Communications
in Computer and Information Science   503

Hongzhi Wang   Haoliang Qi   Wanxiang Che
Zhaowen Qiu   Leilei Kong   Zhongyuan Han
Junyu Lin   Zeguang Lu (Eds.)

Intelligent Computation in Big Data Era

International Conference of Young Computer Scientists,
Engineers and Educators, ICYCSEE 2015
Harbin, China, January 10-12, 2015, Proceedings

Springer

Volume Editors

Hongzhi Wang
Harbin Institute of Technology, China
E-mail: wangzh@hit.edu.cn

Haoliang Qi
Heilongjiang Institute of Technology, Harbin, China
E-mail: haoliangqi163@163.com

Wanxiang Che
Harbin Institute of Technology, China
E-mail: car@hit.edu.cn

Zhaowen Qiu
Northeast Forestry University, Harbin, China
E-mail: qiuzw@yahoo.cn

Leilei Kong
Heilongjiang Institute of Technology, Harbin, China
E-mail: sevenkll@hotmail.com

Zhongyuan Han
Heilongjiang Institute of Technology, Harbin, China
E-mail: hanzhongyuan@gmail.com

Junyu Lin
Harbin Engineering University, China
E-mail: linjunyu@hrbeu.edu.cn

Zeguang Lu
Zhongkeyunhai Company, Harbin, China
E-mail: 13704511101@qq.com

ISSN 1865-0929   e-ISSN 1865-0937
ISBN 978-3-662-46247-8   e-ISBN 978-3-662-46248-5
DOI 10.1007/978-3-662-46248-5
Springer Heidelberg New York Dordrecht London

Library of Congress Control Number: 2014959445

Typesetting: Camera-ready by author, data conversion by Scientific Publishing Services, Chennai, India
Printed on acid-free paper
Springer is part of Springer Science+Business Media (www.springer.com)

Preface

As the program chairs of the International Conference of Young Computer Scientists, Engineers and Educators 2015 (ICYCSEE 2015), it is our great pleasure to welcome you to the proceedings of the conference, which was held in Harbin, China. The goal of this conference is to provide a forum for young computer scientists, engineers, and educators. Even though ICYCSEE is a new conference, it has already had an impact on the community.

The call for papers of this year's conference attracted 200 paper submissions. After the hard work of the Program Committee, 61 papers were accepted to appear in the conference proceedings. The main topic of this conference is big data which is one of the hot research fields. The accepted papers cover a wide range of topics related to big data such as big data management, big data learning, big data harvest, big data communication, big data analysis, big data education, and big data applications.

We would like to thank all the Program Committee members for their hard work in completing the review tasks. Their collective efforts made it possible to attain quality reviews for all the submissions within a few weeks. Their diverse expertise in each individual research area has helped us to create an exciting program for the conference. Their comments and advice helped the authors to improve the quality of their papers and gain deeper insights.

On behalf of the Program Committee, we would also like to thank the two keynote speakers. Prof. Yan Huang, from University of North Texas, gave a talk titled "Data-Driven, Dynamic, and Smarter Urban Transportation Systems." Mr. Xiaosheng Tan, the CTO, VP and CPO (Chief Privacy Officer) of the Qihoo 360 Technology Co. Ltd., discussed "Big Data Based Threat Detection."

Great thanks should also go to the authors and participants for their tremendous supports in making the conference a success. We thank Dr. Lanlan Chang from Springer, whose professional assistance was invaluable in the production of the proceedings.

Besides the technical program, this year the ICYCSEE offered different experiences to the participants. We welcome to you the north-east China to enjoy the beautiful scene of ice and snow in Harbin. We hope you enjoy the conference proceedings.

December 2014

Hongzhi Wang
Haoliang Qi
Wanxiang Che

Organization

The International Conference of Young Computer Scientists, Engineers and Educators (ICYCSEE) 2015 (http://icycsee.hljit.edu.cn) took place in Harbin, China, January 10–12, 2015, hosted by the Heilongjiang Institute of Technology.

General Chairs

Hongzhi Wang   Harbin Institute of Technology
Li Dong   Heilongjiang Institute of Technology

Program Chairs

Haoliang Qi   Heilongjiang Institute of Technology
Wanxiang Che   Harbin Institute of Technology

Educational Program Chairs

Zhaowen Qiu   Northeast Forestry University

Industrial Program Chairs

Yanjuan Sang   Gopha Group

Demo Chairs

Zhongyuan Han   Heilongjiang Institute of Technology

Local Chairs

Zeguang Lu   ZhongkeYunhai Inc.
Leilei Kong   Heilongjiang Institute of Technology

Register Chairs

Junyu Lin   Harbin Engineering University

Financial Chairs

Hui Gao   Harbin Huade University

Program Area Chairs

Artificial Intelligence   Guanglu Sun, Harbin University of Science and
                                    Technology
Big Data   Jinbao Wang, Harbin Institute of Technology

Program Committee

Bach, Nguyen   Carnegie Mellon University, USA
Bu, Jiajun   Zhejiang University, China
Bao, Zhifeng   University of Tasmania, Australia
Bigi, Brigitte   Laboratoire Parole et Langage Aix en Provence,
                                    France
Bond, Francis   Nanyang Technological University, Singapore
Ben, Guosheng   Institute of Computing Technology, Chinese
                                    Academy of Science, China
Blache, Philippe   Laboratoire Parole et Langage Aix en Provence,
                                    France
Bu, Yingyi   UCI, USA
Banchs, Rafael   Human Language Technology Institute for
                                    Infocomm Research A*Star, Singapore
Cai, Shu   University of Southern California, USA
Chan, Paul   Human Language Technology Institute for
                                    Infocomm Research A*Star, Singapore
Chen, Hsin-Hsi   National Taiwan University, Taiwan
Cao, Xiaochun   Institute of Information Engineering, Chinese
                                    Academy of Sciences, China
Castelli, Eric   Maryland Institute College of Art, Vietnam
Chang, Chin-Chen   Feng Chia University, Taiwan
Cheng, Chin-Chuan   Academia Sinica, Taiwan
Cen, Ling   Human Language Technology Institute for
                                    Infocomm Research A*Star, Singapore
Caelen-Haumont, Genevieve   Hanoi University of Science and Technology,
                                    Vietnam
Choi, Key-Sun   Korea Advanced Institute of Science and
                                    Technology, Korea
Chang, Tao-Hsing   National Chiao Tung University, Taiwan
Cui, Bin   Peking University, China

Che, Wanxiang	Harbin Institute of Technology, China
Chen, Yidong	Amoy University, China
Cai, Zhipeng	Georgia State University, USA
Chen, Xiaojiang	Northwest University, China
Chen, Wenliang	Human Language Technology Institute for Infocomm Research A*Star, Singapore
Chen, Wenliang	SooChow University, China
Chen, Boxing	National Research Council Canada, Institute for Information Technology, Canada
Cao, Hoang-Tru	Ho Chi Minh City University of Technology, Vietnam
Dong, Tianyang	Zhejiang University of Technology, China
Deng, Yong	Southwest University, China
Do, Quang	University of Illinois at Urbana-Champaign, America
Do, Ngoc-Diep	Hanoi University of Science and Technology, Vietnam
Dong, Minghui	Institute for Infocomm Research, Singapore
Duan, Xiangyu	Human Language Technology Institute for Infocomm Research A*Star, Singapore
Dinh, Dien	University of Natural Sciences, Vietnam
d'Alessandro, Christophe	The Computer Science Laboratory for Mechanics and Engineering Sciences, France
F., Robert	Carnegie Mellon University, USA
Fu, Guohong	Heilongjiang University, China
Guan, Yi	Harbin Institute of Technology, China
Gao, Jing	Buffalo, The State University of New York, USA
Guo, Li	Institute of Information Engineering, Chinese Academy of Sciences, China
Guo, Tao	China Information Technology security Evaluation Center, China
Ge, Tingjian	University of Massachusetts, Lowell, USA
Gao, Jianfeng	Microsoft, USA
He, Lin	Wuhan University, China
Ho, Bao-Quoc	University of Natural Sciences, Vietnam
Huang, Xuanjing	Fudan University, China
Hsieh, Fuhui	Tatung University, Taiwan
He, Zhongjun	Baidu, USA
Hu, Chunming	Beihang University, China
Huang, Gai-Tai	Takming University of Science and Technology, Taiwan
Ho, Tu-Bao	Japan Advanced Institute of Science and Technology, Japan
Han, Yinhe	Institute of Computing Technology, Chinese Academy of Science, China

Hong, Helena	Nanyang Technological University, Singapore
Hu, Yunhua	Microsoft Research Asia, China
Huang, Yun	National University of Singapore, Singapore
Jiang, Bo	Zhejiang Gongshang University, China
Ji, Donghong	Wuhan University, China
Kashioka, Hideki	The National Institute of Information and Communications Technology, Japan
Kong, Fang	Soochow University, China
Khattak, Asad	Kyung Hee University, Korea
Kim, Jong-Bok	Kyung Hee University, Korea
Liu, Qun	Institute of Computing Technology, Chinese Academy of Science, China
Lin, Hongfei	Dalian University of Technology, China
Liu, Bingquan	Harbin Institute of Technology, China
Li, Deng	Middle and Southern University, China
Lu, Jiaheng	Renmin University, China
Liu, Chao-Lin	National Chengchi University, Taiwan
Li, Shoushan	Soochow University, China
Li, Haifeng	Harbin Institute of Technology, China
Luong, Chi-Mai	Institute of Information Technology, Vietnam
Li, Guoliang	Tsinghua University, China
Lam, Olivia	The University of Hong Kong, Hong Kong
L., Daniel	Acadia University, Canada
Liu, YuanChao	Harbin Institute of Technology, China
Liu, Yang	Tsinghua University, China
Le, Thanh-Huong	Hanoi University, Vietnam
Leung, Cheung-Chi	Human Language Technology Institute for Infocomm Research A*Star, Singapore
Liu, Xiaoguang	Nankai University, China
Lu, Dongxin	ZTE Shenzhen Joint Technology Co., Ltd., China
Lu, Xugang	The National Institute of Information and Communications Technology, Japan
Lu, Jun	Soochow University, China
Lai, Tom	City University of Hong Kong, Hong Kong
Li, Peng	Harbin University of Science and Technology, China
Liu, Donghong	Central China Normal University, China
Lang, Jun	Human Language Technology Institute for Infocomm Research A*Star, Singapore
Ma, Shuai	Beihang University, China

Truong, Quang-Dang-Khoa	Nagaoka University of Technology, Japan
Tao, Jianhua	Institute of Automation of the Chinese Academy of Sciences, China
Tran, Do-Dat	Maryland Institute College of Art, Vietnam
Tian, Feng	Institute of Software of Chinese Academy of Sciences, China
Vallee, Natalie	Institute National Polytechnique de Grenoble, France
Vitrant, Alice	The National Center for Scientific Research, France
Vu, Hai-Quan	University of Natural Sciences, Vietnam
Wang, Huiqiang	Harbin Engineering University, China
Wang, Jinbao	Harbin Institute of Technology, China
Wu, Yunfang	Peking University, China
Wang, Hsin-Min	Academia Sinica, Taiwan
Wu, Chung-Hsien	National Cheng Kung University, Taiwan
Wang, Tao	IDG Arts Company, China
Wang, Ronggang	Peking University ShenZhen Graduate School, China
Wu, Xianchao	Baidu Inc., China
Wutiwiwatchai, Chai	National Electronics and Computer Technology Center, Thailand
Wang, Haifeng	Baidu, China
Wu, Wenjun	Beihang University, China
Wa, Siu	Human Language Technology Institute for Infocomm Research A*Star, Singapore
Wang, Changbo	East China Normal University, China
Wang, Xuan	Harbin Institute of Technology Shenzhen Graduate School, China
Wang, Hui	National University of Singapore, Singapore
Wu, Di	Dalian University of Technology, China
Wang, Zhihui	Dalian University of Technology, China
Wang, Cheng	Xiamen University, China
Wu, Di	Sun Yat-sen University, China
Xu, Jiuyun	China University of Petroleum, China
Xie, Xing	Microsoft Research, China
Xiong, Deyi	SooChow University, China
Xue, Yongzeng	Harbin Institute of Technology, China
Yao, Nianmin	Dalian University of Technology, China
Yao, Yu	Northeastern University, China
Ye, Haizhi	Henan Normal University, China
Yen, Ngoc	Hanoi University of Science and Technology, Vietnam

Yang, Muyun	Harbin Institute of Technology, China
Yu, Hao	Fujitsu, China
Yang, Xiaochun	Northeast University, China
Zhang, Rui	Melbourne University, Australia
Zheng, Dequan	Harbin Institute of Technology, China
Zhou, Shijie	University of Electronic Science and Technology of China, China
Zhang, Dongdong	Microsoft Research Asia, China
Zou, Qingguo	Lanzhou University, China
Zhang, Haiyi	Acadia University, Canada
Zhang, Yue	Singapore University of Technology and Design, Singapore
Zhang, Tian	Nanjing University, China
Zhang, Min	Human Language Technology Institute for Infocomm Research A*Star, Singapore
Zheng, Nengheng	Shenzhen University, China
Zhao, Xiaohui	University of Canberra, Australia
Zhao, Hai	Shanghai Jiao Tong University, China
Zhang, Min	Soochow University, China
Zhang, Xiaoheng	The Hong Kong Polytechnic University, Hong Kong
Zhu, Leihuang	Beijing Institute of Technology, China
Zhang, Hui	Institute for Infocomm Research, Singapore
Zhang, Quan	Institute of Acoustics of the Chinese Academy of Sciences, China
Zhang, Wenqiang	Fudan University, China
Zhu, Muhua	Northeastern University, China
Zhu, Qiaoming	Soochow University, China
Zhen, Han	Beijing Jiaotong University, China
Zhou, Yu	Institute of Automation of the Chinese Academy of Sciences, China
Zhu, Weijun	Zhengzhou University, China
Zhang, Baoxian	University of Chinese Academy of Sciences, China
Zheng, Jun	Southeast University, China
Zhang, Hongli	Harbin Institute of Technology, China
Zhang, Zhiqiang	Harbin Engineering University, China
Zhu, Jingbo	Northeastern University, China
Zong, Chengqing	Institute of Automation of the Chinese Academy of Sciences, China

Table of Contents

Big Data Theory

Unstructured Big Data Processing

Machine Learning for Big Data

Big Data Security

Education Track

Industry Track

Demo Track

Theory of the Solution of Inventive Problems (TRIZ) and Computer Aided Design (CAD) for Micro-electro-mechanical Systems (MEMS)

Huiling Yu[1,2], Shanshan Cui[2], Dongyan Shi[3], Delin Fan[2,*], and Jie Guo[2]

[1]Postdoctoral Research Station of Mechanical Engineering,
Harbin Engineering University, 150001, China
yh12016@163.com
[2]Northeast Forestry University, 150040, China
{yh12016,Cuishanshan510122,gj2618}@163.com, dlfan33@aliyun.com
[3]College of Mechanical and Electrical Engineering,
Harbin Engineering University, 150001, China
920537172@qq.com

Abstract. Satellite remote sensing technology[1] is widely used in all walks of life ,which plays an increasingly remarkable results in natural disasters(sudden and major).With more and more launch and application of high resolution satellite, the texture information in remote sensing imagery becomes much more abundant. In the age of big data, for the infrared remote sensor has short life, which annoys many people. In the packaging process, due to a difference in thermal expansion coefficients [2] between the flip-chip bonded MEMS device and the substrate, cooling after bonding can cause the MEMS to buckle. Combine TAIZ theory with the flexure design in CAD to solve the problem. It can be obtained that increasing fold length can reduce warpage. By solving the deformation problem of MEMS devices can facilitate the development of flip chip technology, and make for the further application of the TRIZ theory in the study of remote sensing equipment.

Keywords: TRIZ conflict matrix, coefficient of thermal expansion, pile welding, Fourier's Law.

1   Introduction

MEMS developed with the advancements of Semiconductor, using similar integrated circuit technology. Along with the integration and performance of integrated circuits have been constantly improving in accordance with the Moore's Law [3],MEMS tried to make mechanical components or systems, sensor, actuator and instrument equipment miniaturization. The electronic industry is seeing an increasing interest of MEMS devices across multiple consumer applications. Design and manufacturing

---

* Corresponding Author.

H. Wang et al. (Eds.): ICYCSEE 2015, CCIS 503, pp. 1–7, 2015.

technology of the MEMS [4] chip has been quite mature, but because the research of MEMS packaging technology[5]has lagged far behind, so that many MEMS chip does not have practical application, which limits the development of MEMS. MEMS packaging technology has its own particularity and complexity. The above poses a tough test for Packaging Technology. To prevent the device (due to thermal mismatch) producing too much stress is the key problem in package technology, which leads to decreasing reliability in the processes of dynamic resource encapsulating[6].Reliability of MEMS is essential to its successful application. In order to solve this problem, TRIZ and the flexure design are used in the package technology. This paper regards the LED of the infrared camera as the object of study .Find the optimization of the parameters and the deterioration of the parameters, and then finds out a Principle to solve this problem.

2   TRIZ Theoretical Basis

TRIZ [7] is a method for creating a science. It is oriented to guide people to innovation. Solving the problem is the core problem of the invention, and the design doesn't overcome conflicts which are not innovative. Discovering and resolving conflicts are the force of evolution, which can make the product more idealized. TRIZ analysis tool consist of algorithm, Su-field analysis, conflict analysis and Functional analysis. The tools are used for analyzing, problem model and conversion. In the process of technical conflict, check the conflict matrix [8]. Different control parameters (A value of 1 or 2) will produce different technical conflict (conflict 1 and conflict 2).Choose proper parameters in order to seek out a correct principle. Then apply the conflict matrix and invention principle to form the innovation solution, as is shown in figure 1.

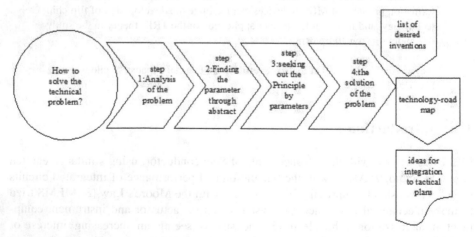

Fig. 1. Conflict analysis process

3   TRIZ Innovative Methods Applied in a Flexible Welding Process

3.1   Description and Analysis of the Bonding Failure in Infrared Camera

Flip chip bonding develops towards bump high density [9] and ultra-fine pitch. The higher the chip power consumption and power density are, the more pronounced effects and surface effects are .The mismatch of heat stress is more severe ,which can cause the MEMS devices to distortion, warping or scratches more easily. Excessive stress concentration can produce the bonding failure. In the package and this welding process, changing the external temperature can cause the release of too much stress due to a difference in thermal expansion coefficients, which causes the package deviation. The main reason for the bonded failure [10] of the infrared camera is LED'S heat dissipation issues, which can have significant effect on the life of the device. After welding process, there is temperature difference between surface and internal as a result of diverse state. This reason can produce the periodic shear stress on the welding surface. Solder crack even the wafer crack would be taken shape caused by the periodic shear stress, eventually result in device failure owning to thermal fatigue. In the chip and the bonding layer, the maximal thermal shear deformation [11] can be estimated by a formula.

$$S = D\Delta\alpha\Delta T / 2 * d \tag{1}$$

In the formula (1), D is chip diagonal size; S is the thermal shear deformation; d is welding layer thickness; $\Delta T = T$ Max-T Min, T max is the solidification of the solder wire temperature, T min is the lowest temperature screening device; $\Delta \alpha$ is the thermal expansion coefficients between the chip and the substrate material. The bigger $\Delta \alpha$ is, the more S becomes. The increase of S will lead to buckle of the device more easily.

3.2   The Solution of the Bonding Failure in Infrared Camera

In this paper, we solve this problem by using the correct material selection, welding method, weld design, flexure design [12], temperature control and TRIZ. Combine parameters of the flexure design with its own problem, and we conclude that changing the folding length, folding spacing and number of folding can reduce the warpag. Fold spacing and fold length [13] can be shown in figure2.Then, we can find out the optimization of the parameters and the deterioration of the parameters in conflict matrix.

In accordance with the above procedure, deal with the welding process issues and seek out the basic factors that need to be solved. The basic elements of flip chip bonding technology [14] innovation methods include the conflict parameters and principles. The conflict parameters for flip chip bonding process refer to having the opposite behavior characteristics. In practice, we can find the optimization of the parameters and the deterioration of the parameters through technical analysis and abstraction. We can seek out the principle by these parameters in conflict matrix. Conflict parameters which are applied in the flip chip bonding process, one part is from descriptions

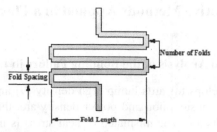

Fig. 2. Parameters for the flexure design

that TRIZ is used in the field of flip welding, and the other part is from an overview on the flip of the welding process. In the actual flip welding, due to the characteristics of the parameter itself, even though one parameter also has the conflict relationship. For example, the speed parameter contains flip welding speed, wire feed speed, welding wire melting speed, cooling speed. When improving the welding wire melting speed, may lead to the deterioration of the welding speed. Thus, innovators can more fully consider the conflict relationship between parameters, in order to get the best innovation solution embodied overall best performance.

In this paper, we solve the bonding failure by changing the folding length. We increase the folding length; meanwhile, the area will increase. So the area is the deterioration parameter. Obviously, an optimization parameter is the stress or pressure. We find out the thermal expansion principle by two parameters in conflict matrix. Specifically, as is shown in table1.

Table 1. Conflict Matrix

		6
		Area of non-moving object
11	The stress or press	10,15,36,37

The identification and resolution of contradictions is a key element of the TRIZ problem solving ethos. Inside all interesting problems and we always find one or more such contradictions. If we can identify these contractions and resolve them, we will have devised a high quality solution to problem-not merely a better compromise, but an innovative solution that breaks free from the existing constraints and provides a step-change towards an ideal system. This contradiction can be expressed as graph 3, which typically takes this form:

In theory, we explain the rationality of the solution by a formula. In the formula (2), σ is the shear stress; Ws is the shear force; A is the cross sectional area. The cross sectional area is proportional to the fold length, on the contrary, σ is inversely proportional to that of the cross sectional area. When we increase the fold length, A (the cross sectional area) will increase, in a result, σ will minish. Therefore, increasing the fold length can be used in flexure design to reduce the warpage[15].

$$\sigma = Ws / A \tag{2}$$

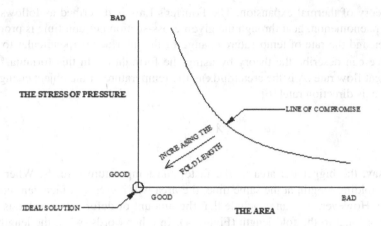

Fig. 3. Flexure design analysis of conflict matrix

The rationality of the method has been proved in theory, By experimentation, we change the folding length so as to gain the amount of deformation value (Table 2)at 50 degrees Celsius.

Table 2. Thermal deformation values measured at 50 ℃

Length (mm)	0	8	16	24	32	40	48	56
Deformation (um)	0	11.026	12.052	10.483	13.703	29.8	13.703	10.483
Length (mm)	64	72	80					
Deformation (um)	12.052	11.026	0					

A graph is gained by above table, through the analysis of this graph, we can arrive at a conclusion that this is an axis of symmetry (As shown in figure 4).

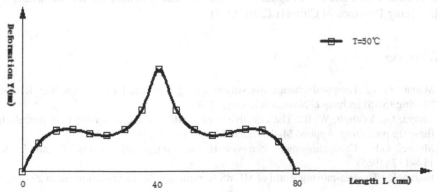

Fig. 4. .Schematic diagram of different body surface thermal expansion values

In the theory of thermal expansion, The Fourier's Law is described as follows, in the thermal phenomenon, heat through the given cross-section per unit time is proportional to area and the rate of temperature change of the interface perpendicular to the direction .We can describe the theory by using the formula (3).In this formula. ϕ is density of heat flow rate ,A is the area,(dt/dx)is the temperature of the object changing along the X axis direction rate[16].

$$\phi = -\lambda A \frac{dt}{dx} \tag{3}$$

That is to say, the bigger the area is, the faster the temperature spreads. When we increase the folding length, at the same time, the area will increase, which can reduce the warpage. However, we can conclude that the amount of deformation value is not always proportional to the fold length (Figure 4). In other words, when the length of the fold length exceeds a value, increasing the fold length can cause the decrease of deformation. So we can change the fold length to solve the bonding failure in infrared camera.

4   Conclusion

Use TRIZ matrix and flexible design to solve the conflict chip bonding failures. It can be concluded that an optimal flexure design will include the longest fold length possible. Besides, that encourages designers to consider the effect from another aspects (hardness, number of folds , fold spacing).By now, reliability is the key to successful application of MEMS devices. However, we know little about the reliability of MEMS devices and MEMS failure modes so far. Therefore, we still have a lot of work to do on how to improve the reliability of MEMS devices through optimization design and advanced technology.

Acknowledgement. Foundation : supported by the Fundamental Research Funds for the Central Universities (DL12EB01-02, DL12CB05) and Heilongjiang Postdoctoral Fund (Grant No. LBH-Z11277) and Natrual Science Foundation for Returness of Heilongjiang Province of China (LC2011C25).

References

1. Mahar, G.A.: Temporal change assessment of agricultural land by Satellite Remote Sensing (SRS) technique. Natural Science (2013)
2. Yanga, C., Youngb, W.-B.: The effective permeability of the under-fill flow domain in flip-chip packaging. Applied Mathematical Modelling (2013)
3. Moore, G.E.: Cramming more components onto integrated circuits. Electronics 38, 114–117 (1965)
4. Bogue, R.: The fast-moving world of MEMS technology. Assembly Automation 29 (2009)

5. Sutanto, J.: Packaging and Non-Hermetic Encapsulation Technology for Flip Chip on Implantable MEMS Devices. Journal of Microelectro Mechanical Systems 21 (2013)
6. van Driel, W.D.: Mechanical reliability challenges for MEMS packages: Capping. Microelectronics Reliability (2007)
7. Hsieh, H.T.: Using TRIZ methods in friction stir welding design. International Journal of Advanced Manufacturing Technology 46, 9–12 (2010)
8. Li, T.S.: Applying TRIZ and AHP to develop innovative design for automated assembly systems. International Journal of Advanced Manufacturing Technology 46, 1–4 (2010)
9. Jan, M.T.: Reliability and Fatigue Analysis in Cantilever-Based MEMS Devices Operating in Harsh Environments. Journal of Quality and Reliability Engineering (2014)
10. Hu, Y., Shen, X.: Research Reviews and Prospects of MEMS Reliability. Integrated Ferroelectrics (2014)
11. Yang, H.J.: Heat transfer in granular materials: effects of nonlinear heat conduction and viscous dissipation. Mathematical Methods in The Applied Sciences 36 (2013)
12. Krijnen, B.: Flexures for large stroke electrostatic actuation in MEMS. Journal of Micromechanics and Microengineering 24 (2014)
13. Lee, Y.C.: Computer-aided design for micro-electro mechanical systems (MEMS). International Journal of Materials & Product Technology (2003)
14. Sutanto, J.: Packaging and Non-Hermetic Encapsulation Technology for Flip Chip on Implantable MEMS Devices. Journal of Microelectro Mechanical Systems 21 (2012)
15. Massoudi, M.: On the heat flux vector for flowing granular materials, Part 1: Effective thermal conductivity and background. Mathematical Methods in the Applied Sciences (2006)
16. Bonetto, F.: Fourier Law: A Challenge to Theorists. World Scientific (2008)

A Distributed Strategy for Defensing Objective Function Attack in Large-scale Cognitive Networks

Guangsheng Feng[1], Junyu Lin[1], Huiqiang Wang[1], Xiaoyu Zhao[1],
Hongwu Lv[1], and Qiao Zhao[2]

[1] College of Computer Science and Technology, Harbin Engineering University,
Harbin 150001, China
{fengguangsheng,linjunyu,wanghuiqiang,zhaoxiaoyu,lvhongwu}@hrbeu.edu.cn
[2] School of Computer and Information Engineering, Harbin University of Commerce,
Harbin 150028, China
zhaoqian@hrbcu.edu.cn

Abstract. Most of existed strategies for defending OFA (Objective Function Attack)are centralized, only suitable for small-scale networks and stressed on the computation complexity and traffic load are usually neglected. In this paper, we pay more attentions on the OFA problem in large-scale cognitive networks, where the big data generated from the network must be considered and the traditional methods could be of helplessness. In this paper, we first analyze the interactive processes between attacker and defender in detail, and then a defense strategy for OFA based on differential game is proposed, abbreviated as DSDG. Secondly, the game saddle point and optimal defense strategy have proved to be existed simultaneously. Simulation results show that the proposed DSDG has a less influence on network performance and a lower rate of packet loss.More importantly, it can cope with the large range OFA effectively.

Keywords: cognitive networks, objective function attack, game model.

1 Introduction

The large-scale cognitive network technology is considered to be a great progress in promoting the reality of cognitive radio networks, but the objective function, i.e., the profit value of network, is crucial to guarantee the network running at an optimal state in such a massive circumstances[1,2]. According to recent studies[3,4,5,6,7], this objective function is quite sensitive to its massive operational data or big data, such that it is easily attacked by malicious users especially in a unstable wireless surroundings. This kind of attack, named Objective Function Attack(OFA), could tamper with the parameters related to the objective function and mislead its cognitive process.

The detail about the OFA could be found in our previous work[8]. Generally speaking, in order to achieve a maximized revenue, an OFA attacker usually

H. Wang et al. (Eds.): ICYCSEE 2015, CCIS 503, pp. 8–15, 2015.

forces the cognitive engine to adjust its objective-function parameters abnormally such that those parameters are deviated from their optimal values. This problem is still an open issue. Traditional scheme to detect OFA is through checking the fluctuation of operational parameters[4,8]. The optimizing process will be terminated forcefully if one or more parameters fluctuate beyond the expectation. Making use of IDS to detect OFA is another approach[5] , which is effective in a small scale network other than the expensive deployment cost. Some recent studies employ swarm intelligence to verify all parameters that could be tampered[4]. Although it's promising in detecting the parameter fluctuations, a heavy computation and communication load will be caused simultaneously. Moreover, a central mechanism is usually required in existing OFA defenses.

   This paper is an extension of our previous work[8]. Considering the fact that those existed OFA defense methods mostly employing central engine to monitor environment and adjust operational parameters will cause a long response delay and non-ideal defense effect, a differential game model is introduced to defending OFA in a distributed way. The rest of this paper is structured as follows: Section 2 constructs a differential game model for defending OFA. Section 3 presents an experimental validation of the proposed approach and discusses the experimental results. Section 4 concludes this paper.

2   System Model

2.1   Objective Function and Its Fluctuation

The objective function is the basis on which the network system could optimize its operational parameters, and the maximum of objective function could be achieved by environment observation and adaptation. Generally, the network operational parameters are usually divided into two categories, one is performance utility R and the other one is security utility S. Therefore, the objective function can be rewritten as (1):

$$f = w_1 R + w_2 S \tag{1}$$

where f represents the objective function, and w_j ($j=1, 2$) represents the weights of parameters R and S, s.t.,$\sum_{j=1}^{2} w_j = 1$.

   When the running environment has changed under a normal circumstance, i.e., no attack exists, the value of its objective function will fluctuate within an acceptable range frequently. Here, we employ fluctuation-checking method [4] to check the OFA. If the fluctuation of objective function does not exceed an upper bound m_j, i.e., equation(2) holds at time t.

$$max \left(\sum_{i=1,i\neq j}^{n} w_i \| f_i^{(t)}(x) - f_i^{(t-1)}(x) \|^{\theta} \right)^{1/\theta} < m_j \tag{2}$$

where $f_i^{(t)}$ is the objective-function value at time t, and θ is the threshold factor.

2.2 DSDG: Modeling OFA and Its Defense Strategy Based on Differential Game Theory

In the interactive process, both the attacker and the defender select their strategies based on a payoff function $J(u, v)$, where u and v are controlling functions for the attacker and defender respectively. The information pattern of the controlling functions is as follows: all the controlling functions adopted by one side before time t are understood completely by the other side, but they do not know the counterpart's following controlling function next time.

Given that both sides are rational, if there are $u^*(t) \in U$ and $v^*(t) \in V$ existed for the attacker and defender respectively such that the inequation (3) holds true for all $u \in U$, $v \in V$, then the optimal strategy is that the attacker selects strategy u^* and the defender selects v^*.

$$J(u, v^*) \leq J(u^*, v^*) = \{\kappa\} \leq J(u^*, v) \tag{3}$$

In other words, once the defender selects strategy v^*, the total loss will not exceed κ whatever the defender selects, and vice versa. Thus, κ is the optimal payoff, i.e., κ is related to the saddle point of this game.

According to above discussions, the network system will continually adjust its operational parameters to maximize its revenue. However, if the adjustment is too much, some potential threats will be incurred. Accordingly, a threat factor $P(t)$ is introduced to limit the parameters adjustment. $P(t)$ is mainly affected by the parameters on a node system, whose expression could be determined from two stages including network initialization as well as suffering an OFA.

In the initialization stage, the threat factor is affected by the continuous adjustment process of performance utility $R(t)$ and security utility $S(t)$, and has an approximate liner relation with them. Thus, the default value or initial value of $P(t)$ could be represented as follows:

$$P(t) = \alpha R(t) + \beta S(t) \tag{4}$$

where α and β denote the impact coefficients for performance and security utility.

In second stage, i.e., a network node under an attack, some parameters will be adjusted to maximize its objective function, but some new threats will be incurred simultaneously. Besides the operational parameters, the node security is also related with its neighbors' threat factors. Thus, $P(t)$ will be various with both itself parameter configurations and its neighbors' while suffering an attack. In consideration of the transmission radius, the threat of a network node mainly comes from its neighbors that could be covered by the transmission power. Due to the exponential decay of wireless channels, the impact degree is decreased exponentially. On this basis, the transmission scope is divided into n sub-areas according to its distance, and then the mean of threat factors is calculated as:

$$\bar{P}_i(t) = \sum_{r=1}^{n} \left(\frac{1}{N} \cdot d^r \cdot \sum_{j=1}^{N_r} P_{r,j}(t) \right) \tag{5}$$

where N_r denotes the node quantity in the rth sub-area, d^r denotes the impact weight of the rth sub-area to the ith node, and $P_{r,j}$ denotes the threat factor of the jth node in the rth sub-area.

An OFA process can be described as follows: an attacker takes strategy $u(\tau)$ to disturb one of the operational parameters in the target node, such as performance utility R. On the contrary, the node system will take strategy $v(\tau)$ to reduce other parameters like security utility S, such that the attacked parameter could be improved for maximizing its objective function. However, if the security level $S(t)$ is lowered to a dangerous level, it is easy to be attacked by malicious users for this target system under such circumstances. Due to this issue, a threat factor $P(t)(0 \le P(t) \le 1)$ is introduced in this paper, which guarantees the target node security level $S(t)$ remaining above a predefined threshold. Moreover, our ultimate goal is to select an optimal strategy to defense the faced attack at the end of interval T. Thus, the process of OFA attack and defense is modeled as a differential game in our work, which is described as in as (6) and (7).

$$\begin{cases} R'(t) & = -u(\tau) + v(\tau) \\ S'(t) & = -sign^* kv(\tau) \\ P'(t) & = \partial v^m(\tau) \\ R(0) & = \xi_1, \quad S(0) = \xi_2 \end{cases} \tag{6}$$

$$J(u, v) = \int_{t_0}^{T_0} \{w_1 R(t) + w_2 S(t) - w_3 \bar{P}(t)\} dt \tag{7}$$

Equations (6) are differential expressions of $R(t)$, $S(t)$ and $P(t)$; $u(\tau)$ and $v(\tau)$ are the strategies adopted by the attacker and its defender respectively. $S(t)$ is the security utility of the node system at time t and $sign = (S_0 - S_t)/|S_0 - S_t|$, where S_0 and S_t are the security utilities before and after suffering attack respectively. Coefficient $k(0 < k < 1)$ denotes the impact degree of $S(t)$ from $v(\tau)$. The security of running environment is reflected by parameters ∂ and m . If the security of fundamental surroundings is low, those two parameters should be increased correspondingly. On the contrary, they should be decreased. $R(0) = \xi_1$ and $S(0) = \xi_2$ denote the initial performance utility and security utility. In Equation(7), $J(u, v)$ is a payoff function of a node system, where $w_j(j = 1, 2, 3)$ are the weights of $R(t)$, $S(t)$ and $\bar{P}(t)$. Equations (6) and (7) conform to the general pattern of OFA and its defense. If the attacker uses an elaborated strategy to impair some parameters in a target node, this node in turn has to adjust other parameters to maintain its current maximum profit. According to this model, if there exists a saddle point $\kappa(\tau)$ in combination (6) and(7), there will be a best strategy existed for defending OFA.

2.3 Proof of Saddle Point

Generally, both attacker and defender have their own optional strategy levels. Let u $(0 \le u \le 1)$ denotes the attacker's strategy level, i.e., attacking power. $u = 1$ represents the most intensive attack, and $u = 0$ means the opposite case. In this way, v ($0 \le v \le 1$) is the defense strategy level.

Without loss of generality, $[t_0, T_0]$ is denoted the duration of an OFA, i.e., the OFA begins at time t_0 and ends at time T_0, in which there are n-time game rounds between the attacker and defender. As our previous work[8], we use a quad $G(U, V, n, T)$ to represent the process between the attacker and defender. Here, $U = \langle u_1, u_2, \cdots, u_n \rangle$ and $V = \langle v_1, v_2, \cdots, v_n \rangle$. In other words, the attacker's strategy is u_i and the defender's strategy is v_i in the ith game round respectively ($i = 1, 2, \cdots, n$). Thus, the whole progress is divided into n sub-processes, and the jth sub-process is continuing for a duration $I_j = [t_0 + \delta \cdot (j-1), t_0 + \delta \cdot j](j = 1, 2, \cdots, n)$, where $\delta = (T_0 - t_0)/n$. Accordingly, the attacker and network try to get the best result, thus the last payoff is approximately equal to (8):

$$J(u, v) = \sup_{\Gamma^{\delta \cdot 1}} \inf_{\Delta_{\delta \cdot 1}} \cdots \sup_{\Gamma^{\delta \cdot n}} \inf_{\Delta_{\delta \cdot n}} J[\Delta_{\delta \cdot 1}, \Gamma^{\delta \cdot 1}, \Delta_{\delta \cdot 2}, \Gamma^{\delta \cdot 2}, \cdots, \Delta_{\delta \cdot n}, \Gamma^{\delta \cdot n}] \tag{8}$$

where $\Delta_{\delta \cdot j}$ and $\Gamma^{\delta \cdot j}$ means the activity or strategy at jth step adopted by the attacker and defender respectively. Thus, if a strategy pair $[\Delta^*, \Gamma^*]$ meets the condition $J[\Delta, \Gamma^*] \leq J[\Delta^*, \Gamma^*] = \{\kappa\} \leq [\Delta^*, \Gamma]$, $[\Delta^*, \Gamma^*]$ is the saddle point of this differential game model. In other words, Δ^* and Γ^* are the optimal strategy for the OFA attacker and its defender respectively. Their detailed definitions could be found in our previous work[8], where we have inferred that there is a Nash Equilibrium from the point of experiment. In this paper, we pay more attention on the theoretical proof.

In a cognitive network, the attacker and its defender game with each other with a form of (6) and (7), and there is a differential game value κ existed , which is accord to the optimal strategy at the current time.

Proof. The differential game equations and payoff function are represented as (6) and (7). Given $(0, t)$ as a game process interval, the related parameters $R(t)$, $S(t)$ and $P(t)$ are calculated as follows:

$$\begin{cases} R(t) = \int_0^t [-u(\tau) + v(\tau)]d\tau + \xi_1 \\ S(t) = \int_0^t -sign * kv(\tau)d\tau + \xi_2 \\ P(t) = \int_0^t \partial v^m(\tau)d\tau \end{cases} \tag{9}$$

Let $\gamma_1 = w_1\xi_1$ and $\gamma_2 = w_1 - w_2 sign * k$, where $\gamma_1, \gamma_2 > 0$. $\bar{P}(t)$ is determined according to (5). To easy analysis, let Ψ denotes :

$$\Psi = \bar{P}_i(t) = \sum_{r=2}^{n} \left(\frac{1}{N_r} \cdot d^r \cdot \sum_{j=1}^{N_r} P_{r,j}(t) \right) + \alpha \xi_1 + \beta \xi_2 \tag{10}$$

Thus, $\bar{P}(t) = d \cdot \int_0^t \partial v^m(\tau)d\tau + \Psi$, and Ψ is a constant.

The payoff function can be transformed to the following equation (11).

$$J(u, v) = \int_{t_0}^{T_0} \{w_1 \int_0^t -u(\tau)d\tau + \gamma_2 \int_0^t v(\tau)d\tau - d \cdot w_3 \int_0^{t_0} \partial v^m d\tau - \Psi + \gamma_1\}dt \tag{11}$$

Without loss of generality, let constant $d = 1/2$ to simplify the expression in the process of deviation. During the process of game, let the node system take the dominant strategy and inferior strategy respectively, and then calculate the value κ^δ and κ_δ as shown in (12).

$$\begin{cases} \kappa^\delta = \dfrac{1}{2} \left(\gamma_2^{m-1} \sqrt{\dfrac{2\gamma_2}{w_3 m \partial}} - \dfrac{\gamma_2^2}{w_3 m^2 \partial} - w_1 \right) T_0^2 + (\gamma_1 - \Psi)T_0 + \sigma \\ \kappa_\delta = \dfrac{1}{2} \left(\gamma_2^{m-1} \sqrt{\dfrac{2\gamma_2}{w_3 m \partial}} - \dfrac{\gamma_2^2}{w_3 m^2 \partial} - w_1 \right) T_0^2 + (\gamma_1 - \Psi)T_0 + \sigma \end{cases} \quad (12)$$

Thus,

$$\kappa = \frac{1}{2} \left(\gamma_2^{m-1} \sqrt{\frac{2\gamma_2}{w_3 m \partial}} - \frac{\gamma_2^2}{w_3 m^2 \partial} - w_1 \right) T_0^2 + (\gamma_1 - \Psi)T_0 + \sigma \quad (13)$$

Consequently, $\kappa^\delta = \kappa_\delta$ is established, i.e., the optimal strategy is existed in the differential game.

Q.E.D.

3 Simulation and Analysis

Performance utility R, security utility S and threat factor P are main parameters, whose weights are w_1, w_2 and w_3 respectively. $R = 0.6$ and $S = 0.85$ are current optimum values with weights 0.4 and 0.3 respectively, which are obtained through learning under normal circumstances. The value of P with a weight 0.3 is thought to be zero at the beginning. The node system under attack will be a period of vulnerability before those parameters resumed. During this period, the node system is susceptible to be attacked. However, we introduce a threat factor $P(t)$ to limit the reduction of $S(t)$, thus the period of vulnerability will be shorted and the risk of suffering from other attacks will also be reduced indirectly. In order to verify above views, our simulations are conducted within a square $10km \times 10km$ region, where 200 nodes are placed randomly.

As shown in figure 1, 20 nodes that have been selected randomly in advance are attacked simultaneously. This figure shows the comparison of defense strategies with a threat factor P or not. At $t = 9$, the parameter R closes to the dangerous level, but the value S is decreased to counter this situation. Then at $t = 15$, the fluctuations of all parameters are resumed to their normal levels. Moreover, the security utility will not be decreased too much during the process of defense due to the threat factor.

Figure 2 shows the comparison of traffic loads between DSDG and PSO. Here, the nodes are divided into 3 categories randomly: normal, attacking, and affected nodes. The packet size is set to be 64B, the traffic is generated automatically, and the affected nodes selected randomly are limited to 30. In PSO, lots of monitoring data are required to collect from the network surroundings, the traffic load is quite high especially when an OFA happens, which has a negative impact on

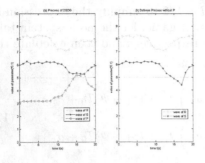 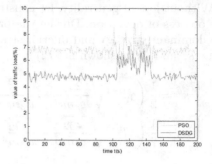

Fig. 1. The defense process   **Fig. 2.** Traffic loads

the network performance. In contrast, our approach only collects threat factor P from its neighbors periodically, thus its negative impact is relatively lesser.

Figure 3 shows the comparison of defense efficiency between two approaches. If the number of affected nodes is less than three, the PSO has a faster convergence speed. Otherwise, the defense efficiency of the proposed approach declines slowly, but the efficiency of PSO is decreased dramatically. The reason is that the parameters of the affected nodes are resumed only in the assistance of all the other nodes in PSO. While there are too many affected nodes, some errors are possibly existed in the collected data from network surroundings, which have negative impact on optimizing the parameters. With the quantity of the affected nodes increasing, more error environmental data are collected by the PSO, which results a larger gap between the reconfigure parameters and the optimal ones. If more than ten nodes are attacked simultaneously, the PSO approach is almost invalid. In the proposed approach, if some nodes are suffering attack simultaneously, their threat factors will be increased correspondingly and then be collected by their neighbors respectively. According to this, their neighbors could infer that the security state is deteriorated. Therefore, their defense process will triggered to response in time.

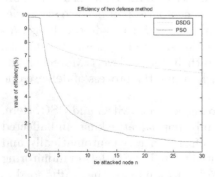 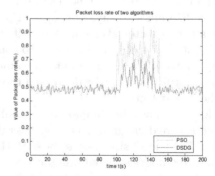

Fig. 3. Efficiency of two methods   **Fig. 4.** Packet loss rate of two methods

Figure 4 shows the comparison of packet loss in attacking period between two methods. In PSO, the security utility of the affected node has been lowered below the risk level before the optimal configurations are calculated. Thus, the cognitive engine is required to collect all other nodes parameters to find a global optimal configuration, which causes a higher rate of packet loss. On the contrary, the proposed method only collects the threat factors from its neighbors, and thus both the traffic load and the packet loss rate are lower than the PSO's.

4 Conclusion

In this paper, we model the OFA defense with some differential equations, and the game process between attacker and defender is described in a distributed way in large-scale network circumstances. Meanwhile, the independent and real-time defense strategy is employed by each network node to avoid the long response delay. Each node optimizes its parameters independently, which has reduced computation and traffic significantly.

Acknowledgments. This work is supported by the Research Fund for the Doctoral Program of Higher Education of China (20122304130002), the Natural Science Foundation of China (61370212), the Natural Science Foundation of Heilongjiang Province (ZD 201102), the Fundamental Research Fund for the Central Universities (HEUCFZ1213, HEUCF100601), and Postdoctoral Science Foundation of Heilongjiang Province (LBH-210204).

References

1. Thomas, R.W., Friend, D.H., Dasilva, L.A., Mackenzie, A.B.: Cognitive networks: adaptation and learning to achieve end-to-end performance objectives. IEEE Communications Magazine 44(12), 51–57 (2006)
2. Fortuna, C., Mohorcic, M.: Trends in the development of communication networks: Cognitive networks. Computer Networks 53(9), 1354–1376 (2009)
3. Parvin, S., Hussain, F.K., Hussain, O.K., Han, S., Tian, B., Chang, E.: Cognitive radio network security: A survey. Journal of Network and Computer Applications 35(6), 1691–1708 (2012)
4. Pei, Q., Li, H., Ma, J., Fan, K.: Defense against objective function attacks in cognitive radio networks. Chinese Journal of Electronics 20(4), 138–142 (2011)
5. Le, N.O., Hern, N.S.J., Soriano, M.: Securing cognitive radio networks. International Journal of Communication Systems 23(5), 633–652 (2010)
6. Reddy, Y.: Security Issues and Threats in Cognitive Radio Networks. In: The Ninth Advanced International Conference on Telecommunications(AICT 2013), Bangkok, Thailand, pp. 84–89 (2013)
7. El-Hajj, W., Safa, H., Guizani, M.: Survey of security issues in cognitive radio networks. Journal of Internet Technology 12(2), 181–198 (2011)
8. Feng, G., Zhao, X., Wang, H., et al.: A Game-Theoretic View on Objective Function Attack and ITS Defense. In: 4th IEEE Conference on Communication Systems and Network Technologies (CSNT), pp. 1152–1156 (2014)

An Energy Efficient Random Walk Based Expanding Ring Search Method for Wireless Networks

Huiqiang Wang, Xiuxiu Wen, Junyu Lin, Guangsheng Feng, and Hongwu Lv

College of Computer Science and Technology, Harbin Engineering University
Harbin, 150001, China
wenxiuxiu@hrbeu.edu.cn

Abstract. Wireless networks generate large amount of data. It is important to design energy efficient data search method since power of wireless nodes is finite. Expanding Ring Search (ERS) is a data search technique exploring for targets progressively, which is widely used to locate destinations or information in wireless networks. The existing studies on improving the energy efficiency of ERS cannot work without positioning systems. In this paper, we combine the technique of random walk with ERS, and propose a random walk based expanding ring search method (RWERS) for large-scale wireless networks. RWERS can work without using positioning systems, and improve the energy efficiency of ERS by preventing each node from transmitting the same request more than once using the technique of random walk. We compare RWERS with the optimal ERS strategy and CERS in networks with various shapes of terrains. The simulation results show that RWERS decreases the energy cost by 50% without decreasing in success rate compared with ERS, and has twice the success rate of CERS when the network is sparse. RWERS can be applied to various shapes of terrains better compared with CERS.

Keywords: wireless networks, big data, Expanding Ring Search, random walk, energy efficient.

1 Introduction

Wireless networks consist of wireless nodes with finite power, which makes it important to design energy efficient search method.

There are a number of works on designing energy efficient search method. Jin, Shi, Jiang, and Zhong estimate the popularity of the searched items, and then use such knowledge to guide search process [1, 2, 3, 4]. Tang, Liu and Zuo introduce the Ant Colony Algorithm into Flooding-based search to decrease its high amount bandwidths and resources in unstructured peer-to-peer networks [6,7,8]. However, peer-to-peer networks are mainly based on the virtual overlay network constructed in the application layer. There is mismatch between the overlay network and the physical network [9], so the works mentioned above, focusing on peer-to-peer networks, are inappropriate for wireless networks.

H. Wang et al. (Eds.): ICYCSEE 2015, CCIS 503, pp. 16–22, 2015.
© Springer-Verlag Berlin Heidelberg 2015

Duy uses the concept of overhearing scheme and proposes a method to cut down the transmissions of the leaf nodes [10]. Shweta extends his works [11]. Rachuri applies the principles of area coverage to ERS and propose a protocol called Coverage based Expanding Ring Search (CERS) for energy efficient and scalable search in WSNs [12]. In CERS, the transmissions are performed such that only a subset of total sensor nodes transmit the search packet to cover the entire terrain area while others listen. Lu proposes Unduplicated Expanding Ring Search methods (UERS), which performs like CERS, but without selections of *relaynodes*[13]. These two works cannot work without positioning systems. Different from their work, we focus on designing search method that works without the knowledge of positions.

We combine the technique of random walk with ERS, and propose a new energy efficient search method (RWERS) for wireless networks. RWERS explores for targets progressively, and prevents each node from transmitting the same request more than once using the technique of random walk to improve the energy efficiency of search process. Furthermore, the nodes located near the source node do not need to transmit the same request every time a search is conducted, and the load is balanced. RWERS can work properly without GPS or other positioning systems, which makes RWERS more scalable, especially for wireless networks composed of low-cost nodes, and can be applied to various shapes of terrains.

The rest of this paper is organized as follows. We present the details of RWERS in Section 2. Section 3 presents results of simulations which verify our conclusions. We conclude our work and state future work directions in Section 4.

2 Protocol Design

2.1 Basic Idea

The basic idea of RWERS is to try to make the nodes that have received the request in previous searches silent in the new search.

When a source node prepares to find a resource, it first floods a request to all the nodes within a small radius, which is the same as the first search of ERS. If the first search fails, the source node sends a random walk message to find a *helper*, a node located at the edge of the area that has been flooded. Then the helper starts the second search on behalf of the source node. Every time a search fails in locating the resource, the source node finds a new *helper* that will start the next search. This continues until the resource is found or all the nodes are requested.

It is hard for a node to decide whether to relay the request when the search is started by a *helper*. However, if the pre-knowledge of the distance (minimum hops) from the source node is known, a node can make decision based on the pre-knowledge, and the problem is solved. Before the source node finds resource for the first time, we assign each node with an *altitude* to help nodes get this pre-knowledge.

2.2 Random Walks Based Expanding Ring Search

Altitude **Assignment**

The source node first sets 0 as its *altitude*, and broadcasts a message with TTL=0. When a node receives the request for the first time, it increases the TTL of the message, and sets the TTL value as its *altitude*. When it receives the message again, it extracts the source address and the TTL from the message, and keeps them in memory (The information got from the duplicated messages tells the *altitude*s of neighbors). This process continues until all the nodes receive the message. Thus, each node is assigned with an *altitude*, and the local information, the *altitude*s of neighbors, is obtained at the same time.

The *altitude* assignment process is only performed once, and the *altitude* information can be used in all the target discoveries afterwards.

RWERS

Let $SSet = \{L_1, L_2, L_3, ..., L_n\}$ be the search strategy of RWERS, where L_i is an integer, representing the *altitude* limitation of the i th search, and $L_i < L_{i+1}$.

The first search
The source node first broadcasts a request. When a node receives the request for the first time, it checks whether it is the target. If so, it sends a reply to the source node; if not, it checks whether its *altitude* is lower than L_1. If the answer is yes, it broadcasts the request; if not, it does nothing. So all the nodes with *altitude* lower than L_1 are requested. If the first search fails, the second search starts.

The ith search ($i \geq 2$)
If the $i-1$ th search fails, *helper* discovery begins. The i th search begins after the *helper* is found.

A random walk message starts its walking from the source node. When a node receives the message, it checks whether it has sent the request or not. If so, it means that the node is not located at the edge of the area that has been flooded, and it cannot be selected as a *helper*, thus the node sends the random walk message to one of its neighbors whose *altitude*s are higher (the *altitude*s of neighbors are obtained in the *altitude* assignment process); if not, the node selects itself to be a *helper*. This *helper* discovery process will be preformed several times until a *helper* is found.

The *helper* broadcasts a request carrying the *altitude* limitation, L_i. Each node checks whether it has received the request before, and whether it is the target. When the answers are all no and its *altitude* is lower than L_i, it broadcasts the request.

If the i th search fails, the next search starts.

3   Simulations

We compare RWERS with ERS, CERS, and Gossip [14] in the networks with various shapes of terrains.

3.1   Simulation Setup

Nodes are uniformly deployed, and the transmission radius of nodes is fixed at 30m. We assume that a single target exists in the network and each non-source node has equal probability to be the target. The strategy for ERS is $\{\lfloor (R-1)/2 \rfloor, R\}$ as it is found to be optimum [13]. The parameter, k, for CERS is fixed at 1 as this value is proved to be most energy efficiency one. The simulation scenarios:

Circle: We consider two networks in the circular terrains with radius being 180m and 360m respectively, and the source node is fixed at the origin, the number of nodes is varied from 200 to 2000 in increments of 200.

Square: We consider a network in the square terrain (360m*360m), and the source node fixed at the center, the number of nodes is varied from 200 to 1200 in increments of 200.

Rectangle: We consider a network in the rectangle terrain (720m*360m), and the source node fixed at the center, the number of nodes is varied from 200 to 1200 in increments of 200.

3.2   Simulation Results

Circle Terrain. Figure 1 shows the variation of the number of requests versus the node density using RWERS, ERS and CERS in a network with small terrain radius (180m). The number of requests can reflect the energy cost. The most energy consuming method is Gossip as we can see from Figure 1. The number of requests for CERS is much smaller than other method, but it can only works under the condition that the positions of nodes are known. Other methods can work properly without knowing locations.

The success rate increases with N as shown in Figure 2. When N is less than 400, the success rate is less than one for each method mainly due to the lack of network connectivity. When N is greater than 600, ERS and RWERS can detect target in probability 1, the success rate of CERS is much lower than other methods. CERS uses a coverage based method to select RelayedNodes, and the decrease of N can strongly effect this process, resulting in low success rate.

We further double the terrain radius and halve the density of previous simulation, and run it again. Figure 3 and 4 show the results. When the terrain radius is doubled, all conclusions are the same as the previous ones except the success rate, which decreases slightly. This is caused by the halving of network density. The results obtained from the two simulations show that RWERS can decrease the number of requests by 50% comparing with ERS, and double the success rate of CERS when the network is sparse, say, less than 0.019/sq.m, in networks with circle terrain.

Square Terrain.
Figure 5 and 6 show the performance of RWERS, ERS, CERS, and Gossip in a network with square terrain. Comparing the results with those in circle terrains, we can find that RWRES still works properly after changing the terrain shape into square, while CERS has a decrease in success rate. The reason is that CERS can only works in circle terrains, and it cannot send request into some regions if the terrain is non-circular shape. [11]

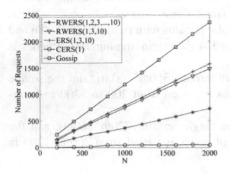

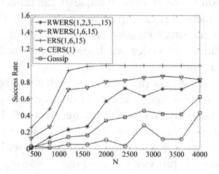

Fig. 1. Average number of request transmitted for finding the target in dense network with circle terrain

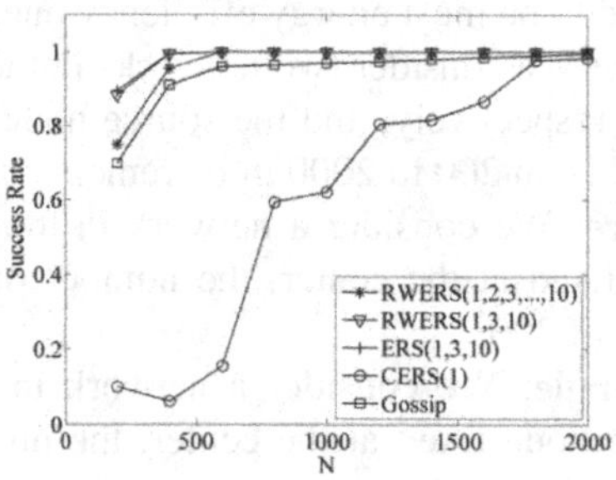

Fig. 2. Average success probability of the target discoveries in dense network with circle terrain

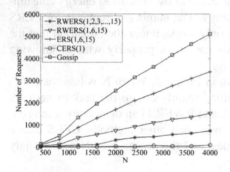

Fig. 3. Average number of request transmitted for finding the target in dense network with circle terrain

Fig. 4. Average success probability of the target discoveries in dense network with circle terrain

Rectangle Terrain.

Figure 7 and 8 show the performance of RWERS, ERS, CERS, and Gossip in a network with rectangle terrain that length-width ratio = 2: 1. When N is greater than 1200, the success rate is about 25% for CERS, 80% for RWERS with Increment = 1, and almost 100% for RWERS with $SSet = \{1,7,15\}$. The results indicate that our method appears random, which is caused by the lack of connectivity of nodes at the same *altitude* in the network with non-circular terrain. But if the strategy is carefully selected, RWRES can still detect the target in probability 1. Thus RWERS can be applied to terrains with various shapes, and has more scalability than CERS.

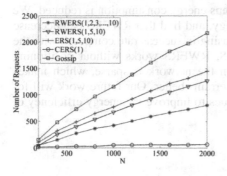 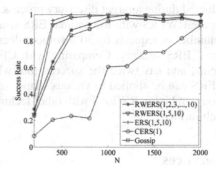

Fig. 5. Average number of request transmitted for finding the target in the network with square terrain

Fig. 6. Average success probability of the target discoveries in the network with square terrain

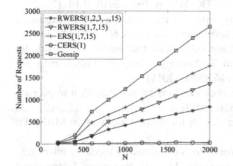 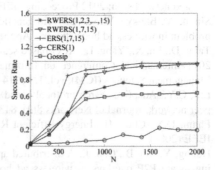

Fig. 7. Average number of request transmitted for finding the target in dense network with rectangle terrain

Fig. 8. Average success probability of the target discoveries in dense network with rectangle terrain

4   Conclusions

ERS is a widely used flooding technique for locating destinations or information in wireless networks. In this paper, we combine the technique of random walk with ERS, and propose a new method (Random Walk based Expanding Ring Search, RWERS) for improving energy efficiency of ERS in wireless networks. By assigning each node with an *altitude*, RWERS can work without the knowledge of positions, which makes it more scalable, especially for wireless networks composed of low-cost nodes. By use of *helpers*, which are found using random walk method, each node transmits the same request only once even though multiple searches may be conducted during a target discovery process, thus energy consumption is reduced. We compare RWERS with the optimal ERS strategy, and find that RWERS can decrease the number of requests by 50% without decreasing in success rate compared with the optimal ERS strategy. Comparing with CERS, RWERS works without positioning systems, and has twice the success rate when the network is sparse, which makes RWERS can be applied to various shapes of terrains better. Our future work will focus on combining RWERS with other techniques to improve the energy efficiency of search further.

References

1. Gaeta, R., Sereno, M.: Generalized probabilistic flooding in unstructured peer-to-peer networks. IEEE Transactions on Parallel and Distributed Systems 22(12), 2055–2062 (2011)
2. Barjini, H., Othman, M., Ibrahim, H., et al.: Shortcoming, problems and analytical comparison for flooding-based search techniques in unstructured P2P networks. Peer-to-Peer Networking and Applications 5(1), 1–13 (2012)
3. Sharifi, L., Khorsandi, S.: A popularity-based query scheme in P2P networks using adaptive gossip sampling. Peer-to-Peer Networking and Applications 6(1), 75–85 (2013)
4. Das Sarma, A., Molla, A.R., Pandurangan, G.: Near-optimal random walk sampling in distributed networks. In: 2012 Proceedings IEEE INFOCOM, pp. 2906–2910. IEEE (2012)
5. Singh, A., Bhukya, W.N.: A hybrid genetic algorithm for the minimum energy broadcast problem in wireless ad hoc networks. Applied Soft Computing 11(1), 667–674 (2011)
6. Tang, D., Lu, X., Yang, L.: ACO-based search algorithm in unstructured P2P Network. In: 2011 International Conference on Information Technology, Computer Engineering and Management Sciences (ICM), vol. 1, pp. 143–146. IEEE (2011)
7. Liu, C.Y.: Adaptive search protocol based on optimized ant colony algorithm in peer-to-peer network. Journal of Networks 8(4), 843–850 (2013)
8. Pham, D.N., Choo, H.: Energy Efficient Ring Search for Route Discovery in MANETs. IEEE (2008)
9. Huang, Y., Jin, B., Cao, J.: A distributed approach to construction of topology mismatching aware P2P overlays in wireless ad hoc networks. In: 14th Euromicro International Conference on Parallel, Distributed, and Network-Based Processing, PDP 2006, p. 8. IEEE (2006)
10. Mishra, S., Singh, J., Dixit, A., et al.: Modified Expanding Ring Search Algorithm for Ad-hoc Networks. International Journal of Computer Science and Information Technologies 3(3), 4067–4070 (2012)
11. Rachuri, K.K., Siva Ram Murthy, C.: On the scalability of expanding ring search for dense wireless sensor networks. Journal of Parallel and Distributed Computing 70(9), 917–929 (2010)
12. Shuaibing, L., Zhezhuang, X., Yinghua, Z.: Unduplicated Expanding Ring Search in Wireless Sensor and Actor Networks (2012)
13. Deng, J., Zuyev, S.: On search sets of expanding ring search in wireless networks. Ad Hoc Networks 6(7), 1168–1181 (2008)
14. do Amaral, M.K., Pellenz, M.E., Penna, M.C., et al.: Performance evaluation of gossip algorithms in WSNs using outage probability. In: 2013 International Symposium on Wireless and Pervasive Computing (ISWPC), pp. 1–5. IEEE (2013)

Reconstructing White Matter Fiber from Brain DTI for Neuroimage Analysis*

Gentian Li, Youxiang Duan, and Qifeng Sun

China University of Petroleum (East China), Qingdao 266580, China
LiGentian@163.com, {yxduan,sunqf}@upc.edu.cn

Abstract. Diffusion Tensor Imaging (DTI), which is the magnetic resonance technology, is applied widely, especially in analyzing the brain function and disease. DTI is a four dimensional image with the principle of diffusion of water molecules, which based on the diffusion characteristic of water. In DTI, each voxel has its own value and orientation, which result in the track of movement of water molecules by the FACT algorithm. It is the tracking of brain white matter fiber, while the tractography is the tracking image of neural fiber bundles according to the principle. All images running are big data processing. The tractrography is important for analyzing the function of brain, creating the brain connectivity and analyzing neuroimage with disease. The paper mainly shows the procedure of data processing by registration that has existed in some researches. However, there is a significant comparison with nonlinear-registration and linear-registration in the paper.

Keywords: Magnetic Resonance, DTI, Tractography, Algorithm.

1 Introduction

With the help of the imaging technology, the medical profession has got great progress in treating previously intractable diseases, for example, the brain diseases. This involves a wide range of imaging, such as Magnetic Resonance Imaging (MRI), Diffusion Weighting Imaging (DWI) and Diffusion Tensor Imaging (DTI). At present, DTI becomes a popular imaging modality for brain analysis in vivo. [1][2] Its fundamental principle is to measure the movement of water in the brain tissue. In DTI image, every voxel encodes the directionality of diffusion, which leads to the possibility of building up a tractography by tracking algorithms. [3]

There are many methods of analyzing brain diseases so far, for example, getting the T1-weighted MRI image or analysis of Tract-based Spatial Statistics (TBSS). It is known that when we give magnetic resonance, water molecules are stimulated and get energy in the location of a magnetic field. [4] And then the energy will be relaxed, producing signals. If the movement of water molecules is tend to freedom, such as in some tissues whose T1 time is longer, the relaxation takes longer time and the signal will be lower. However, in some regions, like fat, whose T1 time is short, the relaxation

---

* Project (No. 14CX02138A) Supported by the Foundamental Research Funds for the Central Universities.

H. Wang et al. (Eds.): ICYCSEE 2015, CCIS 503, pp. 23–30, 2015.

takes shorter time and the signal will be high. Therefore, we can use T1-weighted MRI images to distinguish different tissues, for example, signal will weak in cerebrospinal fluid. Another method is by TBSS, which analyze the differences in Fractional Anisotropy (FA) image that is transformed by DTI image. [5] FA value is the characteristic of diffusion of water molecules (I will introduce how to calculate FA value in the next background chapter). Each voxel in FA image has its own value and we can get the different regions by comparing the whole voxels, so that we classify the normal and patient. [6][7] However, they cannot find any fiber track in the brain, such as Alzheimer's Disease (AD).

In my research, we use the DTI image to build up tractography, which reflects the fibre bundles trending in the brain. [8][9] We aim to classify the healthy persons and the patients by the tracking the change of fiber connections. In my work, I concentrate on analyzing a single subject first, and then extend to analyzing a group of subjects, and found linear-registration is better than nonlinear-registration in tractography by comparing the result of registration.

2   MRI DWI and DTI

MRI is a method which can measure the movement of molecules, mainly water molecules, in vivo and biological tissues. The measurement of water molecules moving is non-damaged. Also, the movement of water molecules is not free in tissues, because it can be affect by many obstacles, and it reflects the structure of obstacles in turn, such as white matter fiber etc. The different characteristics of MRI are due to the different structural tissues at microscopic level. For example, in some typical T1-weighted images, water molecules are stimulated and get energy in the location of a magnetic field. And then the energy will be relaxed, producing signals in MRI. If the movement of water molecules is tend to freedom, such as in some tissues whose T1 time is long, the relaxation takes longer time and the signal will be lower. However, in some regions, like fat, whose T1 time is short, the relaxation takes shorter time and the signal will be high. In T2-weighted images, contrast process will happen. And then, it can be checked and analyzed with the difference of structure both normal and diseased subjects. Therefore, MRI has been having some extremely successful applications especially in clinical application. Mostly, its main clinical application is creating the brain connectivity and the treatment of the intractable brain diseases, for example, finding sub-categories disease, tracking disease progression for curing, studying normal development or aging, and creating brain connectivity to find brain function.

In order to make MRI images have diffusion more sensitively, the solution by adding gradient pulse magnetic field in one orientation, rather than use a same homogeneous magnetic field in the application of DWI. Because the molecules move with different rates at gradient pulse magnetic field, which lead to molecules procession is proportional to the magnetic strength, and then it causes signal attenuation. Moreover, another gradient pulse with the same magnetic strength is put the opposite orientation to refocus the spins. However, the refocusing maybe not perfect for molecules because it moved during the time interval between in two gradient pulse magnetic field, and the signal is decreased. The method was firstly designed for NMR by Stejskal and Tanner. The correspondence equation is

$$\frac{S}{S_0} = \exp[-\gamma^2 G^2 \delta^2 (\Delta - \frac{\delta}{3})D] \tag{1}$$

where S is the signal strength with the gradient pulse magnetic field, S_0 is the signal strength without the gradient pulse magnetic field, γ is the gyromagnetic ratio, G is the amplitude of the gradient pulse, δ is the time of keeping the gradient pulse, Δ is the time interval between the two pulses, D is the diffusion-coefficient.

A coefficient is cited in DWI. It is b parameter, which depend on scanners parameters. The diffusion-coefficient D is replaced by ADC. So, the equation is converted simply in the following:

$$\begin{cases} \dfrac{S}{S_0} = \exp[-b \bullet ADC] \\ b = \gamma^2 G^2 \delta^2 (\Delta - \dfrac{\delta}{3}) \end{cases} \tag{2}$$

where S is the signal strength with the gradient pulse magnetic field, S_0 is the signal strength without the gradient pulse magnetic field, b is the diffusion-weighting factor, ADC is the apparent-diffusion-coefficient, γ is the gyromagnetic ratio, G is the amplitude of the gradient pulse, δ is the time of keeping the gradient pulse, Δ is the time interval between the two pulses, D is the diffusion-coefficient.

According to Eq. (2), every voxel will be calculated in one orientation by the following equation:

$$ADC_{(x,y,z)} = \frac{In(\frac{S_{(x,y,z)}}{S_{0(x,y,z)}})}{b} \tag{3}$$

where (x, y, z) is the coordinate of every voxel in MRI.

In DWI, there are three directional gradient pulses. The three gradient pulses can give you three values of ADC along each of orientation ADC_x, ADC_y and ADC_z. So, an average diffusivity in the voxel can be calculated in the following equation:

$$ADC_i = \frac{ADC_x + ADC_y + ADC_z}{3} \tag{4}$$

where ADC_x, ADC_y and ADC_z are the three directional gradient pulses, and ADC_i is the value of every voxel.

However, we get that the value of every voxel reflects the isotropic diffusion but not the anisotropy of diffusion according to Eq. (4). So, DTI remedy it mostly.

So, finding a good way to describe the anisotropy of diffusion is the important problem. This results in the FA which is the square root of the sum of squares (SRSS) of the diffusion differences, divided by the SRSS of the diffusivities. The equation is in the following:

$$FA = \frac{\sqrt{3((\lambda_1 - \lambda)^2 + (\lambda_2 - \lambda)^2 + (\lambda_3 - \lambda)^2)}}{\sqrt{2(\lambda_1^2 + \lambda_2^2 + \lambda_3^2)}} \tag{5}$$

where λ_1, λ_2 and λ_3 are the three eigenvalues, λ is mean diffusivity.

According to Eq. (5), we get that the FA value is tend to 1 if the λ_1 eigenvalue is very high but λ_2 eigenvalue and λ_3 eigenvalue are lower, which illustrate that diffusion is fast along λ_1 direction and slower in λ_2 and λ_3 directions. It is anisotropy of diffusion. Also, the FA value is tend to 0 if the λ_1 λ_2 and λ_3 eigenvalues are similar, which illustrate the diffusion is isotropic. Therefore, the FA is more accurate.

3   Registration

Before any meaningful analysis could be conducted in a group of subjects, we have to manage some problems existing in DTI images, such as ghosting due to slight motion, complete disappearance of image due to severe motion, etc. Therefore, we employ the Eddy Correction to solve these problems. Moreover, when we get a group of subjects to analysis, we need to transform all the subjects from their native space into a common space for comparison. This process is called image registration. As a commonly used space, the MNI152 [10] space is my choice. This process involves registering the subjects that have been eddy-corrected to MNI152 common space, and then conducting fiber-tracking for all the subjects and comparison. [11]

There are two ways for registration, linear registration and non-linear registration. Linear- registration is that the original image is registered to template image by pan, zoom and rotate processing. It will generate a mapping matrix, which is the transformation process, while the nonlinear-registration is the internal distorted transformation process. It will generate a mapping warp, which is the corresponding transformation process.

In research, we register the FA image to the template image (FMRIB58_FA image). We find that the result is better using nonlinear-registration than linear-registration. The Fig. 1 is the comparison between linear-registration and nonlinear-registration with FA image.

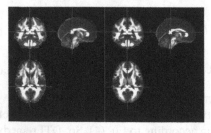

Fig. 1. the comparison between linear-registration and nonlinear-registration with FA image (the left is FA image after linear-registration, while the right is nonlinear-registration. Color red represents the difference.)

Fig. 2. the comparison between linear-registration and nonlinear-registration with tractography (the left is tractography after nonlinear-registration, while the right is linear-registration)

However, the final tractography is better by using linear-registration than nonlinear-registration. According to Fig. 2, we find the left tractography (nonlinear-registration) is shapeless, but the right tractography is shape like the basic brain fiber bundle.

Finally, the linear-registration is recommened in research because the final comparison is tractography.

4   Tracking

The fiber tracking (tractography) is reconstructed and for tract according to previous algorithm of Fiber Assignment by Continuous Tracking (FACT) method. It is stupid for searching algorithm. The Fig. 3 describes the basic principle of tracking.

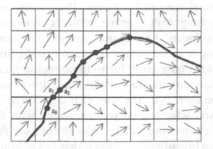

Fig. 3. illustrates that the arrows are principle eigenvector which has the value and orientation. The blue route is the tracked fibre tract. So, find a point that the tracking start. It is seed point a_0 whose FA value is larger than the thresholds. The thresholds contain the range of value and angle. They are user definable in the program, which reflect the range of real neural fibre region. When move along the eigenvector orientation, the tract meets the border of the new voxel at point a_1. If the eigenvector at point a_1 is in the thresholds, the tract orientation will be changed to that of the new voxel. This processing is two side directions along the seed point. Finally, the tract is end at point whose FA value is lower/larger than the thresholds or the angle between two eigenvectors is lower/larger than the thresholds.

5   Process and Results

The data is 10 normal subjects and 10 patient subjects. Table 1 shows the subjects' information.

Table 1. The subjects' information

subjects	number	field /T	TR /ms	TE /ms	directions *repeats	b0
normal	10	3.0	12300.0	68.3	41*1	5
AD	10	3.0	12300.0	68.3	41*1	5

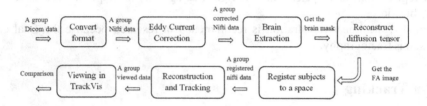

Fig. 4. the processing of group subjects

The Fig. 4 is the processing of group subjects. Firstly, convert all raw DICOM data to NIfTI data using Dcm2nii tool. Secondly, put all the 4D NIfTI data in Eddy Current Correction, and get a group corrected subjects. Then, there are many differences of individual subject processing with a group subjects processing. Because of many differences of different subjects' images, we must correct these images and register these images to a common space. It is very common that the image have some quality problems such as ghosting (phase error due to finite motion), complete disappearance of image (due to severe motion), missing data (upload, download, or conversion error). So, these images must be corrected. The way of image correction is Eddy Correction and Brain Extraction. And when we do some research with these images, we must register these images to a common space in order to compare these images with some difference. So that, we can investigate the local properties of brain tissues in application of some brain disease. The correction and registration are running in FSL. The data is reconstructed and tracked. Finally, we view them with TrackVis and find the difference between the normal and patients.

The Fig. 5 is the comparison of tractography before registration and after registration. The left is 4 tractography randomly selected subjects from 20 subjects before registration. The right is 4 tractography that are same subjects after registration.

The Fig. 6 is the comparison of tractography between normal and patient randomly selected. The subject's tractography A and B are the normal, while the subject's tractography C and D are the Alzheimer's disease patients. We define a sphere whose radius is 2 and location is same, and tractography can pass through it. The viewing is the tractography that pass through the sphere.

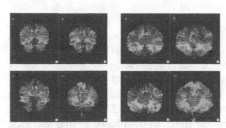

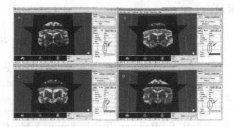

Fig. 5. the comparison of tractography before registration and after registration

Fig. 6. the comparison of tractography between normal and patient

According to Fig. 5, we get some results. Obviously, the tractography before registration locate different location on the space and the size of tractography also are different, for example, picture A and C are tend to higher than picture B and D, and AC are tend to larger than AD. Moreover, picture D is crook in the space. However, the tractography after registration is tidy because they are registered to MNI152 space, which are more benefit and more authoritative for analyzing.

Fig. 6 illustrates that there are some different tractography by comparing the normal and the Alzheimer's disease patient. At the same coordinate, tractography that pass through the sphere are transverse fiber bundles marked red tracks mostly in picture A and B. However, in picture C, tractography that passes through the sphere is very tiny fiber bundles. In picture D, tractography that passes through the sphere is longitudinal fiber bundles marked blue tracks. So we propose that the Alzheimer's disease result from the different fiber bundles in brain, either the lack of necessary neural fiber bundles or the neural fiber deformation.

6 Conclusions

From the results of a group subjects containing the normal and patient, we can find that there are some difference of fiber tracking. For example, there are some fiber tracking interrupted in some regions and the number of fiber tracking is less comparing with the normal subject. Maybe, these interrupted and lost fiber result in the disease. Also, if we find some patients have some symptoms, we can use the result of fiber tracking to analyze the reason that lead to symptoms. So, we can predict some brain disease. Meanwhile, we also find some treatments to help cure. The research is about how to tracking both one subject and a group subjects. We can find some improve at the registration term, which is different from some other application, such as, TBSS. Maybe the results of TBSS are more intuitive for classifying the normal and patient. But, we try to find a right way to describe these differences, for example, the connected strength between of two brain regions, or the number of fiber tracking between of two brain regions. Another hand, finding an accurate way to get completed fiber tracking may be emphasis considered.

References

1. Le Bihan, D., Mangin, J.F., Poupon, C., et al.: Diffusion tensor imaging: concepts and applications. Journal of magnetic resonance imaging 13(4), 534–546 (2001)
2. Dong, Q., Welsh, R.C., Chenevert, T.L., et al.: Clinical applications of diffusion tensor imaging. Journal of Magnetic Resonance Imaging 19(1), 6–18 (2004)
3. Fillard, P., Descoteaux, M., Goh, A., et al.: Quantitative evaluation of 10 tractography algorithms on a realistic diffusion MR phantom. Neuroimage 56(1), 220–234 (2011)
4. Jbabdi, S., Behrens, T.E.J., Smith, S.M.: Crossing fibres in tract-based spatial statistics. Neuroimage 49(1), 249–256 (2010)
5. Smith, S.M., Jenkinson, M., Johansen-Berg, H., et al.: Tract-based spatial statistics: voxelwise analysis of multi-subject diffusion data. Neuroimage 31(4), 1487–1505 (2006)
6. Liu, Y., Spulber, G., Lehtimäki, K.K., et al.: Diffusion tensor imaging and tract-based spatial statistics in Alzheimer's disease and mild cognitive impairment. Neurobiology of Aging 32(9), 1558–1571 (2011)
7. Mori, S., van Zijl, P.: Fiber tracking: principles and strategies–a technical review. NMR in Biomedicine 15(7-8), 468–480 (2002)
8. Zhang, Y., Zhang, J., Oishi, K., et al.: Atlas-guided tract reconstruction for automated and comprehensive examination of the white matter anatomy. Neuroimage 52(4), 1289–1301 (2010)
9. Brett, M., et al.: Using the Talairach atlas with the MNI template. Neuroimage 13(6), 85–85 (2001)
10. Mori, S., Crain, B.J., Chacko, V.P., et al.: Three-dimensional tracking of axonal projections in the brain by magnetic resonance imaging. Annals of Neurology 45(2), 265–269 (1999)
11. Jiang, H., van Zijl, P., Kim, J., et al.: DtiStudio: resource program for diffusion tensor computation and fiber bundle tracking. Computer Methods and Programs in Biomedicine 81(2), 106–116 (2006)

Resolution Method of Six-Element Linguistic Truth-Valued First-Order Logic System

Li Zou[1,2,*], Di Liu[1], Yingxin Wang[3], and Juan Qu[4]

[1] School of Computer and Information Technology, Liaoning Normal University, Dalian, China
[2] State Key Laboratory for Novel Software Technology, Nanjing University, Nanjing, China
[3] Marine Engineering College, Dalian Maritime University, China
[4] Anshan Radio and TV University, Liaoning, China
zoulicn@163.com

Abstract. Based on 6-elements linguistic truth-valued lattice implication algebras this paper discusses 6-elements linguistic truth-valued first-order logic system. With some special properties of 6-elements linguistic truth-valued first-order logic, we discussed the satisfiable problem of 6-elements linguistic truth-valued first-order logic and proposed a resolution method of 6-elements linguistic truth-valued first-order logic. Then the resolution algorithm is presented and an example illustrates the effectiveness of the proposed method.

Keywords: Linguistic truth-valued lattice implication algebra, 6-elements linguistic truth-valued first-order logic, Automated reasoning.

1   Introduction

Automated reasoning one of the most important research directions of Artificial Intelligence, which allows computers to reason completely, or nearly completely, automatically. Words and languages are important aspects for embodying human intelligence. A lattice is an algebraic abstract of the real world, it can be used to describe certain information that may be comparable or incomparable. Human beings usually express world knowledge by using natural language with full of vague and imprecise concepts. Lattice-valued logic system is an important case of multi-valued logic [1].

Nowadays, there exist many alternative methods to linguistic valued based intelligent information processing, such as Martinez et al. proposed a 2-tuple fuzzy linguistic representation model [2-3] and so on. Lattice-valued logic [4],as one kind of multi-valued logic, extends the chain-type truth-valued field to general lattice in which the truth-values are incompletely comparable with each other. Ho discussed the ordering structure of linguistic hedges, and proposed hedge algebra to deal with CWW [5]. Xu gave α-resolution automated reasoning algorithms, established their soundness and completeness in $LF(x)$, which can directly apply to $L_{n\times2}F(x)$ [6]. In 2007, Xu et al. gave the α-resolution field of quasi-regular normal generalized literals in $LP(x)$, and also gave its concrete judging table [7].Recently, many linguistic approaches have been proposed and applied to solve problems with linguistic assessment, e.g., person-

H. Wang et al. (Eds.): ICYCSEE 2015, CCIS 503, pp. 31–37, 2015.

nel management [8], web information processing [9], multi-criteria decision making [10] and fuzzy risk analysis [11-12]. In 2010, He et al. proposed α-lock resolution method based on lattice-valued propositional logic $L_nP(x)$ and established its soundness and weak completeness [13]. Zou et al. proposed a framework of linguistic truth-valued propositional logic and developed the reasoning method of linguistic truth-valued logic system [14-15].

This paper will discuss 6-elements linguistic truth-valued first-order logic system based on 6-elements linguistic truth-valued lattice implication algebra. And will discuss the satisfiable problem of 6-elements linguistic truth-valued first-order logic and propose a resolution method of 6-elements linguistic truth-valued first-order logic. Then will present the resolution algorithm and give an example illustrates the effectiveness.

2   Preliminaries

Definition 1 [16]. Qualitative linguistic hedge variable set is obtained from the partition of linguistic hedge operator set due to the effect of the linguistic hedge operators to the proposition. We can get three classes of the linguistic hedge operators: {strengthen operators, none, weaken operators}. Denote the set of qualitative values by symbol $\{h_+, h_0, h_-\}$. Let h be a qualitative linguistic hedge variable, its qualitative value $[h]$ is defined as follows:

$$[h] = \begin{cases} h_+, & \text{if } h \text{ is a strengthen operator;} \\ h_0, & \text{if } h \text{ has no effect to the truth or there is not hedge operator;} \\ h_-, & \text{if } h \text{ is a weaken operator.} \end{cases}$$

The operation " $\oplus$ ", " $\otimes$ "and " ' "of qualitative linguistic hedge value is defined as follows:

Table1. Operation " $\oplus$ "

$\oplus$	h_+	h_0	h_-
h_+	h_+	h_+	h_0
h_0	h_+	h_0	h_-
h_-	h_0	h_-	h_-

Table 2. Operation " $\otimes$ "

$\otimes$	h_+	h_0	h_-
h_+	h_+	h_0	h_-
h_0	h_0	h_0	h_0
h_-	h_-	h_0	h_+

Table 3. Negative operation

$[x]$	$[x]'$
h_+	h_-
h_0	h_0
h_-	h_+

Let $V = \{h_+T, h_0T, h_-T, h_+F, h_0F, h_-F\}$, $L_6 = (V, \vee, \wedge, \rightarrow)$ where " $\vee$ "," $\wedge$ "operations are shown in the Hasse diagram of L_6 defined as fig 1 and " ' "," $\rightarrow$ "operation are defined as table 4 and table 5 respectively. Then $L_6 = (V, \vee, \wedge, \rightarrow)$ is a lattice implication algebra.

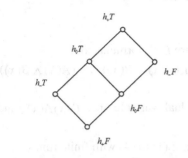

Fig. 1. Hasse diagram of L_6

Table 4. Complementary operator of $L = (V, \vee, \wedge, ', \rightarrow)$

V	V'
h_+F	h_+T
h_0F	h_0T
h_-F	h_-T
h_-T	h_-F
h_0T	h_0F
h_+T	h_+F

Table 5. Implication operator of $L = (V, \vee, \wedge, ', \rightarrow)$

$\rightarrow$	h_+F	h_0F	h_-F	h_-T	h_0T	h_+T
h_+F	h_+T	h_+T	h_+T	h_+T	h_+T	h_+T
h_0F	h_0T	h_+T	h_0T	h_0T	h_+T	h_+T
h_-F	h_-T	h_0T	h_+T	h_-T	h_0T	h_+T
h_-T	h_-F	h_-F	h_-F	h_+T	h_+T	h_+T
h_0T	h_0F	h_-F	h_-F	h_0T	h_+T	h_+T
h_+T	h_+F	h_0F	h_-F	h_-T	h_0T	h_+T

3   Six-elements Linguistic Truth-Valued First-Order Logic

Definition 2. The formal language of 6-elements linguistic truth-valued first-order logic is made up of the following symbols:

1. Individual constants: $a, b, c, \ldots$;
2. Individual variables: $x, y, z, \ldots$;
3. Function of symbols: $f, g, h, \ldots$;
4. The first-order languages: $F, G, H, \ldots$;
5. Conjunction symbols: $\neg, \wedge, \vee, \rightarrow, \leftrightarrow, \ldots$;
6. Quantifier symbols: $\forall, \exists$;
7. Punctuations: , , (,) .

Definition 3. Let x be a individual variable and $P(x)$ be a formula of atom, $v((\forall x)P(x))$ and $v((\exists x)P(x))$ are called $Lvfl_6$ atom where $v \in L_6$.

Definition 4. The formula of $Lvfl_6$ is defined as follows:

1. $Lvfl_6$ atom $hP(x)$ is a formula where $h \in H$;
2. If $A(x)$ is a $Lvfl_6$ formula then $(hA(x)), (\neg A(x))$ are $Lvfl_6$ formulas too;
3.If $A(x), B(x)$ are $Lvfl_6$ formulas then $\neg A(x), (A(x) \vee B(x))$, $(A(x) \wedge B(x))$ and $(A(x) \to B(x))$ are $Lvfl_6$ formulas too;
4. If $A(x)$ is a $Lvfl_6$ formula and x is a individual variable then $(\forall x)A(x)$ and $(\exists x)B(x)$ are $Lvfl_6$ formulas too;
5. All the formulas are the symbolic strings which use (1) to (4) with finite times.

Definition 5. A valuation of $Lvfl_6$ is a propositional algebra homomorphism $\gamma : Lvfl_6 \to L_6$.

Theorem 1. Let $A(x)$ and $B(x)$ be contains the variable x of 6-elements linguistic truth-valued first-order logic formula, the following properties are hold :

$$\gamma((\forall x)A(x) \to (\exists x)B(x)) \geq \gamma((\exists x)(\neg A(x) \vee B(x)))$$

Proof: In 6-elements linguistic truth-valued Propositional logic system, $(\forall x)A(x)$ and $(\exists x)B(x)$ as two atoms:

$$\gamma((\forall x)A(x) \to (\exists x)B(x))$$
$$\geq \neg \gamma(((\forall x)A(x)) \vee (\exists x)B(x))$$
$$= \gamma((\exists x)A(x) \vee (\exists x)B(x))$$
$$= \gamma((\exists x)(A(x) \vee B(x)))$$

Therefore, $\gamma((\forall x)A(x) \to (\exists x)B(x)) \geq \gamma((\exists x)(\neg A(x) \vee B(x)))$.

Theorem 2. Let $P(x)$ contain the variable x of first-order logic formula and c is a constant, the following properties are hold:

$$(1) \quad (\forall x)P(x) \Rightarrow P(x)$$
$$(2) \quad (\exists x)P(x) \Rightarrow P(c)$$

Proof: Let E be a domain of individual.
(1)If $\gamma((\forall x)P(x)) = L_6$. That is to say, for any $x \in E$, let the truth values of $P(x)$ be L_6. So we can get $\gamma(P(x)) = L_6$. Therefore, $(\forall x)P(x) \Rightarrow P(x)$.
(2)If $\gamma((\exists x)P(x)) = L_6$. That is to say, there is a constant $c \in E$, let the truth values of $P(x)$ be L_6. So we can get $\gamma(P(c)) = L_6$. Therefore, $(\exists x)P(x) \Rightarrow P(c)$.

Definition 6. Let $h_1 P(x)$ and $h_2(\neg P(x))$ be two formulas of $Lvfl_6$, and $\gamma((h_1 P(x)) \vee (h_2(\neg P(x)))) \in J_0$ for any γ , then $h_1 P(x)$ and $h_2(\neg P(x))$ are called complementary literals.

Theorem 3. Let G be a formula of $Lvfl_6$ and J be a filter of L_6. G is J-false if and only if there exists a deduction from G which can deduce J-null clause.

Proof: According to 6-elements linguistic truth-valued Propositional logic system, it can be proved easily.

Here we give the resolution method of $Lvfl_6$:

Step 1: Convert the formula into standard form and obtain clauses;

Step 2: If there exists J_0-false then the theorem is proved, stop. Otherwise go to Step 3;

Step 3: Resolution of J_0-complementary literals, the hedge operator of resolvent is the result of the operation of between the former hedge.

Step 4: If we get J_0-false clause, then the theorem is proved, stop. Otherwise go to Step 3.

Example 1

Premise: $(\forall x)(P(x) \rightarrow Q(x))$ and $(\forall x)(Q(x) \rightarrow R(x))$.

Conclusion: $(\forall x)(P(x) \rightarrow R(x))$.

Let $\gamma(P(x)) = h_+T$, $\gamma(Q(x)) = h_-T$ and $\gamma(R(x)) = h_+T$.

Transforming premises into clauses, we obtain

(1) $\neg P(x) \vee Q(x)$

(2) $\neg Q(x) \vee R(x)$.

The negation of the conclusion is

$$\neg((\forall x)(P(x) \rightarrow R(x)))$$
$$= (\exists x)\neg(P(x) \rightarrow R(x))$$
$$= (\exists x)\neg(\neg P(x) \vee R(x))$$
$$= (\exists x)(P(x) \wedge \neg R(x))$$
$$= P(a) \wedge \neg R(a)$$

The resolution proof is as follows:

(1) $h_+(\neg P(x)) \vee h_-(Q(x))$

(2) $h_-(\neg Q(x)) \vee h_+(R(x))$

(3) $h_+(P(a))$

(4) $h_+(\neg R(a))$

(5) $h_0(Q(a))$ from(1)and(3)

(6) $(h_0 \oplus h_+)(R(a)) = h_+(R(a))$ from(2)and(5)

(7) $\square$ from(4)and(6).

Thus, we have established the conclusion.

4   Conclusions

This paper discusses 6-elements linguistic truth-valued first-order logic based on 6-elements linguistic truth-valued lattice implication algebra. In 6-elements linguistic truth-valued first-order logic system discusses the satisfiable problem of 6-elements linguistic truth-valued first-order logic and proposes a resolution method of 6-elements linguistic truth-valued first-order logic. We will discuss the lock resolution and semantic resolution of the 6-elements linguistic truth-valued first-order logic in further work.

Acknowledgments. This work is partly supported by National Nature Science Foundation of China (Grant No.61105059,61175055,61173100), International Cooperation and Exchange of the National Natural Science Foundation of China (Grant No.61210306079), Sichuan Key Technology Research and Development Program (Grant No.2011FZ0051), Radio Administration Bureau of MIIT of China (Grant No.[2011]146), China Institution of Communications (Grant No.[2011]051), and Sichuan Key Laboratory of Intelligent Network Information Processing (Grant No.SGXZD1002-10),Liaoning Excellent Talents in University (LJQ2011116).

References

1. Xu, Y.: Lattice implication algebras. J. Southwest Jiaotong Univ. 89(1), 20–27 (1993)
2. Martinez, L., Ruan, D., Herrera, F.: Herrera, Computing with words in decision support systems: an overview on models and applications. International Journal of Computational Intelligence Systems 3(4), 382–395 (2010)
3. Herrera, F., Martinez, L.: A 2-Tuple fuzzy linguistic representation model for computing with words. IEEE Transactions on Fuzzy Systems 8(6), 746–752 (2000)
4. Xu, Y., Ruan, D., Qin, K.Y., Liu, J.: Lattice-Valued Logic: An alternative approach to treat fuzziness and incomparability. Springer, Heidelberg (2003)
5. Ho, N.C., Long, N.V.: Fuzziness measure on complete hedge algebras and quantifying semantics of terms in linear hedge algebras. Fuzzy Sets and Systems 158, 452–471 (2007)
6. Xu, Y., Liu, J., Ruan, D., Lee, T.T.: On the consistency of rule bases based on lattice-valued first-order logic LF(x). International Journal of Intelligent Systems 21, 399–424 (2006)
7. Xu, Y., Li, X.B., Liu, J., Ruan, D.: Determination of a-resolution for lattice-valued first-order logic based on lattice implication. In: Proc. 2007 International Conference on Intelligent Systems and Knowledge Engineering, pp. 1567–1574 (2007)
8. Pei, Z., Xu, Y., Ruan, D., Qin, K.: Extracting complex linguistic data summaries from personnel database via simple linguistic aggregations. Information Sciences 179, 2325–2332 (2009)
9. Pei, Z., Ruan, D., Xu, Y., Liu, J.: Handling linguistic web information based on a Multi-agent system. Int. J. Intelligent Systems 22, 435–453 (2007)
10. Chen, S.W., Liu, J., Wang, H., Xu, Y., Augusto, J.C.: A linguistic multi-criteria decision making approach based on logical reasoning. Information Sciences 258, 266–276 (2013)
11. Pei, Z.: Fuzzy risk analysis based on linguistic information fusion. ICIC Express Letters 3(3), 325–330 (2009)
12. Pei, Z., Shi, P.: Fuzzy risk anal sis based on linguistic aggregation operators. Int. J. Innovative Computing, Information and Control 7(12), 7105–7118 (2011)
13. He, X.X., Xu, Y., Li, Y.F., Liu, J., Martinez, L., Ruan, D.: α-satisfiability andα-lock resolution for a lattice-valued logic. In: The 5th International Conference on Hybrid Artificial Intelligence Systems, San Sebastian, Spain, pp. 320–327 (2010)

14. Zou, L., Ruan, D., Pei, Z., Xu, Y.: A linguistic truth-valued reasoning approach in decision making with incomparable information. Journal of Intelligent and Fuzzy Systems 19(4-5), 335–343 (2008)
15. Zou, L., Liu, X., Wu, Z., Xu, Y.: A uniform approach of linguistic truth values in sensor evaluation. International Journal of Fuzzy Optimization and Decision Making 7(4), 387–397 (2008)
16. Zou, L., Liu, X., Xu, Y.: Resolution method of linguistic truth-valued proposi tional logic. In: Proc. 2005 International Conference on Neural Networks and Brain, ICNN&B 2005, pp. 1996–1999 (2005)

Graph Similarity Join with K-Hop Tree Indexing

Yue Wang, Hongzhi Wang, Chen Ye, and Hong Gao

Harbin Institute of Technology, Harbin, 150001, China
hitwangyue@gmail.com, {wangzh,honggao}@hit.edu.cn,
sunnyleaves0228@qq.com

Abstract. Graph similarity join has become imperative for integrating noisy and inconsistent data from multiple data sources. The edit distance is commonly used to measure the similarity between graphs. To accelerate the similarity join based on graph edit distance, in the paper, we make use of a preprocessing strategy to remove the mismatching graph pairs with significant differences. Then a novel method of building indexes for each graph is proposed by grouping the nodes which can be reached in k hops for each key node with structure conservation, which is the k-hop-tree based indexing method. Experiments on real and synthetic graph databases also confirm that our method can achieve good join quality in graph similarity join. Besides, the join process can be finished in polynomial time.

Keywords: graph similarity join, edit distance constraint, k-hop tree based indexing, structure conservation, boundary filtering.

1   Introduction

Graphs have become a vital part of data representation in many real applications including biological and chemical information systems [9], pattern recognition [6] and social network [2, 5]. The growing popularity of graph utilization has generated more and more problems relevant to graph data analysis. Among them, graph similarity join problem has attracted much attention, which is used in graph information integration. The input of the graph similarity join problem we study is a graph database while the output is a set of pairs of graphs whose differences are within a predefined threshold. The join on two graph databases can be processed in similar method.

In multiple autonomous graph databases, the graphs may have error or the graphs referring to the same real-world entity may have various structures. Thus Graph similarity join with different constraints is important and essential on account that noisy and inconsistent data exist in the integrated database. In this case, graph similarity join can help to find graph pairs in which the edit distance between the two graphs is smaller than a threshold. We use an example to illustrate such operation. Suppose two corporations whose businesses are mainly about social network systems are going to merge. Each corporation has a graph database storing his user groups in terms of graphs. Then they need to figure out a way to match these graphs in order to integrate two such graphs into one. In these

H. Wang et al. (Eds.): ICYCSEE 2015, CCIS 503, pp. 38–47, 2015.

graphs, we assume each node represents a person with his name as the label and an edge indicates friendship between the two endpoints. Since in real life such graphs can be really big, we only cut a very small portion from two graphs. However, in real world, these minor differences may be caused by different representations in different data sources, thus the two social groups are probably identical. Therefore, these two graphs need to be matched by similarity join.

Among various graph similarity measures, graph edit distance [1, 7] has been widely used in representing distances between graphs. However, the computation of graph edit distance is a NP-Complete problem. In order to solve this problem, some approximate algorithms have been proposed as alternatives. One of the solutions aims to convert this problem to binary linear programming and then compute the bounds [4]. [3] seeks unbounded suboptimal answers with heuristic techniques. These two methods are both low in efficiency and not suitable for large data set. Other algorithms try to transform graphs into different data structures such as indexes. The tree-based q-gram [8] and path-based q-gram [9] methods belong to this category. However, tree-based q-gram method is only effective for sparse graphs with the loose lower bound caused by the great influence of edit operations on common q-grams. Besides, tree-based q-grams are usually short in length, which affects the join efficiency by generating candidates in large size. Path-based q-gram method is more flexible. However, this method only keeps each node's structural information along paths through it and considers each path separately. I believe there is some information loss in the process, which may affect the join quality.

To avoid the drawbacks of these methods, in this paper, we aim to solve the graph similarity join problem effectively with polynomial time and strive to preserve structural information. We first build indexes for each graph and then use the indexes to evaluate the dissimilarity between each graph pairs in the clusters.

In our method, we preprocess the graphs to accelerate the join. Graphs are clustered into several groups according to their node similarities so that the graphs inside each cluster share certain amount of nodes with the same label while those in different clusters seem less similar. In order to perform join efficiently in each cluster, we build a novel index for each graph by finding the nodes which can be reached within k hops for each key node.

In summary, we make the following contributions in this paper:

—In order to apply graph similarity join method with structure conservation, we propose a novel index structure called k-hop tree. We also make use of key nodes in the process to reduce unnecessary expensive comparisons.
—To efficiently find graph pairs that can be joined, clustering strategy is used by grouping the graphs which share certain similarity to some extent. Thus similarity join can be applied inside each cluster to avoid useless comparison.
—We propose an efficiently join algorithm based on k-hop tree indexing.
—We verify that our method can achieve accurate join results in polynomial time through experiments performed on real and synthetic data.

The rest of the paper is organized as follows. Section 2 provides some preliminary information, including definitions and notifications. In Section 3, we introduce the indexing strategy based on k-hop tree. We describe the preprocessing technique as well as the boundary filtering strategy in Section 4. The similarity join algorithm is also proposed in this section. Experimental results and analysis are proposed in Section 5. In Section 6, related work is discussed and analyzed. In Section 7, the conclusion is displayed together with future works.

2   Preliminary and Notations

Problem Statement: The graph similarity join problem we study takes a graph database as input and returns pairs of graphs whose graph edit distance is smaller than a certain predefined threshold. We denote the threshold as θ. Graph similarity join can be classified into two types. One is that there are two graph sets G and H, and similarity join returns one graph from each set, denoted as $\{(g, h) \mid ged(g, h) \leq \theta, g \in G, h \in H\}$. In the other type, similarity join is applied inside a graph set where each graph is distinguished from others by a unique identifier. In this case, the problem is $\{(g, h) \mid ged(g, h) \leq \theta, g \in G, h \in G, g \neq h\}$. Since these two types of problems can be processed in similar method. We focus on the second type in this paper.

Definitions and Notations. In this paper, we focus on the similarity join problem of undirected graphs without self-loops or multiple edges to simplify the discussions. A graph G is described as a quadruple (V, E, IV, IE), where V denotes the set of vertices, E represents the set of edges, IV and IE assign labels to vertices and edges, respectively. The number of vertices and edges in G is denoted by |V| and |E|, respectively. IV(u) represents the label of node u while IE(e(u, v)) represents the label of an edge between node u and v.

The graph edit distance [1, 7] between graph g and h is denoted by $ged(g, h)$. It is the minimum cost of edit operations that are used to transform graph g into graph h. As discussed in [9], the operations include: (1) Insert an isolated vertex into the graph; (2) Delete a vertex with all edges connecting to it from the graph; (3) Change the label of a vertex; (4) Insert an edge between two disconnected vertices ;(5) Delete an edge from the graph and (6) Change the label of an edge.

3   K-hop Tree Based Indexing Strategy

In this section, a novel index based on k-hop tree is introduced to accelerate graph similarity join. In this method, we build index for each graph by finding the nodes which can be reached within k hops for each key node in the graph. This index structure is easy to construct and maintain, which leads to more efficient implementation of similarity join.

There has been a notion in previous works which is called k neighborhood, which is usually used in finding communities in social networks or studying broadcasting in computer networks. Their notion does have different backgrounds from ours, which is used for indexing local structural information in graphs.

Several definitions are introduced to demonstrate how we construct the indexes based on k-hop tree.

Definition 1 (kth Percentile). Given a data set, the value x_i with the property that k percent of the data entries equals to or below x_i is called the kth percentile.

The most commonly used percentiles other than median are quartiles. The first quartile, denoted by Q_1, is the 25th percentile while the third quartile, denoted by Q_3, is the 75th percentile. Midquartile is the average of the 25th percentile and the 75th percentile. That is, midquartile = $(Q_1 + Q_3) / 2$.

Definition 2 (Key Nodes). Given a graph G with n nodes $(v_1, v_2...v_n)$, the graph is traversed and the degree of node v_i are recorded as d_i. Let w denote the midquartile of degrees. That is, $w = (Q_1 + Q_3) / 2$. We define the nodes whose degree is less than $w + 1$ as *key nodes*. Mathematically, $v_i \in$ key nodes, if $d_i < w + 1$.

After defining key nodes, we introduce how to build the k-hop tree indexing.

Definition 3 (K-hop Tree Index). Given a graph G with n nodes $(v_1, v_2...v_n)$, for each key node v_i, we build a tree index for it. The root the tree is v_i, its descendents are those which can be reached in k hops from v_i in the graph.

The k-hop tree indexing generation is provided in Algorithm 1. Firstly, we find key nodes for each graph (Line 2-4). Then in Line 5-7, all the nodes that are within k hops of each key node are found. Indexes are generated in Line 8.

The upper bound of the time complexity of the index generation algorithm is $O(Nn^3)$, where N is the number of graphs and n denotes the maximal number of vertexes in a graph. With the consideration that the size of each graph is not large, the index generation algorithm is efficient. Additionally, the index can be generated offline.

Algorithm 1. Index Generation

```
Input: Data set S, Q₁, Q₃, k;
Output: Index_tree
1 for each graph Gᵢ in S do
2     for each node v in Gᵢ do
3         if degree(v) < (Q₁[i] + Q₃[i])/2+1 then
4             v ∈ key_nodes[i]
5     for each key node u in Gᵢ do
6         W ← nodes within k hops to u
7         Index_node(u) ← W
8         In Index_tree(u), u is root, its children are W
9 return Index_tree
```

---

4   Graph Similarity Join Algorithms Based on K-hop Tree Indexing

The algorithm consists of two parts. One is the preprocessing step, which works as an initial filter to remove non-promising pairs and obtain the candidate set. The other is an edit distance based dissimilarity computation algorithm which is applied on each candidate pair and decide whether they can match.

4.1   Preprocessing

In this subsection, we introduce preprocessing strategy to remove non-promising graph pairs, in order to reduce the data size to be processed.

For each pair of graph G_1 and G_2, with their number of vertexes N_1 and N_2 respectively, we can find the number of common label vertexes, which is denoted by m. There may be repeated labels in a graph, but each unique label is only counted once. The total number of different vertexes in G_1 and G_2 is denoted as p. Then we define dissimilarity between G_1 and G_2 in formula (1).

$$\text{dissimilarity}(G_1, G_2) = (p-m)/p. \tag{1}$$

At the same time, a threshold is predefined to filter out non-promising graph pairs. Usually, the threshold is selected according to the data set. If the data set is randomly constructed, then a bigger threshold is preferred because random graphs can be more different and vice versa. When dissimilarity between two graphs survives the threshold, they are promising candidate pairs and are clustered into one group. We use Example 3 to illustrate the idea of the preprocessing.

Algorithm 2 shows how the preprocessing step works. The input is a graph data set and the output is the clustering results. Firstly, we calculate the number of vertexes with the same label each pair of graphs shares in Lines 2. Then in Line 3, we compute the dissimilarity between each two graphs according to Formula (1). In Lines 4 to 12, the graph pairs whose dissimilarity is no more than the threshold are classified into the same cluster. After finishing clustering, we only need to compute the edit distance between each graph pairs in each cluster. Thus the join algorithm can run very efficiently.

Algorithm 2. Clustering

```
Input: Data set S, filter threshold θ;
Output: Cluster_Set
1 for each two graphs G_i and G_j in S do
2     mat_num(G_i,G_j)= number of common label nodes in G_i, G_j
3     dissimilarity = 1 - mat_num(G_i,G_j)/(|V_Gi|+|V_Gj|)
4     if dissimilarity <= θ then
5         if sign(G_i) = k then
6             (G_i,G_j)is added to Cluster_k
7         elseif sign(G_j) = t then
8             (G_i,G_j)is added to Cluster_t
9         else
10            creat a new cluster Cluster_x in Cluster_Set
11            sign(G_i) = sign(G_j) = x
12            (G_i,G_j)is added to Cluster_x
13 return Cluster_Set
```

We later tested the performance of the clustering technique on synthetic data and found that more than 94% non-promising graph pairs had been removed in the clustering process. That is, we spend relatively small amount of effort doing the clustering and save much effort on the join process since the clustering technique is quite simple and effort-saving. This result is drawn from general data sets. On specific data sets such as chemical compounds where graphs may share the same node labels but have very different topology, the clustering technique is less effective. But as long as not all graphs have very identical node labels, the technique will still work and lead to improvement in efficiency.

4.2 K-hop Tree Indexing Based Join Algorithm

Graph similarity join techniques will be described in this subsection. Before the join, it is supposed that an index has been constructed for each graph according to the method in Section 3 and the graphs are clustered according to the method in Section 4.1. Then in each cluster, we compare each pair of graphs. To accelerate the comparison, the edit distance between two graphs is computed through their index structures. Based on this, the difference rate between each graph pairs is computed, which works as the criteria to decide whether the graph pair can match.

Note that the graphs in each cluster are already in pairwise format after clustering, so we apply comparison on each graph pair in each cluster. We compare graph Gi and Gj in the following steps. We use $ged(Gi,Gj)$ to denote the edit distance between the indexes of Gi and Gj. The framework of the comparison is a sort-merge on the key nodes of the two graphs. To accelerate the comparison of labels, we assign a uniform id to each label and the key nodes of each graph are sorted by the ids of their labels.

(1)   Both G_i and Gj have a pointer to indicate their current key node respectively. For the current key node u in Gi, apply the idea of sort-merge to see whether there are nodes with the same label in Gj. Note that there may be nodes with the same label as u in Gi, in this case, we denote them as $u_1, u_2,...u_d$ and deal with them together.

a)   If there is only one node u in G_i and one node v in G_j with the same label, we then compare the indexes of u and v. The edit distance between indexing trees is computed as the minimum number of edit operations to unitize (make them consistent) the leaf nodes of the two trees.

b)   If there is one node u in G_i and multiple v ($v_1, v_2,...v_m$) in G_j that have the same label, then we compare the indexes between u and $v_1, v_2,...v_m$ respectively. That is, compute edit distance of {index(u), index(v_1)}, {index(u), index(v_2)} and so on. We choose the smallest one as the edit distance.

c)   If there are multiple nodes u ($u_1, u_2,...u_d$) in G_i and one node v in G_j that have the same label, then we compare the indexes between $u_1, u_2,...u_d$ and v respectively and choose the smallest edit distance.

d)   If there are multiple nodes u ($u_1, u_2,...u_d$) in G_i and v ($v_1, v_2,...v_m$) in G_j that have the same label, then we compare the indexes of all possible combinations between u and v and choose the smallest sum of edit distances as our matching.

e)   Otherwise, there is no such label in G_j. Keep the node waiting until all key nodes in G_i have been traversed.

(2)   For each node w that is waiting (according to their order of joining the waiting list), compare it with all unmatched key nodes in G_j, and choose the smallest edit distance. The way of computing edit distance has been stated in step (1). One difference is that we add one to the distance since the two trees have different roots.

(3)   The total graph edit distance $ged(G_i, G_j)$ is the sum of the edit distances determined in previous steps. For each pairwise candidate in each cluster, the graph edit distance is computed.

(4)   Finally, difference rate between each graph pairs, which is denoted as diff_rate(G_i, G_j), is computed with formula (2). If it survives the predefined join threshold, then the two graphs can match with each other. Otherwise, they fail in graph similarity join.

$$diff\_rate(G_i, G_j) = \begin{cases} (2/3) \times ged(G_i, G_j)/(|V_{Gi}| \times |V_{Gj}|), & \text{if } |V_{Gi}| + |V_{Gj}| \leq 10 \\ ged(G_i, G_j)/(|V_{Gi}| \times |V_{Gj}|), & \text{if } |V_{Gi}| + |V_{Gj}| > 10 \end{cases} \quad (2)$$

In the formula above, $|V_G|$ denotes the number of vertexes in G. We heuristically divide the computation of *diff_rate*(G_i, G_j) into two parts on account that when a candidate pair has very small vertex size, very few differences between the two graphs can cause large influence since in this case, the index of a key node almost contains global structural information of the graph rather than local structural information around its neighborhood. Therefore, we apply a coefficient to make some adjustments in order to get more accurate results. The number 2/3 and 10 are both heuristically chosen based on the algorithm we have described.

5   Performance Experiments

In this section, experiments are performed on real and synthetic data set for performance evaluation on our methods. All experiments are performed on a PC with Intel Core Duo 2GHz, 2G main memory and 250G hard disk. The operating system is Windows 7. All algorithms are implemented with C++.

The real graph database is from the Protein Data Bank[1], which contains 600 protein structures. Synthetic data is generated by randomly creating 100 to 500 graphs where the number of vertexes in each graph is randomly generated, the vertex labels are randomly assigned from a predefined label set and edges are randomly created between two vertexes. The average number of vertexes in the set is 28.

5.1   The Impact of Clustering and Edit Distance Computation

In this part, we test the clustering strategy and the edit distance computation algorithm by counting the number of candidates and non-promising candidate pairs.

We use synthetic data for this experiment. The coefficient k is chosen as 2, the filter threshold is set to be 0.40 and the join threshold is set to be 0.15. The result is shown in Figure 3 (a). The two curves in this figure denote the candidate numbers after applying clustering and join respectively. The maximum number of candidate pairs is N(N-1)/2, where N is the number of graphs. Based on this observation, more than 94% non-promising graph pairs have been removed in the clustering process which surely improves the join efficiency.

---

[1] http://www.iam.unibe.ch/fki/databases/
    iam-graph-database/download-the-iam-graph-database

Besides, the number of candidate pairs that survive similarity join is also provided. From the results, we can see that our strategy can effectively filter out unmatched candidate pairs though they survive the initial filter. As the number of graphs grows, the number of candidate pairs also increases smoothly.

5.2   The Impact of Parameter *K*

In this section, the impact of parameter k on the join results is tested. Two groups of experiments are performed on synthetic data.

The first experiment plots the run time versus the number of joined graph pairs with different k in Figure 1(a). We see that the influence of k on run time is small. Besides, the number of matched graph pairs first grows with increasing k and then stays almost unchanged when k reaches 3 or 4. That is because a larger k can usually generate larger indexing structures, thus indexes of different nodes are more likely to be similar. Nevertheless, when k increases, it will reach a point where indexes of each node do not change any more and consequently the number of joined graph pairs keeps almost unchanged.

The second experiment plots run time versus different parameter k and different filter thresholds. The filter threshold is the one applied in the preprocessing step. The experimental result is provided in Figure 1(b). When the filter threshold is small, the run time is also reduced evidently and less graph pairs are joined later. Consequently, when the filter threshold is too small, graph pairs with small vertex sizes will not survive this initial filter even if they can survive similarity join. Therefore, the threshold should be carefully chosen. One possible way would be to keep a relatively small hold-out set and use it to determine the most suitable threshold.

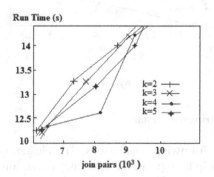

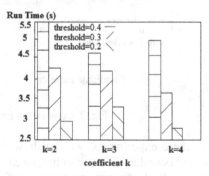

(a)   The run time with various *k*   (b)   Run time with *k* and threshold

Fig. 1. Experimental results of the Impact of k

5.3   Compare with other Methods

We compare our similarity join algorithm with the state-of-the-art algorithms, κ-AT [8] and GSimJoin [9]. Even though graph embedding [10] is used for graph classification and clustering, the problem is from ours and in their experiment part, they only tested their proposed methods on different datasets in terms of classification accuracy, which is totally different from ours. Therefore, we do not compare their methods with ours. The Protein graph database is used for the comparison.

First, the number of candidates after filtering is compared in each method. In κ-AT and GSimJoin, the candidates we use are Cand-2. The outcome of the experiment is shown in Figure 2(a). From the results, it can be observed that κ-AT generates the least number of candidates while GSimJoin and our method generate comparable amount of candidates.

Then, the total run time of the three methods is evaluated, which is displayed in Figure 2(b). κ-AT needs more run time than the other two methods though it has smallest candidate size. The run time of our method falls in between the other two while that of GSimJoin is the least.

In general, our method obtains good efficiency with guarantee on the join quality. Although GSimJoin has the least run time, as we have indicated, path-based q-gram method breaks apart the local structural information since it only keeps each node's structural information along paths through it and then considers each path separately. There can be some information loss in the process, which may affect the join quality. Therefore, its join quality is somewhat doubted. On the other hand, κ-AT is not as efficient as our algorithm.

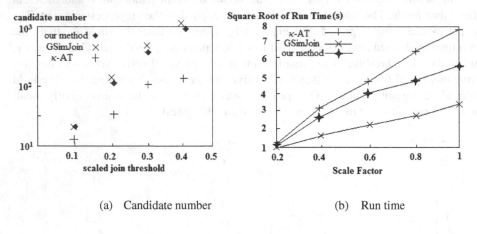

(a)   Candidate number   (b)   Run time

Fig. 2. Comparison Experimental Results

From these experiments, we can draw the following conclusions. The clustering process and boundary filtering technique can help to improve the efficiency of our join algorithm, with the former one more significant. Besides, our algorithm is robust because it is not sensitive to the change of related parameters. Moreover, our method can achieve both high efficiency and join quality at the same time.

6   Conclusions

The wide applications of graph database have generated many data management problems, among which graph similarity join is a popular one which has attracted much attention in the literature. In the paper, we study the graph similarity join problem with edit distance as the constraint. We propose a new idea of k-hop tree indexing strategy for join. After non-promising graph pairs are filtered out, a second filter of boundary filtering technique is used to remove candidate pairs which cannot survive the boundary constraints of similarity join. Generally, our algorithm is performed in the following procedures. First, the preprocessing step is applied by clustering all the graphs. Second, k-hop tree indexing structures are constructed for each key node of the graphs. Then similarity join algorithm is used with boundary filtering to find graph pairs whose edit distance survives the join threshold. Finally, the join result is returned. We have experimentally verified that our similarity join method is efficient and effective. It is also a flexible method without many complicated computations.

Acknowledgements. This paper was partially supported by NGFR 973 grant 2012CB316200, NSFC grant 61472099 and National Sci-Tech Support Plan 2015BAH10F00.

References

1. Bunke, H., Allermann, G.: Inexact graph matching for structural pattern recognition. Pattern Recognition Letters 1(4), 245–253 (1983)
2. Fan, W., Li, J., Ma, S., Tang, N., Wu, Y., Wu, Y.: Graph pattern matching: From intractable to polynomial time. PVLDB 3(1), 264–275 (2010)
3. Fankhauser, S., Riesen, K., Bunke, H.: Speeding up graph edit distance computation through fast bipartite matching. In: GbRPR, pp. 102–111 (2011)
4. Justice, D., Hero, A.O.: A binary linear programming formulation of the graph edit distance. IEEE Trans. Pattern Anal. Mach. Intell. 28(8), 1200–1214 (2006)
5. Ma, S., Cao, Y., Fan, W., Huai, J., Wo, T.: Capturing topology in graph pattern matching. PVLDB 5(4), 310–321 (2011)
6. Qin, J., Wang, W., Lu, Y., Xiao, C., Lin, X.: Efficient Exact Edit Similarity Query Processing with Asymmetric Signature Schemes. In: SIGMOD (2011)
7. Sanfeliu, A., Fu, K.-S.: A distance measure between attributed relational graphs for pattern recognition. IEEE Transactions on Systems, Man, and Cybernetics 13(3), 353–362 (1983)
8. Wang, G., Wang, B., Yang, X., Yu, G.: Efficiently indexing large sparse graphs for similarity search. IEEE Transactions on Knowledge and Data Engineering PP(99), 1 (2010)
9. Zhao, X., Xiao, C., Lin, X., Wang, W.: Efficient graph similarity joins with edit distance constraints. In: ICDE (2012)
10. Conte, D., Ramel, J.-Y., Sidère, N., Luqman, M.M., Gauzère, B., Gibert, J., Brun, L., Vento, M.: A Comparison of Explicit and Implicit Graph Embedding Methods for Pattern Recognition. In: Proceedings of the 9th IAPR-TC15 workshop on Graph-based Representation in Pattern Recognition (GbR 2013), pp. 81–90 (2013)

Efficient String Similarity Search on Disks

Jinbao Wang and Donghua Yang*

The Academy of Fundamental and Interdisciplinary Sciences,
Harbin Institute of Technology, Harbin, 150080, China
{wangjinbao,yang.dh}@hit.edu.cn

Abstract. String similarity search is a basic operation for various applications, such as data cleaning, spell checking, bioinformatics and information integration. Memory based q-gram inverted indexes fail to
support string similarity search over large scale string datasets due to
the memory limitation, and it can no longer work if the data size grows
beyond the memory size. In the era of big data, large string dataset
are quite common. Existing external memory method, Behm-Index, only
supports length-filter and prefix filter. This paper proposes LPA-Index
to reduce I/O cost for better query response time, and LPA-Index is a
disk resident index which suffers no limitation on data size compared to
memory size. LPA-Index supports multiple filters to reduce query candidates effectively, and it adaptively reads inverted lists during query
processing for better I/O performance. Experiment results demonstrate
the efficiency of LPA-Index and its advantages over existing state-of-art
disk index Behm-Index with regard to I/O cost and query response time.

Keywords: We would like to encourage you to list your keywords within
the abstract section.

1 Introduction

String similarity search[1] is a basic operation for a variety of applications, for
instance data cleaning and bioinformatics. With the rapid growth of data volume
generated and stored by human, large scale string datasets have become quite
common. Google has kept a N-Gram dataset consisting of more than 10^{12} tuples
by 2011. In the area of bioinformatics, the GeneBank dataset stores more than
10^6 items and the total size has reached 416GB. The era of big data has come. In
recent years, more and more larger datasets have been reported. It is necessary
to support efficient string similarity search over such large scale datasets.

Existing works on string similarity search are mainly based on q-gram and
memory resident inverted indexes, with which the query processing consists of
two major phases named filtering and verification. They convert each string to
a set of q-grams, and string ids with the same q-gram are organized into an
inverted list. The inverted index includes all the inverted lists whose q-gram
occurs in the string dataset. The size of an inverted index is usually larger than

---

* Corresponding author.

H. Wang et al. (Eds.): ICYCSEE 2015, CCIS 503, pp. 48–55, 2015.

the original dataset. When the dataset grows comparable to the memory size, existing works fail to provide acceptable query performance or even can not work. Behm-Index[2] implements an external memory based inverted index, and proposes an adaptive query processing algorithm based on prefix filter.

This paper targets at better query performance, and combines existing optimization techniques for string similarity search, including length filter, positional filter, prefix filter and asymmetric gram scheme, to build LPA-Index. We present the construction method and implementation details of LPA-Index. During string similarity search, LPA-Index adopts asymmetric gram scheme and adaptively selects a set of grams from the query string for loading inverted lists. Compared to Behm-Index, LPA-Index reduces query cost from two opposite aspects. First, LPA-Index employs positional filter in the index structure to further reduce the number of candidates. At the same time, LPA-Index answers queries with asymmetric gram scheme in disk environments and significantly reduces the cost for loading inverted lists. Experimental results on a real dataset show that LPA-Index achieves a great performance improvement over Behm-Index.

The main contributions of this paper are listed as follows:

1. We design LPA-Index consisting of a memory resident *gram trie* and a disk resident inverted index, which support length filter and positional filter to reduce the query cost in the verification phase.
2. An adpative query processing algorithm is presented for LPA-Index based on prefix filter and asymmetric gram scheme, and the query cost in the filtering phase is significantly reduced in disk environments.
3. Extensive experiments are conducted on a real dataset to evaluate LPA-Index. Experimental results illustrate a significant performance improvement of LPA-Index over state-of-art Behm-Index.

1.1 Preliminaries

Here we present some necessary preliminaries on string similarity search and inverted indexes.

Definition 1. *Edit Distance* *Given two strings s and t, the edit distance between them denoted by $ED(s,t)$ is the minimum number of single character modification to transform one string to the other. The modifications include insertion, deletion and substitution.*

Definition 2. *String Similarity Search (based on edit distance)* *string similarity search $Q(s, \theta)$ consists of a string s and an edit distance threshold θ. It searches in a string dataset S and return all the strings in S, whose edit distance to s is no more than θ, i.e. $\{t|t \in S, ED(s,t) \leqslant \theta\}$.*

String similarity search based on inverted indexes has two major phases: filtering and verification. In the filtering phase, a set of inverted lists are merged to find candidates whose frequence are no less than a specified threshold. In the verification phase, edit distance between candidates and the query string are computed, and the final query results are returned.

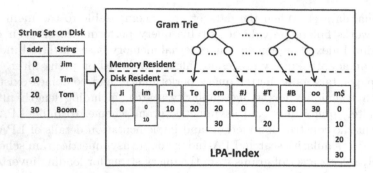

Fig. 1. Structure of LPA-Index

2 LPA-Index

A LPA-Index consists of a main memory resident *gram trie* and a disk resident inverted index, as depicted in Fig. 1. The main memory resident *gram trie* adopts a trie structure to store all the q-grams of the given string dataset S. Each leaf node of the *gram trie* represents a q-gram, and it keeps the disk address of the inverted list for the corresponding q-gram. Fig. 1 shows the four trie nodes which store the disk address of 2-gram "To", "$\#j$", "$\#B$" and "oo".

To enhance the filter effect of LPA-Index with both length filter and positonal filter, the length of a string s and the positions of s's q-grams are taken into accounts when calculating the q-gram set of s. Given $GS = g_0, g_1, ..., g_{|s|+q-2}$, in which g_i is the ith q-gram of s, LPA-Index builds a new gram set denoted $GS^{LPA} = \{(g_0, |s|, 0), (g_1, |s|, 1), ..., (g_{|s|+q-2}, |s|, |s| + q - 2)\}$. During the insertion of $(g_i, |s|, i)$ into the *gram trie*, the numeric values of $|s|$ and i are directly considered as characters following the content of g_i.

In the process of string similarity search which adopts length filter and positional filter, inverted list with the same q-gram and consecutive string lengths and gram positions are often loaded at the same time in the filter phase. To improve the disk reading performance of multiple inverted lists with the same q-gram and consecutive string length and gram positions, LPA-Index stores inverted lists with the same $q - gram$ together, and order them by string length and gram position. In more details, these inverted lists are firstly ordered by string length, and then by gram position. To support large string dataset on disks, LPA-Index stores the disk address of strings in the inverted list other than string id, which is adopted in Behm-Index. There are two major concerns. Firstly, if string ids are kept in inverted list, the map from string ids to disk addresses of strings must be referred in verification phase. The map size grows linearly with the number of strings, and it is not affordable to keep it in memory while dealing with large string dataset. Secondly, an integer string id is not feasible for too large string dataset, although such string datasets are already visible in real life applications. LPA-Index stores 8B disk addresses instead of 4B string ids due to the above concerns. Though more disk space is consumed,

LPA-Index works with much less main memory compared to existing state-of-art Behm-Index. To further reduce the main memory consumption, LPA-Index uses string length segment and gram position segment instead of the original string length and gram position. Consecutive string lengths and gram positions are organized into the same segment, thus the number of entries in the *gram trie* is greatly reduced and returns into control. In the rest of this paper, the string length segment and gram position segment are denoted l_{len} and l_{pos} respectively.

3   Query Processing

Algorithm 1: GRAMSELECTION

Input: string dataset S, LPA-Index *index* for S, query $Q(s, \theta)$, start position
p_s, end position p_e
Output: query gram set $QG(s) \subseteq G(s)$

1 **for** $p_s \leq i \leq p_e$ **do**
2 $\quad$ calculate $c(i)$;

3 **if** $p_e - p_s \leq q$ **then**
4 $\quad$ **return** $i = arg_{p_s \leq i \leq p_e} min\{c(i)\}$

5 **else**
6 $\quad$ $i = arg_{\frac{p_s+p_e}{2}-q+1\leq i\leq \frac{p_s+p_e}{2}+q-1} min\{c(p_s, i-q+1) \times c(i) \times c(i+q-1, p_e)\}$

7 **return** $GramSelection(p_s, i-q+1) \cup \{GS(s)[i]\} \cup GramSelection(i+q-1, p_e)$

LPA-Index stores the length of each invert list in the *gram trie*, and it selects a set of grams from $L(GS(s))$ whose inverted lists are used for query processing while the other inverted lists keep untouched. To satisfy the constraints of asymmetric gram scheme, in which each single-character modification destroys just one gram, LPA-Index selects grams whose positions have no overlap in content. There are multiple selecting strategy for a given query string s, and LPA-Index selects a set of gram $GmSel(s)$ from $GS(s)$ to minimize the following cost model.

$$c(GmSel(s)) = \prod_{g \in GmSel(s)} \frac{|L(g)|}{|S|}. \tag{1}$$

This model comes from the observation that the pruning effect of an inverted list in the filter phase is approximately decided by its length, and the ratio $|L(g)|/|S|$ can be regarded as the number of elements $L(g)$ trends to reserve as candidates. Thus, the model above is an approximation to measure the ratio of the candidates to the total number of strings $|S|$. Next we present the definition of Query Gram Selection, whose goal is to select a set of grams for query processing and to minimize the cost model in equation 1. Algorithm 1 will solve the Query Gram Selection problem. Here we denote $c(i) = |L(GS(s)[i])|/|S|$, which is the cost of selecting the ith gram from $GS(s)$.

Definition 3. QUERY GRAM SELECTION *Given a LPA-Index index, a query string* s. *Compute a gram set* $GmSel(s) \subseteq GS(s)$, $\forall g_i, g_j \in GmSel(s)$, *their position difference in* $GS(s)$ *is no less than gram length* q *and minimize* $c(GmSel(s)) = \prod_{g \in GmSel(s)} \frac{|L(g)|}{|S|}$.

Algorithm 2: LPAQUERYPROCESSING

Input: string dataset S, LPA-Index *index* for S, query $Q(s, \theta)$
Output: string dataset $\{t | t \in S, ED(s, t) \leq \theta\}$
1 $gramList = GramSelection(s, Q(s, \theta), 0, |GS(s)| - 1)$;
2 $prefix = |gramList| - \theta + 1$;
3 read the first $prefix$ inverted lists in $gramList$;
4 combine the read lists into L_{ini} and calculate the frequency of each disk address;
5 $result = \phi$;
6 $\tau = 2$;
7 **while** *there are lists left and reading a next list is beneficial* **do**
8 | read a next list L;
9 | combine L into L_{ini};
10 | remove each element in L_{ini} whose frequency is smaller than τ;
11 | $\tau = \tau + 1$;

12 **for** *all* $addr \in L_{ini}$ **do**
13 | $t = read(addr)$;
14 | **if** $ED(s, t) \leqslant \theta$ **then**
15 | | $result = result \cup \{t\}$;

16 **return** $result$;

The problem QUERY GRAM SELECTION. defined above is polynomial time solvable, and Alg. 1 is able to compute the optimal solution between a starting position p_s and an ending position p_e in $GS(s)$ with dynamic programming technique. The optimal solution of QUERY GRAM SELECTION can be obtained by running Alg. 1 with $p_s = 0$ and $p_e = |GS(s)| - 1$. Given the starting position p_s and ending position p_e, Alg. 1 only selects the gram with minimal cost from $GS(s)[p_s]$ to $GS(s)[p_e]$ if $p_e - p_s < q$. This is because only one gram can be chosen according to the constraint of asymmetric gram scheme. If $p_e - p_s \geqslant q$, then multiple grams should be selected from $GS(s)[p_s]$ to $GS(s)[p_e]$. At this time, Alg. 1 employs the following optimal sub-structure to construct dynamic programming.

$$c(p_s, p_e) = \min_{\frac{p_s + p_e}{2} - q + 1 \leqslant i \leqslant \frac{p_s + p_e}{2} + q - 1} c(p_s, i - q + 1) \times c(i) \times c(i + q - 1, p_e) \quad (2)$$

Theorem 1. *Given a query string* s, *Alg. 1 is able to compute the optimal solution of* QUERY GRAM SELECTION *when specifying* $p_s = 0$ *and* $p_e = |GS(s)| - 1$.

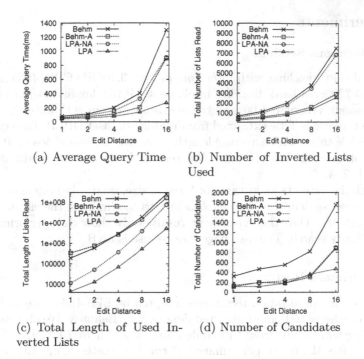

(a) Average Query Time

(b) Number of Inverted Lists Used

(c) Total Length of Used Inverted Lists

(d) Number of Candidates

Fig. 2. Query Performance on Title Dataset

Given a query string s, whose q-gram set is $GS(s)$, the time complexity of Alg. 1 is $O(q \times |GS(s)| \times \log GS(s))$. The space complexity of Alg. 1 is $O(|GS(s)| \times \log GS(s))$.

Alg. 1 computes the query grams, which are used in the filter phase. Then, Alg. 2 invokes Alg. 1 to answer a string similarity search.In the inverted list reading process, Alg. 2 needs to check whether reading an inverted list is beneficial. We denote that the cost to verify a candidate as $c_{verify}(ms)$, the cost of reading the next inverted list L as c_L, and the number of elements with τ-frequency in L_{ini} as num_τ. With the assumption the probability that each string address exists in L is the same, then after reading L we have $num_{tau} \times (1 - |L|/|S|)$ elements filtered. In the verification phase, the cost to verify candidates is reduced by $num_{tau} \times (1 - |L|/|S|) \times c_{verify}$. On the other hand, reading L incurs a cost $c_L = a \times |L| + b(ms)$, where a is decided by the disk data transferring speed and b is decided by the disk random access speed. In this paper, we test our computer with a diversity disk readings and the values of a and b are 10^{-4} and 10 respectively. Verifying a candidate incur a disk random search and an edit distance computation, and the time is dominated by disk random search, whick a around $10ms$. In Alg. 2, a next inverted list is used if $num_\tau \times (1 - |L|/|S|) \times c_{verify} > c_L$.

4 Experiments

4.1 Experiment Settings

Our testbed is PC machine, with Intel Core i5-2400 3.10GHz CPU, 4GB memory and 500GB 7200rpm hard disk. We implement LPA-Index and Behm-Index in JAVA whose jdk version is 1.06, and the operating system is Ubuntu 12.04 LTS. The dataset we use is Title extracted from real-life dataset DBLP. Title contains 2078174 article titles, whose average length is 68. We random choose 100 strings from each dataset as query strings. We test Title dataset with edit distance threshold 1, 2, 4, 8.

Although the datasets we use are not large in size, we invoke system command $echo$ 3 > $\backslash proc \backslash system \backslash vm \backslash drop\_caches$ in java to eliminate the effects of disk buffer in memory. Thus the experiment results only show disk performance of string similarity search. The disk page size we use is 8KB.

4.2 Experiment Results

To illustrate the optimization techniques adopted by LPA-Index, we add asymmetric gram scheme to Behm-Index and denote it by Behm-A. We also eliminate asymmetric gram scheme from LPA-Index and denote it by LPA-NA.

Fig. 2 depicts the query performance of the four methods on Title dataset. As shown in Fig. 2(a), Behm-Index took the longest time to finish a query, followed by Behm-A then LPA-NA. LPA-Index provides shortest average query time. Compared to Behm-Index, LPA-Index shows a maximal speed-up at 5. Fig. 2(b) shows the number of inverted lists read by 100 queries. It is easy to see that LPA-Index and Behm-A use fewer inverted lists. This illustrates that the asymmetric gram scheme adopted by LPA-Index is effective to reduce the number of inverted list loaded. Fig. 2(c) presents the number of inverted list entry read by 100 queries. It can be seen that LPA-Index and LPA-NA read much fewer entries than Behm-A and Behm-Index. This shows the effectiveness of positional filter used by LPA-Index. Fig. 2(d) compares the candidate number of the four methods with 100 queries. From Fig. 2(d), we can see that LPA-Index and LPA-NA have to verify much fewer candidates. This proves that the positional filter is able to reduce the cost of verification phase.

5 Related Work

Inverted index compression are proposed by [4]. These works combine multiple inverted lists or remove partial lists to spare memory space. Incremental update methods of inverted indexes are studied in [5]. [6] proposes asymmetric gram scheme to reduce the number of inverted lists used for query processing. B^{ed}-tree[7] is disk resident tree structured index to support string similarity search. It stores strings in a B^+-tree and provide multiple functions. However, B^+-tree involves many disk random search, which incurs long query time in turn. Behm-Index[2] is the first work known to study inverted index on disk for string similarity search. It combines prefix filter and adaptive query algorithm to answer string similarity search. Compared to Behm-Index, our work support positional filter to reduce candidate number. At the same time, we adopt asymmetric gram scheme and design efficient query gram selection algorithm to significantly improve the query performance.

6 Conclusion

This paper studies string similarity search in disks, and designs LPA-Index, which combines prefix filter, length filter and positional filter and employs asymmetric gram scheme. We present an index construction method and an efficient query algorithm under asymmetric gram scheme. LPA-Index adopts prefix filter and carefully controls the number of inverted lists used for query processing based on asymmetric gram scheme. Experiments are conducted on a real dataset dataset. Performance comparison shows that LPA-Index provide significant performance improvement over the existing state-of-art work Behm-Index.

Acknowledgments. This work is funded by Project (No. 61272046) supported by the National Natural Science Foundation of China; Project supported by the Natural Science Foundation of Heilongjiang Province,China(Grant No. F201317); The Fundamental Research Funds for the Central University (Grant No. HIT.NSRIF.2015065); China Postdoctoral Science Foundation Funded Project(Grant No. 2013T60372, 2014M561351).

References

1. Wagner, R., Fischer, M.: The string-to-string correction problem. Journal of the ACM 21(1), 168–173 (1974)
2. Behm, A., Chen, L., et al.: Answering approximate string queries on large data sets using external memory. In: Proc of IEEE ICDE 2011, pp. 888–899. IEEE Computer Society, Washington, DC (2011)
3. Zobel, J., Moffat, A.: Inverted files for text search engines. ACM Computer Survey 38(2), 6–20 (2006)
4. Chen, L., Jiaheng, L., et al.: Efficient merging and filtering algorithms for approximate string searches. In: Proc of IEEE ICDE 2008, pp. 257–266. IEEE Computer Society, Washington DC (2008)
5. Hadijieleftheriou, M., Koudas, N., et al.: Increamental maintenance of length normalized indexes for approximate string matching. In: Proc of ACM SIGMOD 2010, pp. 429–440. ACM, New York (2011)
6. Jianbin, Q., Wei, W., et al.: Efficient exact edit similarity query processing with the asymmetric signature scheme. In: Proc of ACM SIGMOD 2011, pp. 1033–1044. ACM, New York (2011)
7. Zhenjie, Z., Marios, H., Beng-Chin, O., et al.: Bed-tree: An all-purpose index structure for string similarity search based on edit distance. In: Proc of SIGMOD 2010, pp. 915–926. ACM, New York (2010)

A Joint Link Prediction Method for Social Network

Xiaoqin Xie, Yijia Li, Zhiqiang Zhang, Shuai Han, and Haiwei Pan

College of Computer Science and Technology
Harbin Engineering University, 150001 Harbin, China
xiexiaoqin@hrbeu.edu.cn

Abstract. The popularity of social network services has caused the rapid growth of the users. To predict the links between users has been recognized as one of the key tasks in social network analysis. Most of the present link prediction methods either analyze the topology structure of social network graph or just concern the user's interests. These will lead to the low accuracy of prediction. Furthermore, the large amount of user interest information increases the difficulties for common interest extraction. In order to solve the above problems, this paper proposes a joint social network link prediction method-JLPM. Firstly, we give the problem formulation. Secondly, we define a joint prediction feature model(JPFM) to describe user interest topic feature and network topology structure feature synthetically, and present corresponding feature extracting algorithm. JPFM uses the LDA topic model to extract user interest topics and uses a random walk algorithm to extract the network topology features. Thirdly, by transforming the link prediction problem to a classification problem, we use the typical SVM classifier to predict the possible links. Finally, experimental results on *citation* data set show the feasibility of our method.

Keywords: Social Network, Link Prediction, Topic Model, Random Walk.

1 Introduction

Community-based social network services such as Twitter, Facebook and Myspace have already gathered billions of extensively acting users and are still attracting thousands of newbies each day[1][2]. Effectively modeling user interest and friendship between users, accordingly recommending services and/or suggesting friends are fundamental to all social network services[2]. Link prediction has been recognized as one of the key tasks in social network analysis.

Most link prediction methods depend on such node attributes as user interests, user comments, user-created documents and so on. Furthermore, user behaviors may influence the future link in social networks while the information about a user's behavior is often scattered in both friendship links and interest features. But it is difficult to obtain the concrete node attributes because of the privacy issues. In addition, the openness and dynamics of social network make the information provided by users be not effective or not correct. Most of the present link prediction methods either analyze the topology structure of the social network graph or just concern the user's interests.

H. Wang et al. (Eds.): ICYCSEE 2015, CCIS 503, pp. 56–64, 2015.

These will lead to the low accuracy of prediction. Furthermore, the large amount of user interest information increases the difficulties for common interest extraction. Moreover, most present link prediction approaches ignore the interest relations among users. In order to improve the link prediction performance, we synthetically take into account both the user interest and the network structure features. Then we propose a Joint Link Prediction Method(JLPM) to predict the unknown links between nodes. JLPM transforms the link prediction problem into a classification problem and integrates the above two kinds of features. Parimi et al.[3] use the LDA model to model user interest and construct features for the link prediction problem based on the resulting topic distributions. However, their approach could be inaccurate because that neither user friendship nor user behavior information is taken into account. Similar to their idea, we also use LDA topic modeling to extract user interest topics features to avoid the direct use of user interests. Furthermore, we use a random walk algorithm to extract network topology features.

The contributions of this paper are as follows: (1) Proposing a joint prediction feature model JPFM and corresponding extracting method which considers not only the user interest features but also the network topology features. (2) Proposing a joint method JLPM for social network link prediction. (3) Experimental results on real datasets show that our method can achieve better performance.

The rest of the paper is organized as follows: Section 2 discusses related works. Section 3 gives the problem formulation. Section 4 provides a detailed description of JLPM including the model architecture, feature extraction and model construction. Section 5 presents the experimental design and results. We conclude the paper with a summary and propose the future works in Section 6.

2   Related Works

Link prediction has been an important task for analyzing social networks and also has many applications in a wide range of domains [4]. Hsu et al.[5] have studied the friendship relations of social network prediction and classification problems based on the network structure and user attribute data. They also mentioned interest features. But their study does not capture the semantic relationships between the users' interests[3].   Taskar et al.[6] have studied the use of a relational markov network framework for the task of link prediction. Similarly O'Madadhain[7] uses machine learning methods to analyze the location information network. In these researches the link prediction problem is always transformed into a classification problem, and the features of the network topology and node information are extracted from the training data. Our research is related to theirs, but we use different features. Shuang Hong Yang et al. [2] presented a FIP model that integrates the learning for interest targeting and friendship prediction into one single process. But FIP considers the link prediction problem and interest targeting problem as paralleled parts. In fact, only after a user interest is targeted we could achieve the goal of link prediction. So our research takes a different perspective. We extract user interest as a feature for link prediction. Magdalini Eirinaki et al. [8] presented a trust-aware system for generating personalized user recommendations in social networks. Our research does not concern the trust relationship. Jin Huang et al. [9] address the link prediction problem by aggregating heterogeneous social networks and propose a joint manifold factorization method. They transfer the common group structure knowledge between two related matrices and explore the individual matrix geometric structure. But they mainly focus on the graph structure features and ignore the topic feature of individuals. In addition, their prediction goal is different from ours.

Real social network scale is very large. It is becoming necessary to accurately pre-process these data for extracting hidden patterns. Topic modeling techniques are one of the generative probabilistic models that have been successfully used to identify inherent topics in collections of data, and have shown good performance when used to predict. Latent Dirichlet Allocation(LDA)[10][11] is such a generative probabilistic model. Parimi et al.[3] use the LDA model to model user interest and construct features for the link prediction problem based on the resulting topic distributions. They verify that the LDA model can reduce the link prediction time complexity. However, their approach could be inaccurate because that neither user friendship nor user behavior information is taken into account. Our research is same as theirs when modeling the user interest features. But our JLPM method improves the similarity calculation accuracy of the topic features and further improves the link predictive accuracy. Lichtenwalter et al.[12] proposes the PropFlow algorithm to extract the network topology information in complex networks.

3   Problem Definition

Assume that G=(V, E) denote an undirected social network , where $V=\{v_1, v_2, ..., v_n\}$ means n users, $E=\{e_1, e_2, ..., e_m\}$ represents m friendship links between v_i. An edge $e=(v_i, v_j)$ represents the friendship link at time t(e). The link prediction problem can be formulated as follows:

PROBLEM 1. Given a social network G at time t, a training interval[t, t+1] in a supervised training set for link prediction, the link prediction task is to output a list of edges not present in G[t], but are predicted to appear in the network G[t+1].

Definition 3.1: User Interest Feature (UIF). Each user node v is associated with a vector $\delta_v \in R^T$ of k-dimensional topic distribution ($\sum_z \delta_{vz} =1$). Each element δ_{vz} is the probability of the node on topic z. We call the vector δ_v as user v's UIF.

Definition 3.2: User Interest Feature Matrix IFM = $(b_{ik})_{n \times d}$ where b_{ik} means the probability of node v_i on the kth topic, d is the number of topics.

Definition 3.3: User Interest Topic Similarity Matrix ITSM= $(ts_{ij})_{n \times n}$ where ts_{ij} is the semantic similarity between the user interest feature vector of user v_i and v_j, and n is the number of users.

Definition 3.4: Topology Structure Feature (TSF). In a social network, a user node v_i usually may has a path to node v_j, and this path has a probability weight that a restricted random walk starting from v_i ends at v_j in r steps or fewer using link weights as transition probabilities. The restrictions are that the walk terminates upon reaching v_j or upon revisiting any node including v_i. Formally, each user node v_i is associated with a vector $\gamma_{vi} \in R^T$ of n-dimensional path distribution ($\sum_{vj \in V \cap vj \neq vi} \gamma_{vij} = 1$). Each γ_{vij} is the probability of the v_i linking to v_j. We call the vector γ_{vi} as user v_i's TSF.

Definition 3.5: Topology Structure Feature Matrix TSFM= $(tsfm_{ij})_{n \times n}$ $tsfm_{ij}$ means the probability of node v_i has a path to v_j. n is the number of nodes.

Definition 3.6: Topology Structure Feature Similarity Matrix TSFSM= $(ss_{ij})_{n \times n}$ where ss_{ij} is the similarity between the v_i's topology structure feature and that of v_j.

Definition 3.7: Joint Prediction Feature Model (JPFM). JPFM is a matrix $(PF_{ij})_{n \times n}$, where $PF_{ij} = (ts_{ij}, ss_{ij}, linkFlag)$, and $linkFlag$ represents that whether there is a friend link between v_i and v_j.

PROBLEM 2. Detailed Link Prediction Problem. Given a G from time 1 to t, $G^t = (V^t, E^t, JPFM^t)$, where $JPFM^t$ is the features at time t, the task is to infer those edges e_{ij} which satisfies that the $linkFlag=N$ at time t and $linkFlag=Y$ at time $t+1$.

4   Joint Link Prediction Method For Social Network

In this section we propose a joint link prediction method (JLPM) to incorporate node features and structure feature for better modeling and predicting the friendship links.

4.1   Joint Link Prediction Algorithm

Figure 1 shows the architecture of JLPM. The left part illustrates the input: a social network of 8 users. The right part shows the output of our link prediction: the red bold edges are predicted possible links. The middle two parts are the critical components of our method: feature extraction and link prediction. We need to extract the features of user interests and network topology. We transform the predicting problem into a classifying problem and use both types of features as input of the learning algorithm.

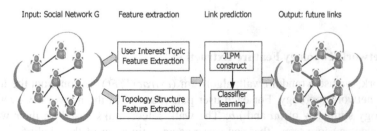

Fig. 1. Joint Link Prediction Method Architecture

The joint link prediction algorithm can be described as follows:

Algorithm 4.1: Joint link predicting

Input: Social network G= (V, E);
Output: The possible link lists;
1: IFM←ExtractUIF(G); // for each user v in V,extracting user interest features:
2: TSFM←ExtractTSF(G);//extracting network topology features
3: A←construct JPFM;
4: use A as input, classifier learning and predicting;
5: ranking the link prediction probability and output the top-k values.

In step 1, the user interest features are extracted. In step 2, the global network topology features are extracted. In step 3, the local interest feature and global topology features are integrated into the unified JPFM matrix. In step 4, we use a classifier to learn the prediction model. In our research, we use SVM classifier. Step 5 outputs the possible future links.

4.2 User Interest Feature Extraction

We use the LDA [10] Topic model to extract the user interest features. The user interest feature extracting algorithm is described as follows:

Algorithm 4.2: ExtractUIF(); //Extract user interest feature

Input: Social network G= (V, E);
Output: user interest features matrix ifm;
1: Call extractTopic();   //get the interest topic probability distribution;:
2: Calculate the interest topic probability distribution;
3: save and return the user interest feature matrix;

We build a LDA model to extract interest topics. And we use the Gibbs sampling method to estimate the parameters. Given two users ua and ub, The topic distribution vector of ua is $TA=(ta_1,... ta_k)$, topic distribution vector of ub is $TB=(tb_1,... tb_k)$, where k is the number of the topics. By calculating the pairwise similarity between user topic vectors, we can get the user interest topic similarity matrix.The similarity calculating formula is as follows:

$$Sim(TA,TB)\frac{\sum\limits_{i=1}^{k} ta_i \times tb_i}{\sqrt{(\sum\limits_{i=1}^{k} ta_i^2)(\sum\limits_{i=1}^{k} tb_i^2)}} \tag{4-1}$$

4.3 Network Topology Feature Extraction

In this work, we use a random walk algorithm ($extractTSF()$) to construct the features based on network topology. For each user pair (ua, ub) in the network, we calculate the topology similarity of ua and ub. The walk selects links based on their weights. This produces a score $tsfm_{sd}$ that can serve as an estimation of the likelihood of new links. For each node v_s in G, we call $extractSingleTSF()$ algorithm to get the topology structure features. To get the TSF, *extractTSF()* will call *extractSingleTSF()* repeatedly for all nodes in G. The input of *extractSingleTSF()* is the social network G, node v_s, and the max length r. And the output of *extractSingleTSF()* is the topology structure feature matrix tsfm$[v_s][v_d]$ for user v_s. If the path length from v_s to v_d is larger than r, then tsfm$[v_s][v_d]$=0; or else the value of tsfm$[v_s][v_d]$ is the link possibility of this path.

4.4   JPFM Constructing

After the above feature extraction works, we have got the social network features similarity matrix ITSM[][] and TSFM[][]. We use a weight sum method as the following formula to synthesize two features. r and are s two adjustable parameters.

$$JPFM[i][j] = r * ITSM[i][j] + s * TSFSM[i][j]$$ (4-2)

5   Experiments

5.1   Dataset

We use a subset of the citation dataset of the *Citeseer*[1] site to test the performance and scalability of our approach. We crawl about 1000 papers in such topics as "association rule mining", "sentiment analysis", "topic models" and "privacy security" and so on. Then we extract the title, abstracts, citation relationship information to form the citation data set. We consider the paper and author as nodes, and consider the citation relationships between papers as the edges.

5.2   Experimental Design

In this work, we use the SVM classifier to analyze data in *Citeseer*. We compare the JLPM with the JLPM-T and JLPM-S methods. JLPM-T denotes that it only consider the topic interest feature and JLPM-S means that it only takes account of the network topology structure features. The following experiments have been performed in our work. 1) Experiment 1: we use the JLPM to predict the friend relationship when selecting the different network scale. We compare the performance of JLPM with that of JLPM-T and JLPM-S in networks with different number of nodes. This experiment will show that how network topology features influence the predicting results. 2) Experiment 2: we use the JLPM to predict the friend relationship when selecting different topic number. We compare the performance of JLPM with that of JLPM-T and JLPM-S under different number of topics. This experiment will show that how individual interest features influence the predicting results. 3) Experiment 3: we observe the performances under different r and s parameter values. 4) Experiment 4: we compare our method with Common Neighbor method, Adamic/Adar method and Katz method [12]. We evaluate the performance of different approaches in terms of the

---

[1] http://citeseerx.ist.psu.edu/index

AUC, namely the area under the receiver operating characteristic curve. AUC is used widely in link prediction researches.

5.3   Results

Figure 2 and 3 are the results of the experiment1 and experiment 2 respectively. As shown in figure 2, we can see that the AUC value of JLPM is between 0.82 and 0.87 and it shows that the JLPM achieve better performance than single JLPM-S or JLPM-S method. The main reason is that the JLPM considers both the node feature and the network structure features of the social network, and this leads to a better prediction results. In experiment 2, as shown in figure 3, it is also obvious that JLPM achieves the highest AUC value when comparing with JLPM-T and JLPM-S method.

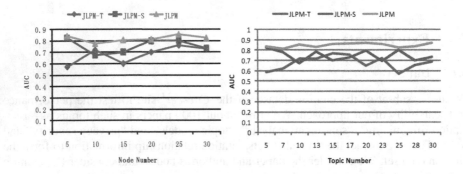

Fig. 2. AUC at different node numbers   **Fig. 3.** AUC at different topic numbers

Figure 4 shows that the AUC values of different r and s parameter values. We can see that with the ascending of the percentage topic feature parameter r, the AUC value is descending. It means that the topic features play a weaker role than the structure feature in link prediction on our test data set.

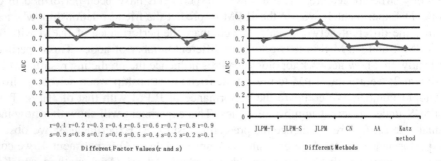

Fig. 4. AUC at different parameter values   **Fig. 5.** AUC of different methods

We compare our methods with the CN, AA and Katz methods. Figure 5 depicts the comparison results. For example, JLPM is superior to CN about 34% in terms of AUC. The results demonstrate the superiority of our proposed JLPM method.

6   Conclusions

We have proposed a joint link prediction method JLPM, which takes advantage of both local user interest data and global network topology structure to predict friendship links in a social network. JPFM model synthesizes the two parts of features. Our JLPM method improves the prediction performance by using LDA to extract user social interests and using a random walk algorithm to extract the network topology features. The experimental results show that the JLPM is effective and practical.

However, JLPM also has some limitations. For example, our approach just takes into account the static image of social network. In the dynamic and evolutionary social networks, how can JLPM extract various features and yield a precise prediction? These are our future works. In addition, more real data sets are needed to testify our method.

Acknowledgements. This research is supported by the Natural Science Foundation of China(No.61202090, 61272184, 61370084) , the Program for New Century Excellent Talents in University No.NCET-11-0829, Natural Science Foundation of HeiLong-Jiang Province(No. F201130), and Fundamental Research Funds for the Central Universities under grant No HEUCF100609.

References

1. Boyd, M.D., Ellison, B.N.: Social Network Sites: Definition, History, and Scholarship. Journal of Computer-Mediated Communication 13 (2007)
2. Yang, S.H., Long, B., Smola, A., et al.: Like like alike — Joint Friendship and Interest Propagation in Social Networks. In: WWW 2011 (2011)
3. Parimi, R., Caragea, D.: Predicting Friendship Links in Social Networks Using a Topic Modeling Approach. In: Huang, J.Z., Cao, L., Srivastava, J. (eds.) PAKDD 2011, Part II. LNCS, vol. 6635, pp. 75–86. Springer, Heidelberg (2011)
4. Hasan, M.A., Zaki, M.J.: A Survey of link prediction in social network. Social network data analytics, vol. 6. Kluwer Academic Publishers (2012)
5. Hsu, H.W., Weninger, T., Paradesi, R.S.M., Lancaster, J.: Structural link analysis from user profiles and friends networks: a feature construction approach. In: Proceedings of International Conference on Weblogs and Social Media, ICWSM 2007, Boulder, CO, USA (2007)
6. Taskar, B., Wong, M., Abbeel, P., Koller, D.: Link Prediction in Relational Data. In: Proc. of 17th Neural Information Processing Systems, NIPS (2003)
7. O'Madadhain, J., Hutchins, J., Smyth, P.: Prediction and ranking algorithms for event-based network data. In: ACM SIGKDD, New York (2005)

8. Eirinaki, M., Louta, M.D., Varlamis, I.: A Trust-Aware System for Personalized User Recommendations in Social Networks. IEEE Transactions on Systems, Man, and Cybernetics: Systems 44(4) (April 2014)
9. Huang, J., Nie, F., Huang, H., Tu, Y.-C.: Trust prediction via aggregating heterogeneous social networks. CIKM2012:1774-1778
10. Blei, D., Ng, Y.A., Jordan, I.M.: Latent Dirichlet Allocation. Journal of Machine Learning Research 3, 993–1022 (2003)
11. Xing, W., Bruce, C.: LDA-based document models for Adhocretrieval. In: Proceedings of the 29th Annual International SIGIR Conference, Washington (2006)
12. Lichtenwalter, R.N., Lussier, J.T., Chawla, N.V.: New perspectives and methods in link prediction. In: Proceedings of the 16th ACM SIGKDD International Conference on Knowledge Discovery and Data Mining, Washington (2010)

HCS: Expanding H-Code RAID 6 without Recalculating Parity Blocks in Big Data Circumstance

Shiying Xia[1], Yu Mao[2,*], Minsheng Tan[1], and Weipeng Jing[3]

[1] University of South China, Hengyang, Hunan, China
[2] Hengyang Normal University, Hunan, China
21236128@qq.com
[3] College of Information and Computer Engineering, Northeast Forestry University, Harbin, China

Abstract. This paper introduces a new RAID 6 expanding method HCS, which is facing the circumstance of big data. HCS expands H-Code manner RAID 6. Two key techniques are used to avoid parity blocks' recalculating. The first one is anti-diagonal data blocks' selection, and the other one is horizontal data migration. These two techniques ensure the data blocks are retained in the same verification zone, that is horizontal verification zone and anti-diagonal verification zone. Experimental results showed that, compared with SDM, which is also a fast expansion method, HCS can reduce 3.6% expansion time and promote 4.62% performance under four traces.

Keywords: Big data, H-Code, RAID expanding, Horizontal coding, Anti-diagonal parity, Horizontal parity.

1 Introduction

In big data circumstance, the gross of data becomes more and more, this requires storage device has the ability of expansion. As a fundamental storage device, RAID [1] [2] is popular used in different scenarios. Different RAID level can be used in different situation, for RAID 0 expand a single hard disk capacity, and RAID 1 offers security storage based on mirror data store. High level RAIDs balance the storage capacity and security, such as RAID 4,5 and 6. RAID 4 and RAID 5 use 1 disk to store the parity blocks in average, which can ensure the failure of 1 disk error. RAID 6 uses two disks to store the parity blocks in average which can ensure the failure of 2 disks error.

With the data increase sharply and quickly, expanding an existence RAID is a critical situation. Expanding a RAID means two tasks, the first one is adding some new disks to the existence RAID, and the second is moving old data to new disks to pursue the balanced data accessing of whole RAID.

---

* Corresponding author.

H. Wang et al. (Eds.): ICYCSEE 2015, CCIS 503, pp. 65–72, 2015.

The key techniques lie in the second task. When moving old data, the first object is online data migration. The storage devices must offer 24×7 service, and when mi-grating the old data, outer user cannot feel it. The second object is balanced data accessing. After data migration, each disk should retain the equal data blocks and parity blocks.

RAID 6 has two kinds of coding, the first one is vertical coding [3] [4] [5] [6] [7] [8] and the second one is horizontal coding [9] [10] [11] [12] [13]. Vertical coding separates the parity blocks into each disk, and horizontal coding uses one disk to store parity blocks. In general, vertical coding has balanced data distribution and horizontal coding does not have balanced data distribution. The H-Code [3] described in this article is a horizontal coding.

H-Code uses two parity stripes to verify the data blocks, so it can restore the whole RAID 6 under two disks' error. H-Code can code n disks, and $n = p + 1$, which p is a prime number. Figure. 1 shows a case of H-Code's two parity stripes. In Fig. 1(a), horizontal parity blocks are stored in a disk. For example, PA is the verification of four blocks which marked as A. Figure. 1(b) shows the anti-diagonal parity stripe. All the A is verified as QA. Anti-diagonal parity blocks do not take part in the horizontal verification and vice versa.

D		D	D	D	PD
C	C		C	C	PC
B	B	B		B	PB
A	A	A	A		PA

(a) Horizontal parity stripe

A	QD	D	C	B	
B	A	QC	D	C	
C	B	A	QB	D	
D	C	B	A	QA	

(b) Anti-diagonal parity stripe

Fig. 1. H-Code coding when $p = 5$

2   Related Works

In recent years, many scientists devote in research the expanding a RAID.

MD-Reshape [14] and GA(Gradual Assimilation) [15] use simple way to expand a RAID. They read all the data of old RAID, then write them to the new RAID. These two methods are time wasting way.

SLAS [16] upgraded GA. It uses reorder window and slide window to control data migration, so it can promote data moving speed. But SLAS is only used in RAID 0.

ALV [17] uses reorder window and slide window to RAID 5. Compared with MD-Reshape, ALV can reduce about 35.33% migration time. Although the time is reduced, ALV also should migrate all the data to new RAID.

FastScale [18] migrates certain data to new disk(s), which can evidently reduce the data migration time, and FastScale is used in RAID 0.

PBM [19] expands the RAID 5 through migrating portion of data blocks and parity blocks. It achieved fast expanding a RAID 5. Legg format the new disk(s) to avoid recalculation of parity blocks of RAID 5 [20], and Corbett avoid recalculation verification by redistributing parity blocks [21].

SDM [22] moves the blocks of a RAID 6 through lest I/O costing. SDM can expand a vertical coding RAID like P-Code [23] and horizontal coding RAID like RDP [24]. But for H-Code, SDM has no expansion solution.

3   Detail Analysis

HCS is a new expansion solution faces H-Code. HCS has two objects in expanding RAID 6. That is,

- No parity blocks recalculation. In H-Code, there are two types of parity blocks that are P and Q. Parity blocks' recalculation will cost significant time. HCS uses zeroing technique to avoid P and Q's recalculation.
- Expanding arbitrary times. After expansion, new RAID 6 is also H-Code coding manner. So new RAID can be expanded the second time.

For illustrating the formulas and introduce HCS, we define some notations as Table 1.

Table 1. Some Notations

Symbol	Meaning
n	Disk count of old RAID
m	Disk count of new added
p	A prime number
P	Horizontal parity block
Q	Anti-diagonal parity block
$E(n, m)$	The process of expansion
(R, D)	A logical address of a block, which R is row number and D is disk number
S_b	Size of a block
S_{so}	Size of stripe in old RAID, the unit is S_b
S_{sn}	Size of stripe in new RAID, the unit is S_b
SN	Count of stripes in old RAID

3.1   Data Migration of One Stripe

In this subsection, we describe our data migration means uses $E(6, 2)$. In $E(6, 2)$, old RAID has 6 disks, and $p = 5$. After migration, new RAID has 8 disks, and $p = 7$. A stripe of H-Code is shows as Fig. 1.

Figure. 2 shows a stripe's data distribution of 6-disk RAID and 8-disk RAID. We numbered the disks and data rows. Comparing two figures, we use two key techniques to achieve data migration. The first one is anti-diagonal data selection, which is all the data blocks above parity Qs will be migrated. In Fig. 2, 3 Ds, 2 Cs and 1 B should be migrated. The other one is horizontal data migration. Horizontal data migration is to fit the target of none parity blocks' recalculation. To detailing horizontal data migration, we use Fig. 3 to illustrate.

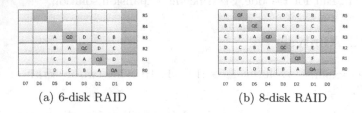

(a) 6-disk RAID   (b) 8-disk RAID

Fig. 2. A stripe data distribution of 6-disk and 8-disk RAID

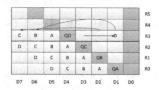

Fig. 3. Horizontal data migration

First, to avoid horizontal parity blocks' recalculation, we move data horizontally. That is to say, all the data blocks have the same row number as before they are moved. Second, to avoid anti-diagonal parity blocks' recalculation, top-right data blocks should move to the same verification zone as before. That is to say, any data block should move to the proper disk to take part in the same verification as before, like $(R3, D1)$ move to $(R3, D6)$, $(R3, D2)$ move to $(R3, D7)$ and $(R3, D3)$ move to $(R3, D1)$. And the third step, all the move-out blocks should be written 0. For example, after migration, $(R3, D3)$ and $(R3, D2)$ should be written 0 to ensure no modify the parity value.

3.2   Data Addressing after Data Migration

After data migration, some data blocks have their new address. This forces we have new data addressing algorithm to ensure right data accessing. The progress of addressing is to translate outer requests' LBA to a blocks' logical address (R, D).

For a LBA, its block number b is:

$$b = LBA/S_b \tag{1}$$

Next step is to find the corresponding stripe number s. For old RAID, a stripe has S_{so} blocks, and for new RAID, a stripe has S_{sn} blocks.

$$S_{so} = n \times (n-2) - 2(n-2) = (n-2)^2 \tag{2}$$

$$S_{sn} = (n+m-2)^2 \tag{3}$$

If $b < (SN \times S_{so})$, b is an old data, otherwise, it's a new data which fill the blank blocks after migration. For old data, b/S_{so} is its stripe number s, and its offset in the stripe is $b mod S_{so}$. The logical address of old data block b can expressed as (R, D), which R can be expressed as equation 4.

$$R = \lfloor (b \ mod \ S_{so})/(n-2) \rfloor \tag{4}$$

For b is unique numbered over parity blocks Q and referred old location D'.

$$D' = (b \ mod \ S_{so}) \ mod \ (n-2) \tag{5}$$

Considering the block b can be right or left of the parity block Q, the real disk number of b is:

$$D = \begin{cases} D' & R \geq D' \\ D'+1 & R < D' \end{cases} \tag{6}$$

Here D is disk number of b before migration. Uses the data migration algorithm, we get the new disk number $D*$ of new disk number.

3.3   Data Recovering

RAID 6 can be recovered after 2 disks' failure. When migrating, this character should be retained. Migration uses whole-stripe-move method, that is to say, moving a stripe cannot be interrupted. If failure occurs, current moving stripe will stop migration and recover its original data distribution.

For not migrated stripe, RAID 6 can use normal recovering progress to recover the data. For migrated stripe, as Fig. 4 illustrated, we give the recovering chain. Figure. 4 shows when two disks $D4$ and $D5$ are failure, the recover chain is:

Fig. 4. Data recovering when $D4$ and $D5$ are failure

$$A(3,5) \rightarrow A(2,4) \rightarrow B(2,5) \rightarrow B(1,4) \rightarrow C(1,5) \rightarrow C(0,4) \rightarrow CD(0,5) \rightarrow QD.$$

4   Experimental Results

We use DiskSim [25] to simulate the H-Code RAID 6 and perform the expansion progress. DiskSim is a block-level disk simulator, we combined the outer requests and migration requests in a queue, which can be sent to the simulator. In the queue, we use a fitted number to control the flow. If the outer requests per second larger than this number, migration requests will slow down, else migration requests will be sent to the queue.

For no expansion solution faces H-Code, we compared the HCS with SDM, which is an expansion solution faces horizontal coding, RDP. $E(6,2)$ and $E(8,4)$ are simulated under four traces, which are Hm0 from Microsoft [26], Home2 [27] from FIU, Financial1 and Financial2 from SPC [28].

4.1   Expansion Time

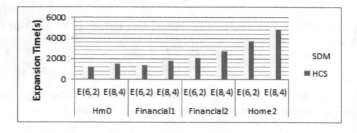

Fig. 5. Expansion time comparison

Figure. 5 shows the different expansion time under four traces. SDM and HCS have almost the same expansion time. Experimental results showed that HCS reduced 3.6% and 3.63% migration time than SDM in $E(6,2)$ and $E(8,4)$ correspondingly. The reason is HCS have not to recalculate the parity blocks but SDM do.

4.2   Performance in Expanding

We use average response time as the metric of performance. From Fig. 6, we can see HCS has less response time than SDM when expanding. Experimental data showed that the response time of HCS less than SDM about 4.62% in average.

5   Conclusions and Further Work

H-Code is a new RAID 6 coding manner. We proposed corresponding expansion method, HCS. HCS use two key techniques to achieve expanding this kind of RAID, which is anti-diagonal data selection and horizontal data moving. These

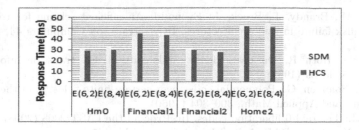

Fig. 6. Performance comparison

two key techniques ensure no parity blocks will be recalculated. Experimental results showed that we can reduce 3.6% migration time compared with SDM in average and promote about 4.62% average response time.

After expansion, data blocks in new RAID are not even distributed. It will cause more access will occur in the parity disk, which is D0 in HCS. But our goal is to ex-pand the RAID multiple times, and the original H-Code manner RAID 6 also not an even-distributed RAID. To trade off two goals, more work should be done in future.

Acknowledgement. This work described in this paper is supported by the Fundamental Research Funds for the Central Universities (DL13CB05) and the Technological innovation talent research project in Harbin (2014RFQXJ132).

References

1. Patterson, D., Katz, G.R.: A case for redundant arrays of inexpensive disks(raid). In: Proceeding of the SIGMOD 1988 (1988)
2. Peter, M., Chen, E.K.L.: Raid: High-performance, reliable secondary storage. ACM Computing Surveys 26, 145–185 (1994)
3. Wu, C., Wan, S.: H-code: A hybrid mds array code to optimize partial stripe writes in raid-6. In: Parallel and Distributed Processing Symposium (IPDPS), pp. 782–793 (2011)
4. Jin, C., Jiang, H.: P-code: A new raid-6 code with optimal properties. In: Proceedings of the 23rd International Conference on Supercomputing, pp. 360–369 (2009)
5. Cassuto, Y., Bruck, J.: Cyclic lowest density mds array codes. IEEE Trans. on Information Theory 55, 1721–1729 (2009)
6. Wu, C., He, X., Wu, G.: HDP code: A horizontal-diagonal parity code to optimize i/o load balancing in raid-6. In: Proceeding of the DSN 2011 (2011)
7. Xu, L., Bruck, J.: X-code: Mds array codes with optimal encoding. IEEE Trans. on Information Theory 45, 272–276 (1999)
8. Xu, L., Bruck, J.: Low-density mds codes and factors of complete graphs. IEEE Trans. on Information Theory 45, 1817–1826 (1999)
9. Corbett, P., English, B., Goel, A.: Row-diagonal parity for double disk failure correction. In: Proceedings of the 3rd USENIX Conference on File and Storage Technologies (2004), pp. 1–14 (2004)

10. Blaum, M., Brandy, J., Bruck, J.: Evenodd: An efficient scheme for tolerating double disk failure in raid architectures. IEEE Trans. on Computers 44, 192–202 (1995)
11. Blaum, M., Roth, R.: On lowest density mds codes. IEEE Trans. on Information Theory 45, 46–59 (1999)
12. Reed, I., Solomon, G.: Polynomial codes over certain finite fields. J. of the Society for Indus. and Applied Math., 300–304 (1960)
13. Plank, J.: The raid-6 liberation codes. In: Proc. of the FAST 2008 (2008)
14. Brown, N.: Online RAID-5 Resizing. Drivers/md/raid5.c in the source code of Linux Kernel 2.6.18, http://www.kernel.org/ (September 2006)
15. Gonzalez, J.L., Cortes, T.: Increasing the capacity of raid5 by online gradual assimilation. In: Proceedings of the International Workshop on Storage Network Architecture and Parallel I/Os - SNAPI 2004, pp. 17–24 (2004)
16. Blaum, M., Brandy, J., Bruck, J.: Slas: An efficient approach to scaling round-robin striped volumes. ACM Transactions on Storage 3 (2007)
17. Zhang, G., Zheng, W., Shu, J.: Alv: A new data redistribution approach to raid-5 scaling. IEEE Transactions on Computers 59, 345–357 (2010)
18. Zheng, W., Zhang, G.: Fastscale: accelerate raid scaling by minimizing data migration. In: Proceedings of the 19th USENIX conference on File and Storage Technologies (2011)
19. Mao, Y., Wan, J., Zhu, Y., Xie, C.: A new parity-based migration method to expand raid-5. IEEE Transactions on Parallel and Distributed Systems 25, 1945–1954 (2014)
20. Legg, C.B.: Method of increasing the storage capacity of a level five RAID disk array by adding, in a single step, a new parity block and n-1 new data blocks which respectively reside in a new columns (December 1999)
21. Corbett, P.F., Kleiman, S.R., English, R.M.: Semi-static distribution technique, US Patent 7,185,144 (February 2007)
22. Wu, C., He, X., Han, J., Tan, H.: Sdm: A stripe-based data migration scheme to improve the scalability of raid-6. In: 2012 IEEE International Conference on Cluster Computing (CLUSTER), pp. 284–292 (2012)
23. Jin, C., Jiang, H., Feng, D., Tian, L.: P-code: A new raid-6 code with optimal properties. In: Proceedings of the 23rd International Conference on Supercomputing (2009)
24. Corbett, P., English, B., Goel, A.: Row-diagonal parity for double disk failure correction. In: Proceedings of the 3rd USENIX Conference on File and Storage Technologies, pp. 1–14 (2004)
25. (DiskSim) University of Michigan and Carnegie Mellon University, http://www.pdl.cmu.edu/DiskSim
26. Narayanan, D., Donnelly, A., Rowstron, A.: Write offloading: Practical power manage-ment for enterprise storage. ACM Transactions on Storage 4, 1–23 (2008)
27. (SNIA) Storage Networking Industry Association,FIU-Home2 block I/O trace (2010), http://iotta.snia.org/traces/414
28. (Storage) Financial1.spc and Financial2.spc., http://traces.cs.umass.edu/index.php/Storage/

Efficient Processing of Multi-way Joins Using MapReduce

Linlin Ding, Siping Liu, Yu Liu, Aili Liu, and Baoyan Song*

School of Information, Liaoning University, Shenyang Liaoning, P.R. China
{dinglinlin,bysong}@lnu.edu.cn,
{spliu,dearboll23,aili07liu}@gmail.com

Abstract. Multi-way join is critical for many big data applications such as data mining and knowledge discovery. Even though lots of research have been devoted to processing multi-way joins using MapReduce, there are still several problems in general to be further improved, such as transferring numerous unpromising intermediate data and lacking of better coordination mechanisms. This work proposes an efficient multi-way joins processing model using MapReduce, named Sharing-Coordination-MapReduce (SC-MapReduce), which has the functions of sharing and coordination. Our SC-MapReduce model can filter the unpromising intermediate data largely by using the sharing mechanism and optimize the multiple tasks coordination of multi-way joins. Extensive experiments show that the proposed model is efficient, robust and scalable.

Keywords: MapReduce, multi-way joins, sharing and coordination.

1 Introduction

Multi-way join is an important and frequently used operation for many big data applications including data mining and knowledge discovery. Since join processing is expensive, especially for large data sets, multi-way join is a costly operation. When processing multi-way joins of big data, a natural solution to ensure the reasonable response time is parallel processing. As a parallel programming model, MapReduce [1] becomes the popular big data programming model for its simplicity, flexibility, fault-tolerance and scalability.

MapReduce is designed to process a single input data set, so multi-way join is not directly supported by MapReduce framework. Although lots of research have been devoted to processing join using MapReduce[2–6], the existing works still have several problems to be further researched. For instance, there are numerous intermediate data to be transferred from Map phase to Reduce phase. When the final output is much smaller than the original input, the numerous unpromising intermediate data would waste the bandwith and I/O. In addition, when processing multi-way joins using several passes MapReduce, the next MapReduce computation cannot start until the previous computation is over.

---

* Corresponding author.

H. Wang et al. (Eds.): ICYCSEE 2015, CCIS 503, pp. 73–80, 2015.
© Springer-Verlag Berlin Heidelberg 2015

In this work, we investigate the problems of multi-way join query processing in MapReduce and analyze the performance and bottlenecks of the existing solutions. An efficient multi-way join query processing model using MapReduce, named Sharing-Coordination-MapReduce (SC-MapReduce) is designed. By using the mechanisms of sharing and coordination, SC-MapReduce can enhance the parallelism and reduce the network cost. In SC-MapReduce, first, a sharing mechanism is proposed to filter the unpromising intermediate data and reduce the network and I/O cost. Then, We design a multi-tasks coordination mechanism for processing multi-way joins. Using this coordination mechanism, the next task do not need to wait for the completion of the previous to start its processing, which will save the waiting time and enhance the parallelism..

The remainder of this paper is organized as follows. Section 2 introduces the related work. Section 3 presents our SC-MapReduce model. The sharing and coordination mechanisms of SC-MapReduce model are given in Section 4. Section 5 reports the experimental results. Section 6 concludes the paper.

2 Related Work

2.1 Problem Statement

Definition 1. *(Two-way Join) Given two data sets $R_1(A_1, S_1)$ and $R_2(A_1, S_2)$ on a common attribute A_1. S_1 and S_2 can be the single attribute or the array of multiple attributes, as shown $R_1(A_1, S_1) \bowtie R_2(A_1, S_2) = (A_1, S_1, S_2)$.*

Definition 2. *(Multi-way Same Attribute Join) Given n (n>2) data sets $R_i(A_1, S_i)(i=1,2,...,n)$ on a common attribute A_1. S_i can be the single attribute or the array of multiple attributes, as shown $R_1(A_1, S_1) \bowtie R_2(A_1, S_2) \bowtie, ..., \bowtie R_n(A_1, S_n) = (A_1, S_1, S_2, ..., S_n)$*

Definition 3. *(Multi-way Different Attributes Join) Given n (n>2) data sets $R_i(A_{i-1}, S_i, A_{i+1})(i=1,2,...,n)$ on common attributes A_i. S_i can be the single attribute or the array of multiple attributes, as shown $R_1(A_0, S_1, A_1) \bowtie R_2(A_1, S_2, A_2) \bowtie, ..., \bowtie R_n(A_{n-1}, S_n, A_n) = (A_0, ..., A_n, S_1, ..., S_n)$*

2.2 Join of MapReduce

There are many existing works of processing joins using MapReduce [4–12]. MapReduce online [5] proposes a pipelined job interaction mechanism to avoid intermediate data materialization. Map-Join-Reduce [9] improves MapReduce runtime to process complex data analysis tasks on large clusters. Paper [7] studies and optimizes multi-way equi-join in MapReduce by selecting a query plan with the lowest input replication cost. Paper [6] is the first one to study all theta-joins and explores optimality properties for them in MapReduce-based systems. Paper [4] studies the problem of processing multi-way theta-joins using MapReduce from a cost-effective perspective. Vernica et al. [8] present an in-depth study of a special type of similarity join in MapReduce.

2.3   Multi-way Joins of MapReduce

The multi-way join contains two types, multi-way same attribute join and multi-way different attributes join. The two-way join is the special case of multi-way join. Processing multi-way join using MapReduce can be implemented by one pass MapReduce computation and multiple MapReduce computations in sequential. Paper [2] analyzes the efficiency of multi-way join from the view of network communication cost. Foto N.Afrati [3] discusses how to identify one pass MapReduce or multiple MapReduce computations.

For one pass MapReduce processing multi-way same attribute or two-way join, Map phase is mainly responsible for labeling the join tuples and identifying the relation of the tuples. The real join operations are implemented in Reduce phase. However, in Shuffle phase, it can transfer the corresponding unpromising intermediate data many times which would increase the I/O and communication cost. For the multiple MapReduce computations processing multi-way different attributes join, each MapReduce computation completes two-way join until all the data sets are processed in sequential. The next MapReduce computation waits for the completion of the previous MapReduce computation, and then starts its computation.

3   SC-MapReduce Framework

3.1   SC-MapReduce Overview

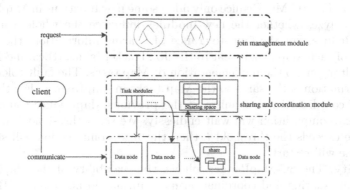

Fig. 1. Architecture of SC-MapReduce Framework

In order to enhance the performance of multi-way joins, in this paper, we propose SC-MapReduce, a Sharing-Coordination-MapReduce framework. Figure 1 shows the whole architecture of SC-MapReduce which briefly inherits the original MapReduce framework, containing the Master node and the Slave nodes. The Master node is mainly responsible for managing the whole system, scheduling and distributing the computation jobs. Furthermore, the join management module and the sharing and coordination module are added in the Master node. The sharing and coordination module contains the sharing mechanism and the coordination mechanism. The real computation is done by the Slave nodes.

Join Management Module. The join management module is in charge of receiving the connecting request from clients and designing the join order of the multi-way join, then generates the execution queue of multi-way join.

Sharing Mechanism. The sharing mechanism mainly processes the communications in one computation course. Each Slave node has a share space to save the sharing information to filter the unpromising intermediate data.

Coordination Mechanism. The coordination mechanism designs for the parallelism of multiple MapReduce computations of multi-way join. When processing multi-way joins, the SC-MapReduce framework can first complete processing a part of join data and wait for the completion of the remaining part from the previous computations, so as to enhance the parallelism of SC-MapReduce.

As shown in Figure 1, when a client submits a multi-way join to the Master node, it first designs the join order of the multi-way join by the join management module. Then, the sharing and coordination module of Master node can obtain the sharing information of the multi-way join and design the implementation of multiple MapReduce computations. Then, the Map and Reduce tasks can use the sharing information to filter the unpromising intermediate data. The multiple computations can also start early to process parts of the inputs. The details of sharing and coordination mechanisms are illustrated in Section 4.

3.2   Availability and Scalability

Availability. The SC-MapReduce only adds some new functions in MapReduce, so it only analyzes whether the new modules influence the whole availability of SC-MapReduce. For the fault-tolerance of sharing information, the sharing information of each Map task is sent to the JobTracker by heartbeat mechanism, so it is simultaneous to the implementation of Map tasks. The fault-tolerance of sharing information is the same to the Map tasks in MapReduce. For the fault-tolerance of coordination mechanism, if the previous MapReduce computation fails, the next computation will wait endlessly. We set a threshold t, when the waiting time exceeds the threshold t, then the next computation will stop and the failed one will restart.

Scalability. The SC-MapReduce follows the scalability of the MapReduce. First, the new sharing and coordination mechanisms do not modify HDFS, so the users can add new Slave nodes naturally. Second, although the new modules are added to the Master node, the sharing mechanism can be realized by only following the implementations of Map tasks. The coordination mechanism only occupies few resource, so the performance of the Master node is not influenced.

4   Sharing and Coordination Module

The sharing and coordination module mainly contains the sharing mechanism and the coordination mechanism. The sharing mechanism is mainly charge of identifying the sharing information to filter the unpromising intermediate data during one MapReduce computation course. The coordination mechanism can generate the parallel implementation of multiple MapReduce computations.

4.1 Sharing Mechanism

In the traditional MapReduce, the implementations of Mappers and Reducers are independent without communication among Mappers, neither among Reducers. Each Mapper has no processing information of the other ones, the same as the Reducers. When the amount of multi-way joins final results is much smaller than the original data, processing the unpromising data would waste the running time and increase the network and I/O cost. Therefore, if the Master node can share some useful sharing information among the Map and Reduce tasks, the unpromising data can be filtered by the sharing information so as to enhance the performance of multi-way joins.

The main course of sharing mechanism is as follows. First, for the input join data sets, the Map tasks process the first few tables according to the join queue obtained by the join management module. Second, each Map task gains and saves its local join attributes, and then sends them to the Master node. Then, after receiving the local sharing information of the Map tasks, the Master node generates the global sharing information and sends it to the other tables. After the Map tasks receiving the global sharing information, they can use it to filter the unpromising intermediate data when processing the other join tables.

However, MapReduce is suit for the single file input. When processing multiple files, the JobTracker will split all the files, then generates the default job queue of Map tasks and distributes the Map tasks to the free nodes. Therefore, we modify some functions of the original MapReduce. SC-MapReduce modifies the scheduling and constructing mechanisms of Map tasks so as to scan the tables according to the join queue. After scanning one join table, the sharing information can be sent, received and distributed among Map tasks and the Master node. The Master node and the Slave nodes all have sharing space to send and receive the sharing information. The JobTracker is in charge of starting and stopping the scheduling queue according to the join queue order.

4.2 Coordination Module

As we know, there are multiple MapReduce computations to complete the multi-way join. The previous MapReduce computation may be the input of the next MapReduce computation, so each MapReduce computation should wait for the completion of the previous computation. However, for the next computation, there are parts of input data which can be processed first without affecting the following data. For example, multi-way join $R_1 \bowtie R_2 \bowtie R_3$, the first MapReduce computation completes the join of $R_1 \bowtie R_2$. The next job realizes join of R_3 and the output of R_1 and R_2. We can beforehand process the Map tasks of R_3 and wait for the completion of $R_1 \bowtie R_2$, and then gain the final results. Therefore, the parallelism of system can be enhanced by optimal coordination mechanisms.

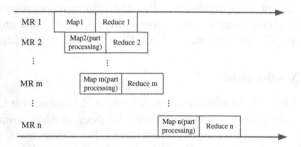

Fig. 2. Coordination Mechanism of SC-MapReduce

The coordination mechanism can coordinate the MapReduce computations. For the multi-way join with n MapReduce jobs, the concurrency is m when there are m MapReduce jobs running at the same time as shown in Figure 2. The processing course is as follows.

(1) Start m MapReduce computations in order and construct task queue for the part of the input data.
(2) For the i MapReduce computation, if its input is completed, then it implements normally. If it needs to wait for the output of the previous MapReduce, it designs the dormant state of task queue after completing the following Map tasks, and then communicates with the Master node.
(3) The Master node restarts the i MapReduce computation when the input of i MapReduce completes according to the implementation of the tasks, and then refreshes the task queue and implements the tasks.
(4) After some MapReduce computations complete, the concurrency is smaller than m. If there are tasks that have not start, the SC-MapReduce adds them into the task queue and starts the new tasks following step 1 to step 3.

5   Experimental Evaluation

5.1   Experimental Setup

The experimental setup is as follows. The experimental setup is a Hadoop cluster running on 9 nodes in a high speed Gigabit network, with one node as the Master node and the Coordinator node, the others as the Slave nodes. Each node has an Intel Quad Core 2.66GHZ CPU, 4GB memory and CentOS Linux 5.6. We use Hadoop 0.20.2 and compile the source codes under JDK 1.6.

We take three tables join as an example to evaluate the SC-MapReduce by comparing with MapReduce in two distributions, uniform and *zipf*. The data in our experiments are synthetic data of network logs. Each record has 100B-10000B and the join attribute is 1B-10B. The log files come from different parts of the network. The percentages of join attributes are 5%, 10%, 65% and 75%, with 10% as the default value. The size of the three tables join is 1G, 5G, 10G and 20G, with 10G as the default value.

5.2   Experimental Results

Figure 3(a) shows the performance of multi-way join running time with changing the join attributes percentage in uniform distribution. We can see that the performance of SC-MapReduce is better than MapReduce. The running time of MapReduce is relative stable for no filtering mechanism and uniform distribution. In SC-MapReduce, the join attributes of 5% and 10% are few, so the global sharing information is small and can filter more unpromising intermediate data. Therefore, the running time is relative short.

Figure 3(b) shows the performance of multi-way join running time with changing the join attributes percentage in *zipf* distribution. In *zipf* distribution, there will be one or few Reduce tasks to process much intermediate data, so the performance in *zipf* distribution is not as well as the uniform distribution. However, the sharing information of SC-MapReduce can also filter numerous unpromising intermediate data.

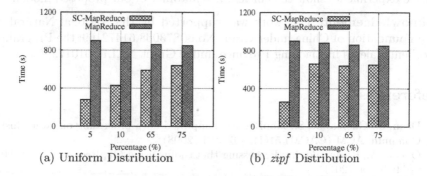

(a) Uniform Distribution   (b) *zipf* Distribution

Fig. 3. Running Time with Changing Percentage

Figure 4(a) shows the performance of multi-way join running time with changing the data size in uniform distribution. With the increasing of data size, the running time of MapReduce and SC-MapReduce both increase, but SC-MapReduce is optimal to MapReduce. The global sharing information of SC-MapReduce can filter large scale of join data which are not the final results.

Figure 4(b) shows the performance of running time with changing the data size in *zipf* distribution. The running time increases with the increasing of the data size in SC-MapReduce and MapReduce. The performance of SC-MapReduce is also better than MapReduce. In *zipf*, the sharing information of SC-MapReduce can be used to filter the unpromising intermediate data too.

6   Conclusions

In this paper, we research the problem of multi-way join query processing based on MapReduce framework. First, we propose an efficient multi-way join query processing model with the functions of sharing and coordination, SC-MapReduce.

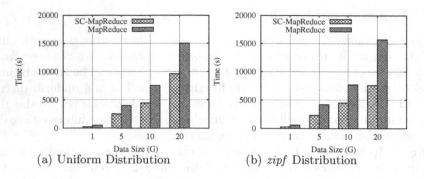

(a) Uniform Distribution   (b) *zipf* Distribution

Fig. 4. Running Time with Changing Data Size

Then, the sharing and coordination module is illustrated in detail. Finally, extensive experiments show the efficiency, scalability of our proposed model.

Acknowledgement. This work was supported by the National Natural Science Foundation of China under Grant No.60873068,61472169; the Program for Excellent Talents in Liaoning Province under Grant No.LR201017.

References

1. Dean, J., Ghemawat, S.: MapReduce: simplified data processing on large clusters. Commun. ACM (CACM) 51(1), 107–113 (2008)
2. Okcan, A., Riedewald, M.: Processing theta-joins using MapReduce. In: SIGMOD, pp. 949–960 (2011)
3. Afrati, F.N., Ullman, J.D.: Optimizing Multiway Joins in a Map-Reduce Environment. IEEE Trans. Knowl. Data Eng (TKDE) 23(9), 1282–1298 (2011)
4. Zhang, X., Chen, L., Wang, M.: Efficient Multi-way Theta-Join Processing Using MapReduce. PVLDB 5(11), 1184–1195 (2012)
5. Pansare, N., Borkar, V.R., Jermaine, C., Condie, T.: Online Aggregation for Large MapReduce Jobs. PVLDB 4(11), 1135–1145 (2011)
6. Okcan, A., Riedewald, M.: Processing theta-joins using MapReduce. In: SIGMOD, pp. 949–960 (2011)
7. Afrati, F.N., Ullman, J.D.: Optimizing joins in a map-reduce environment. In: EDBT, pp. 99–110 (2010)
8. Vernica, R., Carey, M.J., Li, C.: Efficient parallel set-similarity joins using MapReduce. In: SIGMOD, pp. 495–506 (2010)
9. Jiang, D., Tung, A.K.H., Chen, G.: MAP-JOIN-REDUCE: Toward Scalable and Efficient Data Analysis on Large Clusters. IEEE Trans. Knowl. Data Eng (TKDE) 23(9), 1299–1311 (2011)
10. Fries, S., Boden, B., et al.: PHiDJ: Parallel similarity self-join for high-dimensional vector data with MapReduce. In: ICDE, pp. 796–807 (2014)
11. Ma, Y., Meng, X.: Set similarity join on massive probabilistic data using MapReduce. Distributed and Parallel Databases (DPD) 32(3), 447–464 (2014)
12. Lee, T., Bae, H.-C., et al.: Join processing with threshold-based filtering in MapReduce. The Journal of Supercomputing (TJS) 69(2), 793–813 (2014)

A Selection Algorithm of Service Providers for Optimized Data Placement in Multi-Cloud Storage Environment

Wenbin Yao and Liang Lu

Beijing Key Laboratory of Intelligent Telecommunications Software and Multimedia, Beijing
University of Posts and Telecommunications, Beijing, China
yaowenbin@bupt.edu.cn, luliang1024@gmail.com

Abstract. The benefits of cloud storage come along with challenges and open
issues about availability of services, vendor lock-in and data security, etc. One
solution to mitigate the problems is the multi-cloud storage, where the selection
of service providers is a key point. In this paper, an algorithm that can select optimal provider subset for data placement among a set of providers in multi-
cloud storage architecture based on IDA is proposed, designed to achieve good
tradeoff among storage cost, algorithm cost, vendor lock-in, transmission performance and data availability. Experiments demonstrate that it is efficient and
accurate to find optimal solutions in reasonable amount of time, using parameters taken from real cloud providers.

Keywords: cloud storage, multi-cloud, service provider selection, data placement, information dispersal algorithm.

1 Introduction

Along with the rise of big data, cloud storage providers have made great progress,
providing online mass storage to users. Cloud storage has the advantages of strong
availability, good scalability and flexibility, etc. However, there are still some concerns users have about cloud storage, being the obstacles to the growth of it.

(1) *Availability of services.* Although most cloud providers have ensured the reliability and availability of the services highly available through service-level agreement
(SLA), software bugs, user errors, administrator errors, malicious insiders, and natural
catastrophes endangering availability are inevitable and unpredictable [1]. Reports
indicate that many cloud providers appeared a few hours, or even days of service
interruptions [2]. When the SLA is broken, the compensation may only be some payment reduction, which does not help reduce costumers' loss at all [3].

(2) *Vendor lock-in.* When users place a huge amount of data into a particular
cloud provider, it becomes very difficult to switch to another one, because some providers charge for data download, which is definitely the most costly part of data
migration [4]. Meanwhile, so far all of the cloud storage service providers have not
provided uniform service interface in accordance with each other, increasing the difficulty of data migration in the cloud. Thus, users have a "vendor lock-in".

H. Wang et al. (Eds.): ICYCSEE 2015, CCIS 503, pp. 81–92, 2015.

(3) *Data security*. Once data is outsourced in the cloud, users lose complete control over it [8]. Sensitive data may be exposed in front of the cloud managers, and if the cloud appears security attacks, the sensitive data may be damaged permanently or be stolen by malicious attacker, and it will result in huge losses.

One solution to mitigate the above concerns is the *multi-cloud storage* [5], to spread out the data storage among several providers, at a fine granularity rather than a coarse one [6]. Among the implementations of multi-cloud storage, the architecture based on Information Dispersal Algorithm (IDA) [7] is an outstanding one. That is, data objects are split to chunks using IDA, in which a threshold number of trunks are needed to retrieve the original data. The trunks are distributed into available service providers in such a way that no less than a threshold number of providers can take part in successful retrieval of the data, providing better security and reducing vender lock-in. Meanwhile, because the data objects are stored redundantly with IDA, the risks for example the loss in case of a complete failure of parts of service providers are avoided, so the reliability and availability of the data is guaranteed.

Due to the diversity of cloud service providers, to select them in order to achieve the maximum benefits becomes a challenging and important problem in cloud computing. Different cloud service providers supply a variety of services with different SLAs and performance, which are proportional to the cost. However, to users of multi-cloud storage, the process of provider selection and data placement should be as transparent as possible, because what they care most is getting a fine tradeoff among all factors. So we need an algorithm to find out the optimal placement of objects meeting the need of users on multi-cloud environment.

This paper focuses on selecting optimal cloud providers for data objects placement in the multi-cloud storage architecture based on IDA, taking factors like storage cost and data availability, etc. into account. An algorithm is proposed to select a subset of a given cloud providers to store data objects, and to work out the threshold number to retrieve them, with the goal to achieve good tradeoff among storage cost, algorithm cost, vendor lock-in, transmission performance and data availability.

The rest of this paper is organized as follows: In the next section, we discuss the related papers with ours. The optimal cloud service provider selection algorithm is described in Section 3. In section 4, we present the results of experiments using price, availability and other parameters taken from real cloud providers. Finally, we conclude our paper, followed by a list of references.

2   Related Work

Multi-cloud has captured significant attention from both industry and academic in the recent years. Petcu et al. [5] provided a comprehensive overview on classification, needs, obstacles, and future of multi-cloud. Strunk et al. [8] propose π-cloud, a hybrid cloud architecture, which aims to put the user in a position to externalize his IT-infrastructure without losing data control, based on IDA to ensure availability, encryption for secrecy and cryptographic hashing for integrity. They discuss the optimal values of n and k with respect to availability, time cost and space cost. However, they don't provide a solution for obtaining n and k. Another solution, Scalia [9], adapts the placement of data based on its access pattern and subject to optimization objectives, such as storage costs. The goal of the data placement is to minimize the price that the data owner has to pay given a set of customer rules, e.g., availability, durability. Their algorithm needs the rules set by customers. RACS (Redundant Array of Cloud Storage) is a cloud storage proxy that transparently splits an object among multiple cloud providers [6]. This study's goal is also to increase availability and mitigate object lock-in. The idea behind the study is the same as the well-known techniques, RAID5. Their work differs from ours since our objective function is different.

Singh et al. [10] describe a model that the customer divides his data among several service providers available based on his available budget. They provide a linear programming model to complete the data distributing with multiple optimization objectives. The work of Yaser Mansouri et al. [4] is to deal with achieving maximum availability with a given budget as a constraint such that objects are split into chunks replicated across several cloud storage. They use a dynamic algorithm to solve the problem. The way of splitting an object to chunks and replicating across cloud providers in our algorithm is different from theirs.

Though the works above are about provider selection on multi-cloud environment, they have differences in fundamentals, constraints or objective functions from ours. To the best of our knowledge, our work in this paper has never been studied yet.

3   Service Provider Selection Algorithm

In multi-cloud storage architecture based on IDA, the data object is chunked by (k,n)-threshold scheme, and the chunks are stored separately into different service providers. Thus, the service provider selection can be parameterized with n providers to store the chunks and k providers required to restore the original object.

3.1   Problem Description

Consider a set of independent cloud storage service providers represented by $SP = \bigcup_{i=1}^{p}\{sp_i\}$, where sp_i is an individual service provider, and $|SP| = p$.

Definition 1. (Object Placement Solution): Let Φ denote a subset of SP, and let $\Phi_{obj} = \bigcup_{j=1}^{n}\{sp_j\}$ be a placement solution of data object obj, where $sp_j \in SP$, and $\Phi_{obj} \subseteq SP$. Each chunk will be stored in one provider, so the size of Φ_{obj} is n, equal to the amount of chunks.

Definition 2. (Storage Cost): Let $Fu(\Phi_{obj})$ be the cost of uploading all chunks of object obj to service providers in Φ_{obj}. For sp_j, we use $Pput_j$ to denote its price for single data upload request (PUT or POST request in HTTP), and Pup_j to denote the per GB price for transferring data into sp_j. The size of each data chunk can be represented by $|obj|/k$. Thus, we have:

$$Fu(\Phi_{obj}) = \sum_{j=1}^{n} \left\{ Pput_j + Pup_j \times \frac{|obj|}{k} \right\} \tag{1}$$

Let $Fs(\Phi_{obj})$ be the cost of storing all chunks of obj in Φ_{obj} in time T_{obj}. Here we use T_{obj} to denote the stored time of obj, and its value is the average stored time of data objects of its type, got by doing statistics, or equal to the data migration cycle when there doesn't exist statistical data. Let $Pstorage_j$ be the storage price of sp_j for per GB and per month. So, $Fs(\Phi_{obj})$ can be obtained as:

$$Fs(\Phi_{obj}) = \sum_{j=1}^{n} \left\{ Pstorage_j \times T_{obj} \times \frac{|obj|}{k} \right\} \tag{2}$$

The probability of downloading a data chunk from provider sp_j is k/n, as we only need k providers to restore obj. We define the predicted total downloaded amount of obj in the stored time T_{obj} as D_{obj}, use $Pget_j$ to denote the price for single data download request (GET request in HTTP), and $Pdown_j$ to denote the per GB price for transferring data out from sp_j. The cost for downloading chunks to retrieve obj can be predicted as:

$$Fd(\Phi_{obj}) = \frac{k}{n} \times \sum_{j=1}^{n} \left\{ Pget_j \times D_{obj} + Pdown_j \times \frac{|obj|}{k} \times D_{obj} \right\} \tag{3}$$

In summary, the storage cost of Φ_{obj} is given by:

$$F(\Phi_{obj}) = Fu(\Phi_{obj}) + Fs(\Phi_{obj}) + Fd(\Phi_{obj}) \tag{4}$$

Definition 3. (Expected Availability) In SLA, service availability refers to the probability of the service is available upon an agreed period of time. The expected availability for Φ_{obj} is the possibility to retrieve obj from Φ_{obj}, and we use $A(\Phi_{obj})$ to denote it:

$$A(\Phi_{obj}) = 1 - \sum_{\Omega \subset \Phi_{obj}} \left(\prod_{sp_j \in \Omega} ava_j \bullet \prod_{sp_j \notin \Omega} (1 - ava_j) \right) \tag{5}$$

Where $ava_j (\in (0,1])$ is the service availability of sp_j, Ω denotes the subsets of Φ_{obj}, size less than k. When the data object in Φ_{obj} is available, there should be at least k service providers available. We calculate $A(\Phi_{obj})$ by adding up the possibility that there are no more than k available service providers in Φ_{obj}.

Definition 4. (Network Performance) Let net_j be the network transfer performance between our system and sp_j. It determines whether our system can efficiently complete the objects' storage, and it is a reflection of network performance, which can be calculated by measuring the average value of data transfer speed of a certain period of time for several times. The overall network transfer of Φ_{obj} is defined as:

$$N(\Phi_{obj}) = \frac{1}{n}\sum_{i=1}^{n} net_i \qquad (6)$$

Definition 5. (Vendor Lock-in) The lock-in factor $L(\Phi_{obj}) \in (0,1]$ of a placement solution is defined as:

$$L(\Phi_{obj}) = \frac{1}{n} \qquad (7)$$

Definition 6. (Algorithm Cost) The time complexity of IDA is $O(k \cdot \log n)$, meaning that the larger the value of k or n is, the longer the time to split and merge the object needs. The algorithm cost of Φ_{obj} using IDA can be defined as:

$$C(\Phi_{obj}) = k \cdot \log n \qquad (8)$$

Definition 7. (Objective function) The objective of object placement solution selection is to find a subset $\Phi_{obj} \subseteq SP$ for each object so that $F(\Phi_{obj})$, $L(\Phi_{obj})$ and $C(\Phi_{obj})$ is minimized, $F(\Phi_{obj})$ and $N(\Phi_{obj})$ is maximized. We define our objective function in Equation (9):

$$\text{Min } O(\Phi_{obj}) = \frac{\lambda \times \dfrac{F(\Phi_{obj})}{F(\Phi_{obj})_{\max}} + \eta \times \dfrac{C(\Phi_{obj})}{C(\Phi_{obj})_{\max}} + \varphi \times L(\Phi_{obj})}{A(\Phi_{obj}) \times \dfrac{N(\Phi_{obj})}{N(\Phi_{obj})_{\max}}} \qquad (9)$$

Where $\lambda, \eta, \varphi \in [0,1]$ are the weight factors between storage cost, algorithm cost and vendor lock-in, and $\lambda + \eta + \varphi = 1$. They will be assigned by users, such that the values of $F(\Phi_{obj})$, $C(\Phi_{obj})$ and $L(\Phi_{obj})$ will make a trade-off in placement solution selection. We normalize the values of $F(\Phi_{obj})$, $C(\Phi_{obj})$ and $N(\Phi_{obj})$ by having they divided by their maximum value.

3.2 Selection Algorithm

We solve the problem defined by above section with some recursive relations. Assume that we have a provider set SP size of p. We use $\Phi_{i,j}$ to denote a provider subset of i providers, where $|\Phi_{i,j}| = j$, $j \leq i \leq p$. The amount of these subsets like $\Phi_{i,j}$ is C_i^j. Let $Target_{i,j} = \bigcup_{th=1}^{C_i^j}\{\Phi_{i,j}^{th}\}$ represent all these subsets, and then we have a recursive equation as follows:

$$Target_{i,j} = \begin{cases} Target_{i-1,j} \cup Target_{i,j}', & j>1, i \geq j \\ Target_{i,j} = \bigcup_{th=1}^{i}\{\Phi_{i,1}^{th}\}, & j=1, i \geq j \end{cases} \tag{10}$$

Where sp_i is the i-th provider of n providers in SP, and $Target_{i-1,j}'$ represents union set $\bigcup_{th=1}^{C_{i-1}^{j-1}}\{\Phi_{i-1,j-1}^{th} \cup sp_i\}$. We can work out $Target_{i,j}$ by its recursive relation with $Target_{i-1,j}$ and $Target_{i-1,j-1}$ according to Equation (10).

Now we consider the placement solution of obj. If we define $\bar{\Phi}_{i,j}$ as $\Phi_{i-1,j-1} \cup sp_i$, for the placement solution $\bar{\Phi}_{obj,i,j}$ and $\Phi_{obj,i-1,j-1}$ for obj, the storage cost $F(\bar{\Phi}_{obj,i,j})$ can be calculated by the following recursive equation:

$$F(\bar{\Phi}_{obj,i,j}) = \begin{cases} F(\Phi_{obj,i-1,j-1}) + f(sp_i) & i>1 \\ f(sp_i), & i=1 \end{cases}$$

$$f(sp_i) = Pput_i + Pup_i \times \frac{|obj|}{q} + Pstorage_i \times T_{obj} \times \frac{|obj|}{q} + \frac{q}{j} \times Pget_i \times D_{obj} + Pdown_i \times \frac{|obj|}{j} \times D_{obj}$$

$$\tag{11}$$

Where $q \in [1,j]$ is the threshold value for $\bar{\Phi}_{obj,i,j}$ to retrieve obj. Also, the expected availability $A(\bar{\Phi}_{obj,i,j})$ when the threshold value is q can be calculated by Equation (12):

$$A(\bar{\Phi}_{obj,i,j},q) = \begin{cases} A(\Phi_{obj,i-1,j-1},q-1) \times ava_i + A(\Phi_{obj,i-1,j-1},q) \times (1-ava_i), & q>1 \\ 1 - \prod_{y=1}^{i}(1-ava_y), & q=1 \end{cases} \tag{12}$$

The network performance $N(\bar{\Phi}_{obj,i,j})$ can be calculated as:

$$N(\bar{\Phi}_{obj,i,j}) = \frac{i-1}{i}N(\Phi_{obj,i-1,j-1}) + \frac{net_i}{i} \tag{13}$$

Considering the placement solution set $Target_{obj,i,j}$ for storing obj into j providers, for each $\Phi_{obj,i,j} \in Target_{obj,i,j}$ and $q \in [1,j]$, the storage cost, expected availability, network performance of $\Phi_{obj,i,j}$ can be calculated by Equation (11)-(13). The selection algorithm for storing obj into p providers is outlined in Fig. 1.

3.3   An Implementation of Algorithm

In this section, we present a multi-cloud storage architecture in which the service provider selection algorithm is implemented. The mainly related services of the proposed architecture are shown in Fig. 2.

Algorithm 1 Service provider selection algorithm

Input: $obj, SP, \{\lambda, \eta, \varphi\}$
Output: Φ'_{obj}, k
  for each $i \in [1, p]$ **do**
    for each $j \in [1, i]$ **do**
      if $j = 1$ **then**
        $\Phi_{obj} \leftarrow \bigcup_{t=1}^{i} \{sp_t\}, q \leftarrow 1$
        $target[i][j] \leftarrow \{\Phi_{obj}, q, F(\Phi_{obj}), A(\Phi_{obj}), L(\Phi_{obj}), C(\Phi_{obj}), N(\Phi_{obj})\}$
      else
        if $i > j$ **then**
          $target[i][j] \leftarrow target[i][j] \bigcup target[i-1][j]$
        end if
        for each $\Phi_{obj} \in target[i-1][j-1]$ **do**
          $\overline{\Phi}_{obj} = \Phi_{obj} \bigcup sp_i$
          for each $q \in [1, j]$ **do**
            add $\{\overline{\Phi}_{obj}, q, F(\overline{\Phi}_{obj}), A(\overline{\Phi}_{obj}), L(\overline{\Phi}_{obj}), C(\overline{\Phi}_{obj})\}, N(\overline{\Phi}_{obj})\}$ *to* $target[i][j]$
          end for
        end for
      end if
    end for
  end for
  for each $\{\Phi_{obj}, q, F(\Phi_{obj}), A(\Phi_{obj}), L(\Phi_{obj}), C(\Phi_{obj}), N(\Phi_{obj})\} \in \bigcup_{t=1}^{p} target[p][t]$ **do**
    if $O(\Phi_{obj}) > min$ **then**
      $min = O(\Phi_{obj}), \Phi'_{obj} = \Phi_{obj}, k = q$
    end if
  end for
  return Φ'_{obj}, k

Fig. 1. Service provider selection algorithm

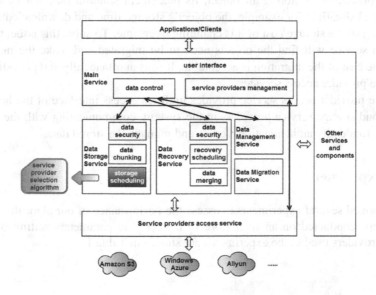

Fig. 2. The multi-cloud storage architecture

A .Main Service

The main service provides a friendly, flexible user interface for the applications and clients to complete data management and service provider management. The data control module processes and forwards users' data storage or access requests. Service provider management module controls the joining, leaving and configuring of providers, taking charge of real-time monitoring of their load state and connection state.

B. Data Storage Service

Data chunking module uses IDA to split data objects, and record the chunks' meta information. Data security module is used to encrypt and signature the objects, in order to ensure the data's confidentiality and integrity. Storage scheduling module is responsible for completing the storage of the data chunks according to the *service provider selection algorithm*, and storing them into different providers at last.

C. Data Recovery Service

When recovering data objects, the data security module takes charge of the decryption and integrity validation. Recovery scheduling module calculates the optimal recovery solutions and gets the data chunks. The data merging module restores the original data with the IDA.

D .Data Management Service

Data management service is used for querying, deleting the stored data objects. Meanwhile, data management service collects object's stored time, object's downloaded amount, service providers' network performance, and other information by doing statistics, providing support for selection of the optimal service providers.

E .Data Migration Service

Upon procedure of storing an object, its placement solution may not be optimal under actual situation, for example, the object's storage time and downloaded amount is changing so the storage cost of it is changing over time. To solve this issue, the data migration service will find the objects need to be migrated and make the migration plan, if the cost of the migration is acceptable. It runs automatically and periodically.

F. Service provider access service

Service provider access service provides uniform access interface of the heterogeneous cloud storage service providers in the system, communicating with the service providers to identity authentication, access and manage the stored data.

4   Experiment

We performed several experiments to assess the performance of our algorithm. Experiments are conducted on an Intel Core i3 laptop. The parameters setting of cloud storage providers used in the experiments are shown in Table 1.

Table 1. Cloud storage service providers' parameters

Service providers	Storage Price	Request Price		Data Transfer Price		Availability	Network Performance
		PUT/POST	GET	In	Out		
Amazon S3 (Tokyo)	0.033	$0.047	$0.0037	$0.00	$0.00	99.99%	0.4409
Amazon S3 (Singapore)	0.03	$0.05	$0.004	$0.00	$0.00	99.99%	0.3612
Windows Azure	0.024	$0.0005	$0.0005	$0.00	$0.04	99.9%	0.5012
Aliyun	0.0292	$0.0017	$0.0017	$0.00	$0.125	99.9%	0.9752
QiNiu	0.0275	$0.0017	$0.0017	$0.00	$0.0833	99.9%	1.2461

In Table 1, the unit of storage price is dollar per GB per month, the unit of request price is dollar per 10,000 requests, the unit of data transfer price is dollar per GB, and the unit of network performance is MB per second. For Amazon S3, we use the parameters of Standard Storage of its three levels. For Windows Azure, we take the parameters of Locally Redundant Storage for block blobs of the four redundancy levels. And if the service providers develop tiered pricing, we just use the lowest level for pricing in our experiments. Average value of data transfer speed is measured by uploading files 200MB in size for 5 times. For Experimental purposes, the size of data object *obj* is 1 GB, stored time is 1 month, and downloaded amount is 1,000 times.

The weight factors play important roles in the selection algorithm, and the values of them show what users are more concerned with. For example, as the value of λ increases, the effect of storage cost of an object placement solution increases. In Table 1, we can see that parameters of Aliyun and QiNiu have advantages over the others, especially the parameters about storage cost. As shown in Table 2, the larger λ is, the greater the likelihood of Aliyun or QiNiu will be a member of the optimal object placement solution is. Meanwhile, when λ become larger in the three weight factors, the objective function value increases, which mean the solution is less competitive.

Table 2. Cloud storage service providers' parameters

Weight Factors	Optimal Solution	Threshold Value	Objective Function Value
{0.5,0.3,0.2}	{Aliyun, QiNiu}	2	0.1774
{0.4,0.4,0.2}	{Aliyun, QiNiu}	1	0.1611
{0.3,0.5,0.2}	{Aliyun, QiNiu}	1	0.1682
{0.5,0.2,0.3}	{Aliyun, QiNiu}	2	0.2142
{0.4,0.2,0.4}	{ Azure, Aliyun, QiNiu}	2	0.2674
{0.3,0.2,0.5}	{Aliyun, QiNiu}	2	0.3108
{0.3,0.4,0.3}	{Aliyun, QiNiu}	1	0.2147
{0.3,0.5,0.2}	{Aliyun, QiNiu}	1	0.1682
{0.1,0.4,0.5}	{Azure, Aliyun, QiNiu}	1	0.3083

In our algorithm, when $k=1$, it means that an object have n replicas and the way to store objects is similar to the way of redundant storage. Because IDA is not used in normal redundant storage, the algorithm cost is zero. If we let $\eta=0$ and fix $k=1$, it is like in the redundant storage architecture. Table 3 is parts of the result of our algorithm when $\lambda=0.6, \eta=0, \varphi=0.4$ and $k=1$, where {Windows Azure, Aliyun, Qi-Niu} is the best solution with the objective function value 0.2096. Our algorithm can make an optimal selection in the redundant storage architecture.

Table 3. Placement solution selection in redundant storage

(NVSC: Normalized value of storage cost, EA: Expected availability, VL: Vendor lock-in, NVNP: Normalized value of network performance, OFV: Objective function value)

Placement solution	NVSC	EA	NVNP	VL	OFV
{Windows Azure}	0.0093	0.999	0.4022	1	1.0094
{Aliyun}	0.0119	0.999	0.7826	1	0.5207
{Aliyun, QiNiu}	0.0229	0.999999	0.8913	0.5	0.2398
{Amazon S3 (Tokyo), Windows Azure}	0.0398	0.9999999	0.378	0.5	0.5923
{Windows Azure, Aliyun, QiNiu}	**0.0322**	**0.999999999**	**0.7283**	**0.333333**	**0.2096**
{Amazon S3 (Tokyo), Amazon S3 (Singapore), Aliyun}	0.0727	0.99999999999	0.4754	0.333333	0.3722

In another experiment, we choose $\Phi_{obj}=\{$Amazon S3 (Tokyo), Amazon S3 (Singapore), Windows Azure, Aliyun, QiNiu$\}$, and make $\lambda=0.5, \eta=0.3, \varphi=0.2$. The influence of threshold value q is shown in Fig. 3. When the placement solution is fixed, the changing of q has no effect on the value of network performance and vender lock-in. We observed that the objective function balances the effects of storage cost, expected availability and algorithm cost of Φ_{obj}. When $q=2$, the objective value of Φ_{obj} reaches its minimum.

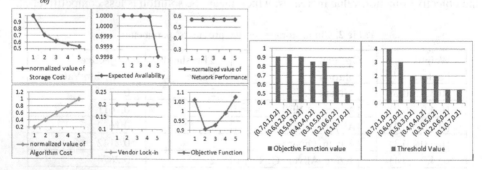

Fig. 3. Values of objective function, etc versus threshold values

Fig. 4. Values of objective function and ver-threshold versus weight factors

Because the value of vender lock-in doesn't change over the threshold value, we make φ be a fixed value when we test the effects of weight factors on objective function value and threshold value. Fig. 4 depicts the result of test for solution {Amazon S3 (Tokyo), Amazon S3 (Singapore), Windows Azure, Aliyun, QiNiu}. Both the value of threshold and objective function decreases when λ is varied from 0.7 to 0.1. If λ decreases, the effect of the algorithm cost increases. So the smaller the value of threshold, the smaller the value of objective function. Our algorithm chooses the solutions with smaller threshold value.

5   Conclusions

We addressed an issue related to placing data chunks of the objects in multi-cloud architecture based on IDA. In order to tackle the issue, we propose a selection algorithm based on recursive relations. The experiments show that the proposed algorithm is efficient and accurate to determine the optimal solution of the objects placements with given weight factors. As our future work, we aim to consider the queries and the processing cost when selecting the cloud providers. Also, the cost of migration from one provider to another due to urgent needs will be studied.

Acknowledgement. This study is supported by the National Natural Science Foundation of China(61370069), the National High Technology Research and Development Program("863"Program) of China (2012AA012600), the Cosponsored Project of Beijing Committee of Education, the Fundamental Research Funds for the Central Universities (BUPT2011RCZJ16) and China Information Security Special Fund (NDRC).

References

1. Kotla, R., Alvisi, L., Dahlin, M.: SafeStore: a durable and practical storage system. In: USENIX Annual Technical Conference, pp. 129–142 (2007)
2. Serious cloud failures and disasters of 2011, http://www.cloudways.com/blog/cloud-failures-disasters-of-2011/
3. Mu, S., Chen, K., Gao, P., Ye, F., Wu, Y., Zheng, W.: µLibCloud: Providing High Available and Uniform Accessing to Multiple Cloud Storages. In: 2012 ACM/IEEE 13th International Conference on Grid Computing (GRID), pp. 201–208. IEEE (2012)
4. Mansouri, Y., Toosi, A.N., Buyya, R.: Brokering Algorithms for Optimizing the Availability and Cost of Cloud Storage Services. In: 2013 IEEE 5th International Conference on Cloud Computing Technology and Science (CloudCom), pp. 581–589. IEEE (2013)
5. Petcu, D.: Multi-Cloud: expectations and current approaches. In: Proceedings of the 2013 International Workshop on Multi-cloud Applications and Federated Clouds, pp. 1–6. ACM (2013)
6. Abu-Libdeh, H., Princehouse, L., Weatherspoon, H.: RACS: a case for cloud storage diversity. In: Proceedings of the 1st ACM Symposium on Cloud Computing, pp. 229–240. ACM (2010)

7. Rabin, M.O.: Efficient dispersal of information for security, load balancing, and fault tolerance. J. ACM. 36, 335–348 (1989)
8. Strunk, A., Mosch, M., Gross, S., Thoss, Y., Schill, A.: Building a Flexible Service Architecture for User Controlled Hybrid Clouds. In: 2012 Seventh International Conference on Availability, Reliability and Security, pp. 149–154. IEEE (2012)
9. Papaioannou, T.G., Bonvin, N., Aberer, K.: Scalia: an adaptive scheme for efficient multi-cloud storage. In: Proceedings of the International Conference on High Performance Computing, Networking, Storage and Analysis, p. 20. IEEE Computer Society Press (2012)
10. Singh, Y., Kandah, F., Zhang, W.: A secured cost-effective multi-cloud storage in cloud computing. In: 2011 IEEE Conference on Computer Communications Workshops (INFOCOM WKSHPS), pp. 619–624. IEEE (2011)

Data-Aware Partitioning Schema in MapReduce

Liang Junjie, Liu Qiongni, Yin Li, and Yu Dunhui

School of Computer Science and Information Engineering,
Hubei University, Wuhan 430062, China

Abstract. With the advantages of MapReduce programming model in parallel computing and processing of data and tasks on large-scale clusters, a Data-aware partitioning schema in MapReduce for large-scale high-dimensional data is proposed. It optimizes partition method of data blocks with the same contribution to computation in MapReduce. Using a two-stage data partitioning strategy, the data are uniformly distributed into data blocks by clustering and partitioning. The experiments show that the data-aware partitioning schema is very effective and extensible for improving the query efficiency of high-dimensional data.

Keywords: Cloud Computing, MapReduce, High-dimensional data, Data-aware partitioning.

The information technology such as next-generation internet, internet of things and cloud computing has been extensively developed and applied, users face many challenges such as massive data or big data. Data exist in the form of high dimensions in many fields. For example, data of financial transactions, multimedia, space and biology can have up to dozens and even hundreds of dimensions. Similarity query is the commonly used method of the high-dimensional space data. Many researchers are focusing on that how to improve the efficiency of similarity query for large-scale high-dimensional data.

To improve the query performance for high-dimensional data, an efficient data access strategy is crucial. The common method for avoiding unnecessary data access is the data partition technology. In the big data era how to perfectly combine the data partition technology with the model of parallel computing becomes a new research direction. In recent years, researches on data partition technology in the environment of cloud computing have attracted much attention [1,2,3,4]. Eltabakh et al. proposed CoHadoop [5] for extending the Hadoop. A file-level property locator is described in NameNode and the relevant data partitions are stored to the same DataNode. However, it is still difficult for the CoHadoop to identify whether the two files are relevant or not. In the paper published by Zaschke on the SIGMOD conference in 2014, a multi-dimensional data storage and index structure named PH tree [6] were described. To reduce storage space, PH tree shares binary description prefixes of the index structure, and support rapid data access during point query and range query. But it is still necessary to further study how to extend the PH tree to the MapReduce framework. Liu et al. published a paper [7] domestically in 2013 proposing a sampling algorithm for determining the space partition function rapidly. The index structure with MapReduce computing model was established, and the K-nearest neighbor join query was realized

H. Wang et al. (Eds.): ICYCSEE 2015, CCIS 503, pp. 93–100, 2015.
© Springer-Verlag Berlin Heidelberg 2015

by using distributed R-tree index. The method is mainly confined to the geospatial knnJ query. The influence on the performance of algorithm brought by multi-dimensional space was not tested.

Aiming at the lacks of data block partition optimization strategy and cannot guarantee that each data partition has the same contribution to the computation in MapReduce framework. In this study, a Data-aware partitioning schema in MapReduce is proposed. The ability of parallel computing and processing of data and tasks on large-scale clusters in MapReduce is utilized. The partition mechanism of the high-dimensional data blocks is optimized based on the high-dimensional data distribution and the characteristics of approximate query. The experiments show that the data-aware partitioning schema has significant advantages in promoting the query efficiency and expansibility of high-dimensional data.

1   Related Work

To overcome the problems in high-dimensional space effectively, the author proposed a high-dimensional index mechanism [8] based on bitcodes in early years. In the mechanism, the advantages in representation of approximate vectors and transform approach of one-dimensional vector are comprehensively used. Moreover, one-dimensional distance is used to represent the distance relationship between the point object and reference point, and bitcodes are used to approximately show the positional relationship. The high-dimensional vectors are compressed into two-dimensional vector (distance, bitcode), which is called bitcode representation for short. Then all the point objects are organized by using the unique B+ tree index. The range to be searched can be greatly reduced through two layers of filtering during indexing to realize the rapid KNN query.

Table 1. Symbol Description

Symbol	Meaning
$DS = \{P_1, P_2, ..., P_n\}$	Dataset of high-dimensional vector
d	Data dimension
$P(p_1, p_2, ..., p_d)$	high-dimensional description of data point P
$P(D_p, B_p)$	2-dimensional description of data point P, that is bitcode index value, D_p is distance between P and the reference point, B_p is bitcode of P
$C = \{C_1, C_2, ..., C_m\}$	Set of clusters
$O = \{O_1, O_2, ..., O_m\}$	Center point set of clusters
$AS = \{A_1, A_2, ..., A_{2^d}\}$	Set of partitions in a cluster
$A(b_1, b_2, ..., b_d)$	The bitcode of a partition

In this study, the bitcode representation is used to encode the high-dimensional vectors and filter the candidates during query. In data analytic environment, data sets are relatively stable.

To simplify the problem, the following reasonable assumptions are made.

2   Data Partitioning

The MapReduce framework is composed of a Master node for providing metadata service and many Slave nodes for storage. This design is suitable for parallel computing and task processing of large-scale data. To adapt to the mechanism, data preprocessing is necessary before the indexes are established. In this way, data can be divided into several data blocks as uniformly as possible, and then all of them can contribute to the final result.

2.1   Cluster Partitioning

The first thing of cluster partitioning is to carry out clustering analysis on the data set DS. Without losing generality, $K-means$ clustering method is usually used to perform global analysis on data in the d -dimensional space. The data cluster $C=\{C_1, C_2, ..., C_m\}$ and corresponding center of mass $O=\{O_1, O_2, ..., O_m\}$ can be obtained.

2.2   Partitioning

After the data clustering and partitioning, each data cluster is further partitioned according to the distribution characteristics of the data. Then the partition $AS=\{A_1, A_2, ..., A_{2^d}\}$ of data clusters can be obtained. For the sake of convenience, the center of mass of data clusters are taken as the reference points for computing the bitcode index values of partitions and point objects. Other points can also be taken as reference points.

Theorem 1. (Dimensions of Partitions) Suppose that cluster size is denoted by $Size(C)$, and the size of point object is denoted by $Size(P)$. The size of query result is denoted by K, then dimensions used to partition a cluster are defined as follows:

$$d' = \left\lfloor \log_2 \frac{Size(C)}{Size(P) \times K \times c} \right\rfloor \quad (d' \le d)$$

where c is a constant ($c \ge 1$). It is used to guarantee that the number of these point objects, which are contained in the partitions, is at least c times of K. Then requirement for the number of objects in query result can be basically met by determining only one candidate partition for each candidate cluster. Hence computation complexity for searching many partitions is avoided. Let's take 20-dimensional vector space in experimental data for example. The size of data set DS is 540MB, and the size of point object is 104Byte. The DS is divided into 4 clusters which are A, B, C and D according to K-means clustering algorithm, where the size of cluster B is 250MB. Suppose that K=10 and c=20. The dimension for partitioning the data cluster B is 13 according to theorem 1. Cluster B is divided into 2^{13} partitions.

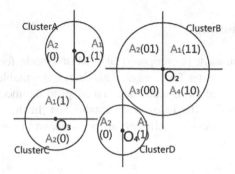

Fig. 1. Schematic diagram of data partition

Definition 1. Bitcode of a Partition $A(b_1, b_2, ..., b_{d'})$ Suppose that the center of mass of the cluster where partition A is located is denoted by $O(o_1, o_2, ..., o_{d'})$. Since all the point objects in the same partition have the same bitcodes [8], the bitcode of partition A can be obtained by that of any point $P(p_1, p_2, ..., p_{d'})$ in the partition.

$$b_i = \begin{cases} 1 & p_i \geq o_i \\ 0 & otherwise \end{cases} \quad (1 \leq i \leq d')$$

It can be seen from definition 1 that the bitcode of a partition represents the positional relationship between the partition and reference point of cluster. Any partition in the data cluster can be represented by the unique bitcode string with the length of d'. The two-dimensional vector space is taken as an example. The representation of bitcode of each partition is shown as Figure 1.

3   Data Storage

To combine the principle of relevance and parallelism during data storage, the data-aware partitioning schema places all the Blocks of a cluster data into a Slave node. One Slave node can store the data of many clusters. Moreover, the data of all partitions in a cluster are uniformly distributed into each block of the cluster. After candidate partitions are scanned during query, parallel computing can be executed by using multiple Map tasks. In this way, parallelism and query efficiency are improved.

In the data-aware partitioning schema, a locator table is designed on the Master node to store the distribution information on the Slave nodes of each cluster. As shown in Figure 2 and Figure 3, the locator able indicates that cluster A and cluster D are stored on Slave1. Cluster B is stored on Slave 2, and cluster C is stored on Slave 3. Therefore, data block A1 and A2 of cluster A and data block D1 of cluster D are stored on Slave1. Data block B1, B2 and B3 of cluster B are stored on Slave2. Data block C1 and C2 of cluster C are stored on Slave3. Other three Slave nodes store the backup of the clusters.

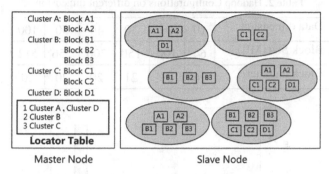

Fig. 2. The logical storage structure of data partition

4 Experimental Results and Analysis

4.1 Experimental Environment

In the experiment, an internal network was constructed with 4 HP blade servers to constitute a Hadoop cluster. One of the nodes was taken as Master, and the other three nodes were taken as Slaves. Each node on the network accesses each other through a 100M network card. CPU of the server of Master node: Inter (R), Xeon (R), E5620, 2.4GHz and 4.4 Core. Memory: 8GB, Disk: 300G*8. CPU of the server of Slave node: Inter(R), Xeon (TM), 3.00GHZ and 4 Core. Memory: 1GB. Disk: 146.8G*2. Each server was installed with OS: 64bit, CentOS6.2, Hadoop of Version 1.0.3 and Eclipse of Version 4.3.1. The default parameter configuration block of Hadoop is 64M, and the number of backups is 3.

4.2 Experimental Result and Analysis

The factors influencing the run time of experiment are data size, data dimension, K value of query and number of servers and so on. Therefore, when the influence on run time brought by a certain factor is studied, other factors are fixed. Then the reliability of experimental data can be assured. Index construction is the pretreatment processes in this study, following a write-once read-many access model. Thus, the time of index construction is not included in the total time.

Scalability. The scalability of the data-aware partitioning schema is studied from two aspects.

(1) Effect of data dimension: It is shown in Fig.3 that the data size is 20 million and K=20, and the number of cluster nodes is 4. The change of execution time of KNN query is observed when the data dimensions are 10, 20, 30, 40, 50 and 100, respectively. It can be seen from the experimental result that when the number of servers are fixed, the execution time increases as the data dimension increases. The amplitude of

increase is relatively stable with the level of 43.7%. There is no sharp increase of execution time as the dimension increases.

Table 2. Hadoop Configurations on different dimension

Data dimension	10	20	30	40	50	100
Block Size(MB)	64	100	150	200	256	512
Block Number	16	21	21	20	20	20

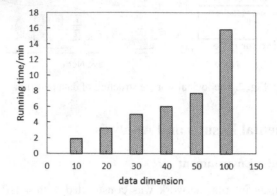

Fig. 3. Scalability (Data Dimension)

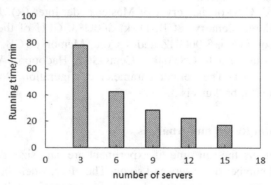

Fig. 4. Scalability (Servers)

At the same time, the experimental data indicates that the increase of execution time slows down (the increase rates of execution time from 10 dimensions to 50 dimensions are 71.3%, 56.8%, 18.8% and 27.8%) as the dimension continually increases. This suggests that the data-aware partitioning schema is very robust to high-dimensional data. Clustering and partitioning are included. These techniques greatly decrease the search range of data. Moreover, bitcode index values of data are compressed. The cost of calculating the distance between high-dimensional vectors is decreased. This is why the dimension does not greatly affect the execution time.

Especially, to decrease the influence brought by the cost of data transmission under the MapReduce framework in the experiment, the number of blocks after data partitioning with different dimensions is ensured to be consistent by adjusting the parameter configuration of Hadoop. The experimental result can better reflect the influence on query time brought by dimensions(Table2).

(2) Effect of number of cluster nodes: It is shown in Fig.4 that the data size is 20 million, and K=20. The change of execution time of KNN query is observed when the numbers of Slave nodes are 3, 6, 9, 12 and 15, respectively. It is found from the experimental result that when the data dimension are fixed, the execution time obviously decreases as the number of servers increases. The average amplitude of decrease is about 1/3. As the number of cluster nodes increases, data are uniformly distributed to more servers through the data-aware partitioning schema for calculation and task processing. The ability of data processing is increased, and the query time is decreased.

The two methods above show that the data-aware partitioning schema has effective scalability.

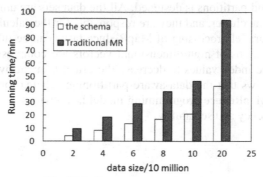

Fig. 5(a). Comparison on data size

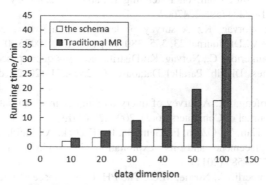

Fig. 5.(b). Comparison on data dimension

Comparison between the Data-aware Partitioning Schema and Traditional MapReduce. The advantages of the data-aware partitioning schema to MapReduce without data partitioning are checked. It is shown in Fig.5 that K=20 and the number of cluster nodes is 4, and the data dimension is 20 in Fig.5(a), and the data size is 20 million in Fig.5(b). KNN query performance is, as shown in Fig. 5(a) and Fig. 5(b), rather independent of data size and data dimension, the data-aware partitioning schema is consistently faster than traditional MapReduce. This suggests that the data storage with locality and relevance can better adapt to similarity query requirement with large-scale high-dimensional dataset. Moreover, dimension reduction encoding method based on the data distribution characteristics in high-dimensional space have obvious effect for promoting the query efficiency.

5   Conclusion

For the MapReduce parallel processing mechanism of data and tasks, a data-aware partitioning schema for high-dimensional data that is appropriate for the MapReduce framework is proposed. Targeting at the distribution characteristics of large-scale high-dimensional vector space, the requirements for the approximate query and the programming pattern of MapReduce, an optimization strategy for data partitioning based on clusters and partitions is designed. All the data sets are uniformly distributed into each block in the cluster, and they are responsible for the calculation. In this way, the advantage of parallel processing of MapReduce tasks can be fully utilized. Moreover, the representations of high-dimensional vectors are compressed by using two-dimensional bitcode index values to decrease the effect brought by high dimensions. The experiment shows that the data-aware partitioning schema for high-dimensional data based on the MapReduce programming model has obvious advantage of promoting the query efficiency and scalability.

References

1. Zhang, C., Li, F.F., Jestes, J.: Efficient parallel kNN joins for large data in MapReduce. In: Proc. of the 15th Int'l Conf. on Extending Database Technology (EDBT), pp. 38–49 (2012), doi:10.1145/2247596.2247602
2. Doulkeridis, C., Norvag, K.: A survey of large-scale analytical query processing in MapReduce. The VLDB Journal 23, 355–380 (2014), doi:10.1007/s00778-013-0319-9
3. Vlachou, A., Doulkeridis, C., Norvag, K.: Distributed top-k query processing by exploiting skyline summaries. Distrib. Parallel Database 30, 239–271 (2012), doi:10.1007/s10619-012-7094-2
4. Yingjie, S., Xiaofeng, M.: A survey of query techniques in cloud data management systems. Chinese Journal of Computers 36(2), 209–225 (2013)
5. Eltabakh, M.Y., Tian, Y., Özcan, F., Gemulla, R., Krettek, A., McPherson, J.: CoHadoop: flexible data placement and its exploitation in Hadoop. Proc.VLDB Endow. (PVLDB) 4(9), 575–585 (2011)
6. Zaschke, T., Zimmerli, C., Norrie, M.C.: The PH-Tree: a space-efficient storage structure and multi-dimensional index. In: Proceedings ACM SIGMOD International Conference on Management of Data, pp. 397–408 (2014), doi:10.1007/s00778-013-0319-9
7. Yi, L., Ning, J., Luo, C., Wei, X.: Algorithm for Processing k-Nearest Join Based on R-Tree in MapReduce. Journal of Software 24, 1836–1851 (2013), doi:10.3724/SP.J.1001.2013.04377
8. Junjie, L., Yucai, F.: BC-iDistance: an Optimized High-Dimensional Index for KNN Processing. Journal of Harbin Institute of Technology 6(15), 856–861 (2008)

An ICN-Oriented Name-Based Routing Scheme

Sun Yanbin, Zhang Yu, Zhang Hongli, Fang Binxing, and Shi Jiantao

School of computer science and technology,
Harbin institute of technology, 150001 Harbin, China
{sunyanbin, zhangyu, zhanghongli,
fangbinxing, shijiantao}@nis.hit.edu.cn

Abstract. Information-Centric Networking (ICN) treat contents as the first class citizens and adopt content names for routing. However, ICN faces challenges of big data. The massive content names and location-independent naming bring scalability and efficiency challenges for content addressing. A scalable and efficient name-based routing scheme is a critical component for ICN. This paper proposes a novel Scalable Name-based Geometric Routing scheme, SNGR. To resolve the location-independent names to locations, SNGR utilizes a bi-level sloppy grouping design. To provide scalable location-dependent routing for name resolution, SNGR proposess a universal geometric routing framework. Our theoretical analyses guarantee the performance of SNGR. The experiments by simulation show that SNGR outperformances other similar routing schemes in terms of the scalability of routing table, the reliability to failures, as well as path stretch and latency in name resolution.

Keywords: name-based routing, routing scalability, Information-Centric Networking.

1 Introduction

Information-Centric Networking (ICN) evolves to a buzzword for future Internet architecture research due to the increasing demand for scalable and efficient content distribution in the Big Data Era. ICN focuses on contents that humans care about rather than where contents are. In ICN, contents are identified by unique names and are retrieved by name-based routing, i.e., finding a path to a content object rather than a host.

However, name-based routing faces the challenges of big data, which brings serious scalability and efficiency problems for two reasons. First, the massive namespace of contents could be several orders of magnitude larger than that of network devices. Google reported that it had indexed approximately 10^{12} URLs [2]. Second, whenever a content moves, the routing system may update routing tables or resolution tables to trace the content. Moreover, if content names are completely location-independent, then the topological aggregation of names is nontrivial, and the efficient routing is difficult in lack of location information.

In order to make ICN scalable and efficient in the big data era, a name-based routing scheme should fulfill the following requirements.

- **Location independence**. A content name is not necessarily bound to its location or owner, and thus routing by location-independent names should be supported.

H. Wang et al. (Eds.): ICYCSEE 2015, CCIS 503, pp. 101–108, 2015.

- **Scalability.** The size of routing table should scale with the size of contents and the size of network.
- **Low stretch.** The ratio of the length of routing path to the length of shortest path should be low.
- **Reliability.** The routing system should still work in the case of failures on a limited set of nodes or links.
- **Anycast support.** For the same content, the routing scheme should be capable of finding a nearby replica among multiple replicas distributed in various locations.

In most of name-based routing schemes for ICN, the location independence of content naming is achieved through either an explicit or implicit name resolution service. Those using explicit name resolution, such as DONA[10], PSIRP [11] and NetInf [6], first resolve a content name to a locator, and then obtain the content from the corresponding location. In the DHT-based name resolution schemes[6], [11], [13], scalability is achieved through hierarchy, but the top level DHT cannot guarantee scalability and low stretch. Those using implicit name resolution, such as CCN/NDN [8], directly announce names into the routing system and do routing by names, but may generate lots of updates whenever contents move.

Besides those for ICN, other previous scalable routing schemes cannot satisfy some of above requirements. VRR [5] and Disco [15] are based on compact routing and do not support anycast. Some DHT-based geometric routing schemes [12], [14] cannot guarantee the greedy embedding of topology and so are lack of low stretch and reliability. Some geometric routing schemes with greedy embedding [7], [9] can avoid the above shortcomings and naturally support anycast, but are not location-independent.

In this paper, we explore a universal name-based routing design for greedily embedded geometric routing schemes and present a Scalable Name-based Geometric Routing scheme, SNGR. SNGR consist of two components: a bi-level sloppy grouping design and a geometric routing framework (GRF). The bi-level sloppy grouping design is a specific implementation of name resolution on logical topology. GRF is a universal framework. Generally, any geometric routing scheme can be modified based on GRF and provide location-dependent routing for name resolution.

The rest of the paper is organized as follows. Section II discusses related work. Section III describes the SNGR scheme. While Section IV presents our theoretical analyses, Section V shows evaluation results. Finally, the paper is concluded in Section VI.

2 Related Works

We briefly reviewed related works in two categories: general scalable routing schemes and name-based routing schemes for ICN.

General Scalable Routing Schemes. Virtual Ring Routing (VRR)[5] applies DHT on physical network with compact routing. The average state of node is $O\left(\sqrt{n}\right)$, but the upper bounds of the state and the path stretch are not guaranteed. Disco [15] combines compact routing with consistent hash, guaranteeing the state of node $\widetilde{O}\left(\sqrt{n}\right)$ and the worst-case stretch 7 for the first packet. However, landmark nodes may be bottlenecks. In some cases, reliability of Disco needs to be improved. PIE[7] is a geometric routing scheme. Each node only stores next-hop nodes and uses greedy forwarding for routing. However, PIE is location-dependent routing. Though a name-based routing scheme Sgern[16] is proposed, it is only a specific instance based on PIE and without specific design for name resolution.

Name-Based Routing for ICN. In CCN/NDN [8], contents are directly routed by their names. CCN/NDN also suffers from routing table scalability issue because the size of routing table is related to the number of contents. DONA[10] adopts a tree-like hierarchical topology. The number of routing entries in a router increases as the hierarchy of that router goes up. Some methods [6], [11], [13] are based on hierarchical DHTs. Though they provide local name resolution, routing schemes in independent domains and the top level domain cannot guarantee the scalability and the low stretch. αroute [3] proposes a alpha-based DHT and maps the overlay topology into the underlay topology. However, the distribution of resolution entries is unbalanced.

3 SNGR Routing Scheme

3.1 Overview of SNGR

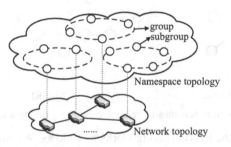

Fig. 1. Structure of SNGR

SNGR (shown by Fig.1) consists of two levels: a logical namespace topology and a physical network topology. The namespace topology adopts a bi-level sloppy grouping strategy to achieve a scalable name resolution system which resolves a content name to multiple locations. The network topology adopts a geometric routing framework, GRF, to provide a location-dependent routing scheme for name resolution.

The bi-level sloppy grouping strategy divides names into multiple groups logically. Within a group, names can be subdivided. Routers are assigned into groups and subgroups according to their names, and are responsible for the resolution of content names that fall into the same group and subgroup.

GRF, which is a universal geometric routing framework, is proposed to provide the location-dependent routing scheme. Unlike traditional geometric routing, GRF brings *vicinity* [15] to geometric routing and adopts a new routing strategy which combines greedy routing and source routing.

Anycast and in-network caching can also be supported by SNGR, they can be our future work. Next, we will describe the components of SNGR in a bottom-up fashion.

3.2   Geometric Routing Framework

Geometric routing [7,9] greedily embeds a topology into a metric space and adopts greedy forwarding for routing. Here, we focus on building a universal geometric routing framework (GRF) for all geometric routing schemes. To support name resolution of SNGR, GRF introduces vicinity and a new routing strategy.

Vicinity. The definition of vicinity is not fixed. Vicinity of a router is a set of routers which are around the router. The vicinity of a router s can be the l closest routers to s (N-vicinity). Based on groups in the namespace topology, the vicinity of s can also be the l' closest routers to s in each group (G-vicinity). The choice of the vicinity is determined by routing requirements.

Both vicinity and neighbors are stored in routing tables. Each router learns the shortest paths to its vicinity via the standard path vector routing protocol.

Hybrid Routing Strategy. With vicinity in routing tables, GRF adopts a hybrid routing strategy which combines global greedy routing with local source routing. The greedy routing is used to greedily select a router (neighbor or vicinity router) from a routing table to forward packets. The source routing is used to communicate with the selected router through the shortest path.

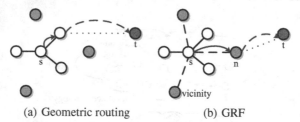

(a) Geometric routing   (b) GRF

Fig. 2. Routing schemes of general geometric routing and GRF

As shown in Fig.2, the hybrid routing works as follows. When a router s receives a packet, it greedily selects a router n from its routing table and guarantee that the sum of $D(s, n)$ (the shortest path length) and $d(n, t)$ (coordinate distance) is minimal. If the selected router is a neighbor, then the packet is directly forwarded. Otherwise, the packet is send to the selected router by the source routing. Each router that receives the packet will repeat the process until the packet reaches the destination.

3.3   Name Resolution System

Here, we propose a bi-level sloppy grouping strategy for the name resolution system.

On the namespace topology, all routers are first divided into multiple groups. Each router obtains a unique binary string by hashing the router name using a consistent hash function, such as SHA-1. The binary string is used to determine the group of the router. Let G denote a group. G consists of a set of routers for which the first k ($k < \log n$, usually set to $\lfloor \log(\sqrt{n}) \rfloor$ or $\lfloor \log(\sqrt{n/\log(n)}) \rfloor$, where n is the size of network) bits of binary strings are the same. Thus, all routers are divided into 2^k groups.

Each group can be further divided into multiple subgroups (SGs) according to the first $k+1 \sim k+h$ (h can be adjusted dynamically) bits of binary strings. All routers in a group are divided into 2^h subgroups. The variable h can be dynamically adjusted. The dynamic size of the subgroup is helpful to reduce the resolution latency and to enhance the robustness of name resolution system.

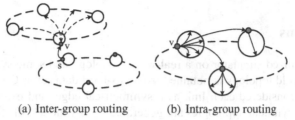

(a) Inter-group routing   (b) Intra-group routing

Fig. 3. Routing on namespace topology

The routing on the name resolution system can be divided into two parts (Fig.3): an inter-group routing and an intra-group routing. The inter-group routing is used for communications between two groups, which can achieved by vicinity. The intra-group routing is used to implement communications within a group.

To implement the intra-group routing, each router stores a DHT routing table including two types of routing entries. (1) Entries in the first type are the coordinates of all routers in the local subgroup, and provide routing within the subgroup by underlying routing. (2) Entries in the second type are used to provide routing between two subgroups. Similar to the G-vicinity, each router stores the m nearest router coordinates in every other subgroup according to the coordinate distance.

A content is registered to all routers in a subgroup, so the content registration and resolution is to find the group and subgroup for the content. The group and subgroup can be obtained via the longest prefix match between two binary strings. The content registration and resolution takes no more than three hops. The first two hops guarantee the shortest path and the shortest coordinate distance, respectively. Thus, resolution router are always close to the requester, which is helpful to avoid the long path.

4   Theoretical Analyses

Node state. The state of a router is the number of entries in a router. There are three tables in each router: a name resolution table, a geometric routing table and a DHT routing table. The average number of name resolution entries is $O(C/(2^{k+h}))$ in SNGR, where C is the number of content objects and 2^{k+h} is the number of subgroups. When $(k + h) = \lceil \log(n) \rceil$ (n is the size of network), $O(C/(2^{k+h}))$ is optimal. That is, each node in a group denotes a subgroup, and the number of name resolution entries is $O(C/n)$.

When the number of resolution entries is optimal, the size of routing tables of SNGR and GRF are related to the number of groups. For SNGR with N-vicinity, it has been proved that the number of routing entries is $O(\sqrt{n \log(n)})$ [15]. For SNGR with G-vicinity, the the numbers of routing entries of GRF and SNGR can be $O(\sqrt{n})$, when the group number and subgroup number are $O(x)$ and $O(n/x)$, respectively

Path stretch. According to PIE[7], the path stretch upper bound of GRF is $O(\log(n))$. However, our experiments indicate that path stretch of GRF is better than that of general geometric routing schemes.

If the path stretch upper bound of GRF is $O(k)$, then the path stretch upper bound of name resolution is $O(2k + 1)$. The conclusion can be proved by the triangle inequality. Thus, the path stretch upper bound of name resolution in SNGR is $O(\log(n))$.

5 Evaluations

We built a centralized simulator on a real-world topology and some synthetic topologies. The real-world topology is obtained from AS-level data set of CAIDA[1] in January, 2012. We considered each link as a symmetrical edge, and extracted the main component. The synthetic topologies are generated by BA model [4]. The initial node number is set to 3, and the average node degree is set to 4. Since Disco is a name-based routing scheme for hosts, we set the subgroup number to be maximum and set the routing table size to be $\widetilde{O}\left(\sqrt{n}\right)$, and evaluated SNGR in comparison with Disco.

Node state. The size of resolution table is determined by the number of subgroups, which will be discussed later. Here we focus on the routing tables.

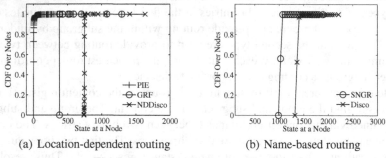

(a) Location-dependent routing   (b) Name-based routing

Fig. 4. Cumulative distribution of node state

Fig.4(a) and Fig.4(b) show the cumulative distributions of node states on the real-world topology. All schemes have balanced distributions of states except few routers with high degrees. PIE has the least node state since each router in PIE only stores its neighbors. GRF and SNGR are better than NDDisco and Disco, respectively. Fig.5 shows the average node states on the synthetic topologies (notice the logarithmic x-axis scale). The slopes of lines for all routings are about 1/2 except PIE, which is consistent with the theoretical analysis.

Path Stretch. Fig.6 shows the average path stretches on synthetic topologies. Obviously, the average path stretches of GRF and SNGR are less than 1.2 and 1.5, respectively. GRF (GRF-N and GRF-G) is better than PIE and NDDisco, SNGR is better than Disco. The path stretch of GRF with N-vicinity (GRF-N) is better than that of GRF with G-vicinity (GRF-G), and so is SNGR. Fig.7(a) and Fig.7(b) are the cumulative distributions of path stretches on the real-world topology. The path stretches of GRF and SNGR are no more than 2.5 and 4, respectively. Especially for GRF-N, the upper bound path stretch is 2.

Reliability. We measured the reliability of routing when nodes or links fail without extra strategies. A proportion of nodes (1% ~ 10%) are randomly set to be failed on

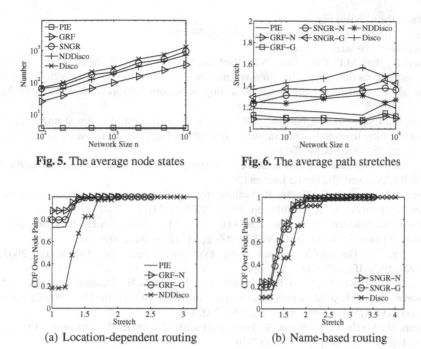

Fig. 5. The average node states

Fig. 6. The average path stretches

(a) Location-dependent routing

(b) Name-based routing

Fig. 7. Cumulative distribution of path stretch

the real-world topology. We found out that success ratios of GRF and SNGR are higher than others. Even in the case that 10% of nodes fail, the routing success ratios are still up to 85% in GRF and SNGR.

6   Conclusions

In this paper, we proposed a scalable and efficient name-based geometric routing (SNGR). SNGR combines name resolution with geometric routing. SNGR guarantees scalable routing tables and resolution tables, low path stretch and high reliability. We have proved that the node states of SNGR and GRF can be both $O(\sqrt{n})$, and the path stretch upper bounds of GRF and name resolution are both $O(\log(n))$. Our experimental results are better than the theoretical analyses. Our work can also be used to address current issues of the Internet, and be applied to some content-oriented applications.

Acknowledgments. This research is supported in part by the National High-Tech Development 863 Program of China under Grant No. 2012AA012506, the National Science and Technology Pillar Program of China under Grant No. 2012BAH37B00, and the National Natural Science Foundation of China under Grant No. 61202457.

References

1. The ipv4 routed/24 as links dataset-jue (2012),
   http://www.caida.org/data/active//ipv4_routed_topology_
   aslinks_dataset.xml

2. We knew the web was big,
   `http://googleblog.blogspot.com/2008/07/`
   `we-knew-web-was-big.html`
3. Ahmed, R., Bari, M.F., Chowdhury, S.R., Rabbani, M.G., Boutaba, R., Mathieu, B.: αroute: A name based routing scheme for information centric networks. In: INFOCOM 2013 (2013)
4. Barabási, A.L., Albert, R.: Emergence of scaling in random networks. Science 286(5439), 509–512 (1999)
5. Caesar, M., Castro, M., Nightingale, E.B., O'Shea, G., Rowstron, A.: Virtual ring routing: network routing inspired by dhts. In: ACM SIGCOMM Computer Communication Review, vol. 36, pp. 351–362. ACM (2006)
6. Dannewitz, C.: Netinf: An information-centric design for the future internet. In: GI/ITG KuVS Workshop on The Future Internet (2009)
7. Herzen, J., Westphal, C., Thiran, P.: Scalable routing easy as pie: A practical isometric embedding protocol. In: ICNP 2011, pp. 49–58. IEEE (2011)
8. Jacobson, V., Smetters, D.K., Thornton, J.D., Plass, M.F., Briggs, N.H., Braynard, R.L.: Networking named content. In: CoNEXT 2009, pp. 1–12. ACM (2009)
9. Kleinberg, R.: Geographic routing using hyperbolic space. In: INFOCOM 2007, pp. 1902–1909. IEEE (2007)
10. Koponen, T., Chawla, M., Chun, B.G., Ermolinskiy, A., Kim, K.H., Shenker, S., Stoica, I.: A data-oriented (and beyond) network architecture. In: ACM SIGCOMM Computer Communication Review, vol. 37, pp. 181–192. ACM (2007)
11. Lagutin, D., Visala, K., Tarkoma, S.: Publish/subscribe for internet: Psirp perspective. In: Future Internet Assembly, pp. 75–84 (2010)
12. Landsiedel, O., Lehmann, K.A., Wehrle, K.: T-dht: topology-based distributed hash tables. In: P2P 2005, pp. 143–144. IEEE (2005)
13. Liu, H., De Foy, X., Zhang, D.: A multi-level dht routing framework with aggregation. In: SIGCOMM workshop on ICN, pp. 43–48 (2012)
14. Newsome, J., Song, D.: Gem: Graph embedding for routing and data-centric storage in sensor networks without geographic information. In: SenSys 2003, pp. 76–88. ACM (2003)
15. Singla, A., Godfrey, P., Fall, K., Iannaccone, G., Ratnasamy, S.: Scalable routing on flat names. In: CoNEXT 2010, p. 20. ACM (2010)
16. Sun, Y.B., Zhang, Y., Zhang, H.L., Fang, B.X.: A scalable geometric routing on location-independent names. In: NDCC 2014 (to be published, 2014)

Partial Clones for Stragglers in MapReduce

Jia Li, Changjian Wang, Dongsheng Li, and Zhen Huang

National Laboratory for Parallel and Distributed Processing, School of Computer Science,
National University of Defense Technology, Changsha, China, 410073
{josephlijia,c_j_wang,lds1201,maths_www}@163.com

Abstract. Stragglers can temporize jobs and reduce cluster efficiency seriously. Many researches have been contributed to the solution, such as Blacklist[8], speculative execution[1, 6], Dolly[8]. In this paper, we put forward a new approach for mitigating stragglers in MapReduce, name Hummer. It starts task clones only for high-risk delaying tasks. Related experiments have been carried and results show that it can decrease the job delaying risk with fewer resources consumption. For small jobs, Hummer also improves job completion time by 48% and 10% compared to LATE and Dolly.

Keywords: MapReduce, mitigating stragglers, task clones.

1   Introduction

Straggler is a critical problem in MapReduce because it can decrease MapReduce performance significantly. Lots of research efforts [1, 2, 3, 6, 7, 8] have been devoted to the solution of stragglers in MapReduce. Some important methods have been proposed, such as Blacklist [6], speculative execution [1, 4], Dolly [6], GRASS [7]. In Blacklist, failed and slow nodes will be put into a blacklist and no new tasks will be assigned to them. Speculative execution launches multiple copies for straggler task and selects the earliest one as result. Dolly clones each task to mitigate stragglers for small jobs. The earliest copy will be adopted.

The above-mentioned approaches are effective for stragglers, but there exist some problems:

- Stragglers rarely appear in a node persistently and they are evenly distributed in cluster [1]. It will cause a waste of resources to put some nodes into blacklist.
- Speculative execution needs some time to detect a straggler task. Waiting means that the job will be delayed.
- Dolly launches multiple copies for each task, and it occupies so many resources.

In this paper, we propose a new approach for stragglers in MapReduce, named Hummer. In Hummer, tasks with high delaying risk will get clones and the risk of stragglers in a job will decrease obviously when resources are limited.

The main contributions of this paper are as follows:

- A new approach for mitigating stragglers is proposed. It can mitigate stragglers in a job by cloning high-risk tasks.
- A risk model of stragglers for MapReduce jobs is established. According to the model, a job delaying risk can be computed.

H. Wang et al. (Eds.): ICYCSEE 2015, CCIS 503, pp. 109–116, 2015.
© Springer-Verlag Berlin Heidelberg 2015

The rest of this paper is organized as follows. Section 2 gives an overview of related works. Section 3 introduces the risk models in Dolly and Hummer. Section 4 presents the design of Hummer. Section 5 shows the experimental results of Hummer. Section 6 concludes the paper.

2   Related Works

Stragglers are slow tasks that complete obviously far behind most of the tasks in job [1], which widely exist in large-scale data intensive computing systems. Although the number of stragglers in a job is small [2, 5, 6, 7], they can delay the job completion seriously and result in a low cluster efficiency [3, 9, 10, 12, 13]. Methods to mitigate stragglers are mainly divided into two categories: active method and passive method. LATE [4] and Mantri [1] are the most popular ones in passive methods (§2.1, §2.2), also called speculative execution ones. Dolly [6] is an active one (§2.3).

2.1   LATE

LATE would detect straggler task. After finding stragglers, it will launch multiple copies for them, and then dispatch these tasks to the nodes to run. When selecting nodes to execute copies of tasks, select the fastest nodes to run. LATE would set a maximum number of copies to prevent thrashing.

2.2   Mantri

Mantri is another approach to detect straggler which concerns about global resources. It adopts a conscious trade-off resource allocation scheme. For stragglers, Mantri estimates the remaining time for this task and the time to run a new copy. It uses a continuous observation and avoids redundant copies [17]. But Mantri also needs a certain time to detect stragglers and leads to a waste of time and job delaying.

2.3   Dolly

To avoid waiting time in speculative execution, Dolly [6, 14] adopts an active way to mitigate stragglers. All tasks will get clones at the beginning of the job. For each task, Dolly selects the fastest copy as result. For the short, real-time jobs [6, 20], Dolly improves the job average response time by 46% and 34%, compared to LATE and Mantri.

However, the resources for cloning in cluster are limited (typically 5 to 10% of the total cluster resources) [6]. Dolly occupies too many resources in cluster and few jobs can get resources to run. Studies show that, there are only few stragglers in a job (typically 9% to14% of job' tasks) [2, 5, 6]. Obviously, there is no need to clone all tasks. Thus, Dolly causes a serious waste of resources.

3   Risk Model

3.1   Job Delaying Risk Model in Dolly

Dolly considers job delaying is due to straggler tasks. Thereby, Dolly model takes probability of each task delaying as fixed value p (turn to be 10% to 14%). In Dolly, c represents the number of copies for a task, and n represents the number of tasks in a job. The risk of job delaying is [11, 15, 18, 19]:

$$\Pr ob_{Dolly} = 1 - (1 - p^c)^n \tag{1}$$

Figure 1 shows how does Dolly eliminate stragglers actively. There exit two main shortages in Dolly model:

- Regard the delaying risk of each task as fixed value is not correct. Task delaying risk is often related with the risk of the node being slow;
- Dolly launches clones for all tasks in a job. Since resources are limited, it will lead to a waste of clone resource.

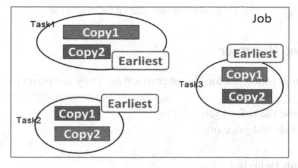

Fig. 1. Dolly launches multiple copies for all tasks

3.2   Job Delaying Risk Model in Hummer

Considering the shortcomings of Dolly, we propose a trade-off approach: Hummer. Hummer launches clones for part of tasks in one job to mitigate stragglers proactively. One task dispatched to high risk node (stragglers appear more often) turns out to be a straggler with higher risk. In Hummer model, different nodes have different risk of stragglers. Figure 2 shows how Hummer launches clones.

We assume that task delaying risk is equal node delaying risk. There are N nodes in cluster, and each node has different computing ability [16]. Each node has different risk of straggler. We calculate each node delaying risk as P_i ($1 \leq i \leq N$) according to historical information.

For a task-i with its clones, its delaying risk is $1 - q_{i1}q_{i2} \ldots q_{ic}$. The risk of task-$i$ delaying without clones is $1 - q_i$. q_i represents the risk of task-i being straggler, $q_i \in$

$\{P_1,P_2,...P_N\}$. q_{ij} denotes the delaying risk of copy-j of task-i, $q_{ij} \in \{P_1,P_2,...P_N\}$, $1 \le i \le n$, $1 \le j \le c$. We get the Hummer model as follows [11, 19]. The risk of job delaying with Hummer:

$$Prob_{Hummer}=1-(1-q_{11}q_{12}\cdots q_{1c})(1-q_{21}q_{22}\cdots q_{2c})$$
$$\cdots(1-q_{m1}q_{m2}\cdots q_{mc})(1-q_{m+1})(1-q_{m+2})\cdots(1-q_n) \tag{2}$$

In Hummer, we don't consider task delaying risk as fixed value any more, but related with the node task is running on.

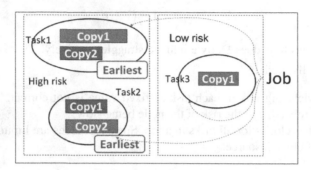

Fig. 2. Hummer clones part of tasks in a job

4   Design of Hummer

There are still some problems for Hummer to solve. They are mainly divided into two categories:
- How we know the risk of job delaying;
- Which tasks should get clones.

4.1   Risk of Job Delaying

Every time Δ, JobTracker will re-calculate the risk of node delaying. That is to say, we update the risk of node delaying every time Δ. We use Hummer model in Section 3.2 to get the job delaying risk.

When a job comes to the head of job queue, JobTracker makes plan for the dispatching of job' tasks. Then JobTracker calculates the risk of each node delaying and regards the risk as the risk of the task delaying which runs on it. Job delaying risk is as follows.

$$Prob_{Hummer}=1-(1-q_{11}q_{12}\cdots q_{1c})(1-q_{21}q_{22}\cdots q_{2c})$$
$$\cdots(1-q_{m1}q_{m2}\cdots q_{mc})(1-q_{m+1})(1-q_{m+2})\cdots(1-q_n) \tag{3}$$

4.2   Tasks Getting Clones in One Job

Which tasks should get clones is another problem to solve. According to 4.1, we can know task delaying risk and job delaying risk. For one job, we compare each task delaying risk with a threshold ρ. If task delaying risk ≥ ρ, we clones this task and launches multiple copies to run on the specified nodes. Otherwise, the task will be scheduled by JobTracker's command.

5   Experimental Evaluation

We evaluate Hummer using traces from 100 nodes in our experiment. We deploy our prototype on a 30-node cluster. Data trace is from a Hadoop cluster called OpenCloud [21], which is a research cluster at Carnegie Mellon University (CMU). The traces include 20-month logs, containing 52675 successful jobs, 4423 failed jobs and 1857 killed jobs. There were 78 users in total that submitted jobs in the logs.

5.1   The Average Job Completion Time

As table 1 shows, we get job bins by the number of tasks. Figure 3 shows the job completion time improvement of Hummer.

From the results, we can see that Hummer is more suitable for small jobs. Hummer reduces the average completion time by 46% and 18% compared to LATE and Dolly.

Table 1. Job bins by the number of tasks

Bin	1	2	3	4
Tasks	0-50	51-100	101-200	201-500

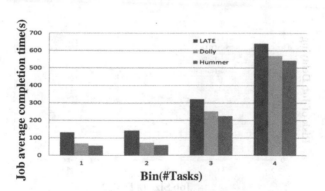

Fig. 3. Average Job Duration in Hummer and Dolly

5.2   Resources Consumption

The most significant advantage of Hummer is that it saves cluster cloning resources. We find that the speed of resources consumption in Hummer is 33% slower than Dolly, as Figure 4 shows.

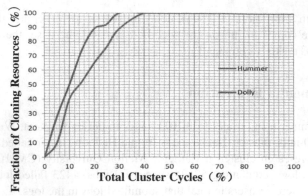

Fig. 4. Speed of Cloning Resources Consumption in Hummer and Dolly. Cloning resources is 5% of the total

5.3   Risk of Job Delaying.

As Figure 5, y-axis represents job delaying risk, and x-axis is the number of tasks in one job.

We can see that as the number of tasks in one job increases, the risk of job delaying also increases. LATE increases fastest. Dolly almost tends to be stable, nearly zero. Hummer is very close to the Dolly [6]. Dolly launches clones for all tasks in a job, so Hummer performs not so well as Dolly in decreasing job delaying risk. But Hummer decreases the risk of job delaying obviously compared to LATE [4].

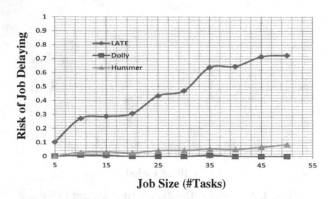

Fig. 5. Risk of job delaying in LATE, Dolly and Hummer

6   Conclusions

This paper presents a proactive approach to solve the straggler problem, called Hummer, which just select partial tasks of a job to launch copies.

For which tasks to choose, Hummer makes use of the node differences. Hummer calculates the risk of each node appearing straggler and updates the risk of each node delaying periodically. Hummer selects tasks which are most possible to be stragglers to launch multiple copies.

Experimental results show that Hummer reduces the job average completion time by almost 46% and 18% compared to LATE and Dolly for small jobs.

Acknowledgements This work is sponsored in part by the National Basic Research Program of China (973) under Grant No.2014CB340303, the National Natural Science Foundation of China under Grant No. 61222205, the Program for New Century Excellent Talents in University, and the Fok Ying-Tong Education Foundation under Grant No. 141066.

References

1. Ananthanarayanan, G., Kandula, S., Greenberg, A., Stoica, I., Harris, E., Saha, B.: Reining in the Outliers in Map-Reduce Clusters using Mantri. In: Proc. of the USENIX OSDI (2010)
2. Kwon, Y., Balazinska, M., Howe, B., Rolia, J.: SkewTune: Mitigating skew in MapReduce applications. In: Proc. of the SIGMOD Conf., pp. 25–36 (2012)
3. Dean, J., Ghemawat, S.: Mapreduce: Simplified data processing on large clusters. In: Proc. of the USENIX OSDI (2004)
4. Zaharia, M., Konwinski, A., Joseph, A.D., Katz, R., Stoica, I.: Improving MapReduce Performance in Heterogeneous Environments. In: Proc. of the USENIX OSDI (2008)
5. Kwon, Y., Balazinska, M., Howe, B., Rolia, J.: SkewTune in action (demonstration). Proc. of the VLDB Endowment 5(12), 1934–1937 (2012)
6. Ananthanarayanan, G., Ghodsi, A., Shenker, S., Stoica, I.: Effective Straggler Mitigation: Attack of the Clones. In: Proc. of the USENIX NSDI (2013)
7. Ananthanarayanan, G., Hung, M.C.-C., Ren, X., Stoica, I., Wierman, A., Yu, M.: GRASS: Trimming Stragglers in Approximation Analytics. In: Proc. of the 11th USENIX NSDI (2014)
8. Kwon, Y., Balazinska, M., Howe, B., Rolia, J.: A Study of Skew in MapReduce Applications. In: Proc. of the Open Cirrus Summit (2011)
9. Chen, Y., Alspaugh, S., Borthakur, D., Katz, R.: Energy Efficiency for Large-Scale MapReduce Workloads with Significant Interactive Analysis. In: Proc. of the ACM EuroSys (2012)
10. Barroso, L.A.: Warehouse-scale computing: Entering the teenage decade. In: Proc. of the ISCA (2011)
11. Resnick, S.: Heavy-tail phenomena: probabilistic and statistical modeling. Springer (2007)
12. Cirne, W., Paranhos, D., Brasileiro, F., Goes, L.F.W., Voorsluys, W.: On the Efficacy, Efficiency and Emergent Behavior of Task Replication in Large Distributed Systems. Parallel Computing 33(3), 213–234 (2007)
13. Hadoop, http://hadoop.apache.org/

14. Ananthanarayanan, G., Ghodsi, A., Shenker, S., Stoica, I.: Why Let Resources Idle? Aggressive Cloning of Jobs with Dolly. In: Proc. of the HotCloud (2012)
15. Ousterhout, K., Wendell, P., Zaharia, M., Stoica, I.: Sparrow: Distributed, Low-Latency Scheduling. In: Proc. of the SOSP (2013)
16. Ghodsi, A., Zaharia, M., Shenker, S., Stoica, I.: Choosy: Max-Min Fair Sharing for Datacenter Jobs with Constraints. In: Proc. of the EuroSys (2013)
17. Zaharia, M., Borthakur, D., Sarma, J.S., Elmeleegy, K., Shenker, S., Stoica, I.: Delay scheduling: a simple technique for achieving locality and fairness in cluster scheduling. In: EuroSys 2010: Proceedings of the 5th European Conference on Computer Systems, pp. 265–278. ACM, New York (2010)
18. Gittins, J.C.: Bandit Processes and Dynamic Allocation Indices. Journal of the Royal Statistical Society. Series B (Methodological) (1979)
19. Sonin, I.: A Generalized Gittins Index for a Markov Chain and Its Recursive Calculation. Statistics & Probability Letters (2008)
20. Dean, J.: Achieving Rapid Response Times in Large Online Services., http://research.google.com/People/jeff/latency.html
21. Ren, K., Kwon, Y., Balazinska, M., Howe, B.: Hadoop's Adolescence: An Analysis of Hadoop Usage in Scientific Workloads. In: Proc. of the VLDB (2013)

A Novel Subpixel Curved Edge Localization Method

Zhengyang Du[1], Wenqiang Zhang[1], Jinxian Qin[2], Hong Lu[2], Zhong Chen[3], and Xidian Zheng[3]

[1] Shanghai Engineering Research Center for Video Technology and System, School of Computer Science, Fudan University, P.R. China
[2] Shanghai Key Lab of Intelligent Information Processing, School of Computer Science, Fudan University, P.R. China
[3] Shanghai Electric Group CO., LTD. Central Academe

Abstract. With the high-speed development of digital image processing technology, machine vision technology has been widely used in automatic detection of industrial products. A large amount of products can be treated by computer instead of human in a shorter time. In the process of automatic detection, edge detection is one of the most commonly used methods. But with the increasing demand for detection precision, traditional pixel-level methods are difficult to meet the requirement, and more subpixel level methods are in the use.

This paper presents a new method to detect curved edge with high precision. First, the target area ratio of pixels near the edge is computed by using one-dimensional edge detection method. Second, parabola is used to approximately represent the curved edge. And we select appropriate parameters to obtain accurate results. This method is able to detect curved edges in subpixel level, and shows its practical effectiveness in automatic measure of products with arc shape in large industrial scene.

Keywords: Machine vision, curved edge detection, subpixel, parabola approximation.

1 Introduction

With the high-speed development of digital image process technology, machine vision technology has been widely used in automatic detection of industrial products. A large amount of products can be treated by computer instead of human in a shorter time. In the process of automatic detection, edge detection is one of the most commonly used methods. But with the increasing demand for detection precision, traditional pixel-level methods are difficult to meet the requirement. To achieve higher precision, higher camera resolution is an effective method, but it is restricted by hardware development and high cost. At present, subpixel edge detection algorithm is the most commonly used method.

Traditional two-dimensional subpixel methods are usually generalized directly from one-dimensional methods. Such methods may not perform well when locating curved edges, especially in irregular situation.

H. Wang et al. (Eds.): ICYCSEE 2015, CCIS 503, pp. 117–127, 2015.

In this paper, a novel subpixel curved edge localization algorithm is proposed. First, one-dimensional method is used to compute the target area ratio near the curved edge. Then, parabola is used to approximately represent parts of the edge. Through parameter selection, edges can be located in high precision. By using this method, subpixel level results can be achieved when locating curved edges. And the time efficiency is proven to meet the demand of big data processing. There is an important practical significance in automatic measurement of product in large industrial scene.

2   Traditional Edge Detection Algorithm

Edge detection is one of the most important research direction in image process field. Lots of researches are based on edge detection. In this chapter, existing edge detection algorithms will be introduced by dividing into edge detection operator and subpixel edge detection.

2.1   Edge Detection Operator

When people observe a target, they regard areas with different colors as edges. Edge detection operators are mathematical representations of the color variety. There are lots of operators, such as Roberts operator, Sobel operator, Prewitt operator, Laplacian operator, Canny operator, Nalwa-Binford operator[1], Sarkar-Boye operator[2] and so on.

As the earliest gradient operator, Roberts operator uses local difference method to compute the difference between adjacent pixels in diagonal directions to represent the gradient. Sobel operator is divided into a vertical operator and a horizontal operator to compute gradient in vertical and horizontal directions, and the larger result is selected. Prewitt operator uses local average difference method to find the edge, it reduces the effect of noise by average the differences. Laplacian operator performs well when detecting isolated points or ends of lines, but it is sensitive to noise. Such operators are all gradient operators in local region, and can be computed by convolution. All these gradient results are used to find edges by setting threshold. A commonly used method is OSTU.

Canny puts forward the three optimal criterions of edge detection operator, and proposes the Canny operator. Canny first smoothes the image, then uses two dimensional first order differential to compute the gradient map, and find possible edge points by using non-maxima suppression algorithm. Finally Canny uses recursive method to obtain pixel-level results.

Such traditional operators are all detecting edges in pixel-level precision. If subpixel results are required, more improvement is necessary.

2.2   Subpixel Precision Edge Detection Algorithm

Subpixel precision edge detection algorithm can be divided into three categories: moment, interpolation and fitting method.

As the first moment method, gray moment method[3] models the edge pixels by using the first three order gray moment to achieve the precise result. Methods, such as spatial moment[4], orthogonal moment[5] and Zernike moment, are proposed in succession. Recently, improved moment methods[6] are still under study.

The core idea of interpolation[7,8] is to interpolate the gray value and the derivative of gray value of edges. According to different interpolation functions, interpolation methods can be divided into quadratic interpolation, polynomial interpolation and so on.

Both moment method and interpolation are sensitive to noise, especially noisy points near edges. Fitting methods[9] are better in this aspect. They are a type of algorithms which uses models to fit the edge. The model may be color model or shape model. Fitting is relatively slower than moment methods and interpolation. However, with the high-speed development of computing ability, speed may become a less important factor than precision.

3   Curved Edge Localization Algorithm

The algorithm of this paper solves the precise localization problem of curved edges in horizontal and vertical directions. It can be used to measure the size of targets with curved edges.

First, the target area ratio of pixels near the edge is computed by using one-dimensional edge detection method. Second, we use parabola to approximately represent the curved edge. And appropriate parameters are selected to obtain accurate results.

3.1   One-Dimensional Edge Detection

In one-dimensional images, pixels can be represented as one-dimensional function form. In the ideal case, function near the edge is like the line in Fig. 1. The edge only influences one pixel. But in practical, it appears like discrete points in Fig. 1. Color changes smoothly near the edge.

One-dimensional edge detection algorithm is using the discrete points to evaluate the break point of the line.

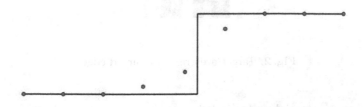

Fig. 1. One-dimensional function

Lots of methods can solve this problem, such as gray moment method and interpolation. In experiments of this paper, interpolation based on Sobel operator is used to achieve subpixel precision results of one-dimensional edges.

3.2   Target Area Ratio

One-dimensional methods can be extended to two-dimensional situation if the edge is a horizontal or vertical straight line. It cannot solve curved edge problems. But one-dimensional methods can still be used to compute the target area ratio near the edge.

Take Fig. 2(a) for example, each block represents a pixel. Preparing for the localization steps, obtaining the target area ratio of each column of pixels is necessary. That is to compute the area ratio of region below the green line within the colored pixels. Before computing the target area ratio, we need to obtain the fuzzy position and the direction of the edge. And extract pixels along the direction to compute the target area ratio.

According to the imaging principal of camera, if the areas of black regions in Fig. 2(b) and Fig. 2(c) are equal, it means that they will perform similar color in images. And the opposite can also be true.

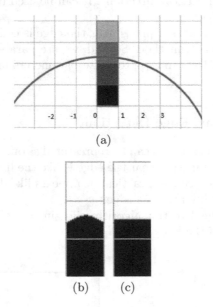

(a)

(b)   (c)

Fig. 2. Target area ratio of curved edge

Assume that a straight edge is being processed, such as Fig.2(c). Subpixel level one-dimensional edge detection algorithm can be used to precisely locate the edge. Target area ratio can be computed in following steps.

Assume that the ith column edge localization result is y_i(the Y-axis value of the edge). The target area ratio of the edge pixel can be represented as $y_i - \lfloor y_i \rfloor$. Considering the region near the edge, the Y-axis value of the lower boundary is Y_0, the height of the region is H, the target area ratio of ith column can be represented as $(y_i - Y_0)/H$.

According to the imaging principal mentioned above, results achieved by using this method can also be effective in situations like Fig. 2(b).

3.3   Curved Edge Detection

In the previous section, target area ratios of all edge pixels have already been obtained. We need to represent them in a unified form. Like Fig. 3, the target area ratios of 5 adjacent columns are represented in the same coordinate system.

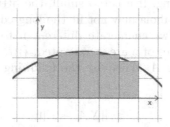

Fig. 3. Curved edge discrete integral graph

The ratio of the ith column is represented as $\left(x_i, y_i^{'}\right)$. Columns near the edge form a function. The Y-axis value means target area ratio of a single column, and the X-axis means the coordinate value.

In this paper, parabola is used to be the unified form to approximately represent the edge.

Assume that the parabola function is $y = ax^2 + bx + c$. if the edge matches the function, the target area ratio of the ith columns can be computed as follow:

$$\int y\, dx = \int ax^2 + bx + c\, dx = \left. \frac{a}{2}x^3 + \frac{b}{2}x^2 + cx \right|_{x_{i-1}}^{x_i} \tag{1}$$

$$= \frac{a}{2}\left(x_i^3 - x_{i-1}^3\right) + \frac{b}{2}\left(x_i^2 - x_{i-1}^2\right) + c\left(x_i - x_{i-1}\right).$$

We use n columns near the edge to evaluate, so we define:

$$X = \begin{pmatrix} x_1^3 & x_1^2 & x_1 \\ x_2^3 - x_1^3 & x_2^2 - x_1^2 & x_2 - x_1 \\ & \cdots & \\ x_n^3 - x_{n-1}^3 & x_n^2 - x_{n-1}^2 & x_n - x_{n-1} \end{pmatrix} ; Y = \begin{pmatrix} y_1^{'} \\ \cdots \\ y_n^{'} \end{pmatrix} ; A = \begin{pmatrix} \frac{a}{3} \\ \frac{b}{2} \\ c \end{pmatrix}. \tag{2}$$

According to (1),

$$XA = Y \tag{3}$$

X is formed by the first three order differences of X-axis, Y is the target area ratio computed above, A is formed by the parameter of the function. Among them, X and Y are already known, there are 3 unknown values in A. If n is not less than 3, A can be worked out. If $n \geq 3$, least squares estimation can be used to estimate the result.

Through the parabola function, the precise localization result of the curved edge can be achieved.

3.4 Parameter Selection

In the previous section, there is an adjustable parameter n, different n may influence the result of fitting. If n is too small, the fitting process may involve error. If n is too large, parabola may not fit the long edge well.

To solve this problem, we enumerate different n to find the best one. Assume that m different values of n are enumerated, $\{n_1, n_2, , n_m\}$, use each n_i to compute the corresponding A_i. Then use the following formula to evaluate the error:

$$err_i = ||XA_i - Y|| \tag{4}$$

Select A_i with the minimum normalized error to be the final result.

3.5 Conclusion

To conclude, our edge localization algorithm is operated as follows to locate a single edge:

- Compute the target area ratio in each column(or row) of pixels by using one-dimensional edges,
- Represent the ratios in the same coordinate system to form the discrete integral graph.
- Enumerate different n, and compute the corresponding parabola fitting result by using the method mentioned in section 3.3,
- Compute the corresponding parabola fitting error with different n by using the method mentioned in section 3.4,
- Find the least error and select the corresponding fitting result to be the localization result.

4 Experiment

In this chapter, the diameters of circular targets are measured to verify the algorithm.

4.1   Simulation

Data. The source images are synthesized by using the square aperture sampling model[11]. An ellipse is included in each image. The gray value of elliptical shape edge is obtained based on square aperture sampling, then the image is convoluted with a Gauss filter and noise is added. The size of the ellipse is 120*60 pixels.

One-Dimensional Edge Detection. The precision of one-dimensional edge results is very important. Synthetic images are used to test the precision of interpolation algorithm based on Sobel operator. The experiment results in different environments are shown in Table. 1.

Table 1. One-dimensional edge detection results with different SNRs

SNR	Average error (pixel)	Standard deviation (pixel)	Max error (pixel)
60dB	0.00314	0.0083	0.002648
50dB	0.00599	0.0128	0.003732
40dB	0.00565	0.0145	0.004752
30dB	0.03363	0.0488	0.010162
20dB	0.14813	0.4236	0.11919

Ellipse Experiment. The edge of ellipse is located by using our method. The experiment results in different environments are shown in Table. 2.

Table 2. Experiment results with different SNRs

SNR	Average error (pixel)	Standard deviation (pixel)	Max error (pixel)
60dB	0.098305	0.051464	0.2707
50dB	0.077569	0.027527	0.3018
40dB	0.091567	0.06989	0.3659
30dB	0.188238	0.146521	0.4021
20dB	0.224266	0.861261	8.6466

Discontinuous Edge Situation. Discontinuous ellipse edges are located by using our method. The discontinuous point is set in different positions. And the SNR of the images is 60dB. The experiment results are shown in Table. 3.

Table 3. Experiment results with different SNRs

Distance between the target and the discontinuous point	Average error (pixel)	Standard deviation (pixel)	Max error (pixel)
0 pixel	0.187302	0.069839	0.3705
1 pixel	0.185524	0.051098	0.2919
2 pixels	0.109696	0.058125	0.2148
3 pixels	0.080194	0.022224	0.1767

4.2   Experiment

Data. The source images with 408*408 pixels resolution are shot with an exposure time of 500μs. The number of the images is 20. Each image is like Fig. 4, it includes circular targets with a diameter of 40mm, and a calibration board which is used to calibrate the size of pixels. The position and direction of targets change in each image.

Pixel Size Calibration. In this experiment, calibration boards are used to compute the pixel size. The side length of each block is 20mm. By using subpixel level Harris operator, the length of sides can be precisely obtained. The ratio of 20mm to the pixel length is the required pixel size.

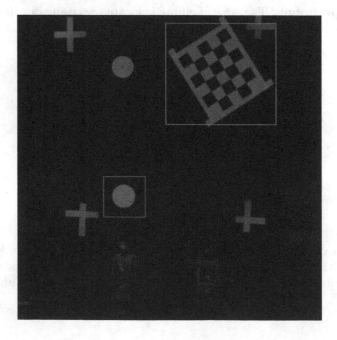

Fig. 4. Experiment data

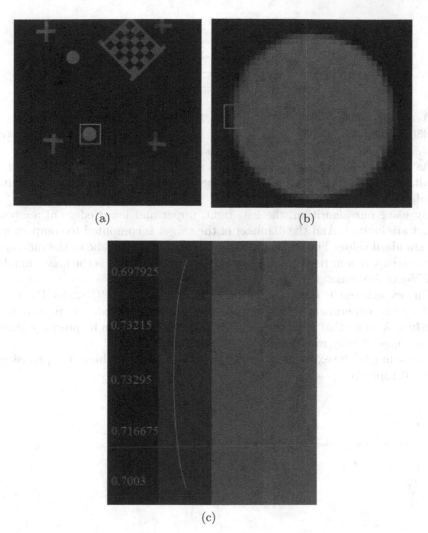

Fig. 5. Localization process and result

Through the experiment, we know that the average length of sides is 15.23738pixels, the standard deviation is 0.355055pixels. So the pixel size is 1.31256mm/pixel.

Experiment Result. The localization result of a single image is shown as Fig. 5. (a) is the original image, fuzzy localization is executed on the original image. The fuzzy localization result is enlarged as (b), and our algorithm is executed on the region in red. The result is shown as (c), which is again enlarged from (b). Each block represent a pixel. The red numbers show the value of target area ratio of each row. The blue curve show the fitting result of the edge.

Table 4. Ellipse experiment

	Average error	Standard deviation	Max error
Image error (pixel)	0.073361	0.040839	0.13094
Actual size error (mm)	0.096291	0.053604	0.17187

According to the section 3.4, let n be 3, 5, 7, and 9. And the fitting results are 0.733075, 0.721075, 0.70655, and 0.6867. The minimum error is obtained when n=3. So 0.733075 is selected to be the final result.

As no standard value is available for a single edge, it is hard to evaluate our result of a single edge. In this section, the error will be analyzed by measuring the diameters of circular targets.

By using our algorithm, the left, right, upper and lower edges of a circular target are located. And the diameter of the target is computed to compare with the standard value. The result is shown as Table. 4. Fig. 6 shows the differences between experiment results and the standard values. X-axis is the image number, and Y-axis is diameter (mm).

The experiment is implemented on Core i7-2630QM CPU@2.00GHz. By repeating the experiment, the run time of locating each edge is proven to be 0.094ms. Assume that 100fps is required, over 100 edges can be precisely located in an image. It may meet most industrial demands.

According to the experiment, our algorithm is able to achieve the precision of about 0.13pixels.

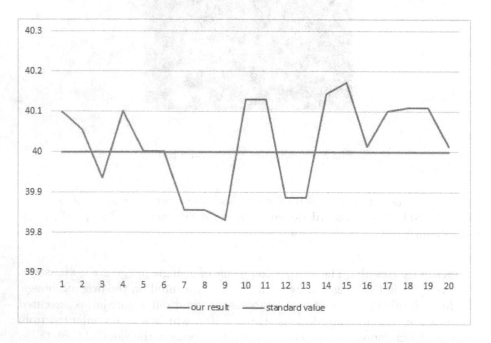

Fig. 6. Result analysis

Comparing with the Zernike moment method[10], the precision of the Zernike moment method on our data is 0.2903pixels.

5   Summary

Our algorithm is proposed to locate curved edge in subpixel precision. The target area ratio of pixels near the edge is computed by using one-dimensional edge detection method. And parabola is used to approximately represent the curved edge and select appropriate parameters to obtain accurate results. Our algorithm is fast enough to meet the industrial demand, and is more precise than other methods.

Acknowledgement. This work was supported in part by the National Natural Science Foundation of China (No. 61170094), Shanghai Committee of Science and Technology (14JC1402202 and 14441904403), and 863 Program 2014AA015101.

References

1. Nalwa, V.S., Binford, T.O.: On detecting edges. IEEE Transactions on Pattern Analysis and Machine Intelligence (6), 699–714 (1986)
2. Sarkar, S., Boyer, K.L.: Optimal infinite impulse response zero crossing based edge detectors. Image Understanding (54), 224–243 (1991)
3. Tabatabai, A.J., Mitchell, O.R.: Edge location to subpixel values in digital imagery. IEEE Transactions on Pattern Analysis and Machine Intelligence 26(2), 188–201 (1984)
4. Lyvers, E.P., Mitchell, O.R.: Subpixel measurements using a moment based edge operator. IEEE Transactions on Pattern Analysis and Machine Intelligence 11(12), 1293–1309 (1989)
5. Ghosal, S., Mehrotra, R.: Orthogonal moment operators for subpixel edge detection. Pattern Recognition 26(2), 295–306 (1993)
6. Qiucheng, S., Yueqian, H., Qingchang, T., Chunjing, L., Ming, L.: A robust edge detection method with sub-pixel accracy. Optik 125(14), 3449–3453 (2014)
7. Xiaobo, W., Xiaoxin, Z.: Improving Resolution of CCD Camera in Dimensional Measurement by Polynominal Interpolation. Chinese Journal of Scientific Instrument 17(2), 154–158 (1996)
8. Qingli, L., Shaojun, Z., Zhongfu, L.: A Improved Subpixel Edge Detecting Algorithm Based on Polynomial Interpolation. Journal of University of Science and Technology Beijing 25(3), 280–283 (2003)
9. Ye, J., Fu, G., Poudel, U.P.: High-accuracy edge detection with blurred edge model. Image and Vision Computing 23(5), 453–467 (2005)
10. Jiwen, C., Jiubin, T.: Algorithm for edge subpixel location based on Zernike moment. Optical Technique 31(5), 779–782 (2005)
11. Changchun, L., Ning, D., Wenhe, L., Haihua, C.: A Method of Making Synthetic Image for Evaluating Edge Location Algorithm. Manufacuring Information Engineering of China 37(19), 52–55 (2008)

Maximal Influence Spread for Social Network Based on MapReduce

Qiqi Shi, Hongzhi Wang, Dong Li, Xinfei Shi, Chen Ye, and Hong Gao

Harbin Institute of Technology
shiqiqi.hit@gmail.com, {wangzh,lee,honggao}@hit.edu.cn,
sunnyleaves0228@qq.com

Abstract. Due to its importance, influence spread maximization problem for social network has been solved by a number of algorithms. However, when it comes to the scalabilities, existing algorithms are not efficient enough to cope with real-world social networks, which are often big networks. To handle big social networks, we propose parallelized influence spread algorithms. Using Map-Reduce in Hadoop as the platform, we proposed Parallel DAGIS algorithm, a parallel influence spread maximization algorithm. Considering information loss in Parallel DAGIS algorithm, we also develop a Parallel Sampling algorithm and change DFS to BFS during search process. Considering two or even more hops neighbor nodes, we further improve accuracy of DHH. Experimental results show that efficiency has been improved, when coping with big social network, by using Parallel DAGIS algorithm and Parallel Sampling algorithm. The accuracy of DHH has been improved by taking into account more than two hops neighbors.

Keywords: Social Network, Map-Reduce, Influence spread.

1   Introduction

A social network, the graph of relationships and interactions within a group of individuals, plays a fundamental role in the spread of information, ideas, and influence among its members. An idea or innovation will spread such as the use of cell phones among college students, the adoption of a new drug within the medical professions. It can either die out quickly or make significant inroads into the population. Thus influence Spread, the extent to which people are likely to be affected by decisions of their friends and colleagues, is a crucial problem.

In recent work, motivated by applications of marketing, Domingos and Richardson [1, 2] posed a fundamental algorithmic problem for social networks. The definition is that with data on a social network and the extent to which individuals influence one another, the goal is to choose the few key individuals to be used for seeding this process to maximal the influence in a given time. One of the applications of this problem is to market a new product that we hope will be adopted by a large fraction of the network.

Several solutions for this problem have been proposed. Domingos and Richardson [2] first present a probabilistic solution. Kempe etc. [3] first formulate this problem to a

H. Wang et al. (Eds.): ICYCSEE 2015, CCIS 503, pp. 128–136, 2015.
© Springer-Verlag Berlin Heidelberg 2015

discrete optimization problem. They proved this problem is NP-hard. Leskovec etc. [4] presents an optimal greedy algorithm which is called cost-effective lazy forward (CELF) framework using the sub-modular properties of influence spread to reduce the calculation for the spread estimation of nodes. In [5], Chen etc. present NewGreedy-ICalgorithm based on the greedy algorithm. When a new seed is selected, the algorithm just transverses the graph randomly once, and it uses the equivalent model as the IC model. They also present a heuristic algorithm which is called Degree Discount. In [6], Shi etc. speed up basic greedy algorithm using the DAG and sampling, and present a more accurate degree discount heuristic algorithm.

The algorithms above are effective and efficient when dealing with small amounts of data. However, when the amount of data increases, the scalability of these algorithms makes them unsuitable for big social networks which now often contain billions of nodes. For example, in [6], it takes a long time to cope with 262k nodes and 1.2M edges by using DAGIS and Sampling algorithm.

With such motivation, in this paper, we attempt to solve the problem of large-scale social network. Because it is common to use parallel mechanism to accelerate computation, we focus on parallel algorithms for influence spread maximization problem. Map-Reduce is a well-known parallel framework, which is suitable for a cluster with hundreds of machines. It has been successfully applied to a number of data mining algorithms such as decision-tree algorithm [7], neural network [8] and frequent pattern mining [9]. Thus in this paper, we use Map-Reduce as the framework and design Map-Reduce-based Improved Greedy Algorithm and Multistep Degree Discount Heuristic. We distribute the calculation of influence spread to different task trackers in Parallel DAGIS and Parallel Sampling algorithm and take into account two-hop and even ten-hop neighbor nodes to improve the accuracy of DDH.

Our contributions in this paper are summarized as follows.

(1) We propose Parallel DAGIS algorithm for influence spread maximization problem in which the calculations of all nodes' influence spread are parallelized. Experimental results show nearly liner speedup of new algorithm under big networks.

(2) We propose Parallel Sampling algorithm in which we parallelize the calculation of all nodes' influence spread and the sampling process. Furthermore, we change unidirectional DFS to bidirectional BFS during search process. Experimental results show that the proposed method can also achieve around linear speedup.

The remaining part the paper is organized as follow. In Section 2, we develop Parallel DAGIS algorithm by parallelizing calculation of each node's influence spread and solve the workflow dependence problem. In Section 3, we propose Parallel Sampling algorithm and also change DFS to BFS which helps scaling up. In Section 4, we extend to two-hop neighbor Degree Discount Heuristic algorithm and discuss the ten-hop neighbors algorithm. Experimental results of the algorithms are presented in Section 5. Finally in Section 6 we draw conclusions of this paper.

2   Parallel DAGIS Algorithm

In this section, we first describe Parallel DAGIS Algorithm. All nodes' influence spreads are computed in parallel. In this way, the efficiency of DAGIS [6] Algorithm is increased especially when dealing with big data. Then we discuss some issues in the implementation of the algorithms, including the use of global variables and workflow dependency. We discuss them respectively.

2.1   Parallelization of Each Node

Define $\delta(A)$ as the expected number of nodes which are active in the final state using the IC or LT model, given A as the initial active seeds set. The influence spread of a set of nodes A is measured as follows.

$$\delta(A) = \sum_{v \in A} \delta^{V-A+v}(v)$$

The influence spread of A is the sum of the influence spread of each node $v \in A$ on sub-graph induced by $V - A + v$. Therefore, the calculation of $\delta(v)$ can be parallelized by each node. Figure 1 illustrates this parallel process.

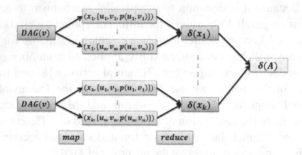

Fig. 1. parallelization of $\delta(A)$ from multi-sources

In **Figure 1**, x_j represents nodes to be calculated in set A. (u_i, v_i) represents SCC in DAG(v) and $p(u_i, v_i)$ represents the weight of the edge (u_i, v_i). Map function transfers the primitive data to Key/Value format and distribute them among different task trackers. Then Reduce function calculates the influence spread of x_j by operating on the Value set with the same Key value. Finally, we sum $\delta(x_j)$ up to get $\delta(A)$. Algorithm 4 shows this process.

Algorithm 4. Parallel DAGIS

Map()
1 **For** each edge $e_i(u_i, v_i)$
2     $u_i \in x_j$;
3   key=x_j;
4   value=$(u_i, v_i, p(u_i, v_i))$;
5   emit(key, value);
Reduce()
1 **For** each node x_j
2   key=x_j;
3   value=$\sum_{x_j \in p[x_i]} IS(x_j) \times w(x_j, x_i)$;
4   emit(key, value);

2.2   Implementation Issues

Then we discuss some issues in the implementation of the proposed algorithm, including the use of global variables and workflow dependency.

2.2.1   Global Variables

Parallelized tasks probably overlap. Figure 2 shows an example.

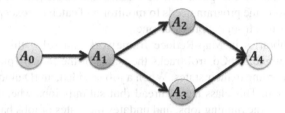

Fig. 2. Duplicate Task

In this example, if A_1 has been calculated when we calculate the influence spread of A_2's precursors, the calculation of A_1 should be avoided when we calculate the influence spread of A_3's precursors. If the thread for calculating the A_1 is running, the thread for A_3 has to wait for the finishing of A_1 thread. If A_1 are not calculated, the running thread for A_3 should calculate A_1 immediately. Therefore, we store A_1 as a global variable to avoid repeating calculation.

A natural idea is to use POSIX threads, semaphores and locks in the parallelization program. However, they do not work well because of communications and IO operations between clusters which are essential in large data results in synchronization, asynchronization and mutex problem and thus maximum speedup cannot be achieved. Hence we use global variables in Map-Reduce.

However, Hadoop in current version does not support global variables and does not advocate the use of global variables. This excessively confines the design of some programs because there are a number of programs using global variables. The reason why Hadoop does not support global variables is that Hadoop tasks are distributed among many compute nodes in the cluster. If it supported read and write operations of global variables, nodes need to communicate with each other which decreases the parallelization. Therefore, we can implement read and write operations of global variables by other means. If user-specified Mapper and Reducer are inner classes of main class, static members defined in the main class can be used as global variables. We can access the global variables by using ":: global". Information about global variables can be stored in the XML files or a configuration file which supports the global variables.

2.2.2   Workflow Dependency

Concurrency problem occurs when more than one task is running in the Map-Reduce workflow. How can we keep these tasks in a normalized order? Can they form liner task sequences or DAG? Those problems can be transformed to a topology sorting problem. Given orders of tasks pairs in a set, we need to judge whether there is a task sequence which violates no premise orders. Meanwhile, there must exist topology order in a DAG and the algorithm is based on generation of $DAG(v)$ from $G_1(v)$. Therefore, the sequence of DAG is same as the workflows of Map-Reduce.

We solve such problem in Map-Reduce as follows. For the liner workflows, the easiest way of unlocking mutex is to carry out a new workflow after the previous one has been finished. And the later workflow has to wait until previous one has finished. If the previous one fails, IO exception will be thrown and subsequent tasks in the pipes cannot run. Therefore, the program needs to monitor and catch the exception, eliminate the intermediate results from the previous one.

Related class libraries in Map-Reduce framework can solve this problem. Take JobControl for example. JobControl tracks the states of the jobs by placing them into different tables according to their states. When a job is added, an ID unique to the group is assigned to the job. This class has a thread that submits jobs when they are ready, monitors the states of the running jobs, and updates the states of jobs based on the state changes of their depending jobs states.

3   Parallel Sampling Algorithm

The parallel DAGIS algorithm above is based on spanning DAG. It loses some information of original graph $G_1(v)$. To avoid the loss of information, we develop sampling strategy on original graph instead of spanning DAG.

In this section, we describe Parallel Sampling Algorithm. It has three steps. First we parallelize the calculation of each node's influence spread. Then we parallelize the process of Sampling. Finally we further change DFS to BFS. In this way, the accuracy of algorithm has been improved and the efficiency of Sampling Algorithm has increased especially when dealing with big data. Then we discuss these three phases.

Step 1. Parallelization of Each Node The phase is the same as the first phase of DAGIS parallelization and the compute of different initial seeds can be parallelized.

Step 2. Parallelization of Sampling Process In the Sampling algorithm, the process of searching samples can be parallelized because each one is independent.

In this step, assuming that samples of some node are larger than T, the statistics of this node's samples are not needed in the following search. Therefore, we add the number of times of searching samples to Value in Map function and add acceptance result of the sample during Reduce function. *Figure 3* illustrates this process. Let i be the number of times of searching samples and r represents whether the sample obtained in one sampling is accepted or not.

Then we further map the second level nodes. We divide each data delivered to machine x_1 into two shares: the first is used for the compute during the extension from the initial nodes to the left; the second is used for the compute during the extension from the initial nodes to the right. Correspondingly, the final reduce process should be changed to three stages: first stage $reduce_1$ computes respectively the left and right extension influence; the second stage $reduce_2$ combines them and computes $\delta(x_1) = \delta_1 \times p_1 + \delta_2 \times (1 - p_1)$. Finally, in the third stage $reduce_3$,

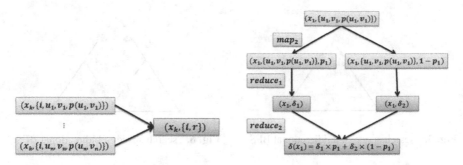

Fig. 3. parallelization of sampling process   Fig. 4. parallelization from two child nodes

$\delta(x_1) + \delta(x_2) + \cdots + \delta(x_k)$ is computed to get the final result $\delta(A)$. Figure 4 illustrate this process and Algorithm 5 shows its flow.

Algorithm 5. *Parallel* Sampling

Map1()
1 **For** each edge$e_i(u_i, v_i)$
2    $u_i \in x_j$;
3   key=x_j;
4   value=$(u_i, v_i, \boldsymbol{p}(u_i, v_i))$;
5   **emit**(key,value);

Map2()
1 **For** each $(x_i,\{u_i, v_i, \boldsymbol{p}(u_i, v_i)\})$
2   key$_1$=(x_i, p_i);
3   value$_1$=$(u_i, v_i, \boldsymbol{p}(u_i, v_i))$;
4   **emit**(key$_2$,value$_2$);
5   key$_2$=$(x_i, 1 - p_i)$;
6   value$_2$=$(u_i, v_i, \boldsymbol{p}(u_i, v_i))$;
7   **emit**(key$_2$,value$_2$);

Reduce$_1$()
1 **For** each (x_i, p_j)
2   Calculate δ_{ij};
3   key=x_i;
4   value= (δ_{ij}, p_{ij});
5   **emit**(key,value);

Reduce$_2$()
1 **For** each node x_i
2    $\delta(x_i) = \Sigma \delta_{ij} \times p_{ij}$;

Step 3. Change to BFS Note that the primitive algorithm is a DFS search from initial seeds. However, DFS often expands unnecessary branches too much. Therefore, DFS should be changed to BFS. **Figure 5** shows ordinary BFS process like a "triangle" extension process.

   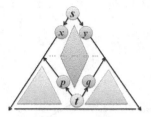

Fig. 5.   search space of unidirectional BFS   **Fig. 6.**   search space of bidirectional BFS

What's more, the primitive algorithm is a unidirectional search from initial seeds which is of low efficiency and should store at least each level node which can probably result in data overflow. Hence we propose the bidirectional search process. In this way, we only need to search the diamond region to avoid the search of two triangular regions below and thus search space will decrease a lot and the algorithm is accelerated. **Figure 6** illustrates the search space of bidirectional BFS.

The bidirectional search process is as follows:

- For reverse BFS extension, a queue is used to store node data by levels.
- After a forward direction BFS, BFS should be extended reversely.
- After a positive search, we should find out whether there are same nodes in the reverse queue. Similarly, after the reverse search, the same work is done.
- The selection of positive or reverse search is as follows. The direction containing fewest nodes to search is chosen. It can decrease the total queue space, speed up pruning and thus accelerate search process.
- If the weights of each edge are different, we can use precedence queue to search at first the node with the least cost to accelerate search.

The pseudo code of such BFS algorithm is shown in **Algorithm 6**.

Algorithm 6. *BFS*

---
1 que_1, que_2 are priority-queues;
2 que1. push_back(s) ; que2. push_back(t) ;
3 **while** (! que1. empty∥! que2. empty)
4   **if** (que1. size () <que2. size ()&&que1. size ()>0) **then**
5   v=que1. pop (); flag=1;
6   **else**
7   v=que2. pop (); flag=2;
8 **for** each u extend from v
9   **if** (flag==1 && u not in que_2) then
10   que1. push_back ();
11   **else if** (flag==2 && u not in que_1) **then**
12   que2. push_back ();

---

4   Experiments

4.1   Experiments Setup

All algorithms are written in Java language in the Hadoop 2.2.15 platform. The running environment is a cluster with eight PCs, each of which has Intel i5 3.1 GHZ CPU, 4GB main memory and 500GB disk. We use a real social network to test our algorithms. Dataset Amazon, the Amazon product co-purchasing network [11], which is dated on March 2, 2003, where nodes are products and a directed edge from s to t means product s is often purchased with product t. It has 262k nodes and 1.2M edges.

4.2   Experiments of Parallel DAGIS and Sampling

In this section, we justify the efficiency of Parallel DAGIS algorithm and Parallel Sampling algorithm. Programs run in different numbers of computers which play the role of nodes in the cluster and calculate the speedup ratio of running time before parallelization to that after parallelization. Figure 7 shows the result. The x-axis is the run time with second as the unit.

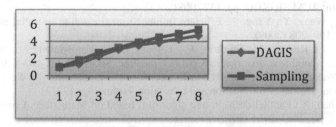

Fig. 7. speedup-ratio of parallel DAGIS and Sampling Algorithm

As the figure shows, the parallelization algorithm has good performance in speed up. The more compute nodes are, the bigger speedup is. When eight compute nodes are used, the speedups of DAGIS and Sampling are 4.7 and 5.4. Meanwhile, the speedup of Sampling is larger than DAGIS because based on the Map-Reduce in the DAGIS parallel algorithm, Sampling is further parallelized, initial nodes are transferred from root to the left and right child nodes of root and DFS is replaced by BFS which improve the algorithm's performance.

5   Conclusions

Existing algorithms of influence spread maximization problem for social network are not efficient enough to cope with real-world social networks, which are often large networks. To handle big social networks, we propose Parallel DAGIS algorithm using Map-Reduce in Hadoop as the platform. Considering information loss in Parallel DAGIS algorithm, we also develop Parallel Sampling algorithm. We further improve accuracy of DHH, considering two or even more hops neighbor nodes. Experimental results show that the efficiency has been improved, when it comes to big social network using Parallel DAGIS algorithm and Parallel Sampling algorithm and the accuracy of DHH has improved by taking into account more than two hops neighbors.

These algorithms are meaningful to the real world especially to support the market strategic decision. We aim at applying these algorithms to the real social networks such as spreading advertisements to the maximum people. In this way, the efficiency of marketing will increase a lot and thus they improve the development of economy.

Acknowledgements. This paper was partially supported by NGFR 973 grant 2012CB316200, NSFC grant 61472099 and National Sci-Tech Support Plan 2015BAH10F00.

References

1. Richardson, M., Domingos, P.: Mining knowledge-sharing sites for viral marketing. In: SIGKDD, pp. 61–70 (2002)
2. Domingos, P., Richardson, M.: Mining the Network Value of Customers. In: Seventh International Conference on Knowledge Discovery and Data Mining (2001)
3. Kempel, D., Kleinberg, J., Tardos, E.: Maximizing the spread of influence through a social network. In: ACM SIGKDD, pp. 137–146 (2003)
4. Chen, W., Wang, Y., Yang, S.: Efficient influence maximization in social networks. In: KDD, pp. 199–208 (2009)
5. Leskovec, J., Krause, A., Guestrin, C., Faloutsos, C., VanBriesen, J., Glance, N.S.: Cost-effective outbreak detection in networks. In: Proc. SIGKDD, pp. 420–429 (2007)
6. Shi, X., Wang, H., Li, J., Gao, H.: Influence Spread Maximization in Social Network. International Journal of Information and Education Technology (2013)
7. Lu, Q., Chen, X.: Parallel decision-tree algorithm based on Mapreduce. Computer Application 2469(9), 2463–2465 (2012)
8. Zhu, C., Rao, R.: The improved bp algorithm based on Map-Reduce and genetic algorithm. In: CSSS 2012 (2012)
9. Chen, G., Yang, Y., Zhang, Y.: MapReduce-based balanced mining for closed frequent itemset. In: ICWS 2012 (2012)
10. Hadoop Infererence, http://apache.etoak.com/hadoop/

Implementation of Dijkstra's Token Circulation
on Sensor Network

Zhiqiang Ma, Achuan Wang[*], and Jifeng Guo

Northeast forestry university, Heilongjiang, Harbin China
mazq_nefu@163.com, tony16269@sina.com, gjf1000@126.com

Abstract. Sensor networks can consist of large number of sensors. Often, sensors networks use low cost units and thus a subject to malfunctions that can bring the system to inconsistent states. After deployment, the system can be situated in places that are hard to reach and therefore manual reboot operations are undesirable and even unfeasible. Therefore, it is imperative to consider the eventual recovery of arbitrary fault when designing sensor networks. Dijkstra's algorithm is an important foundation of self-managing computer system and fault-tolerance computing system in distributed systems, since it allows a distributed system to recover from arbitrary starting state within a finite time. The arbitrary starting state ca model arbitrary failure (as long as the code segment stays correct). Another key advantage of Dijkstra's asynchronous algorithm is that no global clock is needed. This project tests an implementation of Dijkstra's algorithm using snapshotting techniques that we developed in an earlier work. These sensors can initiate from any state but they come into a consistent one after several cycles of running. We demonstrate the usefulness of our testing technique.

Keywords: Dijkstra's algorithm, Sensor network, consistent state.

1   Introduction

This project focuses on the consistency problem of sensor networks, which are distributed systems and sometimes also work as fault tolerance systems. Sensor networks are consisting of large number of sensors that are always suffered from problems such as power failure or natural radiations that may interferes their data transmission. We propose a self-stabilizing algorithm that helps these systems to recover from different inconsistent states to a consistent one, that is to say, if a distributed system is self-stabilized, it recovers or starts from arbitrary states, and then guaranteed to converge to a legitimate state and remain in a legitimate set of states thereafter. This is a big advantage for fault tolerance systems in distributed systems since they are consisting of loosely connected machines, which do not share a global storage, and each of them keeps a partial view of global state.

Dijkstra's algorithm[1][2][6][10][11][12] is firstly introduced in seminar paper of Edsger Dijkstra in 1974, and it is not designed for fault-tolerance systems at the first

---

[*] Corresponding Author.

H. Wang et al. (Eds.): ICYCSEE 2015, CCIS 503, pp. 137–143, 2015.

beginning, but then this algorithm quickly becomes an important foundation of self-managing computer system and fault-tolerance computing system [8][9], since it helps a distributed system to recover from arbitrary state in a finite time after a failure. Beside of this, one of the advantages is that Dijkstra's algorithm does not depend on any strong assumptions but in contrast, some previous algorithms only work when there exists global clocks together with known upper bounds in distributed systems[7].

It is obvious to see that sensor networks are distributed systems since they are consist of large numbers of wired or wireless sensor nodes. There are many reasons for sensors to power off, such as when they deplete their batteries, unstable electric current or damaged by other irresistible natural disasters. We design redundant mechanisms to ensure the damaged sensors to be replaced or repaired when error occurs, but it is also important to maintain the consistency of the whole distributed system, this is the reason that we implement the Dijkstra's algorithm.

In this project, we focus on an implementation of Dijkstra's algorithm on Tmote Sky sensors. This implementation evolves five nodes and a deterministic leader definition instead of a leader election process. In the first step, a deterministic leader is set to node 1 by default, and then each of them picks up a random number that acts as their state in the first cycle, finally all nodes reach a consistent state after several cycles of running and then keep the state in the future.

'Preliminaries' will provide some fundamental on Tmote sky and Contiki, 'Design Strategies' shows related details on how to design the program, 'Implementation and Evaluation' illustrates how to implement the program and 'Conclusion' shows the conclusion that obtained from previous chapter, finally we will make a short discussion.

2   Preliminaries

The system consists of a simulator; in this case, we simulate 5 sensors in Cooja. From section 2.1, we introduce basic knowledge for test bed, Contiki, Cooja and checkpoint process.

2.1   Tmote sky

This implementation is designed to work on Tmote Sky [3] nodes, which is a broadly used energy-saving wireless sensor node, it could also support varies standard ADC interfaces or SPI/I2C interfaces. Tmote Sky use MSP430 series CPU and mostly commonly are F1161 and F5438, among them, F1161 has lowest energy consuming feather in MSP430 series but its computing performance becomes much lower than F5438. Msp430 can make usage of tool chains provided by TI, but those tool chains cannot support full features of this CPU. Tmote sky support both Contiki and Tinyos embedded system. Tmote Sky support extremely low power consumption and highly integrated antenna, it also has become an eco-friendly device and will broadly used in many areas. Tmote sky will be used as both base station (sink node) and ordinary sensors in this research.

2.2   Contiki

Contiki [4] was developed by SICS (Swedish Institute of Computer Science). Comparing with Tinyos, Contiki is better in support for IP connections. Until now, Contiki has published its 2.6 edition by providing many new features. Although Contiki is lightweight, but it support multi-threading (multi-tasking) and build-in TCP/IP stack, it even provide a web browser, which is maybe the smallest in the world.

Contiki is an open-source and multi-threading operating system, which is easy to port to any sensor platforms. Comparing to other embedded operating systems, Contiki gain a great advantage to support its own file system, and it use external flash as its storage media. One good thing for CFS is when updating code base to large amount of sensors, as it is not easy to retrieve all the nodes then upgrade them, then deliver the updated files to their external disk then trigger the upgrade is a smarter method. According to other file systems, CFS cannot use either large-scale data structure or memory cache, both due to the limited memory size. And also the data block size and the read/write frequency have been designed to match the requirement of sensor platforms.

Proto thread is the process controlling model inside Contiki, which is a lightweight threads and running without the support of per-process stacks. Proto thread fulfills the limited hardware requirement from sensors, as it do not have complex state machines, also, it do not support full-featured multi-threading operations. One side effect of proto thread is the lack of support for local variables when switch context between threads or thread blocking, as there is no stack to maintain variable states, but variables can be defined into 'static' to keep their values throughout the whole process.

2.3   Cooja

Cooja is a simulator in Contiki, which is a cross-layer simulation tool that handles simulation from all three levels. But one node can only simulate one layer instead of three. This might little reduce the practical value to some extent.

Cooja combines plug-ins and interfaces together in order to provide a flexible expanded capacity. Plug-ins can be divided into several categories:' *mote plug-in'*, which provides directly functions, such as memory operation, serial input/output and shell input/output; *'simulation plug-in'* provides functions that only take effect within one simulation. *'Cooja plug-ins'* does not depend on any simulations; actually they are general plug-ins for all instances, such as 'control panel', 'simulation visualize' and the timeline.

3   Design Strategies

3.1   Methodology

The purpose for this project is to provide a solution which resolve consistency problems for sensor networks. This design is based on Dijkstra's algorithm and mutual

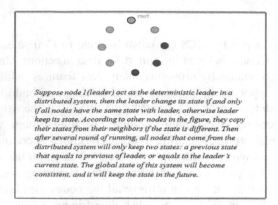

Suppose node 1(leader) act as the deterministic leader in a distributed system, then the leader change its state if and only if all nodes have the same state with leader, otherwise leader keep its state. According to other nodes in the figure, they copy their states from their neighbors if the state is different. Then after several round of running, all nodes that come from the distributed system will only keep two states: a previous state that equals to previous of leader, or equals to the leader's current state. The global state of this system will become consistent, and it will keep the state in the future.

Fig. 1. Dijkstra's algorithm

exclusion. Sensor nodes start from arbitrary states then come into a consistent state by passing one token ring between each of them.

As Fig.1. shows that according to any system that consists of N nodes, each node will change their value between 0 to N+1, all nodes will restricted to N+1 state. And each node other than leader can only have one predictable action in the future: change to the state that equals to leader.

3.2   Design

This project implements a Dijkstra's algorithm on two Tmote sky sensors, which do not share memory but only communicate with broadcast messages. The purpose is tried to reach a consistent state for both two sensors after several cycles of communication, and then keep the consistent state in the future.

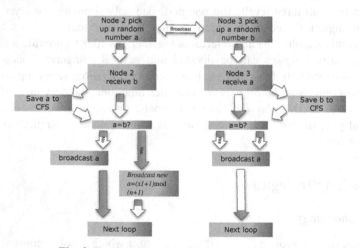

Fig. 2. Detailed design for Dijkstra's algorithm

As we could read in Fig.2., since leader is deterministic during the two sensors, so node 2 is the default leader. Both two sensor firstly select a number as their initial state randomly. Then they start first round of broadcast to send out their states to the other nodes.

When they receive broadcast message, the state comes from message will then extracted and save into a file for further operation. According to the leader node, it will continue broadcast the old state until the state received from node 3 becomes same with its own. Then leader node change to a new state follows the formula 'new state=(x+1) mod (n+1)'. For example, if the old state of leader is 34, then next state will become to '(34+1) mod (2+1) =2'. So according to a sensor network composed by N sensors, the whole system will maintain an N+ 1 state in a consistent state. For example, for a system that consists of 2 nodes, there are totally 3 states: 0, 1, and 2. this is also mean that in most of 3 times of broadcast, this system will come in to a consistent state. According to nodes other than leader, they will change their states follow their neighbor, so if the system contains 2 nodes in this case, node 3 will follow the state of the leader. For example, if the leader's old state is 34, and then node 3 will change its state to node 3 regardless of its original state.

3.3   Implementation and Evaluation

In this project, the communication between different nodes is implemented with rime broadcast. Nodes firstly create a new broadcast channel in the start, and then pad the rime packet with data, while in this case, the packet will pad with numbers that indicates the states of nodes, and then the packet can be sent through the defined broadcast channel. When receive through the call back function, data firstly copy from the packet to a buffer then save into files. The reason for using a file to store states instead of global variables is because the file can keep the data even sensors power off.

It is worthy to mention that the algorithm work with 'SUSE' [5] in this project, which is a new developed tool that can provide a variable level diagnosis to sensor network. It is important to restore the state of rime stack by 'NETSTACK_MAC.off();', since the rime stack will be affected by checkpoint process during diagnosis, and then the broadcast will malfunctioned to send out random data without reset as mentioned before.

When running on sensors that places in a reasonable distance, it cost 4 seconds to finish one cycle of running on both sensors, only 1 seconds for broadcast and receive, and 3 seconds work for an event timer.

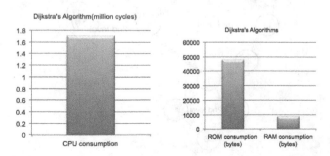

Fig. 3. CPU and memory consumption for Dijkstra's Algorithm

As it shows in the Fig.3., CPU consumption for transfer their state in one round of running is 1.7050251 million cycles. And the RAM consumption is 8666 bytes on both sensors. According to the ROM cost, since the codes running with default make files from Contiki, the consumption of ROM is 47880 bytes.

4   Conclusion

Dijkstra's is one algorithm that can implement self-stabilize in distributed systems. This project develops a self-stabilized distributed system on two sensor nodes by implementing the Dijkstra's algorithm on them. Each node can start with a random number then after at most N round of running, the system comes into a consistent state and keeps the state in the future.

5   Discuss and Future Work

Dijkstra's algorithm can help nodes come into a consistent state in a distributed system within limited round of running, it can also help the system comes into a consistent state when some nodes recover from failures. But one disadvantage is more nodes means more time to reach consistent.One possible improve is to build a two-level (nested) distributed system which top-level nodes is composed by leader nodes. Then if the whole system starts from a chaos state, and if node N is a leader node in top level, then as soon as N has modified its state follows its leader, second level could start to run Dijkstra's algorithm immediately if N is their leader.

As Fig.4. shows, top level firstly start the algorithm, then each second level system start their algorithm as soon as their leader has modify their states that equals to leader in top-level leader. This improvement can work with both deterministic leader and a leader election algorithm. But the time complexity and algorithm complexity have to be considered carefully when compare to traditional Dijkstra's algorithm before decision. As it shows in figure 5, it needs 14 round of running to reach stabilize from tradition Dijkstra's algorithm, but only 7 round of running in this improvement.

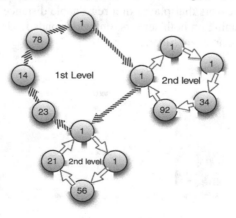

Fig. 4. Nested Dijkstra's Algorithm

Acknowledgements. This work was supported by the Fundamental Research Funds for the Central Universities No. (2572014CB27) and the science and technology project of Heilongjiang Province Education Department No. (12533029).

References

1. Dijkstra, E.W.: A note on two problems in connexion with graphs. Numerische Mathematik 1, 269–271 (1959)
2. Cormen, T.H., Leiserson, C.E., Rivest, R.L., Stein, C.: Section 24.3: Dijkstra's algorithm. In: Introduction to Algorithms, 2nd edn., pp. 595–601. MIT Press and McGraw–Hill (2001) ISBN 0-262-03293-7
3. Moteiv Corporation. Tmote Sky: Datasheet (June 2, 2006)
4. Dunkels, A., vall Gron, B.: Thiemo Voigt Swedish Institute of Computer Science.Contiki - a Lightweight and Flexible Operating System for Tiny Networked Sensors
5. Zhang, X.: Distributed Diagnostics Tool for Sensor Networks (2013)
6. Liu, J., Ma, T., Ma, S.: Systems Engineering-Theory & Practice 31(6), 1153–1157 (2011) (in Chinese)
7. Wang, H.: Science of Surverying and Mapping 39(2), 52–54 (2014) (in Chinese)
8. Han, C., Yi, S., Yang, Y.: Journal of Southwest JiaoTong University 46(2), 252–258 (2011) (in Chinese)
9. Liu, D., Hou, M., Wu, Z., Qu, H.: Computer Science 38(7), 96–99 (2011) (in Chinese)
10. Jiang, D., Dai, L.: Computer Engineering and Applications 47(31), 209–211 (2011) (in Chinese)
11. Zheng, S., Cao, J., Lian, X.: J. Tsinghua Univ. (Sci. & Tech.) 49(11), 1834–1837 (2009) (in Chinese)
12. Duan, Q., Zhou, Y., Chen, Y.: Chinese Journal of Sensors and Actuators 23(11), 1610–1616 (2010) (in Chinese)

Co-simulation and Solving of Clapper Type Relay Dynamic Characteristics Based on Multi-software

Guo Jifeng, Ma Yan[*], Ma Zhiqiang, Li He, and Yu Haoliang

Northeast Forestry University, Heilongjiang, Harbin China
{gjf1000,lihe123}@126.com, myan@vip.163.com,
{zhiqiang1980,haoliang1994}@163.com

Abstract. Based on multi-commercial finite element analysis (FEA) software co-simulation calculating method, the electromagnetic system model was built for solving static and dynamic characteristics of a clap-type rely. Using the Fortran programming language, the solving of differential equation and the calculating of electromagnetic torque interpolation was realized, therefore the MEM coupling system static/dynamic characteristics of the relay was obtained. The validity and accuracy of this method has been confirmed by results of experiments. The conclusions which obtained are valuable in optimizing the clap-type rely production.

Keywords: Co-simulation of Multi-software, Dynamic characteristics, Relay, FEA.

1   Introduction

The clapper type rely is the most widely used electrical switching element, Its performance, life and reliability directly affect the safe and reliable operation of the various facilities, equipment [1-2]. Therefore, the optimization design of the electromagnetic system, mechanical structure, and the improvement of work performance and reliability has have been popular research topics in the industry. However, the effective and accurate calculation of static and dynamic characteristics   is the necessary condition for effective optimization design [3-5].

In this paper, combined with the application of the mechanical dynamics software ADAMS, the finite element analysis software FLUX3d and Fortran programming language, and through the twice interface development by ADAMS, the use of dynamic simulation data bidirectional iterative calculation, the simulation calculation of mechanical-electric-magnetic coupling system can be realized. In the meantime, the static and dynamic characteristics of the clap-type relay and the ways to get them can be dug out. The validity and accuracy of the presented method has been confirmed by the results of experiments.

---

[*] Corresponding author.

H. Wang et al. (Eds.): ICYCSEE 2015, CCIS 503, pp. 144–150, 2015.

2   The Mathematical Model of the Clap-type Rely

2.1   The Electromagnetic and Mechanical System Structure

The main components of a clap-type rely include the coil framework, icon core, yoke, armature, reed, movable contact, static contact, conductive insert, reaction spring and power coil etc., as shown in fig.1. Due to their magneto conductivity, the yoke, armature and iron core make up the electromagnetic system of the relay. The reaction spring, providing the force of restoring to initial position for the armature, as well as the armature and iron core make up the mechanical power system (Fig.2 is a real photo of a clap-type relay).

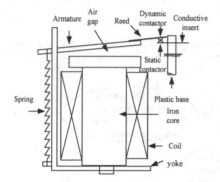 

Fig. 1. Structure diagram of clap-type rely   **Fig. 2.** Real photo of clap-type rely

2.2   Mathematical Model

The armature of the clap-type relay rotates around the shaft according to the following equation under the action of the electromagnetic force moment.

$$J\frac{d\omega}{dt}+T_k+T_f=T_m \tag{1}$$

$$\omega=\frac{d\theta}{dt} \tag{2}$$

The moment of inertia of the armature is J; Rotation angle of armature is θ. Angular velocity of armature is ω.

T_m is the force torque of the electromagnetic suction, T_k is the load torque provided by the spring reaction force and T_f is the frictional torque.

The initial condition is $\omega|_{t=0}=\omega_0=0, \theta|_{t=0}=\theta_0$. The Electromagnetic system voltage balance equation is

$$\frac{d\psi}{dt}=U-iR \tag{3}$$

In this equation, ψ means the coil flux, U means the coil voltage, i mean the electric current and R means the resistance. Considering the formulas, the state equations of dynamic characteristics of clapper electromagnetic relay is built.

3   Multi-software Co-simulation

Based on the virtual prototype ideas, this paper implements the simulation of the clapper electromagnetic relay machine-electric-magnetic couple field through the secondary development finite element analysis software (ANSYS, FLUX3D and ADAMS) and its topological structure is shown in Figure.3.

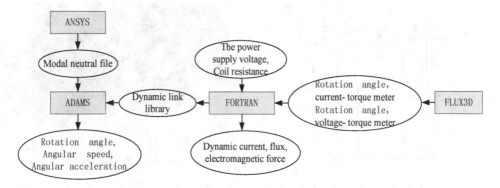

Fig. 3. Co-simulation topological structure

The simulation of dynamic characteristics of clapper electromagnetic relay is a coupling calculation of mechanical and electromagnetism. It includes the calculation of electromagnetism system as well as the dynamic simulation of mechanical system, so it is hard to be solved in single commercial software. What's more, with the existing commercial software development direction toward specialization, it becomes a trend in the field of simulation to complete a complex task through the multi-software collaboration [6-8].

The magnetic structure of relay is irregular, and we construct its magnetic system parametric model using finite element analysis FLUX3D as a solver as Figure 4 shows. When the solution area is much greater than the size of the model, the magnetic potential of the area boundaries can be approximated as 0. This problem is the first kind of boundary condition and that satisfies the equation $\phi(x, y, z)|_{\Gamma 1} = \overline{\phi}(x, y, z) = 0$.

In FLUX3D, the above boundary conditions can be determined by the establishment of infinity boundaries. Then we can calculate to obtain static suctions $F_m(\theta, i)$ and family of flux curves $\psi(\theta, i)$ under different coil current and angle armature. Interpolating with the current, the flux, the displacement of moving core and other parameters, we obtained dynamic electromagnetic torque.

The spring leaf of relays is flexible body, so we should establish modal neutral file in the finite element software. Using ANSYS software, we generate the modal neutral file (*.mnf) of the spring leaf in the form of data streams and import it into ADAMS. The armature, the yoke, the moving core and the dynamic and static contact model attached to the spring leaf are rigid and the structures are regular, so they can be directly built in ADAMS. The reaction spring of relays can be directly replaced with the spring module of ADAMS. After given reasonable property to each entity and reasonable assembled, we load dynamic driving torque in the "center of mass" position.

Programming with FORTRON and the *"sfofub"* of ADAMS as a template, we solute the electromagnetic torque real-time and compile it to make a *.dll* file. Then link it to the simulation sub-step calculation in ADAMS real-time. According to the electromagnetic torque that input in different times, ADAMS solves the multi-body dynamics equations and gets the physical dynamics of virtual prototyping such as the angle, the angular velocity, the angular acceleration of the armature, etc. While programming, the supply voltage E and the coil resistance R are required. Using the variable step size Runge-Kutta method to solve the circuit equations in (1), the response data of structural dynamics such as $\dot{\theta}$, $\ddot{\theta}$ and electromagnetic suction curves can be obtained from ADAMS after the simulation. The simulation model built in ADAMS is shown in Figure.4.

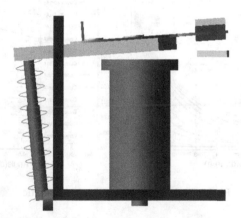

Fig. 4. Mechanical systems ADAMS model

4   The Simulation Solution and the Experimental Verification

4.1   The Simulation Solution

The clapper type of electromagnetic relay of HELLA Company is used to act research object. Its rated voltage is DC24V. Its coil number of turns is 4100. The DC resistance of the relay is 347 Ω, and the wire diameter is 0.250 mm. The moving core, the armature and the yoke are made by 10 steel. Its magnetization curve is shown in Figure.5.

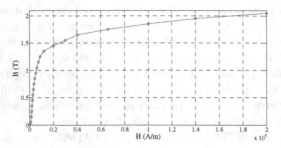

Fig. 5. 10# steel magnetization curve

To calculate to get the relationship in different electromagnetic absorption torque and armature coil electricity flow Angle of the relationship between the gas gap flux and the relationship between armature angles. To calculate different electromagnetic absorption torque and armature coil electricity flow Angle of the relationship between the gas gap flux and the relationship between armature angles, the results are shown in figure 6.

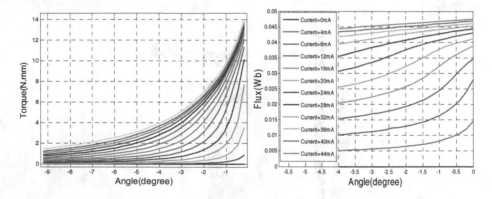

Fig. 6. Static electromagnetic characteristic

4.2   Experimental Verification

Figure.7 is the comparison chart of coil current results between simulation and experiment. The measured values for the oscilloscope for sampling the voltage meter relay closure process. From the contrast curve, the simulation values and measured values within a certain range are consistent. Before the current reaches the maximum, and the process of current better consistency simulation curve and measured curve, the maximum decline stage and reaches the minimum after the rise of the process, the deflection of the two, in the simulation model are the main reasons why not consider the actual movement of the armature component by friction and air resistance, as well as the armature and yoke iron in suction location close degree of joint and the measured object there is a difference. But from the overall trend in terms of both is more consistent.

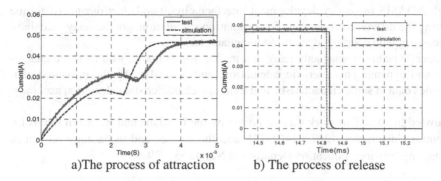

a)The process of attraction   b) The process of release

Fig. 7. Corporations of simulation and experimental results

Action in the process of displacement measurement by laser triangulation, high speed laser displacement sensor is chosen in accordance with the company's German m ILD2220, measurement platform as shown in figure.8. Resolution of 0.3 mu m, measurement range is 10 mm, measuring rate of up to 20 KHz. See the Figure.8.

Figure.9 shows the curves comparison between the armature rotation from dynamic simulation and the income curve. An observation that the armature action time is about 4ms, the experimental data show that the armature is closed after the vibration. Figure in the simulation of very little value in the process of closing the difference with the measured values, the difference is mainly composed of the measured in friction and air resistance.

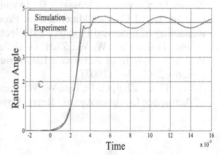

Fig. 8. Laser testing device   **Fig. 9.** Armature corner position changes

5   The Conclusions

Clapper Type electromagnetic relay electromagnetic simulation model is established, and the finite element software FLUX3D simulation analysis of the static characteristic of relay, it is concluded that the excitation current and armature Angle with the electromagnetic torque and flux of the relationship

Put forward Clapper Type electromagnetic relay is more dynamic and static characteristics of joint simulation software, the use of Fortran programming language, through ADAMS secondary development interface circuit and electromagnetic torque of solving the differential equation of interpolation calculation, got the relay machine - electric - magnetic coupling dynamic/static characteristics.

Clapper Type electromagnetic relay at a model, for example, won its static characteristic and dynamic characteristic calculation, and is verified by the measured characteristics data of static characteristic and dynamic characteristic calculation accuracy.

Acknowledgments. This work was supported by the Fundamental Research Funds for the Central Universities（2572014CB27）and the science and technology project of Heilongjiang Province Education Department No. 12533029

References

1. Li, X.: Characteristics Analysis and Detection Technology of the Second Pick-Up of Electromagnetic Relay, pp. 18–25. Hebei University of Technology (2010)
2. You, J., Cai, Z., et al.: Polarized Relay Calculation Model Based on Distribution Parameters of Nonlinear Permagnet Model. Low Voltage Electrical Appliances 21(2), 8–12 (2014)
3. Xue, Y., Zhou, Z.: Clapper type electromagnetic relay dynamic characteristic testing technology research. Low Voltage Electrical Appliances 17(6), 64–68 (2013)
4. Ymaguchi, T., Kawase, Y., Sato, K., et al.: Trajectory Analysis of 2-D Magnetic Resonant Actuator. IEEE Transactions on Magnetics 45(3), 1732–1735 (2009)
5. Wang, X.-B., Lin, H.-Y., Fang, S.-H., et al.: Analysis of Displacement Subsection PWM Control and Dynamic Charamic for Closing Process of Permagnent Contactor. Proceedings of the Csee 30(3), 113–118 (2010)
6. Salais, D., Carvou, E., Jemaa, N.B.: Opening Speed Effect on Arc Duration and Extinction Gap for Usual Contact Materials. In: Proceeding of the 2nd International Conference on Reliability of Electrical Products and Electrial Contacts, pp. 73–77 (2007)
7. Wang, H.-L., Fan, W.-W., Zhai, G.-F., et al.: The Development of Mechanical Virtual Prototype System of Aerospace Electromagnetic Relay. Mechanical and Electronic Components 27(4), 3–7 (2007)
8. Wang, H.-L., Zhai, G.-F., Fan, W.-W.: Research on Key Assembly and Adjust Parameteters of Aerospace Electrom Agnetic Relay. Low Voltage Electrical Appliances 26(5), 5–8 (2008)

iHDFS: A Distributed File System Supporting Incremental Computing*

Zhenhua Wang[1,2], Qingsong Ding[1,2], Fuxiang Gao[1,2], Derong Shen[1,2], and Ge Yu[1,2]

[1]College of Information Science and Engineering, Northeastern University, Shenyang, China
[2]Key Laboratory of Medical Image Computing, Northeastern University, Ministry of Education, Shenyang, China
{wangzhenhua,gaofuxiang,shenderong,yuge}@ise.neu.edu.cn,
dingqingsong@126.com

Abstract. Big data are always processed repeatedly with small changes, which is a major form of big data processing. The feature of incremental change of big data shows that incremental computing mode can improve the performance greatly. HDFS is a distributed file system on Hadoop which is the most popular platform for big data analytics. And HDFS adopts fixed-size chunking policy, which is inefficient facing incremental computing. Therefore, in this paper, we proposed iHDFS (incremental HDFS), a distributed file system, which can provide basic guarantee for big data parallel processing. The iHDFS is implemented as an extension to HDFS. In iHDFS, Rabin fingerprint algorithm is applied to achieve content defined chunking. This policy make data chunking has much higher stability, and the intermediate processing results can be reused efficiently, so the performance of incremental data processing can be improved significantly. The effectiveness and efficiency of iHDFS have been demonstrated by the experimental results.

Keywords: incremental computing, distributed file system, big data, HDFS.

1 Introduction

In recent years, big data applications have become ubiquitous. In the big data environment, the traditional data management technology represented by relational database has been unable to effectively manage these data. How to quickly access big data become the core challenge, then the management technology of big data represented by Hadoop [1] has emerged.

Many big data applications have the characteristics of increment and iteration. Big data processing tasks always run repeatedly when the data changes very small. The characteristic of data incremental change in big data indicates that it can greatly improve performance by using the incremental mode to process big data.

---

* This work is supported by the Fundamental Research Funds of the Central Universities (N130304002).

H. Wang et al. (Eds.): ICYCSEE 2015, CCIS 503, pp. 151–158, 2015.

The core of Hadoop is a distributed file system HDFS and a computing framework MapReduce. Typically, when we process big data in Hadoop platform, data is stored in the HDFS distributed file system and we use MapReduce framework for processing. The existing Hadoop big data management technology don't consider the incremental feature of big data, as a result, it has large processing cost and poor timeliness.

In order to improve the processing efficiency of incremental big data, this paper proposes a distributed file system iHDFS which supports incremental computing, providing basic guarantee for incremental processing of big data. This distributed file system iHDFS is an extension to HDFS. The iHDFS reserves compatibility with HDFS and provides the same interfaces and semantics as HDFS. The main idea of iHDFS is to make the input data of MapReduce, namely the data stored in HDFS, divided into chunks based on content. The content defined chunking regards the text content as boundary, and the change of a small part of data doesn't alter the boundaries of all the chunks. When incremental processing, we just need to compute the changed chunks and get the previous running results from the storage server for the unchanged chunks. This can reduce unnecessary repeated computing and reuse intermediate results, improving the timeliness of big data processing.

2   Related Works

The big data generally uses the distributed storage technology. GFS [2] is a large scale distributed file system that provides massive storage for Google cloud computing. It combines with Chubby, MapReduce [3] and Bigtable very closely and locates in the bottom of all the core technologies. Literature [2] elaborates GFS from its generated background, characteristics, system framework, performance testing and other aspects in detail. HDFS of Hadoop is the open source implementation of GFS.

Currently, mainstream researches and solutions on the incremental computing optimization of big data include: (1) Nectar [4] increases the cache servers in order to cache the intermediate results of DryadLINQ, then, for each query, the intermediate results can be get from the cache server by splitting and contrast of SQL. (2) Nova [5] regards the data processing process as a continuous collection and continuous computing process of data. (3) Percolator [6] proposed by Google uses incremental indexing filter instead of MapReduce to analyze the frequent changing data sets, so that the return speed of the search results are more and more near real-time. (4) McSherry et al. in Microsoft designed a data stream system Naiad [7], supporting incremental iterative computing. The above methods implement the incremental computing of big data, but they aren't compatible with MapReduce and need to adopt a new programming model. Incoop [8] has tried to make incremental processing for Hadoop internal architecture, implementing incremental computing under MapReduce framework.

3   Overview of iHDFS

Considering the reliability, high effectiveness, scalability and the open source characteristics of Hadoop platform, this paper builds an incremental processing framework of big data based on HDFS and MapReduce model, and it can implement the performance advantage of the incremental computing. Fig. 1 is the architecture of the incremental processing of big data, including the incremental distributed file system, the incremental parallel computing model and the storage server. The incremental distributed file system iHDFS is an extension to HDFS and it is the bottom facility of the incremental processing of big data. The incremental parallel computing model is an extension to MapReduce, reading data from iHDFS. The storage server is used to store the intermediate results and it can be implemented by the distributed caching system. The key in this paper is studying the distributed file system iHDFS which supports the incremental computing.

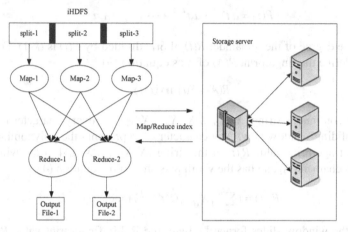

Fig. 1. The architecture of incremental processing for big data

In order to overcome the shortcoming of the fixed-size chunking algorithm in HDFS and improve the efficiency of the incremental processing, this paper presents a content defined chunking algorithm to ensure that the chunks keep stable as far as possible when the content of file changes, minimizing the impact of the data updating and reducing the repeated computing. The content defined chunking policy is implemented by Rabin fingerprint algorithm. The content defined chunking includes coarse-grained content defined chunking and fine-grained content defined chunking. Generally speaking, we use the coarse-grained content defined chunking at first when incremental processing. If data in the file has not changed, then just reuse the results. If the data in the file has changed, then use fine-grained content defined chunking at this moment.

4   Rabin Fingerprint Algorithm

Rabin fingerprint algorithm was proposed by professor Rabin in Harvard University, it has the characteristics of high computational efficiency and good randomness, and it can make the effect of data modification on continuously fingerprint sequence under a local range, commonly used to make quick comparisons and identify duplicated data.

In iHDFS, the implementation of chunking algorithm uses Rabin algorithm [9] to calculate the data fingerprints.

The calculation of Rabin fingerprint is on the finite field $GF(2^n)$. Suppose $S([b_1,...,b_m])$ is a binary string containing m binary characters, then according S, the corresponding (m-1) degree polynomial can be constructed, as is shown in equation (1) (where t is infinitive element).

$$S(t) = b_1 t^{m-1} + b_2 t^{m-2} + \cdots + b_{m-1} t + b_m \tag{1}$$

Give a polynomial P(t) with k degree, as is shown in equation (2).

$$P(t) = a_1 t^{k-1} + a_2 t^{k-2} + \cdots + a_{k-1} t + a_k \tag{2}$$

Then the degree of the remainder R(t) of S(t) divided by P(t) is (k-1). For the given string A, define the fingerprint R(A) of A as equation (3).

$$R(A) = S(t) \bmod P(t) \tag{3}$$

For the continuous string $[X_1, X_2, \ldots X_w, X_{w+1}, X_{w+2}, \ldots]$ (every character is 8 bits), the width of sliding window is w. Any character X_i represents the polynomial $X_i(t)$, then the Rabin fingerprint value $R_i(t)$ of the string $[X_i, X_{i+1}, \ldots, X_{w-i+1}]$ in the window which starts with character X_i and has the width w is shown in equation (4).

$$R_i(t) = (\sum_{j=1}^{w} X_{i+j-1}(t) t^{8w-j}) \bmod P(t) \tag{4}$$

When the window slides forward a byte, the Rabin fingerprint value $R_{i+1}(t)$ of the sliding chunk string $[X_i, X_{i+1}, \ldots, X_{i+w}]$ which starts with character X_i and has the width w is shown in equation (5).

$$R_{i+1}(t) = (\sum_{j=1}^{w} X_{i+j}(t) t^{8w-j}) = R_i(t) t^8 - X_i(t) t^{8w} + X_{w+i}(t) \bmod P(t) \tag{5}$$

5   Content Defined Chunking

5.1   Coarse-grained Content Defined Chunking

Most data sources are composed of some files with different sizes, coarse-grained content defined chunking regards these files as units to divide the data sources into chunks, and each file is a chunk regardless of their sizes. When computing at the first time, it first performs fingerprint processing for each file by using fingerprint functions and gets the fingerprint values of each file. Then each file performs MapReduce computing to get the corresponding computing results. Finally, we should make the fingerprint values of each file and the corresponding computing results form a mapping and store the mapping in the storage server. This processing is easy for query and reuse of results when we make incremental processing.

When we make incremental processing, we still regard these files as units to divide the data sources into coarse-grained chunks and calculate the fingerprint values of each files, meanwhile we compare the fingerprint values with those have been calculated in the storage server. If the query in the storage server has the same fingerprint value, this means we don't need to make MapReduce processing for this file, we just need to find the previous operation result in the storage server through the mapping of fingerprint values. If we can't find any same fingerprint value in the storage server, we need to make MapReduce processing for this file. Meanwhile, we should store the mapping composed by the file's fingerprint value and their calculation results in the storage server for the next calculation.

5.2   Fine-grained Content Defined Chunking

The so-called fine-grained content defined chunking is to divide into chunks based on content for each file. It divides the files into non-overlapping chunks with different sizes according to the content of the files. The content defined chunking algorithm uses a fixed-size window to slide in the files. Firstly it calculates the content's digital fingerprint of the files in the window (use Rabin fingerprint algorithm in iHDFS), then it determines whether the digital fingerprint satisfies a certain condition for determining whether the sliding window boundary is the demarcation of the chunk.

In order to improve the efficiency of incremental processing, the fingerprints of each chunk and their corresponding results of data processing should be stored at the first time. When incremental processing, we still use the sliding window mechanism for file chunk, the window slides to the rear from the head of file by one byte step. Before sliding, the Rabin algorithm should be used to calculate the fingerprint value of data in the window and we should determine if the boundary condition is satisfied.

When we find the same fingerprint value in the storage server, this means that the chunk is the same as it processed at the first time (i.e. the data has not changed), then the chunk needn't to make MapReduce processing and we can find the result of the first time processing from the mapping directly; If you can't find any same fingerprint value in the storage server, this means that the chunk has changed, we need to calculate its Rabin fingerprint again. Meanwhile, after the MapReduce processing, we need to put the mapping composed by the fingerprint value of this chunk and its computing result in the storage server. The process is shown in Fig.2

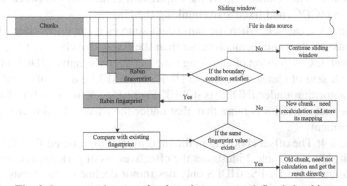

Fig. 2. Incremental processing based on content defined chunking

6   Experiments

6.1   Experimental Environment

The experiments use Ubuntu-13.10 operating system as the experimental test platform and Hadoop version is Hadoop-0.20.2. We use Memcached distributed caching system as a storage server and programming language is Java. Hadoop cluster operation mode is the pseudo-distributed mode.

We use a series of application data from the areas of machine learning, natural language processing, pattern recognition, document analysis and so on as the experimental data set. Assess the processing performance of the incremental computing through the application data of the data-intensive data (word count, co-occurrence matrix) and computing-intensive data (K-means clustering, nearest neighbor classifier).

The relevant experimental parameters are showed in Table 1.

Table 1. Experimental parameters

Parameters	Range of value	Default
Minimum chunk size	16MB~64MB	40MB
Maximum chunk size	64MB~128MB	64MB
Fixed chunk size	64MB	64MB
Rabin fingerprint window	32B~64B	48B
Rabin fingerprint offset	16MB~60MB	16MB

6.2   Experimental Results and Analysis

Experiment 1: Relationship between data update ratio and duplicated chunk ratio.

Fig. 3 shows four distributions of the duplicated chunk ratio under different update ratio for two kinds of data. From these representative data we can find the duplicated ratio of chunk is at least about 65% by using the content defined chunking algorithm in the distributed file system, this means the chunking algorithm has strong adaptability and ensures the stability of the chunk.

Experiment 2: The relationship of the duplicated chunk ratio between content defined chunking (CDC) and fixed-size chunking (FSC).

From Fig. 4 we can see, when the same data sources change the same, the content defined chunking algorithm is much better than the fixed-size chunking algorithm.

Experiment 3: Comparison of chunking time between the naive HDFS and iHDFS.

The default size of chunk under naive HDFS is 64MB, and the offset of the Rabin fingerprint algorithm under iHDFS is 40MB. From Fig. 5, we can find that that the running time under iHDFS is more than that under the naive HDFS in the same operating environment.

Experiment 4: The offset choosing in the incremental distributed file system

The result shown in Table 2 illustrates the effectiveness of performance optimization. Compared to the naive HDFS, iHDFS only has minor decline of the throughput, this is because we have used the Rabin fingerprint algorithm for content defined chunking.

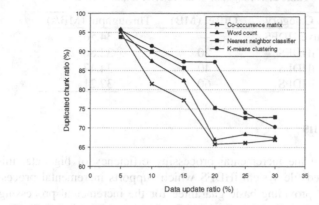

Fig. 3. Relationship between data update ratio and duplicated chunk ratio

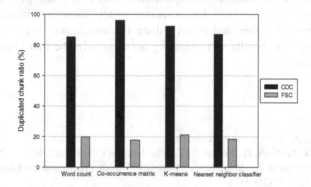

Fig. 4. Comparison between content defined chunking and fixed size chunking

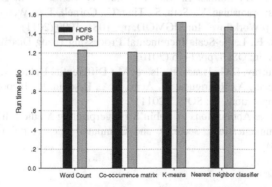

Fig. 5. Chunking time comparison between naive HDFS and iHDFS

Table 2. Throughput of naive HDFS and iHDFS

Category	Offset (MB)	Throughput (MB/s)
naive HDFS	—	34.56
iHDFS	20	32.86
iHDFS	40	34.47
iHDFS	60	32.21

7   Conclusions

In order to improve the incremental processing efficiency of big data, this paper presents a distributed file system iHDFS which supports incremental processing on the basis of HDFS, providing basic guarantee for the incremental processing of big data. Making the content defined chunking for the data stored in the distributed file system reduces the sensitivity of data changing and enhances the stability of data changing. Meanwhile it maintains compatibility with HDFS and provides interfaces and semantics the same as HDFS. Experimental results show that the proposed distributed file system is feasible, and it can reduce the unnecessary repeated computing, reuse the intermediate results of computing and improve the timeliness of big data incremental processing.

References

[1] Hadoop Project, http://hadoop.apache.org
[2] Ghemawat, S., Gobioff, H., Leung, S.-T.: The Google File System. SOSP, 29-43 (2003)
[3] Dean, J., Ghemawat, S.: MapReduce: Simplified Data Processing on Large Clusters. In: OSDI, pp. 137–150 (2004)
[4] Gunda, P.K., Ravindranath, L., Thekkath, C.A., Yu, Y., Zhuang, L.: Nectar: Automatic Management of Data and Computation in Datacenters. In: OSDI, pp. 1–8 (2010)
[5] Olston, C., Chiou, G., Chitnis, L., Liu, F., Han, Y., Larsson, M., Neumann, A., Rao, V.B.N., Sankarasubramanian, V., Seth, S., Tian, C., Cornell, T.Z., Wang, X.: Nova: Continuous Pig/Hadoop Workflows. In: SIGMOD, pp. 1081–1090 (2011)
[6] Peng, D., Dabek, F.: Large-Scale Incremental Processing Using Distributed Transactions and Notifications. In: OSDI, pp. 1–15 (2010)
[7] McSherry, F., Murray, D.G., Isaacs, R., Isard, M.: Differential Dataflow. CIDR (2013)
[8] Bhatotia, P., Wieder, A., Rodrigues, R., Acar, U.A., Pasquin, R.: Incoop: MapReduce for Incremental Computations. In: SOCC (2011)
[9] Broder, A.Z.: Some Applications of Rabin'S Fingerprinting Method. In: In Sequences II: Methods in Communications, Security, and Computer Science, pp. 143–152. Springer, New York (1993)

Chinese Explanatory Segment Recognition as Sequence Labeling

Yu He, Da Pan, and Guohong Fu

School of Computer Science and Technology, Heilongjiang University Harbin 150080, China
heyucs@yahoo.com, pandacs@live.cn, ghfu@hotmail.com

Abstract. How to mine the underlying reasons for opinions is a key issue on opinion mining. In this paper, we propose a CRF-based labeling approach to explanatory segment recognition in Chinese product reviews. To this end, we first reformulate explanatory segments recognition as a labeling task on a sequence of words, and then explore various features from three linguistic levels, namely character, word and semantic under the framework of conditional random fields. Experimental results over product reviews from mobilephone and car domains show that the proposed approach significantly outperforms existing state-of-the-art methods for explanatory segment extraction.

Keywords: opinion mining, explanatory segment recognition, product reviews.

1   Introduction

With the explosive growth of the user-generated opinionated texts on the web over the past years, there has been a dramatic proliferation of research concerned with opinion mining [1][2]. Most previous studies on opinion mining have focused on opinion extraction[3][4] and sentiment classification[5][6], little research has been done on mining the underlying reasons for opinions. In this regard, this paper is primarily concerned with explanatory segment recognition (ESR) in online product reviews.

More recently, the pioneer studies by (Kim et al., 2013) introduce two novel problems on explanatory opinion mining, namely explanatory sentence extraction which aims to rank opinionated sentences in terms of explanatoryness[7], and compact explanatory opinion summarization which aims to extract within-sentence explanatory text segments from input opinionated texts[8], in order to help users better understand the detailed reasons of sentiments and opinions.

However, explanatory opinion mining is still at its infancy stage. There are many challenges along the way. On the one hand, lacking manually-labeled datasets is one of the major bottlenecks that supervised machine learning methods faced. To break this bottleneck, two supervised methods, namely TF-IDF weighting and hidden Markov models have exploited to explanatory sentence extraction and explanatory segment extraction, respectively [7][8]. Although these methods provide an unsupervised solution to explanatory information extraction, their performance is still not satisfactory in comparison to supervised methods. On the other hand, compared to regular

H. Wang et al. (Eds.): ICYCSEE 2015, CCIS 503, pp. 159–168, 2015.

news texts, user-generated reviews are usually written in a free or loose style. Intuitively, the number of reasons of sentiments on a specific product feature may be very limited. It is obviously very challenging to model these various explanatory expressions in real product reviews.

To address the above problems, in this paper we propose a CRF-based approach to ESR in Chinese product reviews. Thus, the challenge we faced from candidate segments getting and ranking to feature selection. Besides the general features in sequence labeling task, such as words and part-of-speech (POS) tags, we also introduce topic words as semantic features for ESR labeling due to the fact that most explanatory expressions are observed to have a close relationship with a fixed topic. Experimental results show that the proposed approach significantly outperforms the state-of-the-art baselines in discussion.

2   Task Description

2.1   Explanatory Opinionated Sentences

Before going further to the problem formalization, we need to clarify the notion of explanatory opinionated sentences and explanatory segments within product reviews.

Table 1. Categorization of opinionated sentences

No.	Types	Opinionated sentences	Explanatory segments
1	Non-explanatory opinionated sentences	这个手机的电池还不错 (*The battery of this mobile phone is good*)	No
2	Simple explanatory opinionated sentences	这部手机很好！我最喜欢它的大屏幕. (*This mobile phone is great! I love its big screen.*)	大屏幕 (*big screen*)
3	Complex explanatory opinionated sentences	这个手机的屏幕分辨率很高，看不到任何像素点。 (*The screen resolution of this mobile phone is very high, and I cannot see any pixels.*)	分辨率很高(*The resolution is very high*) 看不到像素点(*Cannot see any pixels*)

Non-explanatory Opinionated Sentences. As the first example in Table 1 shows, if an opinionated sentence only presents the positive or negative sentiment orientation of the opinion holder to a given target without any explanations or specification for the reasons of the sentiment, then it is a non-explanatory opinionated sentence.

Explanatory Opinionated Sentences. If an opinionated sentence presents the positive or negative sentiment orientation of the opinion holder to a given target, and at the same time, explain or specify the reasons of the sentiment, then it is an explanatory opinionated sentence, as illustrated by the second and third examples in Table 1.

Explanatory Segments. An explanatory segment is actually a consecutive sequence of words in an explanatory opinionated sentence that forms an explanation for the reason of sentiments. For example, 大屏幕 (*big screen*) in the second example in Table 1.

2.2   Task Formulation

In the present study, we reformulate explanatory segments recognition as a labeling task on a sequence of words with "B-I-O-S" 4 tags forms. In this study, we transform our corpus into the character-based corpus, and then merge the lexical chunks and topic chunks in Chinese ESR.

Table 2. An example: chunk representation of explanatory segment

Words	POS tag	Character	Lexical chunk tag	Topic chunk tag	Explanatory Segment tag
手机	n	手	B-n	B-t12	O
		机	I-n	I-t12	O
的	u	的	S-u	S-no	O
屏幕	n	屏	B-n	B-t14	O
		幕	I-n	I-t14	O
		分	B-n	B-t9	B
分辨率	n	辨	I-n	I-t9	I
		率	I-n	I-t9	I
高	a	高	S-a	S-t9	I

As illustrated in Table 2, lexical chunk tags and topic chunk tags follow a similar format like S1-S2, where S1 refers to the position patterns of a token in chunk formation, and S2 denotes the respective categories of words that contain the token. Note that S2 in topic chunk refers to the topic number, and S2 is "no" if current word is irrelevant to any topic. The detailed descriptions on features will be shown in Section 3.

2.3   CRFs for Chinese ESR

We choose CRFs as the basic framework for ESR in that CRFs have proven to be one of the most effective techniques for sequence labeling tasks [9][10]. Compared with other methods, CRFs allow us to exploit numerous observation features as well as state sequence based features or other features to explanatory segment labeling.

Let $X = (x_1, x_2, ..., x_T)$ be an input sequence of Chinese words, $Y = (y_1, y_2, ..., y_T)$ be a sequences of explanatory segment tags as defined in Section 2.2. From a statistical point of view, the goal of segment labeling is to find the most likely sequence of segment tags $\hat{Y}$ for a given sequence of character tokens X that maximizes the conditional probability $p(Y \mid X)$. CRFs modeling uses Markov random fields to decompose the conditional probability $p(Y|X)$ of a sequence of segments tags as a product of probabilities below.

$$p(y|x) = \frac{1}{Z(x)} \exp(\sum_{i=1}^{T} \sum_{j} \lambda_j f_j(y,x,i)).$$ (1)

where $f_j(y,x,i)$ is the jth feature function at position i, associated with a weight λ_j, and $Z(x)$ is a normalization factor that guarantees that the summation of the probability of all sequences of segments tags is one, which can be further calculated by

$$Z(x) = \sum_{y} \exp(\sum_{i=1}^{T} \sum_{j} \lambda_j f_j(y,x,i)).$$ (2)

3   Feature Selection

The major challenge in sequence labeling task is feature selection. In this paper, we do not use some deeper linguistic features, like phrase chunks, functional chunks and so on, because these features are difficult to acquire with accuracy in online user-generated product reviews. Instead, we explore features for ESR at word-form and semantic levels.

3.1   Form Features

To avoid errors caused by Chinese words segmentation, we use Chinese characters as the basic units for labeling, while the word form and POS tags are retained, as show in Table 2.

At character level, we exploit character forms in a fixed window of 3 characters, including the current character c_i, the preceding character c_{i-1}, and the follow character c_{i+1}, and their respective POS tags t_i, t_{i-1}, and t_{i+1}. Table 3 details the feature template at character level.

Table 3. Character-level features

No.	Feature	Definition
L1	c_i	The current character
L2	$c_{i-1}c_i$	The current character and the preceding character
L3	c_ic_{i+1}	The current character and the following character
L4	$c_{i-1}t_i$	The preceding character and the current word's POS tag
L5	t_ic_{i+1}	The current word's POS tag and the following character
L6	$t_{i-1}c_i$	The current character and the preceding word's POS tag
L7	c_it_{i+1}	The current character and the following word's POS tag

At word level, we consider some possible combinations of the current character, the preceding character, the following character and their relevant lexical tags as features for ESR. The feature template at word level is given in detail in Table 4. Where, w_i, w_{i-1} and w_{i+1} denote the category tags of the word containing characters c_i, c_{i-1}, and c_{i+1}, respectively.

Table 4. Word-level features

No.	Feature	Definition
P1	c_iw_i	The current character and word tag
P2	$c_{i-1}w_{i-1}c_iw_i$	The preceding/current character and word tag
P3	$c_iw_ic_{i+1}w_{i+1}$	The current/ following character and word tag
P4	$w_{i-1}c_iw_i$	The preceding word tag and the current character and word tag
P5	$c_iw_iw_{i+1}$	The current character and word tag and the following word tag

3.2 Semantic Feature

We found that there is a strong correlation between explanatory information and opinion targets. For example 屏幕 (*screen*) is the opinion target of explanatory evaluation 分辨率高 (*resolution is high*). But opinion target extraction is a very hard problem in that opinion targets may be implicit or co-referred in online product reviews. Furthermore, opinion targets are difficult to generalized. For example, the three opinion targets 屏幕 (*screen*), 显示屏 (screen) and 触摸屏 (touch screen) are very similar in semantic, but they are different opinion targets in fact.

Now the problem is how to represent the semantic of a word. HowNet[11] is a general strategy. HowNet is a Chinese semantic lexicon with the similar structure with WordNet. However, HowNet mostly includes common words and does not cover the domain specific words in online product reviews. To avoid this problem, we use LDA[12], which is an unsupervised machine learning techniques and can be used to identify latent topic in large-scale document collection or corpus. Actually, LDA utilizes the bag of words model, which considers a document as a word frequency vector, and thus transfers the text information into digital information for modeling. As such, each document is represented as a probability distribution of some topics, while each topic represents many words constituted a probability distribution.

Thus, we have also built a large corpus of mobilephone and car reviews to achieve LDA models. In the training process, α and β are set to 0.5 and 0.05, respectively, and the maximum number of iteration is set to 1000. Fig.1 shows a part of results in mobilephone field.

Topic3:	键盘(*keyboard*), 手感(*feel*), 触屏(*touch screen*), 适合(*fit*), 拿(*take*)......
Topic11:	万(*ten thousand*), 清楚(*clear*), 像素(*pixel*), 800W, 500W, 1200......
Topic14:	天(*day*), 电池(*battery*), 左右(*around*), 小时(*hour*), 次(*times*)......

Fig. 1. A part of LDA models in the mobilephone domain

Then, we can add these topic features to CRFs. Table 5 presents the feature template at semantic level. Where, s_i, s_{i-1} and s_{i+1} denote the topic category tags of the words that involve characters c_i, c_{i-1}, c_{i+1}, respectively.

Table 5. Semantic-level features

No.	Feature	Definition
F1	$c_i s_i$	The current character and semantic chunk tag
F2	$c_{i-1} s_{i-1} c_i s_i$	The preceding/current character and semantic chunk tag
F3	$c_i s_i c_{i+1} s_{i+1}$	The current/ following character and semantic chunk tag
F4	$s_{i-1} c_i s_i$	The preceding semantic chunk tag and the current character and semantic chunk tag
F5	$c_i s_i s_{i+1}$	The current character and semantic chunk tag and the following semantic chunk tag

4   Experiments

4.1   Experimental Setup

To show the generality of the proposed method, we created two corpora of product reviews from two different domains, namely mobilephone and car. Both corpora are manually annotated with explanatory segments by a pair of tags, like "<EXP>" and "</EXP>". The datasets are further divided into training sets and test sets. Table 6 presents the basic statistics information of these datasets.

Table 6. Statistics of the experimental corpus

	Mobilephone	Car
Training	1259	1083
Test	840	723
All	2099	1806

Furthermore, we adopt two typical evaluation methods in information extraction, namely strict evaluation and lenient evaluation[13]. In strict evaluation, an explanatory segment recognized by the system are considered as correct if it is exactly identical to the corresponding gold standard counterpart by human judgment, while in lenient evaluation, an explanatory segment proposed by the system is counted as correct if it covers parts of its counterparts in the gold standard dataset. Furthermore, the performance in explanatory segment extraction is reported in terms of precision (P), recall (R) and F-score (F).

4.2   Baseline Methods

The proposed CRF-based method for explanatory segment recognition is compared with two baseline methods, namely the unsupervised HMMs [8] and the chi-square based method. The two methods perform explanatory segment recognition in two steps. First, all possible explanatory segment candidates are generated from a given opinionated sentence, and then two scoring techniques, viz HMMs and chi-quares, are employed to score the explanatoryness of each candidate, respectively. To obtain syntactic clues for explanatory segment candidate generation, we use LTP, a Chinese language technology platform[14]. Details of the HMM-based explanatory extraction method can be seen in [8]. Unlike the above HMM-based method, the chi-square method use the sum of chi-square values $\chi2$ of each word w_k ($i{\leq}k{\leq}j$) in the segment candidate E as its explanatoryness score, namely

$$Score_{\text{chi-square}}(E) = \sum_{k=i}^{j} \chi^2(w_k).$$ (3)

4.3   Experimental Results

To evaluate our method, we have conducted three experiments over the datasets in Table 6.

(1) Effects of features at different levels on ESR
Our first experiment intends to examine the effects of different features at different linguistic levels on Chinese ESR. This experiment is conducted with a single-pass strategy, which performs explanatory segment labeling in one pass. The results are presented in Table 7 and Table 8.

Table 7. Results for different level features under strict evaluation

Feature	Mobilephone			Car		
	P	R	F	P	R	F
L1	0.266	0.141	0.184	0.240	0.101	0.143
L1-L3	0.517	0.319	0.395	0.437	0.283	0.343
L1-L5	0.520	0.328	0.402	0.442	0.289	0.349
L1-L7	**0.533**	**0.340**	**0.415**	**0.445**	**0.296**	**0.356**
P1	0.317	0.172	0.223	0.339	0.150	0.208
P1-P3	0.461	0.271	0.341	0.397	0.230	0.291
P1-P5	**0.489**	**0.299**	**0.371**	**0.418**	**0.254**	**0.316**
F1	0.330	0.178	0.231	0.308	0.145	0.197
F1-F3	0.469	0.281	0.351	**0.389**	0.228	0.288
F1-F5	**0.476**	**0.292**	**0.363**	0.383	**0.234**	**0.291**

As can be seen in these tables, combing a variety of contextual features can improve the performance of Chinese ESR. Take the evaluation of character-level features L1-L3 as an example. In strict evaluation, the F-score is 0.395 while using character unigram and bigram features only, and the F-score will be increased by nearly 2 percents while integrating contextual words with their corresponding POS tags. However, in lenient evaluation, the effect is relatively limited. Also, we can see that word-level and semantic-level features are not as good as the character-level features.

Table 8. Results for different level features under lenient evaluation

Feature	Mobilephone			Car		
	P	R	F	P	R	F
L1	0.543	0.648	0.591	0.475	0.49	0.472
L1-L3	0.674	0.710	0.691	0.730	0.654	0.690
L1-L5	0.666	**0.721**	**0.692**	0.733	**0.667**	0.699
L1-L7	**0.680**	0.702	0.691	**0.744**	0.664	**0.702**
P1	0.558	0.685	0.615	0.542	0.532	0.537
P1-P3	0.614	**0.696**	0.653	0.660	0.562	0.607
P1-P5	**0.625**	0.691	**0.656**	**0.690**	**0.576**	**0.628**
F1	0.564	0.667	0.611	0.549	0.560	0.555
F1-F3	0.635	0.720	0.675	0.664	0.574	0.616
F1-F5	**0.653**	**0.722**	**0.686**	**0.693**	**0.609**	**0.648**

(2) Effects of different feature combinations on ESR

We integrate these features from 4 aspects, namely character-level + word-level(L1-L7+P1-P5), character-level + semantic-level(L1-L7+F1-F5), word-level + semantic-level(P1-P5+F1-F5) and all three levels(L1-L7+P1-P5+F1-F5). And the results are presented in Table 9 and Table 10.

Table 9. Results for combining different level features under strict evaluation

Feature	Mobilephone			Car		
	P	R	F	P	R	F
L1-L7+P1-P5	**0.541**	0.349	**0.425**	**0.435**	**0.300**	**0.355**
L1-L7+F1-F5	0.525	0.348	0.419	0.425	0.298	0.351
P1-P5+F1-F5	0.493	0.311	0.381	0.405	0.263	0.319
ALL	0.539	**0.351**	**0.425**	0.403	0.283	0.322

Table 10. Results for combining different level features under lenient evaluation

Feature	Mobilephone			Car		
	P	R	F	P	R	F
L1-L7+P1-P5	0.688	0.716	0.702	0.753	0.637	0.690
L1-L7+F1-F5	**0.711**	**0.740**	**0.725**	**0.764**	**0.648**	**0.701**
P1-P5+F1-F5	0.659	0.712	0.684	0.710	0.586	0.642
ALL	0.709	0.729	0.719	0.768	0.621	0.687

From above tables, we can see that incorporating character-level features with word-level features can obtain the best overall F-score under strict evaluation in mobilephone and car fields. However, under lenient evaluation in Table 10, semantic-level features have a great effect to improve F-score of L1-L7. As we have discussed above, topic features are good way to express semantic in ESR.

(3) Comparison to baselines

Our third experiment intends to compare the proposed ESR method to the two baseline methods, which has introduced in section 4.2. The relevant results are given in Table 11 and Table 12.

Table 11. Comparison results for different methods under strict evaluation

Methods	Mobilephone			Car		
	P	R	F	P	R	F
HMM[8]	0.079	0.058	0.067	0.051	0.043	0.047
Chi-squares	0.123	0.091	0.105	0.073	0.060	0.066
Ours best	**0.541**	**0.349**	**0.425**	**0.435**	**0.300**	**0.355**

Table 12. Comparison results for different methods under lenient evaluation

Methods	Mobilephone			Car		
	P	R	F	P	R	F
HMM[8]	0.557	0.192	0.285	0.350	0.126	0.188
Chi-squares	0.552	0.334	0.416	0.451	0.254	0.325
Ours best	**0.711**	**0.740**	**0.725**	**0.764**	**0.648**	**0.701**

It can be seen from Table 11 and 12 that the proposed method consistently performs better than the two baselines. In case of the mobilephone dataset under the lenient evaluation, our method can enhance the F-score by more than 44 and 31 percents, compared to HMMs and chi-squares respectively, illustrating in a sense the effectiveness of the proposed method.

5 Conclusions

In this paper, we proposed a CRF-based labeling approach to explanatory segment recognition in Chinese product reviews. In particular, we have explored features for ESR at three levels, namely characters, words and semantic, and thus examined their respective effects on Chinese ESR through experiments over product reviews from mobilephone and car domains. We show that using multiple level features under labeling strategy can result in performance improvement. While our current method proved to be more effective than the baseline methods like the HMM-based method, it needs a large hand-labeled corpus for training, which is time consuming and inefficient. For future work, we plan to explore some unsupervised or semi-supervised methods to ESR.

Acknowledgments . This study was supported by National Natural Science Foundation of China under Grant No.61170148 and No.60973081, the Returned Scholar Foundation of Heilongjiang Province, Harbin Innovative Foundation for Returnees under Grant No.2009RFLXG007, and the Graduate Innovative Research Projects of Heilongjiang University under Grant No. YJSCX2014-017HLJU, respectively.

References

1. Pang, B., Lee, L.: Opinion mining and sentiment analysis. Foundations and Trends in Information Retrieval 2, 1–135 (2008)
2. Liu, B.: Sentiment analysis and subjectivity. In: Handbook of Natural Language Processing, pp. 627–666 (2010)
3. Zhai, Z., Liu, B., Xu, H., Jia, P.: Constrained LDA for grouping product features in opinion mining. In: Huang, J.Z., Cao, L., Srivastava, J. (eds.) PAKDD 2011, Part I. LNCS, vol. 6634, pp. 448–459. Springer, Heidelberg (2011)
4. Chen, Z., Mukherjee, A., Liu, B.: Aspect extraction with automated prior knowledge learning. In: Proceedings of ACL 2014, pp. 347–358 (2014)
5. Fu, G., Wang, X.: Chinese sentence-level sentiment classification based on fuzzy sets. In: Proceedings of COLING 2010, pp. 312–319 (2010)
6. Maas, A., Daly, R., Pham, P., Huang, D., Ng, A., Potts, C.: Learning word vectors for sentiment analysis. In: Proceedings of ACL 2011, pp. 142–150 (2011)
7. Kim, H., Castellanos, M., Hsu, M., Zhai, C., Dayal, U., Ghosh, R.: Ranking explanatory sentences for opinion summarization. In: Proceedings of SIGIR 2013, pp. 1069–1072 (2013)
8. Kim, H., Castellanos, M., Hsu, M., Zhai, C., Dayal, U., Ghosh, R.: Compact explanatory opinion summarization. In: Proceedings of CIKM 2013, pp. 1697–1702 (2013)
9. Jakob, N., Gurevych, I.: Extracting opinion targets in a single-and cross-domain setting with conditional random fields. In: Proceedings of EMNLP 2010, pp. 1035–1045 (2010)
10. Lafferty, J., Andrew, M., Fernando, P.: Conditional random fields: Probabilistic models for segmenting and labeling sequence data. In: Proceedings of ICML 2001, pp. 282–289 (2001)
11. Liu, Q., Li, S.: Word similarity computing based on how-net. Computational Linguistics and Chinese Language Processing 2(7), 59–76 (2002)
12. Blei, D., Ng, A., Jordan, M.: Latent Dirichlet Allocation. The Journal of Machine Learning Research 3, 993–1022 (2003)
13. Zhao, J., Xu, H., Huang, X., Tan, S., Liu, K., Zhang, Q.: Overview of Chinese Opinion Analysis Evaluation 2008. In: Proceedings of COAE 2008, pp. 1–20 (2008)
14. Che, W., Li, Z., Liu, T.: LTP: A Chinese language technology platform. In: Proceedings of COLING 2010 (Demonstration Volume), pp. 13–16 (2010)

An Unsupervised Method for Short-Text Sentiment Analysis Based on Analysis of Massive Data

Zhenhua Huang, Zhenrong Zhao, Qiong Liu, and Zhenyu Wang

School of Software Engineering South China University of Technology
zhhuangscut@gmail.com

Abstract. Common forms of short text are microblogs, Twitter posts, short product reviews, short movie reviews and instant messages. Sentiment analysis of them has been a hot topic. A highly-accurate model is proposed in this paper for short-text sentiment analysis. The researches target microblog, product review and movie reviews. Words, symbols or sentences with emotional tendencies are proved important indicators in short-text sentiment analysis based on massive users' data. It is an effective method to predict emotional tendencies of short text using these features. The model has noticed the phenomenon of polysemy in single-character emotional word in Chinese and discusses single-character and multi-character emotional word separately. The idea of model can be used to deal with various kinds of short-text data. Experiments show that this model performs well in most cases.

Keywords: sentiment analysis, short text, emotional words, massive data.

1 Introduction

Contemporarily, sentiment analysis applications become widely applied in various forms of short texts like short movie reviews, Twitter, short product reviews. It is called short-text sentiment analysis in this paper. For social networks, such as Twitter or Weibo, allows no more than 140 words. Reviews no more than 150 words are referred to short text in this paper. Short-text sentiment analysis has many usages. Malhar Anjaria[1] used large-scale data from Twitter users to predict results of democratic elections. Jichang Zhao[2] use massive data of weibo to analyze abnormal emotional event.

One prominent issue in emotional analysis is to increase the accuracy as to get more meaningful applications about sentiment analysis. This paper analyzes massive data of short text and mines properties and expressions of the short text (microblogs, movies reviews and product reviews). Short texts have the following characteristics: (1)Colloquial expressions links and emoticons are widely used. (2)Expression of emotions is more intuitive and emotional information is easier to expose which gives the possibility of high accuracy. And Single-character emotional word are rather common in Chinese short-text, like "keng","niu","shui","zan", more in line with colloquial expression which are concentrated signals of users' feelings. They have different meanings when used alone and should be talked separately. This paper proposes a method to deal with the problem of polysemy of single-character emotional word in

H. Wang et al. (Eds.): ICYCSEE 2015, CCIS 503, pp. 169–176, 2015.
© Springer-Verlag Berlin Heidelberg 2015

Chinese expressions. In a large-scale data, 200,000 emotional microblogs, statistics show that words containing emotional tendencies a total of 756,808, with the proportion of 92.3% texts and on average each text contains 3.78 words with emotional tendencies. The single-character emotional words appear 12,556 times. Random sampling of 1000 emotional words, getting single-character emotional words has emotional tendencies is only 58.4%. So it needs to be discussed particularly.

The task of this paper is to try to capture and quantify the principal information and combine the information together and finally use them to predict this text's emotional tendency to get a high accuracy in short-text sentiment analysis. Commonly used classifiers like SVM, KNN, Naïve Bayes are also discussed in this paper. Results prove that the model can achieve a high recall in most cases in short-text sentiment analysis.

2   Related Works

Short-text sentiment analysis can be carried out with supervised or unsupervised methods. Turney[3] applied an unsupervised method of PMI to do sentiment analysis and the results were between 66% and 84%. Paltoglou[4] employed a method based on emotional words, applied to some less relevant datasets from online discussions, tweets and social network comments, etc. Results showed a better performance than what can be observed in other classifiers. Li Dong[5] employed Adaptive Multi-Compositionality (AdaMC) layer to train recursive neural model and results showed that the method increased the highest accuracy rate from 85.4% to 88.5%. But rarely scholars talk about language expression and characteristics of texts in details.

Xu Lin Hong et al [6] classified emotion as seven categories and built emotional lexicon, which contains polarity and strength. We apply method of combining How-Net and PMI proposed by the paper[7], with artificial selection and expansion, construct a thesaurus containing more than 23,000 emotional words with polarity and strength. On this basis, buzzwords and colloquial words are added. Now the thesaurus contains more than 30,000 words and is currently known as the most comprehensive emotion thesaurus.

For semantic similarity computation, Liu Qun[8] proposed a method to calculate the semantic similarity by applying the concept of the original word similarity. The model applies Liu's method and applies a quick method of dividing unknown words to known words to calculate unknown words' semantic similarity.

3   Proposed Model

The model contains four parts. The first three parts get a emotional information set of a text and the model will use the set to predict the polarity of the text. If it doesn't works, the fourth part will begin. Each part will be discussed in details in the following parts.

3.1   Preparation

Firstly, divide short texts into multiple sub-clauses. For each sub-clause s_i, divide it by word segmentation system and will get various word unit u_{ij}, which each word unit contains body w_{ij} and part of speech p_{ij} . Then a short text can be expressed as:

$$T = \sum s_i = \sum_{i=1}^{n} \sum_{j=1}^{m} u_{ij} = \sum_{i=1}^{n} \sum_{j=1}^{m} < w_{ij}, p_{ij} >$$

Process word unit set U as follows, the model is aimed to get an emotional information set U' which can reflect the emotional tendency of short text.

3.2   Part 1: Multi-character Emotional Words

In the word unit set U, some words express certain emotions; some express certain kind of emotion when get together. Words or phrases able to express emotion of text referred to as the emotional element in this paper. The purpose of the process is to find out emotional elements set. There is an inadequacy of the existing Chinese word segmentation system in identifying unknown words. Some multi-character words which contain emotional information would be separated. Traditional method of importing user dictionary could not solve it when dictionary becomes large. We apply a method of "sliding window" to deal with this situation, setting window size = 4, which means the size of a combination of the words unit within 4 steps. Each combination should be in the sub-clause otherwise it will automatically go to the next sub-clause. When the word unit has matched a word in thesaurus, it will be put into the set U', illustrated as condition 1; When a word cannot match the emotional words, the end of the window will slide forwards and combine word units, one step a time. If a combination matches emotional word, the combination will be put into the set U' as condition 2. The processed word units are marked for no more process in future. For judgment of negative or emotion-turning, the same "sliding window" approach is used.

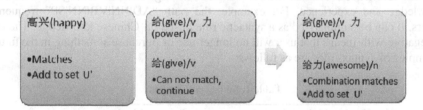

Fig. 1. Condition 1in the left, Condition 2 in the right

For example, when a phrase is "yi yu zhong di", word segmentation system may divide it into "yi/m yu/ng zhong/f di/ude1". Each word after divided no longer contains any emotion, but the method of "sliding window" can identify the original emotional orientation. The phrase is an idiom means "get key point" with positive emotional tendency ".

3.3   Part 2: Single-Character Emotional Words

The article discusses the single-character emotional words separately. In Chinese, the single-character words often have different meanings. If we regard them as emotional word directly, error will often occur. A Chinese single-character emotional word will constitute a new word or phrase with a fixed meaning when it combines with other characters or words. For example, "niu" in itself could be a noun or adjective, it can be combined with other character to show means of cows or bulls. Similarly, "poor" can express "bad", but also has meaning of "less" and "almost". For this particular phenomenon, we have found that when single-character emotional words to express meaning of emotional tendencies, accompanied by expression of the "modifiers", the "modifier part" can be an adverb, adverb phrase, modal particle or sentence punctuation. The experience in a dataset of sampling 2000 data shows that the accuracy of this method up to 92% but the accuracy of the normal text is 58.4%. Steps of the process are as follows: 1) Matching single-character emotional words. There are commonly used 55 positive emotional words and 68 negative emotional words. 2) Find modifiers. Lookup modify signals for the single-character word. 3) Negative judgment and strength calculation.

3.4   Part 3: Syntax Rules

In Chinese, user can still express emotions without emotional words. For example, "Since the melamine scandal broke, never to drink this brand of milk any more." Syntax rules in this part are aimed to match such expressions. And sometimes emotional words will show reverse meaning in certain grammar rules. Each rule can be represented as: rule = (id, rule, polarity, strength, negative).

Before creating rules, firstly we obtain text syntax path using Stanford Parser and make statistics. Choose the frequent syntax path and design rules based on knowledge and experience of Chinese. The method is similar to Popescu[9] and Zhao Yanyan[10]. Need to be clear that we use Stanford Parser just to assist build rules and the rules do not need to include emotional words. We need to design rules according to the Chinese knowledge and experience. For example: the path "AD|NN|VB|NN|SP" frequently occurs. It can be summarized as a syntactic rule based on Chinese knowledge. The rule is negative with means that user will no longer use or purchase something in the future. Examples of syntax rules are as follows:

Table 1. Some syntax rules

absolutely denial words + v (neutral or positive)+n
away (boycott，etc.) from +n
without positive/negative words + any more.
n+ is not what I have thought + positive adj

3.5   Part 4 : Semantic Similarity

Although a comprehensive emotional lexicon and several rules are applied, there are still some cases cannot be matched by the model as expressions are diverse. When the first three parts cannot determine the polarity of the text, semantic similarity will be calculated to predict emotions. Words of "n, d, a, v" four types are reserved, and each will be taken the maximum value with standard set. The equations are as follows:

$$S\_H(w, pos/neg) = \max\{\sum_{i=1}^{n} sim(w, commendatory_i / derogatory_i)\} \quad (1)$$

$$Ploarity(txt) = \max\{\sum S\_H(w_i, pos), \sum S\_H(w_i, neg)\} \quad (2)$$

Equation (1) is to calculate similarity with polarity of a word w and equation (2) is to judge the polarity of a text. When a word is unknown in Hownet, divide the word into several known words and choose the maximum of similarity between words. For example "smile point" = "smile"&"point", choosing max value of {Sim ("smile ", w), Sim ("point", w)} as the similarity between word "(smile point)" with word w.

3.6 Supervised Classifiers

Texts are divided into words, constitute words bag and give each word an id. Then convert texts in the form of vectors as$\{w_1, w_2, \ldots, w_n\}$. Generate consistent format Arff file, a type of document which can be recognized by Weka[11]. Then Weka is employed to choose classifiers to make experiments. The classifiers in experiments are SVM, Bayes Net, Naïve Bayes and KNN.

4   Experiments and Analysis

4.1   Microblogs

To test the model this paper proposed. This paper firstly does experiments on sample test data about microblog of COAE2013 which is open now. 1000 positive and 1000 negative annotated texts are selected. Experimental results are as follows:

Table 2. Results of COAE Microblogs

Classifiers	Precision		Recall		F-Measure		Time/Per
	pos	neg	pos	neg	pos	neg	
BayesNet	0.867	0.843	0.838	0.871	0.852	0.857	58.4ms
NaiveBayes	0.857	0.792	0.771	0.871	0.812	0.83	48.1ms
KNN	0.881	0.806	0.785	0.894	0.83	0.848	6.6ms
SVM	0.926	0.901	0.898	0.928	0.912	0.914	24.6ms
Model	0.935	0.917	0.915	0.936	0.925	0.926	48.2ms

The "Model" represents the proposed method. For KNN, when K = 3, the classification to achieve the best results. Data in the table shows that in the four supervised classification method F-Measure of Naïve Bayes is lowest and SVM performs best.

Bayes Net is first used and between the Naïve Bayes and SVM. Bayes Net has achieved much better results than Naïve Bayes illustrates that the connection between the words are closely, rather than independent. In the short text like microblog with relatively insignificant statistical regularity, better results can be achieved based on linear classifiers such as SVM. The proposed method can also achieve a higher accuracy rate, slightly higher than the SVM method. The relative colloquial expression and the direct expression of emotions occur frequently in short texts makes great contributions to good performance.

For this dataset, single-character emotional words appear 391 times, 250 times effective in positive set and 201 times, 162 times effective in negative set. It shows that the single-character emotional words that do not express emotions occupy a considerable part. If single-character emotional words are regarded as emotional words directly, F-Measure in average will decrease about 3% in the proposed model. So single-character words cannot be ignored in the short-text sentiment analysis.

For processing time, it contains time of modeling and analysis. It can be seen that the processing time is acceptable. The processing time of the proposed model is 48.2ms per text and it is practical. Theoretically, the running time of the model is proportional to the length of a text.

4.2   Weibo Comments and Movie Reviews

Then we use annotated 3000 positive and 3000 negative texts of movie reviews and Weibo comments in the hit movie Tiny Time 3.0 to test the model. Although Weibo comments and movie reviews are different kinds of short texts, they are similar in many ways. For Weibo comments, there is a particular phenomenon that there are many emoticons in microblog. Emoticons have great emotional directivity. We classify emoticons and add strength and polarity information and regard them as the emotional words. In machine learning methods, add an additional dimension to represent the existence and polarity of emoticons. The experimental results are as follows:

Table 3. Results of reviews and comments of Tiny Time 3.0

Classifiers	Precision		Recall		F-Measure		Time/Per
	pos	neg	pos	neg	pos	neg	
BayesNet	0.898	0.943	0.946	0.892	0.921	0.917	356.5ms
NaiveBayes	0.937	0.787	0.743	0.95	0.829	0.861	158.3ms
KNN	0.833	0.714	0.651	0.87	0.731	0.784	51.5ms
SVM	0.947	0.872	0.86	0.952	0.901	0.91	194.7ms
Model	0.913	0.923	0.924	0.912	0.918	0.917	53.6ms

Results show that supervised methods except KNN reached a very high rate of return call for the short-text sentiment analysis. In this dataset, Bayes Net performs best in F-Measure, exceeding the SVM method and slightly higher than the model. Associations between emotional words are quite closely.

4.3   Product Reviews

As for sentiment analysis of product review, we get 2000 positive reviews and 2000 negative reviews from an e-commerce website. All the selected reviews are less than 150 words to make sure they are short texts. Results are as follows:

Table 4. Results of product reviews

Classifiers	Precision		Recall		F-Measure		Time/Per
	pos	neg	pos	neg	pos	neg	
BayesNet	0.929	0.875	0.867	0.934	0.897	0.903	117.8ms
NaiveBayes	0.851	0.819	0.859	0.871	0.83	0.838	81.3ms
KNN	0.881	0.806	0.785	0.894	0.83	0.848	27.5ms
SVM	0.929	0.862	0.85	0.935	0.888	0.897	73.9ms
Model	0.913	0.923	0.924	0.912	0.918	0.917	57.4ms

It can be seen that recall of the model is relatively low in negative product reviews dataset. The reason is that some users evaluate products from multiple respects, including the good and bad respects. The positive reviews are relatively straightforward, tend to point out the advantages of goods and services from the various respects, so that their emotional tendencies express more directly. Therefore, emotional tendency of positive product reviews are more easily captured and maintained a high recall.

4.4   Comparison

As can be seen, the time to train model and process of SVM, Bayes are increasing sharply when data become large. When in large-scale data, the time consumption of SVM and Bayes Net are too large to be useful. While the model is relatively stable. Commonly used classification need model according to the different data sets, but the model proposed can be applied to various short texts. Naïve Bayes is a practical way in massive data. The model performs better than Naïve Bayes and the model suit different kinds of data. Style of emotional expression vary on different users, some are straightforward, while some are subtle. In total, most users tend to choose straightforward expression. And the characteristic make great contribution to the accuracy of model.

5   Conclusion and Future Work

This paper discusses sentiment on short text in details. We propose a high-acurrate model and compare it with four classifiers, and the results show that our model performs well in short-text sentiment analysis. The contributions of this paper are in the following: 1) A more comprehensive and abundant Chinese emotion thesaurus with emotion-al intensity is built. 2) The characteristics of short text and expression are found to have great impact on sentiment analysis based on characteristic in massive data. Emotional words are proved to do a great job in sentiment analysis. 3) We propose a high accurate model for short-text sentiment analysis, which has repeated practicality and scalability, and achieves a high accuracy in classification. And we will design more rules and apply the model in other kinds of short text in the future. And we will make experiments in large-scale mixed data of various kinds of short texts.

Acknowledgment. This work is supported by Guangdong Province Student Innovative Project (1056112105), National Student Innovative Project (201210561102).

References

1. Anjaria, M., Guddeti, R.M.R.: Influence factor based opinion mining of Twitter data using supervised learning. In: Proceedings of the Sixth International Conference on Communication Systems and Networks (2014)
2. Zhao, J., Dong, L., Wu, J., et al.: Moodlens: an emoticon-based sentiment analysis system for chinese tweets. In: Proceedings of the 18th ACM SIGKDD International Conference on Knowledge Discovery and Data Mining, pp. 1528–1531. ACM (2012)
3. Turney, P.D.: Thumbs up or thumbs down?: semantic orientation applied to unsupervised classification of reviews. In: Proceedings of the 40th Annual Meeting on Association for Computational Linguistics, pp. 417–424. Association for Computational Linguistics (2002)
4. Paltoglou, G., Thelwall, M.: Twitter, MySpace, Digg: Unsupervised sentiment analysis in social media. ACM Transactions on Intelligent Systems and Technology (TIST) 3(4), 66 (2012)
5. Dong, L., Wei, F., Zhou, M., et al.: Adaptive multi-compositionality for recursive neural models with applications to sentiment analysis. In: Twenty-Eighth AAAI Conference on Artificial Intelligence (AAAI), AAAI (2014)
6. Linhong, X., Hongfei, L., Yu, P., Hui, R., Jianmei, C.: Constructing the Affective Lexicon Ontology. Journal of the China Society for Scientific and Technical Information 2, 006 (2008)
7. Wang, Z.-Y., Wu, Z.-H., Hu, F.-T.: Words Sentiment Polarity Calculation Based on How-Net and PMI. Computer Engineering 38(15), 187–189, 193 (2012)
8. Liu, Q., Li, S.: Word similarity computing based on How-net. Computational Linguistics and Chinese Language Processing 7(2), 59–76 (2002)
9. Popescu, A.M., Etzioni, O.: Extracting product features and opinions from reviews. In: Mooney, R.J. (ed.) Proc. of the HLT/EMNLP 2005, pp. 339–346. Association for Computational Linguistics, Morristown (2005)
10. YanYan, Z., Qin, B., Che, W.X., et al.: Appraisal Expression Recognition Based on Syntactic Path. Journal of Software 22(5), 887–898 (2011)
11. http://www.cs.waikato.ac.nz/ml/weka/

Normalization of Homophonic Words in Chinese Microblogs

Xin Zhang, Jiaying Song, Yu He, and Guohong Fu

School of Computer Science and Technology, Heilongjiang University
Harbin 150080, China
zhang_xin_nlp@hotmail.com, jy_song@outlook.com,
heyucs@yahoo.com, ghfu@hlju.edu.cn

Abstract. Homophonic words are very popular in Chinese microblog, posing a new challenge for Chinese microblog text analysis. However, to date, there has been very little research conducted on Chinese homophonic words normalization. In this paper, we take Chinese homophonic word normalization as a process of language decoding and propose an n-gram based approach. To this end, we first employ homophonic–original word or character mapping tables to generate normalization candidates for a given sentence with homophonic words, and thus exploit n-gram language models to decode the best normalization from the candidate set. Our experimental results show that using the homophonic-original character mapping table and n-grams trained from the microblog corpus help improve performance in homophonic word recognition and restoration.

Keywords: Microblog analysis, text normalization, homophonic words, n-gram.

1 Introduction

With the rapid development of Web 2.0 technology and social media in recent years, a large number of user generated contents (UGCs) appear in the network. In comparison with formal news texts, social media texts have a free writing style and usually contain a large amount of informal or non-stantard expressions such as homophonic words, abbreviations, spelling errors and so forth. Homophonic word is used as a figure of speech which makes use of the same or similar pronunciation of words in order to aroused association and imagination [1]. Due to the fact that the use of homophonic words makes the language expression novel and vivid [2], homophonic words are very popular in Chinese microblog. The major task of Chinese homophonic word normalization is thus to identify specific context of homophonic word and to restore it into the corresponding original word. Obviously, correct normalization of homophonic words is of great significance for processing or understanding Microblogs and other social media.

In recent years, text normalization is caused widely attention and made great progress. And appearing different normalization methods such as noise channel model [3], rules or dictionary method [4][5], finite state automata [6][7], machine translation

H. Wang et al. (Eds.): ICYCSEE 2015, CCIS 503, pp. 177–187, 2015.
© Springer-Verlag Berlin Heidelberg 2015

model [8][9[10], etc. While research of text normalization mostly focus on text forms such as English social media [4][5], news text [6][10], text message or instant messaging [7][8][9]. We mainly focus on dealing with some non-standard words such as numbers and abbreviations [4-10]. Chinese microblog homophonic normalization seldom research so far.

Compared to other types of text normalization problems, the research of Chinese microblog homophonic words faces many problems in fact. Firstly, the writing style of user-generated microblog is freedom. The generation of homophonic words is complex and diverse such as Chinese character homophonic word, foreign language homophonic word, number homophonic word and admixture homophonic word [2]. Second, homophonic word, as one of the typical features of a network of language, get widespread attention in the field of linguistics [1][2]. But homophonic recognition and restoration issue has not aroused enough attention in the field of natural language processing because we lack relevant linguistic resource, homophonic dictionaries, etc. In addition, according to the characteristics of the Chinese language itself, Chinese text does not have explicitly word boundaries. And the lack of morphological changes of Chinese words brings new difficulties for the recognition and restoration of homophonic words to some extent.

In this paper, we view Chinese homophonic normalization issue as a language decoding process and propose a homophonic recognition and restoration method based on n-gram language models. Therefore, we construct a microblog homophonic corpus and then construct two homophonic normalization knowledge, homophonic - original word and character mapping table according to the mapping relationship on pronunciation. Finally, we select an optimum homophonic normalization from homophonic candidate set by using n-gram language models. Furthermore, we are also exploring the influence of different language models and different homophonic corpus through the experiments. Preliminary experimental results show that the use of microblog language models and homophonic-original character mapping table knowledge helps to improve the identification and restoration performance.

The rest of the paper is organized as follows. In section 2 we briefly summarize the relevant research. In section 3 we analyzed the characteristics of Chinese network homophonic words. In section 4 we introduce normalization method of homophonic word proposed in this paper. We present our experiment results and detailed analysis in section 5. Section 6 concludes this paper with future work.

2   Related Work

In recent years, text normalization research of English, French and other Western languages has great progress. There are different normalization methods and these methods have been widely applied in different fields. Choudhury et al. (2007)[3], taking character as Granularity, propose a text normalization method based on the noise channel model combining morpheme with phoneme feature. Han et al. (2011)[4] mainly focus on recognition errors of Out-of-Vocabulary (OOV) in the text and obtain the right candidate according to the similarity of pronunciation. Hassan and Menezes(2013)[5] propose a dictionary-based text normalization method and obtain normalization dictionary by combining Random Walk framework with n-grams trained from large-scale non-labeled corpus. Sproat et al. (2001)[6] focus on numbers, abbreviations and other non-standard words existed in the English text and propose a normalization method of non-standard based on a finite state automaton. And it has been applied to English speech synthesis. Beaufort et al. (2010) [7] combine noise channel model with rule-based iterative finite state machine to study text normalization issue in French. Recently, Aw et al.(2006)[8] and Bangalore et al. (2002)[9] use machine translation methods to explore SMS text normalization issue.

Compared with Western languages, there is less related research of Chinese text normalization. For resolving number, sign and other nonstandard words speech synthesis problem constructed by non-Chinese characters, Fu et al. (2006)[10] propose a rule-based Chinese normalization method. In addition, Fu et al. (2007)[11] also use the hidden Markov method to explore the issues of Chinese abbreviations corpus extension or restoration. As for the normalization of numerous homophonic words in Chinese microblog, few people research it specifically. Most view homophonic normalization and typo normalization as the same problem to analysis[12]. But, these correction processing strategies are mostly based on the word to start. It is always unable to deal with true homophonic words (e.g., 稀饭-喜欢, 油菜花-有才华, 杯具-悲剧, 大虾-大侠, etc.) and legato homophonic words (e.g., 表-不要, 酱紫-这样子, etc.). The early research about homophonic word is mainly focus on homonyms disambiguation problem. There are typical studies that Chang (1993)[13] propose disambiguation method based on Chinese pinyin syllable and other features, and Lee et al.(1997)[14] propose a repair method which is similar to the voice recognition. While, many homophonic words not only have different pronunciation, the meaning is different. For example, 灰机(飞机), 河蟹(和谐) , etc. These homophonic words can obviously not be resolved through both methods described above.

3   Homophonic Words in Chinese Microblogs

3.1   Microblog Homophonic Corpus

In order to understand the characteristics of homophonic words in Chinese microblogs, we collect 204 typical homophonic words to construct a homophonic corpus from Chinese microblogs, which consists of 1756 sentences, 24135 words and 2276 homophonic words.

3.2   Types of Homophonic Words in Chinese Microblogs

As shown in Table 1, homophonic words in Chinese microblogs can be divided into four main categories in terms of their origins and writing forms.

Table 1. Types of homophonic words in terms of origins and writing forms

Types	Definition	Subclasses	Example
Hanzi homophonic words (HHWs)	Homophonic words originate from mandarin, dialects or foreign languages, and are written in Hanzi.	HHWs from Mandarin	杯具-悲剧
		HHWs from dialect	木有-没有
		HHWs from foreign languages	马屁山-MP3
Foreign homophonic words (FHWs)	Homophonic words originate from foreign languages and are written in foreign letters.	-	B- be, C- see, UR- you are, CU- See you
Numeral homophonic words (NHWs)	Homophonic words originate from Chinese or foreign languages, and are written in Arabic numeral forms.	NHWs from Chinese	0-你, 1- 一/要/有, 2-爱, 5-我/无/呜,
		NHWs from foreign language	2-to/too, 4-for
Mixed homophonic words (MHWs)	Homophonic words are generated with multi-origins and writing forms.	MHWs mixed with English letters and Arabic numerals.	3X-Thanks, P9-啤酒,
		MHWs mixed with Hanzi, English letters and Arabic numerals.	哈9-喝酒, 4人民-为人民

Table 2. Sub-categories of HHWs in Chinese microblogs

Classification criteria	Sub-types	Examples	Distribution	
			Number	%
The number of homophonic characters in HHWs	Mono-character HHWs	帅锅(哥), 妹纸(子)	1673	73.51
	Multi-character HHWs	神马(什么), 鸡冻(激动)	603	26.49
The corresponding relationship between homophonic and original characters in HHWs	One-to-one HHWs	木(没)有	1879	83.35
	One-to-many HHWs	灰(非)常, 灰(飞)机	379	16.65
Whether a HHW are both a normal word and a homophonic word.	Real word homonym	杯具(悲剧)	175	7.69
	Non-real word homonym	介个(这个)	2101	92.31
Whether the homophonic pronunciation is the same with its original	Homonym	狠久(很久)	131	6.02
	Non-homonym	肿么(怎么)	2145	93.98

With regard to the popular use of HHWs in Chinese microblogs, in the present study we focus on the normalization of HHWs. As shown in Table 2, HHWs can be classified into different sub-categories according to different classification criteria.

As can be seen from Table 2, most HHWs are non-homonym, which originates mainly from Chinese dialects. Table 3 presents some of the most common homophonic syllables of non-homonym and their distributions.

Table 3. Most common homophonic syllables in non-homonym HHWs

Types of syllables	Homophonic syllables	Original syllables	Percentage
	zh/ch/sh	z/c/s	12.903%
Initial	n	l	7.258%
	f	h	4.839%
	i	ie	6.452%
Vowel	ei	ui	3.225%
	un	en	1.613%

4   The Method

In fact, homophonic word normalization involves two sub-tasks, namely homophonic word recognition and restoration. In the present study we reformulate these two sub-tasks as a language decoding process and thus solve them in one pass..

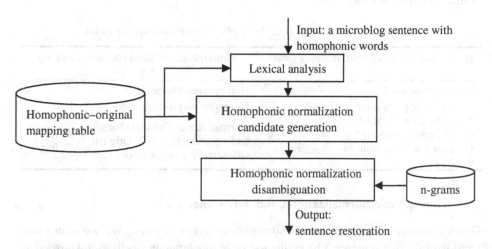

Fig. 1. Overall framework of the system for Chinese homophonic word normalization

As shown in Figure 1, the proposed system for Chinese homophonic word normalization consists of three main modules.

- **Lexical analysis.** The main task of lexical analysis module is to split microblog sentence which may exist homophonic words into word strings.
- **Homophonic normalization candidate generation.** In order to construct the candidate of homophonic normalization we scan the result of lexical analysis module by taking use of homophonic normalization knowledge to indentify the potential homophonic words and restore them as the original word. Potential homophonic words, the original word candidates and other words in the sentence constitute a word lattice. We call it homophonic candidate lattice. Every path of the lattice is a complete sentence.
- **Homophonic normalization disambiguation or decoding.** The main task of homophonic normalized decoding is to find and output a best (highest probability) path from homophonic candidate lattice by using Viterbi algorithm and the n-gram language model. If the normalized sentence is the same with the original one, there are no homophonic words in the sentence. Otherwise, homophonic words are existed in the sentence. By comparing the output with the input sentence can we obtain homophonic -original word mapping pairs

4.1   Homophonic-Original Mapping Table

In order to generate a set of normalization candidates for each homophonic word, we exploit two homophonic normalization knowledge, namely the homophonic-original word mapping table and the homophonic-original character mapping table. As shown in Table 4, this knowledge shows us the correspondence between the homophonic and original word (character).

Table 4. Entry structures of the homophonic-original mapping tables

Homophonic-original word mapping table		Homophonic-original character mapping table	
Homophonic word	碎觉	Homophonic character	灰
Homophonic pronunciation	sui4jiao4	Homophonic pronunciation	hui1
Normalization: Original word1 / pronunciation1 ǀ Original word2 / pronunciation2 ǀ ...	睡 觉 /shui4jiao4	Normalization: Original character1 / pronunciation1 ǀ Original character2 / pronunciation2 ǀ ...	非/fei1 ǀ ㄟ/fei1

4.3   Homophonic Normalization Candidate Generation

Given a sentence likely to contain homophonic words $s=w_1w_2...w_n$, we scan every word w_i $(1{\leq}i{\leq}n)$ in sentence S by taking use of homophonic normalization knowledge. If w_i is a homophonic word, we generate the corresponding original word set {w_{i1}, w_{i2}, ..., w_{imi}}. The generated normalization candidates are stored in a word lattice structure, as shown in Fig, 2.

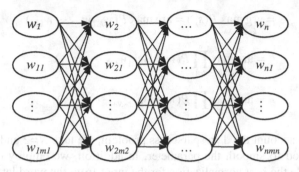

Fig. 2. Word lattice for homophonic normalization candidates

It is noteworthy that normalization candidate generation depends on the homophonic normalization knowledge in use. In case of using the homophonic-original word mapping table, we directly construct normalization candidates for each potential homophonic word in input by referring the mapping table. In case of using the homophonic-original character mapping table, we first judge whether the current word contains homophonic characters by referring the mapping table. If yes, we get all potential original characters for each homophonic character in the word from the mapping table and thus generate its original words by exhaustive combination of the relevant original characters. Fig. 3 illustrates the word lattice of homophonic normalization candidates for the sentence 灰机经常出现晚点的情况 (The airlines are often delayed).

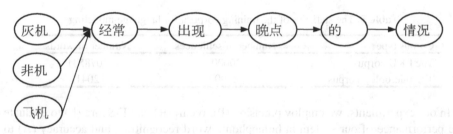

Fig. 3. The word lattice of homophonic normalization candidates for the sentence 灰机/经常/出现/晚点/的/情况

4.3   Homophonic Normalization Disambiguation

Given a sentence $s=w_1w_2...w_n$ with n words, it may have more than one homophonic candidates. As shown in Fig. 2, there may be multiple paths $S = \{s_1, s_2, ... , s_K\}$ from the beginning to the end in homophonic normalization candidate lattice. The main task of homophonic normalization disambiguation based on n-gram language model is to find a best path $\hat{s}$ in the homophonic normalization lattice so that making the probability maximum.

$$\hat{s} = \arg\max_{k=1,\cdots,K} P(s_k) = \arg\max_{k=1,\cdots,K} P(w_{k,1} w_{k,2} \cdots w_{k,n})$$

$$= \arg\max_{k=1,\cdots,K} \prod_{i=1}^{n} P(w_{k,i} \mid w_{k,1} w_{k,2} \cdots w_{k,i-1}) \tag{1}$$

$$\approx \arg\max_{k=1,\cdots,K} \prod_{i=1}^{n} P(w_{k,i} \mid w_{k,i-N+1} \cdots w_{k,i-1})$$

Taking into account the data sparseness problem, we take $N=2$ and use linear interpolation method to smooth the parameter. In addition, we employ the Viterbi algorithm to decode the best normalization for the input from the word lattice.

5   Experimental Results and Analysis

5.2   Experiments Settings

We take Table 1 corpus as test set. In order to investigate the impact of different language model dealing microblog homophonic normalization, we use three different training corpus, the segmented corpus of Peking University's "People's Daily", our microblog segmented corpus based on standards developed by Peking University specification and mix of these two corpus. So we have three different language models: news language model, microblog language model and mix language model. Table 6 shows the statistics of training corpuses.

Table 5. The statistics of the training corpora for language modeling

Corpus type	Number of sentences	Number of words
The PKU corpus	20000	1078531
The microblog corpus	8190	120417

In our experiments, we employ precision (P), recall (R) and F-Score (F) to evaluate the performance of our system in homophonic word recognition, and accuracy (A) to evaluate the performance in homophonic word restoration. These measures are defined by formulas (2)-(5), respectively.

$$P = C_R / T_O \tag{2}$$

$$R = C_R / T_I \tag{3}$$

$$F = 2 * P * R / (P + R) \tag{4}$$

$$A = C_N / T_I \tag{5}$$

Where, T_I denotes the total number of homophonic words in the test date set, T_O stands for the total number of homophonic words identified by the system, C_R represents the total number of correctly-identified homophonic words, and C_N is the total number of correctly-restored homophonic words.

5.2 Experimental Results

As mentioned above, we generate homophonic normalization candidates by using homophonic-original mapping tables and utilize three different n-gram language models to decode homophonic normalization. Therefore, the major purpose of our experiments is to verify the effect of these factors on homophonic word identification and restoration. The results are presented in Table 6.

Table 6. Experimental results for homophonic word normalization using different homophonic mapping tables and language models

Homophonic-original mapping tables	LMs	P	R	F	A
Homophonic-original word mapping table	LMs from News texts	0.9014	0.8875	0.8944	0.8704
	LMs from microblog texts	0.8974	0.8801	0.8886	0.8730
	Hybird LMs	0.9023	0.8884	0.8953	0.8730
Homophonic-original character mapping table	LMs from news texts	0.6937	0.9596	0.8053	0.8908
	LMs from microblog texts	0.6939	0.9607	0.8058	0.8919
	Hybrid LMs	0.6939	0.9607	0.8058	0.8919

We have following observations from the results in Table 7. (1) Compared with the word-based restoration mapping table, homophonic recognition accuracy decrease if we use homophonic - original character mapping table. But the recall increase greatly. This is partly explained the use of homophonic normalization knowledge in word of units can ease the shortage of homophonic normalization coverage problem. (2) Under the same conditions, the microblog language model performance of homophonic recognition and restoration is better than the news language model on the whole, although microblog corpus is much lower than the size of the news training corpus. This reflects differences between non-formal microblog language and for-mal news languages to some extent.

To better understand experimental results the Table 6, we also present the normalization performance for different types of HHWs when using character-based mapping table and hybrid LMs. The results are presented in Table 7.

From Table 7, we can see that the performance for real word homophonic HHWs is slightly better than the non-real ones. There are several possible reasons for this. First, it hard to normalize true homophonic word, but small proportion in real corpus and usage is not a lot of change. For example, the word "油菜花" is a true homophonic word which itself is a common noun. Noun usage is relatively not complicated. Second, "油菜花" is the homophonic form of "有才华" and used as verb-object phrase. It is easy to identify because of the great difference between these two usages. The result of one-to-many homophonic normalization is better than one-to-one's. We analyzed the main reasons as follow: For the one-to-many homophonic character "灰", it may correspond to two standard forms "飞(灰机)" and "非(灰常)". These experiments are realized using the homophonic-original word mapping table, so one-to-many homophonic HHWs can be well resolved in word of units.

Table 7. Normalization results of various types of homophonic words

Types of HHWs		Homophonic word recognition			Homophonic restoration accuracy
		P	R	F	
1	one-word homophonic word	0.8624	0.8891	0.8754	0.8507
	Multi-word homophonic word	0.9653	0.9408	0.9529	0.9392
2	One-to-one	0.8834	0.8949	0.8891	0.8671
	One-to-many	0.9191	0.9516	0.9351	0.9097
3	True homophonic word	0.8947	0.9329	0.9134	0.8947
	Untrue homophonic word	0.8879	0.9003	0.8941	0.8713
4	Homonym	1.0000	0.8017	0.8899	1.0000
	Non-homonym	0.8837	0.9082	0.8991	0.8677

6   Conclusion and Future Work

In this paper, we have presented an n-gram based method to homophonic word normalization in Chinese microblogs. Preliminary results show that using character-based homophonic knowledge and microblog language models helps improve the performance in homophonic word recognition and restoration. Although this method achieved good results in experiments, the performance of the proposed method, especially the performance of homophonic recognition is highly dependent on the coverage of homophonic normalization know-ledge. While the use of the homophonic-original character mapping table can only alleviate this problem to some extent. In the future research, we will study the acquisition and representation of homophonic normalization knowledge more detailed to explore the inherent basic law of Chinese homophonic word to improve the recognition performance of future applications. In addition, the disambiguation of homophonic normalization is extremely important for homophonic normalization. We will expand microblog corpus size and research new restoration models to improve the disambiguation performance of homophonic normalization.

Acknowledgments. This study was supported by National Natural Science Foundation of China under Grant No.61170148 and No.60973081, the Returned Scholar Foundation of Heilongjiang Province, and Harbin Innovative Foundation for Returnees under Grant No.2009RFLXG007, respectively.

References

1. Tan, X., Pu, K., Shen, M.: Chinese Dictionary of Rhetoric. Shanghai Dictionary Publishing House, Shanghai (2010)
2. Zhou, J., Yang, Y.: Features of Network homophonic words in generation and development. Journal of Jianghan University: Humanities Science Edition 31(3), 30–35 (2012)
3. Choudhury, M., Saraf, R., Jain, V., Mukherjee, A., Sarkar, S., Basu, A.: Investigation and modeling of the structure of texting language. International Journal of Document Analysis and Recognition 10(3-4), 157–174 (2007)
4. Han, B., Baldwin, T.: Lexical normalization of short text messages: Makn sens a# twitter. In: Proceedings of the 49th Annual Meeting of the Association for Computational Linguistics: Human Language Technologies, pp. 368–378 (2011)
5. Hassan, H., Menezes, A.: Social text normalization using contextual graph random walks. In: Proceedings of the 51st Annual Meeting of the Association for Computational Linguistics, pp. 1577–1586 (2013)
6. Sproat, R., Black, A.W., Chen, S., Kumar, S., Ostendorf, M., Richards, C.: Normalization of non-standard words. Computer Speech & Language 15(3), 287–333 (2001)
7. Beaufort, R., Roekhaut, S., Cougnon, L., Fairon, C.: A hybrid rule/model-based finite-state framework for normalizing SMS messages. In: Proceedings of the 48th Annual Meeting of the Association for Computational Linguistics, pp. 770–779 (2010)
8. Aw, A., Zhang, M., Xiao, J., Su, J.: A phrase-based statistical model for SMS text normalization. In: Proceedings of the 44st Annual Meeting of the Association for Computational Linguistics, pp. 33–40 (2006)
9. Bangalore, S., Murdock, V., Riccardi, G.: Bootstrapping bilingual data using consensus translation for a multilingual instant messaging system. In: Proceedings of the 19th International Conference on Computational Linguistics, pp. 1–7 (2002)
10. Fu, G., Zhang, M., Zhou, G., Luke, K.: A unified framework for text analysis in Chinese TTS. In: van de Snepscheut, J.L.A. (ed.) Trace Theory and VLSI Design. LNCS, vol. 200, pp. 200–210. Springer, Heidelberg (1985)
11. Fu, G., Luke, K., Webster, J.: Automatic expansion of abbreviations in Chinese news text: A hybrid approach. International Journal of Computer Processing of Oriental Languages 20(2&3), 165–179 (2007)
12. Liu, L., Wang, S., Wang, D., Wang, P., Cao, C.: Automatic Text Detection in Domain Question Answering. Journal of Chinese Information Processing 27(3), 77–83 (2013)
13. Chang, C.: Corpus-based adaptation mechanisms for Chinese homophone disambiguation. In: Proceedings of the Workshop on Very Large Corpora, pp. 94–101 (1993)
14. Lee, Y., Chen, H.: Applying repair processing in Chinese homophone disambiguation. In: Proceedings of the 5th Conference on Applied Natural Language Processing, pp. 57–63 (1997)

Personalized Web Image Retrieval Based on User Interest Model

Zhaowen Qiu[1,*], Haiyan Chen[1], and Haiyi Zhang[2]

[1] Institute of Information and Computer Engineering, Northeast Forestry University, China
qiuzw@nefu.edu.cn
[2] Jodrey School of Computer Science, Acadia University, Canada
haiyi.zhang@acadiau.ca

Abstract. The traditional search engines don't consider that the users interest are different, and they don't provide personalized retrieval service, so the retrieval efficiency is not high. In order to solve the problem, a method for personalized web image retrieval based on user interest model is proposed. Firstly, the formalized definition of user interest model is provided. Then the user interest model combines the methods of explicit tracking and implicit tracking to improve user's interest information and provide personalized web image retrieval. Experimental results show that the user interest model can be successfully applied in web image retrieval.

Keywords: user interest model, personalized, interest learning, web image retrieval.

1 Introduction

Information retrieval services provided by traditional search engines do not take into account the user's differences lead to different users enter the same keyword, the same search results returned. But in reality, due to differences in background knowledge, interests and other aspects of the different needs of users tend to be different. With the rapid development of information technology at the same time, people in trouble among the massive data, which does not distinguish between the user's information retrieval will consume a lot of search time and reduce the efficiency of information retrieval.

In order to get a more accurate search results in line with the actual needs of different users, personalized Web image retrieval has become a research hotspot. The essence of personalized for different users with different service strategy to provide different services, and personalized Web image retrieval is interested in the initiative to learn and record the user based on the user feedback on the search results, suggesting that the user's interest demand [1]. So consider the differences in the user's search, you can greatly improve retrieval efficiency. To solve the above problems, this paper presents the user interest model, support the user's personalized Web image retrieval, in order to improve retrieval efficiency.

---

* Corresponding author.

H. Wang et al. (Eds.): ICYCSEE 2015, CCIS 503, pp. 188–195, 2015.

User interest model [2] is a personalized image retrieval core, which is used to store and manage user interest information. By collecting user feedback, establish initial user interest model, and then through a long learning, and constantly update the user interest model, and ultimately get the user's interest tends to provide personalized search services for different users. Currently, most studies of user interest model is still rare in the image field. Hsuetal[3] The user interest model used in medical image retrieval, but only for this type of image; Lietal[4] According to the user's interest to adjust the visual characteristics of the underlying weights, but ignores the Figure like the high-level semantic information. Through the establishment of user interest model, the use of explicit and implicit method of combining learning and to consider user interest by the time the impact of this factor, and constantly improve the user's interest in information provided by the user interest model personalized web image search service for users [5].

1.1   Image Semantic Feature Extraction

Low-level visual features of images by color, texture and shape features characteristic composition. Among them, the color characteristics of HSV color space based on 20 non-uniform quantization algorithm [6], a total of 20 dimensions. Four statistic texture feature extraction using GLCM, select the features that the problem: Contrast, consistency, relevance pixel gray scale and as entropy feature vector [7], at 00,450,900, 1350 extracting the four directions, four texture, composed of 16-dimensional feature vector [8]. Feature extraction using the seven-dimensional shape of Hu's invariant moments. In this paper, a method based on SVM semantic association to obtain semantic information of the image, as shown in Figure 1 specific method.

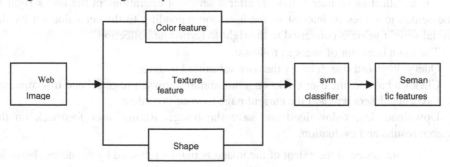

Fig. 1. Extraction of image high-level semantic feature

Based on SVM [9] semantic correlation method is to extract semantic information through the establishment of low-level image features to high-level features of the visual mapping. First calculate the underlying visual image feature vector, and use them as SVM input vectors, then the image class to learn to build low-level visual features of the image is mapped to the high-level semantics. Semantic correlation method, the semantics of a class associated with each one as a separate classification of the two, that is, for ei`ther a semantic category, the positive training examples are all the images included in the class, and is not a counter-example all other classes of image class.

2   User Interest Model

User interest model describes the user's interests and hobbies, plays a central role in personalized retrieval system, is the foundation of personalized service. User interest model is mainly visited by statistical and learning to get the user information, the need to constantly update [9].

User interest model (User interest model, UIM) is defined as six group: $M = (U, S, Q, F, T, \delta,)$ Among them, $U = (u_1, u_2, ..., u_n)$ is a collection of user; $S = (S_1, S_2, ..., S_i, ..., S_n)$ is a collection of interest to the user seman-

tics; $S_i = (s_{i1}, s_{i2}, ..., s_{im})$,it show that M i represents the user interest semantic;

$Q = (Q_1, Q_2, ..., Q_i, ..., Q_n)$ is interested in the semantics of the initial weights corresponding set of users, $Q_i = (q_{i1}, q_{i2}, ..., q_{im})$,it represents the initial weights for the user i of interest semantic of m;

$F = (F_1, F_2, ..., F_i, ..., F_n)$ is that who interested in the right to terminate the corresponding user set value semantics. $F_i = (f_{i1}, f_{i2}, ..., f_{im})$,it represents that user i value the right to terminate the m interested in semantics;

$T = (T_1, T_2, ..., T_i, ..., T_n)$ record that users interested in the semantics of the time it first appears, Important parameters that affect the long-term interests of the users, $T_i = (t_{i1}, t_{i2}, ..., t_{ij}, ..., t_{im})$,it represents that Users interested in the semantics of m i for the first occurrence of the time.

δ is a collection of user behavior, after a series of operations in the user's right to the semantic values of interest to the user corresponding to the set value set by the initial weight value is converted to the right to terminate collection.

The main behavior of users as follows:

Query: keyword search query the user submitted images;

Click on browse: the user clicks on a thumbnail, browse image-related information;

Favorites: Users will add the current page to your favorites;

Download: Users download and save the image; Rating: user feedback on the search results and evaluation.

Users interested in the extent of the image is mainly reflected by the access behavior of users, depending on the access behavior of users, for users interested in semantic weights updated accordingly, as shown in Table 1.

Different users have different semantics for the same degree of interest, and therefore the evaluation made by different feedback δ , weights based on the user interest semantic semantics will be the degree of interest of users is divided into A, B, C, D shown, E5 a rating scale, specific criteria for the classification are shown in Table 2,the lager δ weight, show more interest to users of the semantic category, corresponds to the higher degree of interest.

A user interest corresponding to the highest degree, E corresponds to the lowest degree of user interest.

Table 1. Semantics weight value influence from user visiting action

user access behavior δ	Image semantic weights (q_{ij}, f_{ij})
Query	$q_{ij} = f_{ij} + 0.1$ $\quad f_{ij} = q_{ij}$
Click on Browse	$q_{ij} = f_{ij} + 0.2$ $\quad f_{ij} = q_{ij}$
Favorites	$q_{ij} = f_{ij} + 0.3$ $\quad f_{ij} = q_{ij}$
Download	$q_{ij} = f_{ij} + 0.4$ $\quad f_{ij} = q_{ij}$
Evaluation	$q_{ij} = f_{ij} + \alpha \times 0.1$ $\, f_{ij} = q_{ij}$

Table 2. User interest degree to semantics

The degree of user interest	A	B	C	D	E
Weights α	0	1	2	3	4

3   Users Interested in Learning

Due to the volatility of the user's interest has, therefore need to access the user's behavior, the use of interest corresponding learning algorithm to adapt to constantly update the user's corresponding changes [10]. To address this problem, we first establish the initial user interest model based on the information and the user's first visit to the behavior of user-submitted, and user access behavior using explicit and implicit learning method combining constantly improve the user interest model.

3.1   Explicit Learning

Explicit learning refers users to choose their own interest or the semantics of the retrieved images for feedback and evaluation, so as to achieve the purpose of learning [11]. According to the semantics of the user is interested, you can get the user rating matrix, using an M × N matrix to represent, as shown in equation (1). M wherein M represents the row number of users, N of the N columns represent the semantics of a user element represents a semantic score

$$R(M,N) = \begin{pmatrix} R_{1,1} & \cdots & R_{1,j} & \cdots & R_{1,N} \\ \cdots & \cdots & \cdots & \cdots & \cdots \\ R_{i,1} & \cdots & R_{i,j} & \cdots & R_{i,N} \\ \cdots & \cdots & \cdots & \cdots & \cdots \\ R_{M,1} & \cdots & R_{M,j} & \cdots & R_{M,N} \end{pmatrix} \qquad (1)$$

The value of $R_{i,j}$ is 0,1,2,3,4, the larger the value of $R_{i,j}$, the more that the user interest semantic. Users in image retrieval process, you can explicit feedback information, that after a user browse images, images of scores presented to illustrate the extent of the image of the user interested. We define the rating scale is divided into five, as shown in Table 3-2. Through the study of user feedback information, the system to dynamically update the user interest information.

In practice, the user's interest is constantly changing, and each user generally only a small part of all the information on the entries were evaluated, all due to the $R(M,N)$ become an extremely sparse matrix. User rating for sparse matrix problems, Based on the user's implicit interest in learning to solve.

3.2 Implicit Learning

Since the display user data collected to learn there may be some false information or non-real feedback and evaluation, we can analyze the behavior of the user's access to the user's interest in implicit learning, in order to constantly improve the user interest model, to provide users personalized service [12].

Users interested in implicit learning is automatic, so the system can automatically update the user interest model based on user access behavior. Users to perform queries, click Browse, when collections, download and other actions, the user interested in semantics corresponding weights will increase. When utilizing user interest model retrieval of image information, priority will weights greater semantic class returned to the user. User access behavior reflects the degree of interest to users of the resource, the impact of user access to semantic degree of interest shown in Table 1. According to the behavior of users can get access to the following scoring matrix:

$$
R(M,N) = \begin{pmatrix} f_{1,1} & \cdots & f_{1,j} & \cdots & f_{1,N} \\ \cdots & \cdots & \cdots & \cdots & \cdots \\ f_{i,1} & \cdots & f_{i,j} & \cdots & f_{i,N} \\ \cdots & \cdots & \cdots & \cdots & \cdots \\ f_{M,1} & \cdots & f_{M,j} & \cdots & f_{M,N} \end{pmatrix} \tag{2}
$$

When a user interested in for a resource, the user when browsing this page will consume more time and will often repeat visits to the page, so we can use the access time and access frequency to calculate the user a certain degree of interest in semantics, and users will be affected interest of time, with the loss of time and gradually weakened, and therefore the degree of user interest can be defined as:

$$
F = (\frac{n}{N} + \frac{t}{T}) \times \frac{l}{L} \times e^{\frac{\log_2(T_l - T)}{hl}} \tag{3}
$$

Where n is the number of visits to the semantic-based nodes, the total number of N-based visits, t is access to the semantic node consumes time, T is the total time of the site visit used, l-oriented semantics visit nodes , L is the total number of points semantic festival website. $e^{\frac{\log_2(T_l-T)}{hl}}$ is amnesia factor,h1 is Life Cycle parameters, generally, One week after exposure to new knowledge will begin to forget, so h1 equals seven, T for users interested in the semantics of the first occurrence time, Tl is the last time the user behavior. Degree of interest in using the above model of equation (1) to be updated, user rating matrix obtained, see equation (4).

$$R(M,N) = \begin{pmatrix} R_{1,1} + F_{1,1} & \cdots & R_{1,j} + F_{1,j} & \cdots & R_{1,N} + F_{1,N} \\ \cdots & \cdots & \cdots & \cdots & \cdots \\ R_{i,1} + F_{i,1} & \cdots & R_{i,j} + F_{i,j} & \cdots & R_{i,N} + F_{i,N} \\ \cdots & \cdots & \cdots & \cdots & \cdots \\ R_{M,1} + F_{M,1} & \cdots & R_{M,j} + F_{M,j} & \cdots & R_{M,N} + F_{M,N} \end{pmatrix} \quad (4)$$

Combining the formula (2) and (4) update the user to obtain the final score matrix, see equation (5).

$$R(M,N) = \begin{pmatrix} R_{1,1} + F_{1,1} + f_{1,1} & \cdots & R_{1,j} + F_{1,j} + f_{1,j} & \cdots & R_{1,N} + F_{1,N} + f_{1,N} \\ \cdots & \cdots & \cdots & \cdots & \cdots \\ R_{i,1} + F_{i,1} + f_{i,1} & \cdots & R_{i,j} + F_{i,j} + f_{i,j} & \cdots & R_{i,N} + F_{i,N} + f_{i,N} \\ \cdots & \cdots & \cdots & \cdots & \cdots \\ R_{M,1} + F_{M,1} + f_{M,1} & \cdots & R_{M,j} + F_{M,j} + f_{M,j} & \cdots & R_{M,N} + F_{M,N} + f_{M,N} \end{pmatrix} \quad (5)$$

4   Results

Experiment, Corel Photo gallery and join the network video material in the library portion of the image, a total of 6000 images, these images into 80 semantic categories, such as mammals, birds, reptiles, plants, vehicles, buildings and so on. Select a picture in each of the semantic class as a representative, and labeling. Taking into account the large amount of computation problems, the experiment only 80 randomly selected semantic class 100 image as a query test set, the image retrieval model based on user interest and user interest model does not use image retrieval done comparative experiments, the test results are shown in Figure 2.

As can be seen from the experimental results, the model-based image retrieval of user interest, and do not use the results of user interest model approach compared to average precision has been significantly improved. After a user submits a query image keywords, correct image appears before the first one is the probability of 10 up to 96%, compared with not using user interest model higher than 6.9%. Visible, user interest model presented in this paper can be more accurate characterization of the user's interest, applied to Web image retrieval, effectively improve the efficiency of the image retrieval.

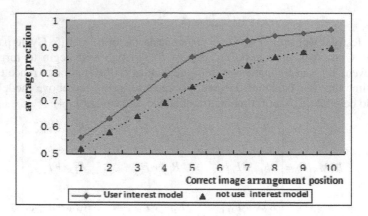

Fig. 2. Image semantics annotation accuracy by using user interest model

5   Conclusions

User interest model is the key technology of personalized image retrieval through user feedback speculated that the user needs to effectively compensate for the semantic gap problem underlying visual features and high-level semantics between. This paper presents an improved user interest model, and gives the formal definition, then the use of explicit and implicit tracking combined method for tracking user interest for learning, and constantly improve the user interest information [13]. Can provide personalized Web image retrieval service for users through the user interest model. The experimental results show that the model of interest to the user to effectively improve the efficiency of the image retrieval. Will further study among users with similar interests hobby migration issues, has studied the use of user interest information to improve the new registered users interested in information retrieval quickly provide personalized service for new users.

Acknowledgment. (Foundation: Supported by the Fundamental Research Funds for the Central Universities(DL10CB01), Heilongjiang Province Natural Science Fund(LC2012C06), Special Funds for Harbin Innovative Talent in Science and Technology(2012RFLXG022)).

References

1. He, L., Jing, Z., Lan, S.: Sun Image retrieval construction method personalized user interest model. Computer Engineering and Applications 45(31), 168–171 (2009)
2. Martin-Bautista, M.J., Kraft, D.H., Vila, M.A., et al.: User profiles and fuzzy logic for Web retrieval issues. Soft Computing 6, 365–372 (2002)
3. Hsu, C.C., Chu, W.W., Taira, R.K.: A Knowledge-Based Approach for Retrieving Images by Content. IEEE Transactions on Knowledge and Data Engineering 8(4), 522–532 (1996)

4. Li, J.H., Liu, M.S., Cheng, Y.: User Interest Model Based Image Retrieval Technique. In: Proceedings of the IEEE International Conference on Automation and Logistics, China, pp. 2265–2269 (2007)
5. Adda, M.: A Formal Model of Information Retrieval Based on User Sensitivities. Procedia Computer Science, 428–436 (2013)
6. Zhaowen, Q., Tian, Z.: A new image color feature extraction method. Harbin Institute of Technology Journal 36(12), 1769–1701 (2004)
7. Zhaowen, Q.: User-oriented Web image retrieval key technology research. Doctoral dissertation. Harbin Institute of Technology (2009)
8. JBC, Based on Web Mining user personalized recommendations Research and realization. Beijing University of Posts and Telecommunications (2013)
9. Xiaohua, L., Lansun, S., Huanqin, L.: Estimation of crowd density based on wavelet and support vector machine. Transactions of the Institute of Measurement and Control 28(3), 299–308 (2006)
10. Park, Y.J., Chang, K.N.: Individual and group behavior-based customer profile model for personalized product recommendation. Expert Systems with Applications 36(2), 1932–1939 (2009)
11. Huang Huadong based personalized search user model study. East China University of Technology (2013)
12. Alberto, D., Pablo, G.: User-model Based Personalized Summarization. Information Processing & Management 43(6), 1715–1734 (2007)
13. Jing, Y.: User interest model and real-time personalized recommendation algorithm. Nanjing University of Posts and Telecommunications (2013)

Detection of Underwater Objects by Adaptive Threshold FCM Based on Frequency Domain and Time Domain

Xingmei Wang[1], Guangyu Liu[2], Lin Li[3], and Shouxuan Jiang[1]

[1]School of Computer Science and Technology, Harbin Engineering University, Harbin, China
wangxingmei@hrbeu.edu.cn
[2]School of Shipbuilding Engineering, Harbin Engineering University, Harbin, China
lgysohumail@sohu.com
[3]School of Information and Communication Engineering,
Harbin Engineering University, Harbin, China
515649273@qq.com

Abstract. According to the characteristics of sonar image data with big data feature, In order to accurately detect underwater objects of sonar image, a novel adaptive threshold FCM (Fuzzy Clustering Algorithm, FCM) based on frequency domain and time domain is proposed. Based on the relationship between sonar image data and big data, Firstly, wavelet de-noising method is used to smooth noise. After de-noising, the sonar image is blocked and each sub-block region is processed by two-dimensional discrete Fourier transform, their maximum amplitude spectrum used as frequency domain character, then time domain of mean and standard deviation, frequency domain of maximum amplitude spectrum are taken for character to complete block k-means clustering, the initial clustering center is determined, after that made use of FCM on sonar image detection, based on clustered image, adaptive threshold is constructed by the distribution of sonar image sea-bottom reverberation region, and final detection results of sonar image are completed. The comparison different experiments demonstrate that the proposed algorithm get good detection precision and adaptability.

Keywords: Sonar image, Character frequency domain, Block k-means clustering, Fuzzy clustering algorithm, Adaptive threshold.

1 Introduction

Sonar image contains three kinds of regions: object-highlight and shadow regions, sea-bottom reverberation region. For sonar image analysis, detection of the sonar image plays a very important role in the later sonar image processing. Its purpose is to extract object-highlight and shadow regions from complex sea-bottom reverberation region. Only under the condition of accurate detection, feature extraction and parameter measurement can be carried on for underwater objects. So underwater objects detection is very valuable for the study of sonar image processing [1]. Recognition, tracking and classification in the ocean depend on whether it has an accurate detection of underwater sonar image [2]. But with the current algorithms, it is very difficult to

H. Wang et al. (Eds.): ICYCSEE 2015, CCIS 503, pp. 196–203, 2015.

improve detection quality of the sonar image, so high quality detection algorithm is urgently needed to explore [3].

The sonar image has serious noise, these noises make sonar image data with complex structure, namely, sonar image data has big data feature. In order to do not appear detection error or excessive detection in detection process, it is vital to denoising before target detection on underwater sonar image. The noise of sonar image concentrate in high frequency region, object-highlight region and shadow region concentrate in low frequency region generally, so sonar image has the character of obvious distribution frequency domain .Wavelet de-noising is suitable for the signal of local and transient, it fully reflect on the signal character of different regions and different resolutions, and it has ability of denoting local signal character in frequency domain and time domain. So the wavelet de-noising method is used to smooth noise of sonar image [4].

FCM can well solve fuzzy calculation problem, this algorithm can be calculated through membership of each pixel and each cluster center, it can alleviate defects resulting of hard clustering [5].

In this paper, adaptive threshold FCM which based on frequency domain and time domain is applied in sonar image. wavelet de-noising method is used to smooth noise, then the sonar image is blocked, three characters of sub-block region are obtained, i.e., the value of mean, standard deviation and maximum amplitude spectrum, completed block k-means clustering, the initial clustering center is determined, after that made use of adaptive threshold FCM to detect sonar image. Adaptive threshold is constructed by s distribution of sonar image sea-bottom reverberation region, and final underwater objects detection results are completed under the environment of sonar image data with big data feature.

This paper is organized as follows: In Section 2, we give original sonar image of wavelet de-noising. In Section 3, we detail block k-means clustering algorithm of sonar image. In Section 4, we present adaptive threshold FCM of the sonar image. Section 5 we discuss detailed experimental results and analysis. Section 6 is the conclusions and future work about the sonar image processing. The last Section is the acknowledgment.

2   Wavelet De-noising of Original Sonar Image

Sonar image is two-dimensional signal generally, by multi-resolution analysis, it is changed into the wavelet domain, then extraction of wavelet coefficient in different scales, multi-resolution decomposition formula of arbitrary two-dimensional signal $X(t) \in L^2(R)$ can be expressed as:

$$x(t) = \sum_k a_j(k)\varphi_{j,k}(t) + \sum_{j=1}^{J} \sum_K d_j(k)\psi_{j,k}(t) \tag{1}$$

Where $\varphi_{j,k}(t) = 2^{-J/2}\varphi(2^{-J}t - k)$ is scaling function, $\psi_{j,k}(t) = 2^{-J/2}$ is wavelet function, $\{\varphi_{j,k}(t)\}$ is orthonormal basis of the scale space V_j, $\{\psi_{j,k}(t)\}$ is orthonormal basis of wavelet space W_j. the decomposition coefficient of $a_j(k)$ and $d_j(k)$ respectively called discrete smooth approximation signals and discrete details signal,

Recursive step by step, sonar image decomposition coefficient can be obtained in different series, Due to the important information distribution of the sonar image is the wavelet decomposition coefficient, so the process of decomposition coefficient can describe sonar image de-noising[6].

After wavelet decomposition of sonar image, the coefficient of very small is noise, the coefficient of relatively great mainly suits for object-highlight and shadow regions, sea-bottom reverberation region. So suitable threshold is set for the parts of high frequency coefficient. A new wavelet coefficient is obtained by threshold function mapping, then make wavelet reconstruction on wavelet coefficient. In the method of threshold de-noising, threshold and threshold function are the most fundamental elements [7].

The most commonly used threshold function is Visu Shrink method which is proposed by Donoho et al. The choice of optimal threshold T is that:

$$T = \sigma\sqrt{2\log N} \tag{2}$$

Where σ is standard deviation of the noise, N is the number of high frequency coefficient in wavelet coefficient.

The experiment result of wavelet de-noising is shown in Fig.1. Fig.1 (a) is the original sonar image (the image size is 162×224). Fig.1 (b) is the de-noising sonar image.

(a) Original sonar image   (b) De-noising sonar image

Fig. 1. Wavelet de-noising of sonar image

From the experiment results, we can get the following conclusions. The method of wavelet de-noising can remove some noise of original sonar image. The sonar image becomes more smoothing. When compared the shadow region and object-highlight region with sea-bottom reverberation region, de-noising sonar image in this paper is more prominent than the original sonar image, and this will bring more conducive to the subsequent underwater objects detection.

3   Block K-means Clustering Algorithm of Sonar Image

3.1   Character of Frequency Domain

The two-dimensional Fourier transform expression is:

$$F(u,v) = \int\limits_{-\infty}^{+\infty} \int\limits_{-\infty}^{+\infty} f(x,y)e^{-j2\pi(ux+vy)}dxdy \tag{3}$$

The amplitude of two-dimensional Fourier transform is that:

$$|F(u,v)| = \sqrt{R^2(u,v) + I^2(u,v)} \tag{4}$$

The phase spectrum of two-dimensional Fourier transform is that:

$$\varphi(u,v) = \arctan \frac{I(u,v)}{R(u,v)} \tag{5}$$

According to the obvious differences of sonar image in object-highlight region, shadow region and sea-bottom reverberation region. Sonar image is blocked, the size of each sub-block is $M * N$, and each sub-block region is processed by two-dimensional discrete Fourier transform, maximum amplitude spectrum is extracted as frequency domain character. Amplitude spectrum formula is:

$$F(u,v) = \frac{1}{MN} \sum_{u=0}^{M-1} \sum_{v=0}^{N-1} f(x,y) \exp[-j2\pi(ux/M + vy/N)] \tag{6}$$

Where $u = 0,1,\cdots M - 1, v = 0,1,\cdots N - 1$. $f(x,y)$ is pixel value of every point in sonar image, and it is real number. Fourier transform is plural form generally.

$$F(u,v) = R(u,v) + jI(u,v) \tag{7}$$

$$|F(u,v)| = \sqrt{R^2(u,v) + I^2(u,v)} \tag{8}$$

3.2 Block K-means Clustering Algorithm

The differences of object-highlight region, shadow region and sea-bottom reverberation region in sonar image are very large [8]. And using block k-means clustering algorithm can more easily detect the object-highlight region and shadow region. Choose the window of the $M * N$, and sonar image is divided into non-overlapped $M * N$ window that a number of B, Three samples of each window are obtained, namely, the value of mean, standard deviation and maximum amplitude spectrum[9].

4 Adaptive Threshold FCM Algorithm of Sonar Image

Clustering centers of k-means clustering and fuzzy clustering are very close. But k-means clustering algorithm is faster convergence speed than fuzzy clustering algorithm. So clustering center of k-means clustering use as initial cluster centers of FCM algorithm, the iteration number of fuzzy clustering algorithm is significantly decreased, and FCM convergence speed is improved [10].

After wavelet de-noising of sonar image, the initial cluster centers is determined by block k-means clustering algorithm, then image is detected by fuzzy clustering algorithm. Based on the image clustered, through gray distribution model of sea-bottom reverberation region, and adaptive threshold is constructed to complete sonar image final detection.

The distribution of sonar image sea-bottom reverberation region can be described by a Gamma distribution $G_\gamma(y;\gamma,\lambda)$.

The probability density function of Gamma distribution is expressed as [9]:

$$W_Y(y;\gamma,\lambda) = \frac{\lambda^\gamma}{\Gamma(\gamma)} y^{\gamma-1} \cdot \exp(-\lambda \cdot y) \tag{9}$$

Where y is the gray value of pixels, γ is shape parameter, λ is scale parameter, $\Gamma(\bullet)$ is function of Gamma.

Shape parameter γ, scale parameter λ, can be estimated by the equation (10) and (11):

$$\hat{\gamma} = \frac{m_1^2}{m_2 - m_1^2} \tag{10}$$

$$\hat{\lambda} = \frac{m_1}{m_2 - m_1^2} \tag{11}$$

Where $m_q = \frac{1}{M}\sum_{i=1}^{M} y_i^q$, $q = 1,2$, $\{y_i\}_{i=1}^M$ is pixel in whole sonar image.

The value of λ is very small on Gamma distribution. So the value of γ is selected as the parameter description of the model. After fuzzy clustering detection of sonar image, the difference of object-highlight region and sea-bottom reverberation region or shadow region and sea-bottom reverberation region is very small, leading to that the object-highlight region and shadow region cannot be separated from sea-bottom reverberation. So the adaptive threshold of the results of fuzzy clustering is proposed to complete the detection of object-highlight region and shadow region [9].

Reference pixel values are set by the results of fuzzy clustering detection, and it is mapped to the initial sonar image. Based on these template, the value of γ_1 is calculate and it is the object-highlight region or shadow region that obey Gamma distribution, the value of γ_2 is calculated and it is the sea-bottom reverberation region that obey Gamma distribution. When increasing the step length of the pixel values, the maximum of $a = \max |\gamma_1 - \gamma_2|$ is obtained, so object-highlight region or shadow region cannot be described by Gamma distribution. And the correct the object-highlight region and shadow region are detected. It is the final detection results of sonar image.

5   Results and Analysis of Experiment

The proposed adaptive threshold FCM based on frequency domain and time domain is tested by the original sonar image.

The first group of experiments is to detect the original sonar image. The experimental results are shown in Fig.2. Fig.2 (a) is the initial cluster centers by block k-means clustering algorithm based on Fig.1 (b), Fig.2 (b) is the result of fuzzy clustering detection, shadow region and sea-bottom reverberation region are hard to distinguish in Fig.2 (b), according to the pixel value of shadow region is above 8, so 8 is selected as the reference pixel value, the step length is 2, and calculate $a = \max |\gamma_1 - \gamma_2|$, the distributed parameter values are shown in Table 1, when the pixel value is 12, a is the maximum, the difference of shadow region and sea-bottom reverberation region is the maximum. Fig.2 (c) is discrete diagram according to Table 1, abscissa is the pixel value of 8 to 24, the step length is 2, ordinate is $a = \max |\gamma_1 - \gamma_2|$. Fig.2 (d) is the remarking result of shadow region and sea-bottom reverberation region when the pixel value is12, a is the maximum. Among them, less than 12 is the shadow region, and greater than or equal to 12 is sea-bottom reverberation region, namely shadow region and sea-bottom reverberation region are distinguished in Fig.2 (b). Fig.2 (e) is the detection result of object-highlight region, shadow region and sea-bottom reverberation region according to Fig.2 (d). Fig.2 (f) is the result of removing isolated region. Fig.2 (g) is the marking result which final detection result maps to the original sonar image.

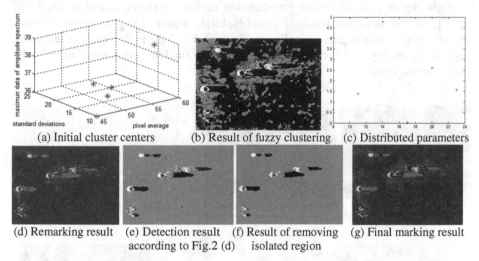

(a) Initial cluster centers   (b) Result of fuzzy clustering   (c) Distributed parameters

(d) Remarking result   (e) Detection result   (f) Result of removing   (g) Final marking result
                       according to Fig.2 (d)   isolated region

Fig. 2. Detection results of group 1 experiments

Table 1. The Distributed parameters (group 1 experiment)

pixel value	8	10	12	14	16	18	20	22	24
γ_1	4.17	3.27	10.67	1.33	5.56	3.76	8.03	4.17	3.27
γ_2	4.80	4.65	5.88	0.12	5.54	6.37	6.45	4.80	4.65
$a = \max \mid \gamma_1 - \gamma_2 \mid$	0.63	1.38	4.80	1.21	0.01	2.61	1.57	0.63	1.38

For comparison, the result of adaptive threshold FCM based on frequency domain and time domain about sonar image is more correct. The comparison experiments demonstrate that the proposed algorithm get good results.

In order to verify the adaptability of the proposed algorithm, Fig. 3 is second group of experiments to detect the original sonar image by the proposed algorithm in this paper with the different type of sonar system. Fig.3 (a) is the original sonar image (the image size is 179×296). Fig.3 (b) is the de-noising sonar image. Fig.3 (c) is the initial cluster centers by block k-means clustering algorithm based on Fig.3 (b), Fig.3 (d) is the result of fuzzy clustering detection, object-highlight region and sea-bottom reverberation region are hard to distinguish in Fig.3 (d), according to the pixel value of object-highlight region is above110, so 110 is selected as the reference pixel value, the step length is 10, and calculate $a = \max \mid \gamma_1 - \gamma_2 \mid$, the distributed parameter values are shown in Table 2, when the pixel value is 180, a is the maximum, the difference of object-highlight region and sea-bottom reverberation region is the maximum. Fig.3 (e) is discrete diagram according to Table 2, abscissa is the pixel value of 110 to 190, the step length is 10, ordinate is $a = \max | \gamma_1 - \gamma_2 |$. Fig.3 (f) is the remarking result of object-highlight region and sea-bottom reverberation region when the pixel value is 180, a is the maximum. Among them, less than 180 is the sea-bottom reverberation region, and greater than or equal to 180 is object-highlight region, namely object-highlight region and sea-bottom reverberation region are distinguished in Fig.3 (d). Fig.3 (g) is the detection result of object-highlight region, shadow region and sea-bottom reverberation region according to Fig.3 (f). Fig.3 (h) is the result of removing isolated region. Fig.3 (i) is the marking result which final detection result maps to the original sonar image.

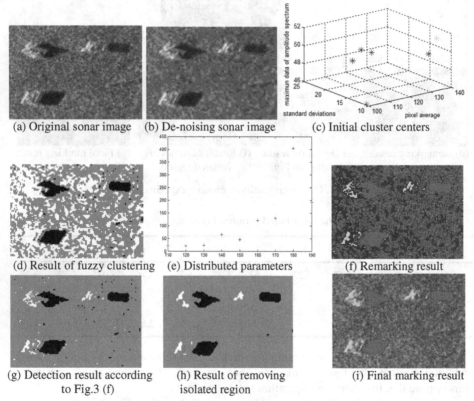

(a) Original sonar image   (b) De-noising sonar image   (c) Initial cluster centers

(d) Result of fuzzy clustering   (e) Distributed parameters   (f) Remarking result

(g) Detection result according to Fig.3 (f)   (h) Result of removing isolated region   (i) Final marking result

Fig. 3. Detection results of group 2 experiments

Table 2. The Distributed parameters (group 2 experiment)

pixel value	110	120	130	140	150	160	170	180	190
γ_1	37	36	41	84	67	142	146	422	206
γ_2	16	16	18	19	21	21	16	19	15
$a = \max | \gamma_1 - \gamma_2 |$	21	20	23	65	46	121	130	403	191

It can be seen from the results of the comparison experiments that the effect of detection will be better when $a = \max |\gamma_1 - \gamma_2|$ is relatively maximum, and at the same time, the algorithm has its adaptability.

6   Conclusions

In this paper, we present a new adaptive threshold FCM based on frequency domain and time domain for detection of sonar image. In order to reduce noise, original image is smoothed by wavelet de-noising. After smoothed processing of sonar image, sonar image is blocked, each sub-block region extracts value of mean, standard deviation and maximum amplitude spectrum to complete block k-means clustering, then FCM is used to detect sonar image, according to results of FCM, adaptive threshold is constructed by the distribution of sonar image sea-bottom reverberation region, the final detection result is obtained under the environment of sonar image data with big data feature.

Acknowledgment. This work was supported by the National Natural Science Foundation of China (41306086), technology innovation talent special foundation of Harbin (2014RFQXJ105) and Fundamental Research Funds for the Central Universities (No.HEUCFR1121, HEUCF100606).

References

1. Liu, G., Bian, H., Shi, H.: Sonar Image Segmentation based on an Improved Level Set Method. In: Proceedings of International Conference on Medical Physics and Biomedical Engineering, pp. 71168–71175 (2012)
2. Ye, X.-F., Zhang, Z.-H., Liu., P.X., Guan, H.-L.: Sonar image segmentation based on GMRF and level-set models. Ocean Engineering 37(10), 891–901 (2010)
3. Wang, X., Liu, G., Li, L., Liu, Z.: A Novel Quantum-inspired Algorithm for Edge Detection of Sonar Image. In: The 33rd Chinese Control Conference, pp. 401–409 (2014)
4. Wu, J.: Wavelet domain de-noising method based on multistage median filtering. The Journal of China Universities of Posts and Telecommunications 20(2), 113–119 (2013)
5. Cinque, L., Foresti, G., Lombardi, L.: A clustering fuzzy approach for image segmentation. Journal of Manufacturing Automation 37(9), 1797–1807 (2004)
6. Chen, Y.T.: Image de-noising and quality measurements by using filtering and wavelet based techniques. AEU-International Journal of Electronics and Communications 68(8), 699–705 (2014)
7. Wei, W.-Y.: New Ostu image segmentation based on intensity stretching on DWT fields. Journal of Northwest Normal University (Natural Science) 45(6), 46–48 (2009)
8. Lei, W., Xiufen, Y., Tian, W.: Segmentation algorithm of fuzzy clustering on sidescan sonar image. Journal of Huazhong University of Science & Technology (Natural Science Edition) 40(9), 25–29 (2012)
9. Ye, X.-F., Wang, X.-M., Zhang, Z.-H., Fang, C.: Study of sonar imagery segmentation algorithm by improved MRF parameter model. Journal of Harbin Engineering Unviersity 30(7), 768–774 (2009)
10. Tian, X.-D., Tong, J.-J., Liu, Z.: Background gray level distribution in sonar images. Technical Acoustics 25(5), 468–472 (2006)

Discovering Event Regions
Using a Large-Scale Trajectory Dataset

Ling Yang, Zhijun Li, and Shouxu Jiang[*]

School of Computer Science and Technology, Harbin Institute of Technology, Harbin, China
{13B903011,lizhijun_os,jsx}@ hit.edu.cn

Abstract. The city is facing the unprecedented pressure with the rapid development and the moving population. Some hidden knowledge can be found to service the social with human trajectory data. In this paper, we define a state-of-the-art concept on fluctuant locations with PCA method and discover the same attribute of fluctuant locations called event with topic model. In the time slice, locations with the same attribute are called event region. Event regions aim to understand the relationship between spatial-temporal locations in the city and to early-warning analyze for the city planning, construction, intelligent navigation, route planning and location based service. We use GeoLife public data to experiment and verify this paper.

Keywords: big data, principal component analysis, topical model, fluctuation location, event region.

1 Introduction

Events are occurring everyday in city, and the mobility of population is increasing the possibility of an event occurs. The features of events ,transient mob, destructive behavior, the state of imbalance and others, harm the city. Therefore, discover events to make the early warning using historical trajectories, which is a high academic value and practical significance. An effective application is of great importance to formulate urban planning requires evaluating a broad range of factors, such as traffic flow, human mobility, point of interests, and road network structures [1].

The events generate relative locations, which are named fluctuation location. Even in the same time segment the fluctuation locations are not necessarily caused by the same events, or that the attribute of fluctuations locations is different. The attribute is the topic in this paper. Fluctuation locations with the same attribute in a time segment is a event region.

We find that the GPS data could overlap strongly in the papers using trajectories data on locations[8][10] and on the relevant inference applications[4][5].Take an example, there will be many people through the same GPS points at the same time. We propose a hypothesis[1] that the appeared GPS points are 'pseudo-sensors' to count numbers through

---

[*] Corresponding author.
[1] Prof. Jiang He is the corresponding author of this paper.

H. Wang et al. (Eds.): ICYCSEE 2015, CCIS 503, pp. 204–211, 2015.

here real-time ."Pseudo-sensors" advantage to conquer the economic costs and space constraints[11].This hypothesis is based on the study of Song[2],which shows humans have a certain trajectory roundedness. The trajectories in the city are more narrow in the scope of space. The hypothesis of "pseudo-sensor" is suit for urban computing.

It is not difficult to detect that the value of fluctuation is not the similar compared with the near location in figure1(b).When an event occurs, the status of interrelated locations will have changed. Events will form a series of fluctuation locations with logical related and time-continuous. To calculate the fluctuation location and to discover the hidden topic with explicit statistical data ,Which are our main contents in the paper. Figure1(a), 10 pseudo sensors in the Zhongguancun North Street count variations in one day. Figure1 (b), the horizontal axis represents the time segment; ordinate indicates the number of statistical values. In order to consider the number of the same location change in a continuous time segment, we subtract the number of continuous time, the formation of the difference value. Fluctuation locations are calculated with PCA on the basis of the differential value in the Figure1(d), the solid circles are the fluctuation locations. Topic model could dig out the attribute of fluctuation location .Location A and D that are caused by the common theme of E1; Location A, B and C are from the same common theme E2 produced in figure(d).Event region 1st is a set of fluctuation location A and D; Fluctuation locations A, B and C in the event region 2nd.

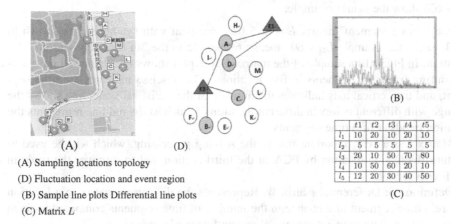

(A)　　　　　　　　(D)

(A) Sampling locations topology

(D) Fluctuation location and event region

(B) Sample line plots Differential line plots

(C) Matrix L

	t1	t2	t3	t4	t5
l_1	10	20	10	20	10
l_2	5	5	5	5	5
l_3	20	10	50	70	80
l_4	10	50	60	20	10
l_5	12	20	30	40	50

(C)

Fig. 1. Scene Description

2　Problem Definitions

Definition 1 (GPS Logs):P is the set of GPS points, p={p1,p2,p3,...,pn}, pi∈P. pi ,including latitude (pi.Lat), longitude (pi.Log) and time (pi.T)

This structure is represented by the following triples: $< p_i.Lat, p_i.Log, p_i.T >$

Latitude change once combined 110.94 km, a division 1.849 kilometers, one second combined 30.8 m; longitude change once co 85.276 km, a division 1.42 km, one second combined 23.69 meters in GPS data specification description. According to the accuracy of this study, we will collect GPS data accurate to three decimal places, we need to form a "location", which is, a "pseudo-sensor" as shown in Table 1.

Table 1. Location Example

LocationID	Latitude	Longitude
24576	39.889	116.341

Trajectory data on the time field has two categories: the date and time. The issue of concern is the location of the moment fluctuations generated. The field of date is deleted. We set a time segment every five minutes and get a total of 288 time segments 24-hour one day. Every five time segments are set to one research unit, namely a time slot. There are 58 time slot in 288 time segments. See figure 2.

Time Slot	S 1					S 25					S58	
Time Segment	t 1			...		t121		...	t125	...	...	t288
Time	00:00:00	00:00:05	...	00:04:55	...	10:05:00	...	10:04:55	...	.....	...	...	.......| 23:59:55

Fig. 2. Example of time, time segment, time slot

Definition 2: "Position - time segment" matrix L. Pseudo sensors count the number in time segments. The location and time segment form a sequence matrix.
Fig1.(C) show the matrix example.

"location-time segment" matrix L is at 5 time segment with synthesis data shown in Fig3. (a).In the example $L_{43} = 60$ means 60 people in the 3th time segment and 4th location. In Fig3. (b) a sample of the matrix L line plot shows the change of the number during five time segments in five locations. The abscissa indicates the time segment, and the vertical axis indicates the number of the statistical scale, line shows the change with different colors in different locations 5.Such as the red line represents the location trends at fifth time segments.

Matrix L is a very important role in the solving processing, which will be used to capture fluctuation location by PCA in the third section and to discover the topic in the fourth section.

Definition 3: Differential matrix W. Represents the difference matrix to a location before a time segment of a change in the number of time segments compared to form a new location-time segment matrix. W matrix forms of expression:

$$W_{ij}=|L_{i(j+1)} -L_{ij}| \qquad (1)$$

Definition 4: Fluctuation location (FL). Fluctuation locations are calculated by the following two characteristics. First, the number dramatically changes in one location at the continuous time segment; Second, the trend of location is different from other locations.

Fluctuation location is the core concept in the paper Unlike the abnormal location. Relevant evidence is discussed in the third section.
Definition 5: Event region. Region is a collection of locations. Event region is a set of fluctuation location with the same attribute. The different causes the different attribute of fluctuation location which is the topic.

$$P(FL_i) = \sum_{i=1}^{T} P(FL_i \mid z_i = j) p(z_i = j) \tag{2}$$

z_i is the hidden variable, i is the time slot, j is the topic identifier.

Event region has three characteristics: First, the connectivity space; Second, the continuity time; Thirdly, the correlation logic. Event region is the target object in the paper. It has an important significance to understand the dynamic characteristic of event region and the correlation between the locations of in the city.

Figure 3 presents the architecture of our method, which consists of two major components: 1) identify the fluctuate location and 2) discovery the event region. We will detail each step of these two components
in the following two sections respectively.

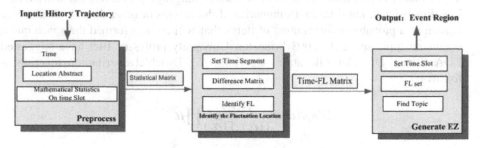

Fig. 3. Framework of discovering the event region

3   Modeling Event Regions

3.1   Calculate the Fluctuation Location with PCA

According to the definition 4, the characteristic of fluctuation locations relatively produce "wave" in continuous-time location. Comparison of different locations to produce "volatility" in the same time segment. In this section we describe the use for Principal Component Analysis (PCA) for anomaly detection. PCA is a widely used dimensionality reduction.PCA exploits the observation that in most explicitly high-dimensional data sets, there is a high implicit correlation between many dimensions (variables) which can be inferred by carrying out an eigen-decomposition of the data covariance matrix[4].

Example of PCA Anomaly Detection: We present a small example to illustrate the use of PCA for anomaly detection. To carry out a PCA analysis we first normalize the L matrix and form the 5×5 L^TL covariance matrix. An eigen decomposition of the covariance matrix show that the eigenvalues in decreasing order are [1.9×10^3, 0. 67×10^3, 0. 02×10^3, 0. 01×10^3, 0].We choose the first eigenvector as the normal subspace and the remaining eigenvectors as the abnormal subspace. All the points are projected onto abnormal subspace and in this space for all points we compute the square of the deviation from the mean. These are [0.4×10^3, 0. 06×10^3, 0.5×10^3, 1. 47×10^3, 0. 49×10^3].Thus the technique correctly identifies locationID 4 as the anomaly. The input matrix is the difference matrix W not matrix L in our method. The covariance matrix eigenvalues vector could be calculated with PCA methods. An eigen decomposition of the covariance matrix show that the eigenvalues in decreasing order are $[0.4 \times 10^3, 0.2 \times 10^3, 0.01 \times 10^3, 0.2]$ All the points are projected onto abnormal subspace and in this space for all points we compute the square of the deviation from the mean. These are

$[0.17 \times 10^3, 0.091 \times 10^3, 0.472 \times 10^3, 0.028 \times 10^3, 0.177 \times 10^3]$.

Thus the technique correctly identifies locationID 3 as the anomaly. The third location is fluctuation location. The fluctuation location is different from the anomaly.

3.2   Discover the Event Region with Topic Model

Topic model is proposed in the field of natural language processing for discovering low-dimensional, multi-faceted summaries of documents or other discrete data. Topic is seen as a probability distribution of items that will be transformed the space into a document topic space. In 2003 Princeton University professor Blei have suggested LDA (Latent Dirichlet Allocation) algorithm [6]. Dirichlet distribution function as follows:

$$Dir(\mu \mid \alpha) = \frac{\Gamma(\alpha_0)}{\Gamma(\alpha_1)..\Gamma(\alpha_k)} \prod_{k=1}^{K} \mu_k^{\alpha_k - 1} \tag{3}$$

where the parameter α is a k-vector. A k-dimension Dirichlet random variable μ can take values in the (k-1)-simplex (a k-vector μ lies in the (k-1)-simplex if $0 \le \mu_k \le 1, \sum_k \mu_k = 1; \alpha_0 = \sum_{k=1}^{k} \alpha_k$, $\Gamma(\cdot)$ is the Gamma function.

For multinomial distribution the conjugate prior is the Dirichlet distribution. Topic model solves the event region to discover the set of fluctuation locations with the same attribute at the same time segment. Our parameters are summarized in Table 2.

Table 2. Trajectory-event analogy to document-topic

trajectory	document
event	topic
fluctuation location	item_W
time slot	item_t
trajectories at same time slot	corpus

Trajectories data could discover the event region with LDA. However, the time slot is a key role for topic. The event region is discovered at one time slot and disappear at two time slot ,which means the evolution of event region. In order to meet the above analysis, we will use references[7] TOT (Topic Over Time) topic model found event region.TOT theme topic model considers the probability distribution is affected by time, time variable obey the Beta distribution. The graphical model is shown in Figure4 [7]. Our notation is summarized in Table 3.

In TOT, topic discovery is influenced not only by word co-occurrences, but also temporal information. TOT model is more t compares with LDA model. Observable variable t which obedience beta distribution, Ψz is Beta distribution. Beta distribution function as follows:

$$Beta(\chi;\alpha,\beta) = \frac{\Gamma(\alpha+\beta)}{\Gamma(\alpha)\Gamma(\beta)}\chi^{\alpha-1}(1-\chi)^{\beta-1} \tag{4}$$

There are two ways of describing its generative process:

1. Draw T multinomial Φ_z from a Dirichlet prior β, one for each event region z;
2. For each trajectory d, draw a multinomial θ_d from a Dirichlet prior α; then for each fluctuate locationω_{di} in trajectory d:

— Draw a event region z_{di} from multinomial θ_d;
— Draw a fluctuate location w_{di} from multinomial φ_{zdi};
— Draw a time slots t_{di} from Beta ψ_{zdi} .

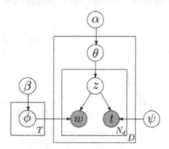

Fig. 4. TOT graph model[7]

Table 3. Notation used in this paper

SYMBOL	DESCRIPTION
T	number of event regions
D	number of trajectories
N_d	number of fluctuation locations token in trajectory d
θ_d	the multinomial distribution of event regions specific to the dth trajectory
Φ_z	the multinomial distribution of fluctuation location ω specific to event region z
Ψ_z	the beta distribution of time slot t specific to event region z
Z_{di}	the event region associated with the ith token in the dth trajectory
ω_{di}	the ith fluctuation location token in dth trajectory
t_{di}	the time slot associated with the ith token in the trajectory d

To calculate event region is transformed into a given fluctuate location and time slot sequence, the conditional probability of the sequence of events regions. For simplicity and speed we estimate these Beta distributions ψ_z by the method of moments, once per iteration of Gibbs sampling. One could estimate the values of the hyperparameters of the TOT model, α and β, from data using a Gibbs EM algorithm [9].Gibbs sampling is a special case of Markov Monte Carlo (Markov-chain Monte

Carlo, MCMC) methods, Every component of the joint distribution of a sampled While keeping the other components unchanged. We begin with the joint probability of a data set, and using the chain rule, we can obtain the conditional probability conveniently .Reference[7] could compute the posterior estimates of θ and Ψ.

4   Experimentation and Evaluation

The trajectories dataset is from reference [3]. This GPS trajectory dataset was collected in (Microsoft Research Asia) Geolife project by 182 users in a period of over five years (from April 2007 to August 2012). A GPS trajectory of this dataset is represented by a sequence of time-stamped points, each of which contains the information of latitude, longitude and altitude. This dataset contains 17,621 trajectories with a total distance of 1,292,951kilometers and a total duration of 50,176 hours. This dataset recoded a broad range of users' outdoor movements, including not only life routines like go home and go to work but also some entertainments and sports activities, such as shopping, sightseeing, dining, hiking, and cycling.

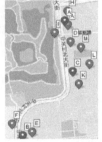

(a) Sample at time slot 11   (b) Sample at time slot 25

Fig. 5. Example of an event region

The example shows the nature of event regions using the true data in Figure 1.Figure 1(a) in a 10 sampling points, Offline calculated at fluctuation locations in figure 5 (a), the non-fluctuation locations are indicated by green. Trajectory data with time slot 11th obtain Figure 5 (a) finds two event regions with red and purple represents. Fluctuation locations with different colors mean the different types, which are event regions. The trajectory data with 25th time slot obtained three event regions in figure 5(b), which are represented by red, purple and blue. There will be different regions in a different time slots. Figure5 (a) and (b) comparison at different time-slot, there will be a different event region. For example,(a) Point B is affected by the same thing in a time slot 11th with points A and C ,which is the same topic. (b) Point B and point R belongs to the new topic by new events. Prior to point A and point C are gone with different time slots. The evolution of events in the event regions is also different. The experiment also confirms the event region is defined for temporal continuity and logical relevance.

5 Conclusion and Future Work

This paper presents a new understanding of urban locations using trajectory data. Our work is based on the basis of relevant studies. We propose a new concept of the fluctuation location unlike the analysis of outliers. We discover the event region at the same time slot on the basis of fluctuation location. The points in experiments change with temporal trajectory, which means the event regions change. It is the evolution of event region. In future work, we completed three major improvements: First, preprocess the data sacristy. Secondly, the PCA can only solve linear problems, how to effectively overcome their own shortcomings improvements. Again, the topic model, the need for the following work.

References

1. Zheng, Y., Capra, L., Wolfson, O., Yang, H.: Urban Computing: concepts, methodologies, and applications. ACM Transaction on Intelligent Systems and Technology (ACM TIST) 5(3) (2014)
2. Song, C., Qu, Z., Blumm, N., et al.: Limits of predictability in human mobility. Science 327(5968), 1018–1021 (2010)
3. Zheng, Y., Xie, X., Ma, W.-Y.: GeoLife: A Collaborative Social Networking Service among User, location and trajectory. IEEE Data Engineering Bulletin 33(2), 32–40 (2010) (Invited paper)
4. Chawla, S., Zheng, Y., Hu, J.: Inferring the root cause in road traffic anomalies. In: IEEE International Conference on Data Mining (ICDM 2012) (2012)
5. Pan, B., Zheng, Y., Wilkie, D., Shahabi, C.: Crowd Sensing of Traffic Anomalies based on Human Mobility and Social Media. ACM SIGSPATIAL GIS 2013 (2013)
6. Blei, D., Ng, A., Jordan, M.: Latent dirichlet allocation. The Journal of Machine Learning Research 3, 993–1022 (2003)
7. Wang, X., McCallum, A.: Topics over time: a non-Markov continuous-time model of topical trends. In: Proceedings of the 12th ACM SIGKDD International Conference on Knowledge Discovery and Data Mining, pp. 424–433. ACM (2006)
8. Häsner, M., Junghans, C., Sengstock, C., et al.: Online Hot Spot Prediction in Road Networks. In: BTW, pp. 187–206 (2011)
9. Heinrich, G.: Parameter estimation for text analysis. Technical report (2005)
10. Mongiovi, M., Bogdanov, P., Ranca, R., et al.: Netspot: Spotting significant anomalous regions on dynamic networks. In: Proceedings of the 13th SIAM International Conference on Data Mining (SDM), Texas-Austin, TX (2013)
11. Peeta, S., Zhang, P.: Counting Device Selection and Reliability: Synthesis Study. Joint Transportation Research Program 332 (2002)

An Evolutional Learning Algorithm
Based on Weighted Likelihood for Image Segmentation[*]

Yu Lin-Sen[1], Liu Yong-Mei[2], Sun Guang-Lu[1], and Li Peng[1]

[1] College of Computer Science and Technology, Harbin University of Science and Technology,
Harbin, China
[2] College of Computer Science and Technology, Harbin Engineering University,
Harbin, China

Abstract. Due to the coupling of model parameters, most spatial mixture models for image segmentation can not directly computed by EM algorithm. The paper proposes an evolutional learning algorithm based on weighted likelihood of mixture models for image segmentation. The proposed algorithm consists of multiple generations of learning algorithm, and each stage of learning algorithm corresponds to an EM algorithm of spatially constraint independent mixture model. The smoothed EM result in spatial domain of each stage is considered as the supervision information to guide the next stage clustering. The spatial constraint information is thus incorporated into the independent mixture model. So the coupling problem of the spatial model parameters can be avoided at a lower computational cost. Experiments using synthetic and real images are presented to show the efficiency of the proposed algorithm.

Keywords: image segmentation, mixture models, spatial constraint, EM algorithm.

1 Introduction

In order to impose spatial constraint on mixture models, different research approaches have been proposed. Most of them utilized Markov Random Field(MRF) as a powerful tool to impose spatial coherence on mixture models. A common method using MRF is to impose spatial smoothness directly on the hidden variable, which indicates the label of image pixels. As Hidden Markov Random Field(HMRF) is not computationally feasible, approximation methods like mean-field[3] are commonly used. The idea of mean-field is based on neglecting the fluctuations of the sites interacting with a considered site, so that the resulting system behaves as one composed of independent variables. Although the computational problem for this method becomes tractable, it requires a large amount of computation.

Another commonly used spatial mixture model based on MRF is Spatially Variant Mixture Model (SVMM)[2]. SVMM imposes spatial constraint on the prior probability. The EM algorithm of SVMM also cannot be obtained in a closed form. The gradient

---

[*]The paper is supported by the National Science Foundation of Heilongjiang province numbered QC2013C060.

H. Wang et al. (Eds.): ICYCSEE 2015, CCIS 503, pp. 212–219, 2015.

projection and quadratic programming algorithm were proposed to solve the constraint optimization problem in the M-step[4][5]. Dirichlet Compound Multinomial-based Spatially Variant Finite Mixture(DCM-SVFMM)[6] assumes that the prior probabilities follow the Dirichlet distribution. The EM algorithm for DCM-SVFMM can acquire close solution; however the computational cost of this method remains quite high.

Different from the above methods, we break down the complex problem into two more easily tractable problems: the first one focus on clustering based on mixture model guided with a certain spatial constraint, and the second problem is how to acquire supervision information from the clustering result by making use of spatial correlation.

The paper is organized as follows. Independent mixture models are introduced briefly in section 2. Based on this model, a weighted spatially constraint mixture model with EM algorithm is given in section 3. And then the proposed evolutional EM algorithm for image segmentation is presented in section 4. Experiment results and conclusion are presented in section 5 and 6 respectively.

2   Independence Mixture Model

Let $X = \{x_1, x_2, \ldots x_N\}$ denote data set of size N, which represents the visual description of an image. We adopt $\pi_i^k = p(k \mid x_i)$ as the prior probability [2]. It represents the probability of the i th pixel which belongs to the k th component model, satisfying the constraints $0 \leq \pi_i^k \leq 1$, $\sum_{k=1}^{K} \pi_i^k = 1$ for $i = 1, 2 \cdots, N$, $k = 1, 2 \cdots, K$. Let $\Pi = \{\pi_1, \pi_2, \cdots, \pi_N\}$ for simplicity.

For a mixture model with K component densities, the probability density function of an observation x_i is expressed by

$$f(x_i \mid \Pi, \Theta) = \sum_{k=1}^{K} \pi_i^k f_k(x_i \mid \theta_k) \tag{1}$$

Where $f_k(x_i \mid \theta_k)$ is a density function parameterized by θ_k, and θ_k is the parameter vector of the k th component density. And we define $\Theta = (\theta_1, \theta_2, \ldots, \theta_K)$ as model parameter vectors. If we do not know an observation x_i is generated by which component density distribution, then Θ is hard to acquire. The problem of incomplete data can be solved by EM algorithm [1]. The observation data x_i for a given pixel depends on the state of the discrete hidden variable z_i^k, $z_i^k = 1$ if x_i originates from the k th component density, else $z_i^k = 0$. Let $Z = \{z_1, z_2, \cdots z_N\}$. If the observation data are independent and identically distributed, now the joint distribution of the complete data $\{X, Z\}$ is given by

$$p(X, Z \mid \Pi, \Theta) = \prod_{i=1}^{N} \left[\pi_i^k f(x_i \mid \theta_k) \right]^{z_i^k} \qquad (2)$$

The log-likelihood function is expressed by

$$L(\Pi, \Theta) = \sum_{i=1}^{N} \sum_{k=1}^{K} z_i^k \left[\log \pi_i^k + \log f(x_i \mid \theta_k) \right] \qquad (3)$$

3   Spatially Constraint Mixture Models and EM Algorithm

3.1   Weighted Likelihood of Mixture Model

The key idea of the proposed approach is that the conditional probability density function for component k of x_i is weighted by a constant w_i^k, which represents the importance to sample x_i in estimating the parameters of component k, where $w_i^k \in [0,1]$. Therefore, the proposed weighted log-likelihood of Gaussian mixture model is:

$$L(\Pi, \Theta) = \prod_{i=1}^{N} \log \left[\pi_k [f_k(x_i \mid \theta_k)]^{w_i^k} \right]^{z_i^k} \qquad (4)$$

In the ideal situation, for example, if we had already known that x_i comes from the k th component model, we should have obtained $w_i^k = 1$ and $w_i^l = 0$ ($l \neq k$). In fact this is also our aim to segment image. If the adjacent pixels have similar weight value, the proposed method can impose the prior knowledge of local spatial correlation of pixels on the mixture model.

3.2   Semi-supervised Learning for the Weighted Likelihood

In the proposed model, we should impose smooth constraint on π_i^k and w_i^k. However it is very hard to obtain close solutions by EM algorithm for such a model. To circumvent the problem, we change unsupervised learning of model parameters into semi-supervised learning. Because π_i^k and w_i^k have same form, we first unified them by using w_i^k. And then we consider w_i^k satisfied the spatial constraint as supervision information. However, acquiring w_i^k is also one of our goals. Because a more accurate w_i^k can be deduced from the semi-supervised clustering result, it should be used to guide to learn the more precise model parameters. The model parameters are thus iteratively updated between acquiring supervision information from clustering result and learning the model parameters guided with the supervision information. The weighted likelihood for the semi-supervised learning is now become:

$$L(\Pi, \Theta) = \prod_{i=1}^{N} \log \left[w_i^k [f_k(x_i \mid \theta_k)]^{w_i^k} \right]^{z_i^k} \qquad (5)$$

The proposed model inherits the advantages of independent mixture models, and meanwhile fuses the spatial correlation of neighbor pixels through the reasonable choice of weights.

3.3 EM Algorithm for Semi-supervised Learning

The E step computes the expectation of the log-likelihood function given the current model parameter estimate $\Theta^{(t)}$:

$$Q(\Theta \mid \Theta^{(t)}) = \sum_{i=1}^{N} \sum_{k=1}^{K} p(k \mid x_i, \Theta^{(t)}) \left[\log w_i^k + w_i^k \log f_k(x_i \mid \theta_k^{(t)})\right] \quad (6)$$

Where, the posterior probability of each pixel $p(k \mid x_i, \Theta^{(t)})$ can be computed:

$$p(k \mid x_i, \Theta^{(t)}) = \frac{[w_i^k f_k(x_i \mid \theta_k^{(t)})]^{w_i^k}}{\sum_{k=1}^{K} [w_i^k f_k(x_i \mid \theta_k^{(t)})]^{w_i^k}} \quad (7)$$

M-step estimates the new parameters using the old ones:

$$\Theta^{(t+1)} = \arg\max_{\Theta} Q(\Theta \mid \Theta^{(t)}) \quad (8)$$

By $\dfrac{\partial Q(\Theta \mid \Theta^{(t)})}{\partial \Theta} = 0$, we can acquire the updated equations of Θ. For Gaussian mixture model, we can get:

$$Q(\Theta \mid \Theta^{(t)}) = \sum_{i=1}^{N} \sum_{k=1}^{K} p(k \mid x_i, \Theta^{(t)}) \, w_i^k$$

$$[(-\frac{1}{2}\log|\Sigma_k|) - \frac{1}{2}(x_i - \mu_k)^T \Sigma_k^{-1}(x_i - \mu_k)] \quad (9)$$

Taking the derivative of $Q(\Theta \mid \Theta^{(t)})$ with respect to μ_k and setting it to zero, we can get:

$$\mu_k^{(t+1)} = \frac{\sum_{i=1}^{N} w_i^k p(k \mid x_i, \Theta^{(t)}) x_i}{\sum_{i=1}^{N} w_i^k p(k \mid x_i, \Theta^{(t)})} \quad (10)$$

Taking the derivative of $Q(\Theta \mid \Theta^{(t)})$ with respect to Σ_k^{-1} and setting it to zero, we can get:

$$\Sigma_k^{(t+1)} = \frac{\sum_{i=1}^{N} w_i^k P(k \mid x_i; \Theta^{(t)})(x_i - \mu_k^{(t+1)}) \cdot (x_i - \mu_k^{(t+1)})^T}{\sum_{i=1}^{N} w_i^k P(k \mid x_i; \Theta^{(t)})} \quad (11)$$

3.4   Acquiring Supervision Information

The evolution begins from a simple clustering algorithm without spatial constraint. The clustering result indicates the origin for each x_i with a certain degree of reliability. Making use of spatial correlation, the initial value of w_i^k can be acquired, which can be utilized to guide the EM clustering with a certain degree of spatial constraint. The initial value of w_i^k is ambiguous at the inception phase of the evolution. Along with the evolution stages, w_i^k becomes unambiguous and approaches a constant.

Because $p(k \mid x_i, \Theta^{(t)})$ in EM algorithm indicates the posterior probability of observation x_i coming from the k th component model. Making use of the prior knowledge that adjacent pixels most likely come from the same component density, we adopt an average filter to smooth the image. The spatial constraint is thus embedded into the prior probability in a reasonable way. Because the information of supervision is deduced from the result of unsupervised learning, the whole algorithm is an unsupervised learning algorithm. We use K-means to initialize model parameters here. The complete algorithm is described in the following section.

4   Evolutional Learning Algorithm

Step 1: Set the stop threshold η for evolutional algorithm; for each stage of EM, set convergence threshold and the number of the maximum iteration times.

Step 2: Use K-means to initialize the model parameters. $ID(i)$ denotes the class label for the i th pixel, and $ID(i) \in \{1, 2, \cdots, K\}$.

Step 3: Initialize $w_i^k = 0, k = 1, 2, \cdots, K$; set $w_i^{ID(i)} = 1$.

Step4: Smooth w_i^k to get $\left[w_i^k\right]^f$, and $\left[w_i^k\right]^f$ is by $w_i^k = \dfrac{\left[w_i^k\right]^f}{\sum_{l=1}^{K}\left[w_i^k\right]^f}$,   Let

$$H_i^{old} = -\sum_{l=1}^{K} w_i^k \log w_i^k.$$

Step 5: Inherit the model parameters from the last stage except w_i^k. And learn model parameters by EM under the supervision of the renewed w_i^k.

1) E-step: compute the posterior probability $p(k \mid x_i, \Theta^{(t)})$ using Eq.(7) for each component k;

2) M-step: update the model parameters according to Eq.(10) and (11) for each component k.

Step 6: After convergence or exceed the maximum iteration times of step 5, compute the entropy of pixels:

$$H_i = -\sum_{l=1}^{K} p(l \mid x_i, \Theta) \log(p(l \mid x_i, \Theta))$$

And acquire supervision information:

1) $p(k \mid x_i, \Theta)$ is smoothed at spatial domain to get $p^f(k \mid x_i, \Theta)$, and then normalize it by $p^f(k \mid x_i, \Theta) = p^f(k \mid x_i, \Theta) \Big/ \sum_{l=1}^{K} p^f(l \mid x_i, \Theta)$;

2) Let $w_i^k = p^f(k \mid x_i, \Theta)$.

Step 7: If $\sum_{i=1}^{N} abs(H_i - H_i^{old})/N < \eta$ then stop, else set $H_i^{old} = H_i$, and go to Step 5 for the next stage evolution.

5   Experiment Results

The proposed algorithm is implemented with MATLAB. We select a gray synthetic image and several color real images to test the efficiency of our algorithm. In the experiments, a 3×3 average filter is adopted to smooth the posterior of image pixels for each stage of EM. The number of the maximum iteration times is set to 10 and the convergence threshold is set to 10^{-4} for each stage of EM algorithm. We compare our method with SVMM[4] and HMRF with Mean field approximation. As for the method of Mean field approximation, we use the software of SpaCEM3 in windows environment. It provides three approximations for HMRF, namely the mean field (MeanF), the simulated field (SimF), and the mode field (Modef) methods. The three algorithms are all initialized by the result of Kmeans.

5.1   Synthetic Image

We first testify our algorithm on a synthetic image. Fig. 1(a) shows the synthetic four class image with gray level of 0,1/3,2/3,and 1. The size of the image is 128×128. We add Gaussian noise to the synthetic image using the MATLAB library function imnoise( ). Fig. 1(b) shows the corrupted images with Gaussian noise of mean 0 and variance 0.01. The noise for each class does not follow Gaussian distribution anymore. The number of cluster is set to 4. The stop threshold of evolution η is set to 0.0001 for our algorithm.

The misclassification ratio (MCR)[3] is adopted here to evaluate the segmentation performance quantitatively. All algorithms are performed on a PC of 2.66 GHz CPU with 2GB RAM. The running time of our algorithm is only 0.66 seconds, significantly less than mean filed approximation method, the shortest running time of them are 34 seconds. The running time of SVMM is longest, it needs 86.7 seconds.

5.2   Real Images

For real world color image, we choose 5 images from the Berkeley image segmentation dataset [7] to testify the proposed algorithm. We resize the original image to 128×192 or 192×128. Fig.2 shows the original image, segmentation results of SimF and our method from top to bottom. We use the original RGB color vector as visual representation and set the number of segments to 7. The stop threshold of evolution η is set to 0.001. The average BDE[10], PRI[8], VOI[9], and GCE[7] for SimF are 15.54, 0.75, 2.67 and 0.34, and that of our method are 14.32, 0.76, 2.47 and 0.31 respectively.

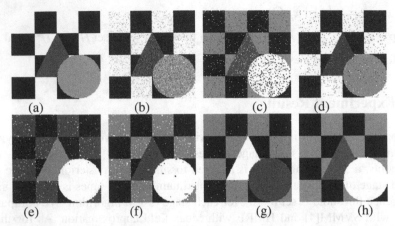

Fig. 1. Segmentation results on synthetic image (a) original image (b) noise corrupted image with Gaussian noise(mean 0, variance 0.01) (c) K-means segmentation (d) SVMM(MCR=3.89%) (e)MeanF(MCR=2.11%) (f) ModeF (MCR=2.93%) (g) SIMF (MCR=0.41%) (h) Proposed method based on the result of (b) (MCR=0.09%)

Fig. 2. Segmentation results of SimF and the proposed method on color images

6   Conclusion

In the paper, an evolutional EM algorithm of the weighted mixture model for image segmentation is proposed. The spatial constraint information is embedded into the weights of the proposed model. As an independent mixture model, the proposed EM can obtain the closed solution, and has advantages such as simplicity and easiness for implementation. Compared with the EM of traditional independent mixture model, the proposed algorithm brings in no extra parameters but the filtering mode and the size of filter window. Although the evolutional algorithm has to execute many EM stages, each stage do not need so many iterations before convergence in fact because of the inheritance of model parameters. The segmentation of our algorithm is comparable to traditional spatial models based on MRF but with less computational burden. Our future work will focus on how to acquire adaptive weights for a more deliberate spatial mixture model.

References

1. Dempster, A.P., Laird, N., Rubin, M.D.B.: Maximum-likelihood from incomplete data via the EM algorithm. Journal of the Royal Statistical Society, Series B (Methodological) 39(1), 1–38 (1977)
2. Sanjay, G.S., Hebert, T.J.: Bayesian pixel classification using spatially variant finite mixtures and the generalized EM algorithm. IEEE Trans. Image Process. 7(7), 1014–1028 (1998)
3. Celeux, G., Forbes, F., Peyrard, N.: EM Procedures Using Mean Field-Like Approximations for Markov Model-Based Image Segmentation. Pattern Recognition 36, 131–144 (2003)
4. Blekas, K., Likas, A., Galatsanos, N.P., Lagaris, I.E.: A spatially constrained mixture model for image segmentation. IEEE Trans. Neural Netw. 16(2), 494–498 (2005)
5. Nikou, C., Galatsanos, N.P., Likas, A.: A class-adaptive spatially variant mixture model for image segmentation. IEEE Trans. Image Process. 16(4), 1121–1130 (2007)
6. Nikou, C., Likas, A., Galatsanos, N.P.: A Bayesian framework for image segmentation with spatially varying mixtures. IEEE Trans.Image Process. 19(9), 2278–2289 (2010)
7. Martin, D., Fowlkes, C., Tal, D., Malik, J.: A database of human segmented natural images and its application to evaluating segmentation algorithms and measuring ecological statistics. In: ICCV, pp. 416–423 (2001)
8. Unnikrishnan, R., Pantofaru, C., Hebert, M.: Toward objective evaluation of image segmentation algorithms. IEEE Trans. Pattern Anal. Mach. Intell. 29(6), 929–944 (2007)
9. Meila, M.: Comparing clusterings: An axiomatic view. In: Proc. Int. Conf. Machine Learning, pp. 577–584 (2005)
10. Freixenet, J., Munoz, X., Raba, D., Marti, J., Cuff, X.: Yet another survey on image segmentation: Region and boundary information integration. In: Proc. Eur. Conf. Computer Vision, pp. 408–422 (2002)

Visualization Analysis for 3D Big Data Modeling

TianChi Zhang, Jing Zhang, JianPei Zhang, HaiWei Pan,
and Kathawach Satianpakiranakorn

College of Computer Science and Technology, Harbin Engineering University, China, 150001
{zhangtianchi,zhangjing}@hrbeu.edu.cn

Abstract. This paper describes an automatic system for 3D big data of face modeling using front and side view images taken by an ordinary digital camera, whose directions are orthogonal. The paper consists of four keys in 3D visualization. Firstly we study the 3D big data of face modeling including feature facial extraction from 2D images. The second part is to represent the technical from Computer Vision, Image Processing and my new method for extract information from images and create 3D model. Thirdly, 3D face modeling based on 2D image software is implemented by C# language, EMGU CV library and XNA framework. Finally, we design experiment, test and record results for measure performance of our method.

Keywords: 3D big data face modeling, Mesh modeling, feature points extraction.

1   Introduction

There is an increasingly rich amount of 3D visual big data available from our daily life. Big Data in 3D computer model research become more and more important. 3D big data modeling is the process of developing a mathematical representation of any three-dimensional surface of an object via specialized software. The product is called a 3D model. It can be displayed as a two-dimensional image through a process called 3D rendering or used in a computer simulation of physical phenomena.

A number of researchers have proposed to create face models from 2D images. Some approaches use two orthogonal views so that the 3D information of facial surface points can be measured [1], [2], [3] They require two cameras which must be carefully set up so that their directions are orthogonal. Zheng(1994)[4] developed a system to construct geometrical object models from image contours. The system requires a turn-table setup. Pighin et al.(2006)[5]developed a system to allow a user to manually specify correspondences across multiple images, and use computer vision techniques to compute 3D reconstructions of specified feature points. A 3D mesh model is then fitted to the reconstructed 3D points[6-8]. Mohamed D(2013)[8] of 3D mesh model with a manually intensive procedure, it was able to generate highly realistic face models.

Because it is difficult to obtain a comprehensive and high quality 3D face database, other approaches have been proposed using the idea of "linear classes of face geometries". Kang and Jones(2002)[9]also use linear spaces of geometrical models to

H. Wang et al. (Eds.): ICYCSEE 2015, CCIS 503, pp. 220–227, 2015.

construct 3D face models from multiple images. But their approach requires manually aligning the generic mesh to one of the images, which is in general a tedious task for an average user. Instead of representing a face as a linear combination of real faces, Liu et al.(2001)[10]represent it as a linear combination of a neutral face and some number of face metrics where a metric is a vector that linearly deforms a face. The metrics in their systems are meaningful face deformations, such as to make the head wider, make the nose bigger, etc. They are defined interactively by artists. Kyu Park(2005) [11] represent the full automate human head modeling by using the 2 view images.

Actually, there are two parameters to evaluate the experiment results, one is the accuracy, the other one is average of time usage. Many methods worked well in average of time usage, especially Mohamed D 3D mesh model, but the average of time usage still need to be raised. Later, the evaluate parameter of the new method will compared with Mohamed D of 3D mesh model.

2   3D Big Data Model

The paper focuses on large scale 3D big data face model from 2D images. Especially, study a methods for design a 3D human face model. The detail is to study the working process with Two-Dimensional Images, Working process with Three-Dimensional Model and do experiment for measure the results from the application.

2.1   Working with Two-Dimensional Images

3D big data visualization still developes going on for looking for the best solution, the problem depend on properties of input and process for solving[12,13]. Firstly, 3D big data face model can be created by manual but it very difficult to determine which features to look like real face. An automatic system for creating facial 3D model has been created for solving this problem. Quality of model depends on the number of vertices, more vertices more slickly. So the feature points extraction step must detect as more as possible.

This part present method for identify key points on both front and side view image as possible. To collect information for creating three-dimensional model. It is divide into 2 subparts consist of frontal and side view feature extraction. The processes with two-dimensional images are automatically calculated to determine the approximate to the user. Because all facial feature points are so many the users can be defined by themselves.

The application has been divided into two main sections consist of working with frontal image and side view image (profile image).

Frontal Image
After user selected 'import frontal image' choice and choose an image. Application will find the face on the image. If it find the face, it will crop only face area and resize the image. Then the image will ready to use for the next step, finding facial feature points. Facial feature points are separated into four groups, show in are primary, secondary, reference form side image and calculated feature points.

The primary feature points are iris, nostrils, mouth corners and middle,they are the most important feature points, they can indicate important areas for looking for the other feature points. These points are the easiest finding and high accuracy. The other feature points are secondary feature points, Ref. from side view points and Calculated feature points. Before application find the primary points. Image must be defined the areas for finding them by feature area locating method. All feature points in front can be shown in Table 1.

Table 1. All feature points can be detected on frontal image

Primary feature points	iris and iris' radius (2points and 1 value)
	nostrils (2 points)
	mouth corners and middle (3 points)
Secondary feature points	eyebrows (6 points)
	eye corners (4 points)
	mouth lips (4 points)
Ref. from side view points	nose bridge, nose tip, nose wings and under nose (5 points)
	under mouth point (1 point)
	chin tip (1 point) and under chin (1 point)
	jaw (2 points)and chin (2 points)
Calculated feature points	Eyelid (6 points)
Summation of all feature points on frontal image	39 points and 1 value

Side View Image

Main purpose of side view feature extraction support the frontal face feature points in deep information and use for reference to find additional feature points. The side view image has feature points less than front view image about half one. Because side view image has less space and feature those can be seen in front view are orthogonal with view of sight, some feature points disappear from side view. So some points must be represented deep rang for two points in front view. All feature points are use sub method in feature extraction on side view image method as shown in Table 2 consist of the feature points can be detected on single image, ref. from front view, calculated feature points and all features.

The two image of front and side view can be projection by chin area. Chin is area between mouth and neck. Positions of shin can be found by draw a vertical line from mouth corner in side view image down ward until as found with chin line. locating the chin points on X-axis with mouth corners on front view and Y-axis with Y value of point on the chin line was found. As shown in Figure 1 and the side view image must be adapted to the frontal image before.

Table 2. All feature points can be detected on side view image

Feature points can be detected on single image	Nose tip point (1 point)
	Eye tail, Mouth corner (2 points)
	Nose bridge, Under nose, Under chin (3 points)
	Eyebrow (3 points)
	Upper mouth lip, Middle mouth lip, Lower mouth lip, Under mouth and Chin tip(5 points)
Ref. from front view points	Upper and Lower eyelid (2 points)
	Nose wing (1 point)
	Jaw and Chin (2 points)
Calculated feature points	Inner eye tail (1 point)
Summation of all feature points on side image	20 points

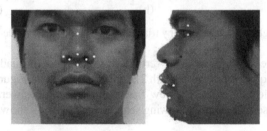

Fig. 1. Show chin detection on front view image projection from side view image

2.2   Working with Three-Dimensional Model

This part creates the three-dimensional model from the information that obtained from two-dimensional images part after finished correcting the feature points. Next, create 3D face model. Application will sent all information from image part to 3D model part. Vertices on 3D space will be defined from a pair of variable, one from front view and the other one from side view. X and Y position will be defined following variable from front view, and Z will be defined following X-value of variable from side view image.

Define Mesh Model

Mesh model is the base of 3D model. It is composed by triangles, call 'polygon'. If it has more feature points, the model will smoother. To draw the triangle, it need to define some points, or vertices, to represent each of the corners of the triangle. Each vertices contain 3 value are positions on 3D space (X, Y and Z). Generally 3D graphic frameworks almost have 3 types of triangle called 'Primitive Type'

Draw Quad with Triangle Function: The Improve Method

Quad descibes the image details better than triangles. In XNA, there are not 'Quad function' same as triangle. The improved method is to use 'Primitive Type' to create Quad. That means a quad is composited by many triangles. By Quad we mean a sequence of points: $T=\{\tau_1<\tau_2<\cdots<\tau_n\}$, where τ_1 is the initial point and is τ_n is the final point, if vertices positions sequence same as $\tau_1>\tau_4>\tau_2>\tau_3$, quad will be incompleted, Figure 2(a) shows two typical triangles. Quad can be created with 2 new Quad functions by use triangle strips or triangle fans. Quad made by triangle strips is shown in Figure 2(b), Quad made by triangle fan is shown in Figure 2(c).

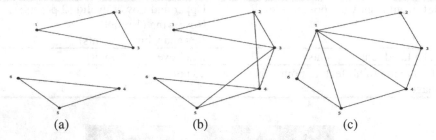

(a)   (b)   (c)

Fig. 2. Two triangles(a), Quad made by triangle strip(b) and Quad made by triangle fan(c)

Quad by triangle strip: create the 4 vertices at the corners of quad. Use triangle strip draw 2 triangles, a quad is composed with 2 triangles, as show in Figure3 (b)

Quad by triangle fan, create the new first point by use the average position of all points. After that use triangle fan following sequence number, as shown in Figure 3(a).

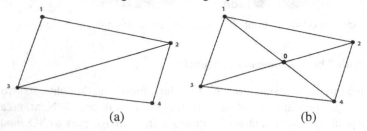

(a)   (b)

Fig. 3. Quads are created by triangle fan (a) and triangle strip (b)

The quad from triangle fan is smoother than another one because each point is not far from other points.From feature extraction part, use 39 points on frontal image and 20 points on side view image to draw the relation between all points on both of the images, the line must not cross or overlap on the other line as shown in Figure 4.

Texturing

A texture coordinate is represented by a two-dimensional (U, V) coordinate, where U is horizontal and V is vertical. The top-left corner of an image is represented by texture coordinate (0, 0), and the bottom-right corner of an image is represented by texture coordinate (1, 1), regardless of the size of the image. To specify a point in the exact middle of a texture, the texture coordinate is (0.5, 0.5). 3D face model after texturing is shown in Figure5.

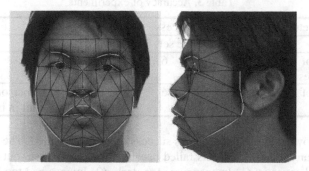

Fig. 4. Show mesh model on image define by feature points

Fig. 5. 3D face model

3   The Experiments

The application is built on Windows 7 Professional by C# language on Microsoft Visual C# Express with EMGU CV, DirectX9 and XNA framework. The independent variable for test image must be Frontal and right views of human face with
white background, clean face(without or least mustache, beard, glasses or unkempt
hair, Normal deportment (no smile, laugh etc.).

There are two parameters to avalue the experiment results,one parameters is the
accuracy, means after the face points are selected judging the seccess rate. The other
parameters to avalue the experiment is the average of time usage, that is, to judge the
using time when select the face points.

This experiment of accuracy use face database from Professor Francesc Tarrés[14],
this database contain facial image of 22 persons in any action, select only straight
face on front and right view facial image, every person have 2 sets of images difference in light source, total 44 sets of images for test with my method.

In group of test images, some test images was out of scope of input. But this experiment allow to use for observe the flexibility of method. The results were in the following Table3.

Table 3. Accuracy of experiments

All sets		success			fail
	All	one set	both sets		
amount of sets of images	44	28	6	22	16
amount of person	22	17	6	11	11 persons (5 persons fail in both sets)

Table 3 shows that there are 28 sets success and 16 sets failed. The average deviation is less than 10. Many test sets failed because the images were out of scope, summary of failed cause is: (1) Images were too dark. (2) Images had too low resolution. (3) Face in images was obscured by hair. (4) Trouble in posture of face.

The purpose of this application is high automatically and it must be able to be valued by feature points average of time usage. Mohamed D has made 3D face modeling and he show the feature extraction from both front and profile images. Many face features were extracted as eye, mouth, nose, shin, and cheek contour and feature deep information from profile image. Mohamed D's method is good and his average of time usage is fast in recent year.

Table 4. Average of time usage competition with Mohamed D method

Step	Time usage (ms) of ours	Time usage (ms) of Mohamed D
First step on frontal image	79.90	180.00
Second step on frontal image	39.00	270.02
First step on side view image	180.94	550.68
After calibrate	1536.45	7360.67
All step	1836.29	8361.35

In order to prove that the methods is accurate and high speed, the improved method uses 44 sets of images for average time usage test competing with Mohamed D's method. Performance on time of the improved method were recorded in Table 4, the total average time for all step is included. Mohamed D's method was 8.36 seconds only the improved method was 1.83 seconds. The experiment results shows that the two parameters of accuracy and average time usage are all better that others.

4   Conclusion

This paper firstly can detect feature points from front and side view of face image. The full feature extraction on images are used for detect the feature and create model. Secondly, each face can be constructed into 3D big data by the step of define mesh model, the improve quad method and texturing. Finally, the experiments results are satisfactory. The average deviation is less than 10%, 3D face model can be create in 1.83 second without corrected by hand.

Acknowledgements. The paper is partly supported by: 1. The Fund of PHD Supervisor from China Institute Committee (20132304110018). 2. The Natural Fund of Hei Longjiang Province (F201246). 3. The National Natural Science Foundation of China under Grant (61272184).

References

1. Akimoto, T., Suenaga, Y., Wallace, R.S.: Automatic creation of 3D facial models. IEEE Computer Graphics and Applications 13, 16–22 (1993)
2. Dariush, B., Kang, S.B., Waters, K.: Spatiotemporal analysis of face profiles: Detection, segmentation, and registration in Automatic Face and Gesture Recognition. In: Third IEEE International Conference on Digital Object Identifier, pp. 248–253. IEEE (1998)
3. Ip, H.S., Yin, L.: Constructing a 3D individualized head model from two orthogonal views. The Visual Computer 12(5), 254–266 (1996)
4. Zheng, J.Y.: Acquiring 3-D models from sequences of contours. IEEE Transactions on Pattern Analysis and Machine Intelligence 16(2), 163–178 (1994)
5. Pighin, F., Hecker, J., Lischinski, D., Szeliski, R., Salesin, D.H.: Synthesizing realistic facial expressions from photographs. In: ACM SIGGRAPH, pp. 19–21. ACM (2006)
6. Hu, Y., Yan, J., Li, W., Shi, P.: 3D Face Landmarking Method under Pose and Expression Variations. IEICE Transactions on Information and Systems 94(3), 729–733 (2011)
7. Hwang, J., Yu, S., Kim, J., Lee, S.: 3D Face Modeling Using the Multi-Deformable Method. Sensors 12, 12–16 (2012)
8. Mohamed, D., Anuj, S., Remco, C.: 3D Face Modeling, Analysis and Recognition, pp. 11–16. Wiley-Blackwell, New Jersey (2013)
9. Kang, S.B., Jones, M.: Appearance-based structure from motion using linear classes of 3-d models 49(1), 5–22 (2002)
10. Liu, Z., Zhang, Z., Jacobs, C., Cohen, M.: Rapid modeling of animated faces from video. The Journal of Visualization and Computer Animation 12(4), 5–8 (2001)
11. Park, K., Zhang, H., Vezhnevets, V.: Image-Based 3D FaceModeling System. EURASIP Journal on Applied Signal Processing, 2072–2090 (2005)
12. Lampitt, A.: Fighting cancer with 3D big data visualization. InfoWorld, 1–5 (2012)
13. Philip, J.: Miner3D:Beatifully Visualizing the big data. Cioreview, 8–12 (2013)
14. Tarrés., F., Rama, A.: Audio Visual Technologies Group,
   http://gps.tsc.upc.es/GTAV/ResearchAreas/UPCFaceDatabase

MBITP: A Map Based Indoor Target Prediction in Smartphone[*]

Bowen Xu and Jinbao Li[**]

School of Computer Science and Technology,
Heilongjiang University, 150080, Harbin, P.R. China
608750@qq.com, jbli@hlju.edu.cn

Abstract. This paper presents MBITP, a novel method for an indoor target prediction through the sensor data which may be the Big Data. To predict target, a probability model is presented. In addition, a real-time error correction technique based on map feature is designed to enhance the estimation accuracy. Based on it, we propose an effective prediction algorithm. The practice evaluation shows that the method introduced in this paper has an acceptable performance in real-time target prediction.

Keywords: target prediction, smartphone, map based, indoor location, probability model.

1 Introduction

With the increasing number of the members in mobile internet, sensor data comes into people vision. More and more sensor data become the Big Data, how to use it to improve our lives is a valuable challenge. Various kinds of services have been emerged. For example, Wang et al. proposed a method which offered useful information by providing context-aware and browsing behavior [1]. Chen et al. proposed a method which analyzed user behavior patterns and predict activity on a particular pattern of behavior in order to give user some recommendations [2]. Nevertheless, these are all for online data analysis and services for outdoor environments. When people go to a place, if they can acquire the information and services before them arriving there, it will enhance the time utilization radio and life satisfaction. We can use GPS module locating to realize it in the outdoors, because of no signal we cannot do it at indoor environment.

Previously, some indoor service applications are used direct or indirect indoor positioning technology. They typically use infrared ray or signal detector to detect relative

---

[*] This work is supported in part by the National Natural Science Foundation of China (NSFC) under Grant No.61370222 and No.61070193, Heilongjiang Province Founds for Distinguished Young Scientists under Grant No.JC201104, Technology Innovation of Heilongjiang Educational Committee under grant No.2013TD012, Program for Group of Science Harbin technological innovation found under grant No.2011RFXXG014.

[**] Corresponding author.

H. Wang et al. (Eds.): ICYCSEE 2015, CCIS 503, pp. 228–236, 2015.

position. Although it is a simple way to use, it costs much manpower to decorate environment. It is a good plan to use in parking lot to detect the car or other bulky goods, but it is not precise to detect the human which has complex behavior and small size.

In order to solve the above challenges, we propose a novel method to locate position and predict target place. This method mainly uses the sensor data locating the smartphone users. It traces the path the user moves, combined with the characteristics of map. Users will get the target prediction after calculation. Service providers could offer unique services based on it.

The main contribution of this paper is as follows. We have established an effective probability model to predict target. Based on it, we propose an effective prediction algorithm. Evaluation in practice shows that the method introduced in this paper has an acceptable performance.

The rest of this paper is organized as follows. We review related works in Section 2, and report the system overview in Section 3. Section 4 overviews Preliminary knowledge, followed by detailed description of indoor prediction algorithm and probability model in Section 5. We conduct evaluation in Section 6; conclude our paper in Section 7.

2   Related Work

In recent years, with the popularity of smart phones, the predicted position based on smart phones has become a hot topic. Zhu X et al. [3], closest to our work, explored how to formulate a method of outdoor location and path tracing without GPS module. Chung J et al. [4] proposed a novel method that utilize different place has different magnetic interference to locate position. Javier J et al. [5] proposed a location technique based on signal propagation model. This method has a higher positioning accuracy in theory, but at indoor environment, it has a low performance because it is hard to find fit signal propagation model in complex environment. In [6], Luo X et al. proposed a novel indoor location system which is based on WIFI RSSI signal. It showed high precision in evaluation, but it costs too much manpower to decorate environment and had low precision in complex scene.

The above mentioned methods need much more preparatory work and the result we get is insufficient to meet the basic needs of location-based services in indoor environments.

3   System Overview

In this section, we briefly introduce our system. Our system can be organized into the framework in Figure 1. The system consists of three components, step detection component, direction detection component and predicted probability model. It consists of four parts of information: accelerometer, gyroscope, map and GPS. After calculating, it outputs where the user walking to.

When the user walks, both the accelerometer and the gyroscope of the phone keep sampling and sending the samples to the step detection algorithm and direction detection algorithm. These algorithms process data in real time. On one hand, it monitors the acceleration data to detect a step and reports new steps to the predicted probability model. On the other hand, it monitors gyroscope data for possible turns. If a turn is detected, then the direction detection algorithm uses the acceleration data to reorient the phone's axes and compute the angular displacement of the turn. It then reports the angular displacement to the predicted probability model. An example of the reported angular displacement is "turn, 90, right". Sometimes errors locate user out of the border, the system would automatically correct user coordinate and change it to the right place. We can provide personalized services to users based on forecast results, but this is not discussed in this article.

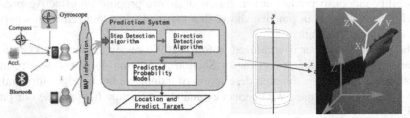

Fig. 1. System Overview   **Fig. 2.** Two coordinate system

4   Preliminaries

In recent work, due to complex human movement, estimating travel distance by taking the double integral of acceleration is not precise. Instead, we use the method which is to count the number of walking steps and then multiply it by the step length. It can be explained by the following formula

$$d = n_{step} \times l_{step} \tag{1}$$

Where d is the movement distance, *nstep* is the number of walking steps and *lstep* is the step length. When walking in a period, acceleration sensor shows a certain law. We adopt the solution in Ref. [7] to process sensor data. It can be calculated as

$$Acc = \sqrt{x_{acc}^2 + y_{acc}^2 + z_{acc}^2} \tag{2}$$

People complete walking with a period consists of two parts, lifting the foot and putting down the foot. Due to the different height and walking custom, these two parts generate a waveform, the sensor values, are not same. For each person, the same threshold value couldn't be found to judge them. In this paper, we take the average value of *Acc* and utilize a correlation coefficient to solve this challenge. It can be showed as

$$\varphi(acc_i) = \begin{cases} 1 & if \ acc_i > u_i + v_i \\ 0 & if \ acc_i < u_i - v_i \\ 2 & else \end{cases} \tag{3}$$

Where acc_i is the acceleration magnitude at time i. Where u_i is $\frac{\sum_{t=1}^{i-1} acc_t}{i-1}$, the average of the series and v_i is the corresponding standard deviation. The two dynamic thresholds $u_i + v_i$ and $u_i - v_i$ mean "up" and "down" respectively. In order to calculate travelling distance, we need to acquire stride length of each person. Different people have different stride length, but in Ref. [8] we can roughly estimate their stride length based on their height. It is shown in the following table.

Table 1. The Ratio between Stride Length and Height

Subjects		The Ratio between Stride Length and Height (%)
10-30 Age	Male	42.36
Group	Female	43.56
40-60 Age	Male	41.17
Group	Female	40.55

There are two coordinates systems, the mobile phone coordinate system and the geographic coordinate system. They are shown in Fig. 2. In fact, the two references are different in most cases. In calculating orientation, we can obtain xyz angular velocity data, but we need Z-axis data. Therefore we take the mobile phone coordinate system transiting to the geographic coordinate system. This problem is also appeared in the recent research [9], where the solution is to find the rotation matrix between the two references by acceleration measurement. It is not fit for pedestrian. It is not able to gain an acceptable rotation matrix because it is changing during walking. The body shape and mobile phone orientation always change during walking, so some noise is recorded and we cannot exclude the noise. Therefore we cannot get a precise rotation matrix.

The angle which is pedestrian turns is same to the Z-axis changes in gyroscope. Our solution is based on two observations. Firstly, when a pedestrian is turning a corner, the value at each axis follows a recurring pattern. Second, when a pedestrian walks in a straight line, the average value in any axis does not fluctuate much. Z-axis direction and the direction of gravity are in a vertical plane. The acceleration sensor value is mainly produced by the gravity. We can use this value to determine the Z-axis direction components of the xyz-axis. No matter how to change the direction of the mobile phone, we also can get the angel changed in Z-axis. Then, the adjusted angular displacement around Z-axis for an incoming turn is calculated as following formula.

$$Z_{displacement} = \frac{d_x \cdot \overline{\beta_x} + d_y \cdot \overline{\beta_y} + d_z \cdot \overline{\beta_z}}{\sqrt{\overline{\beta_x}^2 + \overline{\beta_y}^2 + \overline{\beta_z}^2}} \tag{3}$$

Where d_x, d_y, d_z are the angular displacement of three axis and $\overline{\beta_x}, \overline{\beta_y}, \overline{\beta_z}$ are the average acceleration of three axis. Then we can get the value of the instantaneous acceleration as following formula.

$$Z_{acc} = \frac{\alpha_x \cdot \beta_x + \alpha_y \cdot \beta_y + \alpha_z \cdot \beta_z}{\sqrt{\beta_x^2 + \beta_y^2 + \beta_z^2}} \tag{4}$$

Where $\alpha_x, \alpha_y, \alpha_z$ is the value of the instantaneous angular acceleration. Where $\beta_x, \beta_y, \beta_z$ is the value of the instantaneous linear acceleration.

5   Map Based Indoor Target Prediction Algorithm

To achieve predicting target, we need to analysis these data. So far, some related researches are not focused on human behavior prediction. There are many regularities concealing in human behavior and we can make use of these to infer the human purpose.

In this paper, it proposes a probability model to calculate the probability of target place. We start from the human behavior. When a person is walking to somewhere, he must go with a certain purpose, and he will move toward the target direction on the whole. So, the probability of the place becoming target is in direct proportion to the decreasing distance between person and the place. We can get formula like this. $t_A^n = d_A^n - d_A^{n-1}$. Where d_A^n is the distance t_A^n is the decreased distance and between user and place A and at time n. If the user is thinking something or he is not with a certain purpose, he will walk around a circle or go back forth. In these situations, the value t is not significantly increased. However we should not record and cumulate the value t. So the value t will be cumulated when the corresponding place is in user's vision. In other situation the value t will be deleted. As shown in Fig. 3(a), the place A, B, C are in user's vision and they are belong to a set of viable target points. The place F, G, H, I are not in user's vision, so these value t will be deleted.

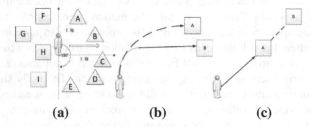

(a)   (b)   (c)

Fig. 3. Target points

Suppose a situation that the user first goes to place A, and then he changed the goal to go to place B.it is shown in Fig. 3(b). If it only rely（relies）on value t, the result will not be changed in a long time. So we introduce another parameter to solve this problem. No matter the user how to walk we can get the line which is between user's position and place A. The user always walks toward their target direction, so we can get the angle θ_0 between forward direction and baseline and the angle θ_A between the line and baseline. We can get other angle about other places. If the θ_0 is similar with the θ_A, place A will be a target place of high probability.

Consider a special situation that place A, place B, and the user are in a line. It is shown in Fig. 3(c). When user walks to them in line, how to distinct which place the user wants to go? From the analysis of the experience, people always select the place which is nearest to them. We assume that place A is the target place in this situation. Therefore, we should introduce a parameter d to make the probability in place A and B different. Where d is the distance between the place and the user's current position. Meanwhile we hope this parameter would not make excessive effect. It is proposed a formula to calculate weight as follow.

$$S_i = \frac{t_i \cdot \cos|\theta_p - \theta_i|}{k \log_{10} d_i} \tag{5}$$

Where t_i is the cumulate sum of the reduce distance between place I and user at different time. The formula is like this: $\sum_{j=0}^n t_I^j$ Where θ_0 is between forward direction and baseline and θ_A is between the line and baseline. Where d is the distance between the place and the user's current position. Therefore we could get the probability of each place becoming target, it is as follow formula.

$$P_i = \frac{s_i}{\sum_{j=1}^{n} s_j} \tag{6}$$

In the end, we formally introduce the proposed algorithm 1.

Algorithm 1: Target Prediction algorithm

input: initial coordinate x, y, Map M, stride length δ ,place P

output: the probability in each place $\Pr\{Pr_1, Pr_2, Pr_3, \ldots, Pr_n\}$

1   start

2    $l_{step} = \delta$; $\theta = 0$;

3   If(sensor is changed)

4   Then $\theta = \theta + \theta'$

5   If(new step)

6   Then step = step + 1; $x = x + l_{step} \cdot \cos\theta$; $y = y + l_{step} \cdot \sin\theta$;

7   If($\{x, y\}$ is out of M)

8   Then x' = the nearest x in the M; y' = the nearest y in the M;

9   For i from 1 to n Do

10        $p_{i_k} = \frac{p_i.y - y}{p_i.x - x}$; $p_{i_d} = \sqrt{(p_i.y - y)^2 + (p_i.x - x)^2}$; $\theta_{p_i} = \arctan(p_{i_k})$;

11   If($|\theta_{p_i} - \theta| < 90$)

12   Then $p_{i_t} = p_{i_t} + |p_{i_d}' - p_{i_d}|$; $S_{p_i} = \frac{p_{i_t} \cdot \cos|\theta_{p_i} - \theta|}{k \log_{10} p_{i_d}}$;

13   Else $p_{i_t} = 0$; $S_{p_i} = 0$;

14   End do

15   For *i* from 1 to n Do

16   Sum = Sum + S_{p_i};

17   End do

18   For *i* from 1 to n Do

19        $Pr_i = \frac{S_{p_i}}{Sum}$

20   End do

21   return Pr;

22 End

---

6   Evaluation

We perform an experiment to assess the accuracy of predictions. Firstly, we introduce the experiment environment. Next we give some brief analyses which are according to the experimental results. Then we analyze some typical situations. In the end, we do some explanations in error results. The mainly equipment is MI 2S mobile phone. It consists of Snapdragon APQ8064 Pro 1741MHz CPU, 2GB RAM and some sensors. In the experiment, the phone can be kept on hand at any angle or fixed in your pocket. In this paper, the experimental scenario is in a corridor platform which is about 120 square meters (15m multiply 8m). it is shown in Fig. 4. We set six interesting places which can attract user. This experiment is mainly done by six volunteers. Let them walk to final target in particular path and in free path. Then our system collects data in real time and gives the probability of each place. We examined whether it is correct by comparing to the actual destination. In order to facilitate the experiment, the initial location of all volunteers is place C.

We give brief analyses as follows. In Fig. 5(a), we let user walk in straight line from C to D. We can get high probability predicting place D, because the walking path is too simple. Then we let user walk in a circle from C to C in Fig. 5(b). Even the curvilinear motion, by measuring the angle of the gyroscope we can accurately obtain the walking path. After finishing three-quarters of a circle, the user move to place C. The probability of place C is 58% and continue rising before arrive at place C. It is shown that we are able to successfully predict the target place before reaching it in a certain distance. Then we do some complex paths test. We let user walk without purpose in a period of time, then let him start to walk to a certain place. We also can get prediction with high probability. In Fig. 6(a) the probability of place A is 1, because there is no other place in user's vision. The system predicated user would not go to other place. In Fig. 6(b), no matter how many laps the user walks, we also get the precise prediction. Next we make a robust test. This test may not accord to human behavior habit, but we can test if this system is stable. We also can get precise prediction, no matter the user walks a lot of laps in Fig. 6(c) or the user walks like snake crawling in Fig. 6(d).

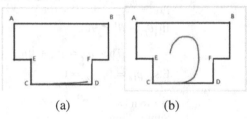

(a)   (b)

Fig. 4. the experimental scene   **Fig. 5.** The result of target prediction

In Fig. 11, the vertical ordinate indicates the prediction probability and the vertical horizontal indicates two meanings. The values on vertical horizontal indicate detected points in the process and the unit on vertical horizontal indicates a user step. We let user walk from C to F and we analyze the entire process. The probability of the place B is gradually increasing when the user is walking through 5 detected points. In the entire process, the user has no tendency to go to the place A, B, C. Therefore the probability of these is zero. The probability of the place B and E is high at first, and then they reduce during the process. Before completion of half the whole distance, we can clearly predict the user to go to F.

As shown in Fig. 7(a). The user walks through a circle from C to F. At point Φ, user is in ready state and he faces to D, so the probability of the place D is high at this time. Next the user walks through a circle, he has a tendency to go B, E, D. Therefore, the probability of these places are respectively raised at first, and then decreased. Then the user starts to move to A and the probability of place A becomes rising. In the end, the result of the highest probability is A. another situation is shown in Fig. 7(b). We let user walk like a snake from C to B. When the user arrives at point 1, the system find that the user maybe go to B, so the probability of B is obviously rising. Then the user arrive at point 2, D is not in user's vision, so the probability of B reduce to zero. The tendency of going to E is obvious. The probability of point E starts to rise. When the user is between point 3 and 4, the tendency of going to B appears again. In the end, when the user at point 5, the probability of B is 84 percent. If the user always changes target, we will not be able to predict precisely. However, when user makes sure his final target, we can give precise prediction as soon as possible.

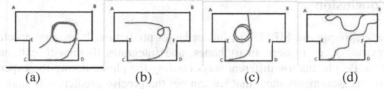

(a)   (b)   (c)   (d)

Fig. 6. The result of target prediction

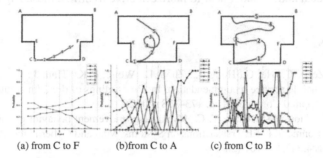

(a) from C to F   (b)from C to A   (c) from C to B

Fig. 7. The probability of different place

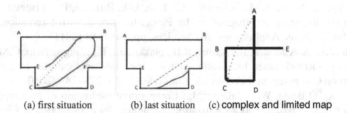

(a) first situation   (b) last situation   (c) complex and limited map

Fig. 8. Error situation

In Fig 7(c), there are two situations which may cause error prediction. The first situation is that the user walks from C to B and then returns to C, but when the user walks back to C the place E is the highest probability. It is explained that the phone is not fixed, and the user turning range is too large. It produces some gyroscope noise, therefore the system cannot gain precise turning angle. To solve it, we can firmly fix a phone in the user body and let user slightly turns. The last situation is that the user walk from C to B, but the result which the system predicts is F. it is explained that at the beginning, there is no data to provide optimizing threshold, and the body shaking makes wrong "step" is recorded. We can collect some data to optimize threshold to avoid this situation happens.

In Fig. 8(c), the user walks only on the black line. If he wants to walk from A to C, he will go through point B or D. Before arriving at B or D, the method which we propose cannot predict C. In the future, we will collect some user data, analyze them and find some measures to overcome this challenge.

7   Conclusion

In this paper, we present MBITP, an indoor target prediction. It uses the accelerometer and gyroscope of modern smartphones, and integrates them with external map information. People can use different ways holding the phone during the experiment. Real-world measurements show that we can get the precise prediction in high probability. We can get the precise prediction in high probability. In the future, we will collect more data and consider how to more effective to utilize these Big Data.

References

1. Wang, J., Zeng, C., He, C., Hong, L., Zhou, L., Wong, R.K., Tian, J.: Context-aware role mining for mobile service recommendation. In: Proceedings of the 27th Annual ACM Symposium on Applied Computing, pp. 173–178 (2012)
2. Chen, T.-S., Chou, Y.-S., Chen, T.-C.: Mining user movement behavior patterns in a mobile service environment. IEEE Transactions on Systems, Man and Cybernetics, Part A: Systems and Humans 42(1), 87–101 (2012)
3. Zhu, X., Li, Q., Chen, G.: APT: Accurate outdoor pedestrian tracking with smartphones. In: 2013 Proceedings IEEE INFOCOM, pp. 2508–2516 (2013)
4. Chung, J., Donahoe, M., Schmandt, C., Kim, I.-J., Razavai, P., Wiseman, M.: Indoor location sensing using geo-magnetism. In: Proceedings of the 9th International Conference on Mobile Systems, Applications, and Services, pp. 141–154 (2011)
5. Diaz, J.J., de A Maues, R., Soares, R.B., Nakamura, E.F., Figueiredo, C.M.S.: Bluepass: an indoor bluetooth-based localization system for mobile applications. In: 2010 IEEE Symposium on Computers and Communications (ISCC), pp. 778–783 (2010)
6. Luo, X., O'Brien, W.J., Julien, C.L.: Comparative evaluation of Received Signal-Strength Index (RSSI) based indoor localization techniques for construction jobsites. Advanced Engineering Informatics 25(2), 355–363 (2011)
7. Jang, H.-J., Kim, J., Hwang, D.-H.: Robust step detection method for pedestrian navigation systems. Electronics Letters 43(14), 749–751 (2007)
8. Sung, B., Yoon, J.: Analysis of stride length and the ratio between height and stride length in 10–60 aged men and women. Korean J. Walk Sci. 8(63-70), 10–60 (2008)
9. Kratz, S., Rohs, M., Essl, G.: Combining acceleration and gyroscope data for motion gesture recognition using classifiers with dimensionality constraints. In: Proceedings of the 2013 International Conference on Intelligent User Interfaces, pp. 173–178 (2013)

A Method of Automatically Generating 2D Animation Intermediate Frames

Zhaowen Qiu[1,*], Haiyan Chen[1], Tingting Zhang[2], and Yan Gao[3]

[1] Information and Computer Engineering, Northeast Forestry University, China
qiuzw@nefu.edu.cn
[2] Heilongjiang Institute of Commerce, China
tuanweizhan@163.com
[3] Software Engineering Institute, East China Normal University, China
ygao@sei.ecnu.edu.cn

Abstract. This paper proposes the automatic generation of the middle frame and the middle frame automatic coloring method of two-dimensional animation process, users simply given starting key frames and end key frames, According to the algorithm proposed in this paper, the system can automatically generate all key frames that in the middle, and based on the starting key frame and termination of key frame color, the generated in the middle of the frame will been automatically chromatically. The experimental results show that, the automatic generation of intermediate frames and the middle frame automatic coloring method of two-dimensional animation is proposed in this paper production process can be successfully used in animation production, greatly improving the efficiency of animation.

Keywords: Intermediate Frames, Automatically Generating, Automatic Color, 2D Animation.

1 Introduction

As early as 1964, American scientist Ken Knowlton [1] produced the first computer animated film, also written in FORTRAN language two-dimensional animation system called BEFLIX, from then on, began a two-dimensional computer animation. 1971, Nestor Burtnyk and Marceh Wein who known as the father of computer animation, that proposed a "computer-generated key frame animation" technique, application of the technology developed MSGEN dimensional animation system[2]. The key frame method is the main method movement of these early computer-generated animation system. 1974, computer animation in the film industry made a breakthrough, Hungar Reter Folders in the Cannes Film Festival with his film "La Fain" won the Prixdn Jnry Award. The main technology used in the film that is a key frame, since the middle of paintings produced by the computer, thus eliminating the need for a large number of mechanical hand-painted production intermediate labor[3]. 1990, the

---

*Corresponding author.

H. Wang et al. (Eds.): ICYCSEE 2015, CCIS 503, pp. 237–241, 2015.

Eleventh Asian Games, television programs, the use of a computer animated head. This is determined by Beijing Science and Education Film Studio and North Jiao tong University co-produced, recorded a two-dimensional animated film on the film marks the first application of computer animation in the film-making [4]. 1995, created by Walt Disney's first fully computer-animated film production, "Toy Story". In computer animation technology development process, we see that computer animation is from the simple to the complex a gradual process of development, and the development of computer animation and cartoons, and the creation of film and television production are closely linked[5]. Since the draw frame interpolation process is very arduous lengthy, it is necessary to draw the outline of the middle of the screen, but also its color, on this basis, this paper proposes a new method to automatically generate an intermediate frame and batch colored, can greatly save manpower, has high practical value.

2   Middle Frame Automatically Generated Technology

The technical editor laplasse on line algorithm based on key graph edit, and according to the feature point matching results, the contour vector (Beizer curve) interpolation, and then obtain the intermediate frames. Specific steps are as follows:

Step one: the foreground and background key frames in different layers in the drawing, the initial key frame and ending key frame editor as an intermediate frame basis, the key can be used to edit the lines of gradient-based deformation Laplace domain editing algorithm [6], This embodiment of the Laplace editing algorithms to establish deformation of each vertex in the adjacent points of the coordinate system:

$$V_i - \frac{1}{d_i} \sum_{j \in Ni} V_j \qquad (1)$$

Where V_i, V_j is the vertex, d_i is the right weight vertex i, N_i is the neighborhood of the vertex i. Drawing process can draw no more than one role in each layer in order to simplify the complexity of the matching algorithm [7].

Step 2: The line image in each layer is enclosed area is closed; This implementation for each frame image, filling technique using seed, closed range search independent, leaf nodes form a tree, for independent single line, linking the first endpoint, forms a closed region; In the leaf nodes of the same layer, if they are in the same closed region, is the formation of upper nodes, use the same method for the other frames, until the formation of a forest; For each node of each frame in the forest, calculate the corresponding region cent of mass and area ;To find the correspondence between the two frames from the root to the leaf node, compared with the cost relation as follows. $w_1 \times area$ of overlap $/ A + w_2 \times area$ of the overlapping $area / B + area$ of the cent of mass distance $w_3 \times AB$, where w_1, w_2 and w_3 are weight coefficients ,the addition of 1.The operation can establish a correspondence between nodes in the same layer for each region of each of its forests and its corresponding frame.

Step 3: By using topological graph matching relationship to establish the initial key frame and end key frames between regions, matching topology map at the same time consider nested relation constraint hierarchy graph and global region shape and local deformation, but nested relation constraint hierarchy graph is not allowed to change the matching region contains relations and hierarchical relationship;

Step 4: The relationship between the matching contour matching region vector;

Step 5: In the closed area of different layer, feature point extraction contour vector, and feature point matching, feature points including the large curvature extreme points and curves, key frame interpolation algorithm of Bezier curve in the parameter domain interpolation;

Step Six: According to the results of the matching feature points of vectored contour lines (Beizer curve) obtained by interpolating the intermediate frames.

3   Automatic Batch Coloring Method

The method to solve the case of two-dimensional animation software currently in use can not be in the lens characters or scene structure is more complex automated batch coloring problems and provide a two-dimensional animation process automatic batch coloring method [8,9].

Step 1: Each of the initial key frames and key frames are drawn to end in a different layer;

Step 2: The image of each layer of the line will been closed sealed area;

Step 3: Matching system to establish the initial key frame and end key frames between regions through the topology map matching;

Step 4: The initial key frames and ending key frames matching relationship with a combination of color information to establish the relationship between the range of color matching;

Step 5: The same color information provided in the region corresponding to each intermediate frame in matching relationship.

This method will be drawn after the closing lines of the enclosed area, calculate the matching relationship between each layer of each region, according to the matching color information to establish the relationship between the combined matching relationship between regions, different areas of the color information is passed to the corresponding region of the middle frame, so that animators color liberated from the heavy session, and thus have more energy into more creative creations.

4   Experiments

In this paper, the process of peaches from small to large, for example, in a .Net environment, automatically generate intermediate frames through programming and batch coloring techniques [10]. First, draw the start frame and end frame of the animation sequence shown in Figure 1.

Select the ten intermediate painting as a demonstration results, and its batch color, shown in Figure 2.

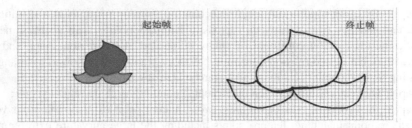

Fig. 1. Start fram and end of fram

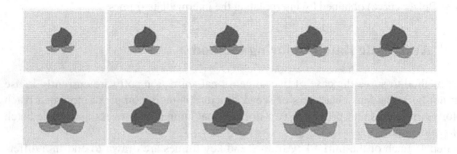

Fig. 2. Automatic generation of intermediate frames

5   Conclusion

The method provided in this paper can replace manual drawing the original middle frame, at the same time, the method of batch coloring is also provided, which can be closed in the region immediately on the needed color, save a lot of time and manpower, so in practical application, this method has high practical value, can increase the production efficiency, to reduce the cost of animation production[11]. Also be pointed out that this method of processing object must be closed areas, when the initial frame and termination time frames are not identical, the middle frame will appear distortion phenomenon, the use of the process there are many restrictions and constraints, how to solve the problems in need of further study.

Acknowledgements. (Foundation: Supported by the Fundamental Research Funds for the Central Universities (DL10CB01), Heilongjiang Province Natural Science Fund (LC2012C06), Special Funds for Harbin Innovative Talent in Science and Technology (2012RFLXG022)).

References

1. Mei Kebing from traditional animation to computer animation (below) - Computer Animation course of development. Publishing and Printing 03, 17–21 (2003)
2. Yang Siyi the history and prospects of four computer animation. Electronic Publishing 05, 45-51 (1995)
3. Chang, C.W., Lee, S.Y.: Automatic cell painting in computer assisted cartoon production using similarity recognition. Journal of Visualization and Computer Animation 8(3), 165–185 (1997)
4. Fill, C.: Designing the 21Century. Journal of Womens Health (2010)
5. Jin-Futang: Status and development of computer animation. Television Technology 09, 8–13 (1995)
6. Bai, W., Yinan, L.: Automatic key frame animation in the middle frame interpolation technology. Xi'an Jiao Tong University 01, 111–115 (1995)
7. Peinan, L., Ruifeng, G., Zhenzhen, D.: Research quaternion spherical Bézier spline interpolation algorithm. Computer Systems 12, 2439–2443 (2009)
8. Liusong, G., Shiqiang, Z., Silver, W.X., Huifang, W.: Smooth gesture planner based on quaternion and B-spline manipulator. Zhejiang University (Engineering Science) 07, 1192–1196 + 1202 (2009)
9. Xiuli, J.: Study based on two-dimensional animation technology components. Huazhong University of Science and Technology (2012)
10. Hua, Y.: Dimensional animated short film. The Old Man and Dog production research and analysis. Zhejiang University (2012)
11. Shun, Y.: Applied research of digital technology in the two-dimensional animation design. Shandong University (2010)

Metric Learning with Relative Distance Constraints: A Modified SVM Approach

Changchun Luo[1], Mu Li[1], Hongzhi Zhang[1], Faqiang Wang[1],
David Zhang[1,2], and Wangmeng Zuo[1]

[1] Computational Perception and Cognition Centre, School of Computer Science and
Technology, Harbin Institute of Technology, Harbin, 150001, China
[2] Biometrics Research Centre, Department of Computing, Hong Kong Polytechnic University,
Hung Hom, Kowloon, Hong Kong
{changchunluo1990,tshfqw}@163.com,
{limuhit,zhanghz0451,cswmzuo}@gmail.com,
csdzhang@comp.polyu.edu.hk

Abstract. Distance metric learning plays an important role in many machine learning tasks. In this paper, we propose a method for learning a Mahanalobis distance metric. By formulating the metric learning problem with relative distance constraints, we suggest a Relative Distance Constrained Metric Learning (RDCML) model which can be easily implemented and effectively solved by a modified support vector machine (SVM) approach. Experimental results on UCI datasets and handwritten digits datasets show that RDCML achieves better or comparable classification accuracy when compared with the state-of-the-art metric learning methods.

Keywords: Metric learning, Mahalanobis distance, Lagrange duality, support vector machine, kernel method.

1 Introduction

Metric learning aims to learn a Mahalanobis distance metric that amplifies the distance between dissimilar samples and shrinks the distance between similar samples. Many machine learning tasks, such as k-Nearest Neighbor (kNN) classification, are heavily relied on the chosen distance metric, where proper distance metric can greatly improve the performance of kNN classification. The learned metric can be used in classification [12], clustering [13], retrieval [14]. Metric learning has been widely applied in many real world applications, such as face verification [4], object classification [12], visual tracking [6], etc.

Most existing metric learning methods are based on pairwise constraints, where the distance between similar samples should be smaller than a upper bound u, while distance between dissimilar samples should be greater than a lower bound l, where $l > u$. The representative pairwise constrained metric learning algorithms include neighborhood component analysis (NCA) [7], information-theoretic metric learning (ITML) [1], maximal collapsing metric learning algorithm (MCML) [3], etc. NCA directly maximizes a stochastic variant of the leave-one-out kNN score on the training set. However, NCA is formulated as a non-convex optimization problem, and its solution may

H. Wang et al. (Eds.): ICYCSEE 2015, CCIS 503, pp. 242–249, 2015.

converge to local optimum. MCML and ITML are formulated as convex semidefinite programming (SDP) problems. MCML tries to collapse all examples in the same class to a single point and push examples in other classes infinitely far away. ITML formulates the problem as that of minimizing the differential relative entropy between two multivariate Gaussians under pairwise constraints on the learned metric.

Considering the complexity of the sample distribution, it is generally unnecessary to require the distances between similar (dissimilar) samples to be lower (higher) than a fixed bound $u(l)$. Different from pairwise constraints, there are a number of metric learning methods based on relative distance constraints. For each triplet $(\mathbf{x}_i, \mathbf{x}_j, \mathbf{x}_k)$ where $\mathbf{x}_i$ and $\mathbf{x}_j$ are similar but $\mathbf{x}_i$ and $\mathbf{x}_k$ are dissimilar, the relative distance constraint restricts that the learned distance $d_{\mathrm{M}}(\mathbf{x}_i, \mathbf{x}_j)$ between $\mathbf{x}_i$ and $\mathbf{x}_j$ should be lower than that $d_{\mathrm{M}}(\mathbf{x}_i, \mathbf{x}_k)$ between $\mathbf{x}_i$ and $\mathbf{x}_k$. By this way, relative distance constraints were introduced to learn a proper Mahalanobis distance metric. Large margin nearest neighbor (LMNN) [2] and BoostMetric [10] are two representative metric learning models based on relative distance constraints. LMNN adopts the k-nearest neighbors (kNN) model to construct a set of triplets, and aims to learn a distance metric that makes the distance $d_{\mathrm{M}}(\mathbf{x}_i, \mathbf{x}_j)$ be separated from $d_{\mathrm{M}}(\mathbf{x}_i, \mathbf{x}_k)$ by a large margin. The LMNN model is formulated as a convex semidefinite programming (SDP) problem, and a solver based on projected gradient has been proposed for solving this SDP problem. Modified LMNN models have been further investigated and applied to multi-task and non-linear metric learning [17, 18]. BoostMetric is a boosting-like algorithm, which adopts a weak learner to learn rank-one positive semidefinite (PSD) matrices and uses an efficient and scalable boosting-based learning process to learn a linear combination of rank-one PSD matrices [10]. Improved BoostMetric methods, e.g., MetricBoost [15] and FrobMetric [19, 20] were further developed and applied to face recognition and image classification. By incorporating two regularization terms in the duality form, Liu and Vemuri [21] suggested a robust metric learning method with relative distance constraints. Schultz and Joachims [22] reduced the distance metric to a diagonal matrix, and used SVM to optimize the proposed metric learning model based on relative distance constraints. In [5], we formulated metric learning as a kernel classification framework, and analyzed the relationship between the state-of-the-art metric learning models.

Most existing metric learning models are formulated as the SDP problems, and a number of optimization algorithms, e.g., projected sub-gradient [2] and Boosting-like [10], have been proposed to solve them. In this paper, we reformulate the metric learning with relative distance constraints as a kernel-based learning model. With this model formulation, we suggest a Relative Distance Constrained Metric Learning (RDCML) model which can be efficiently solved using a modified SVM solver. The experimental results show that RDCML achieves comparable classification accuracy than the competing methods.

The remainder of this paper is organized as follows. In Sect. 2, we introduce the RDCML method and its detailed deduction. In Sect. 3, we present the experimental results on the UCI and the handwritten digits data sets. In Sect. 4, we conclude the paper with some concluding remarks.

2   The RDCML Model

In Mahalanobis distance metric learning, the target is to learn a metric denoted by $\mathbf{M}$, by which the distance between two data points $\mathbf{x}_i$ and $\mathbf{x}_j$ is defined as

$$d_{\mathbf{M}}^2\left(\mathbf{x}_i, \mathbf{x}_j\right) = \left\|\mathbf{x}_i - \mathbf{x}_j\right\|_{\mathbf{M}}^2 = \left(\mathbf{x}_i - \mathbf{x}_j\right)^T \mathbf{M}\left(\mathbf{x}_i - \mathbf{x}_j\right) \qquad (1)$$

To satisfy the non-negativity and triangle inequality of distance metric, $\mathbf{M}$ should be constrained to be a positive semidefinite (PSD) matrix. As a special case, when $\mathbf{M}$ is an identity matrix, the Mahalanobis distance in (1) becomes the Euclidean distance.

2.1   Problem Formulation

Denote $\{(\mathbf{x}_i, y_i) \,|\, i = 1, 2, ..., N\}$ by a training set, where $\mathbf{x}_i \in \mathbb{R}^d$ is the ith training sample, and y_i is the class label of $\mathbf{x}_i$. Here we define $\mathbf{x}_j$ as a hit of $\mathbf{x}_i$ if $\mathbf{x}_j \ (j \neq i)$ and $\mathbf{x}_i$ belong to the same class. Otherwise, we call $\mathbf{x}_j$ a miss of $\mathbf{x}_i$ if $\mathbf{x}_j \ (j \neq i)$ and $\mathbf{x}_i$ belong to different classes. Given a triplet set $\mathcal{T} = \{(\mathbf{x}_i, \mathbf{x}_j, \mathbf{x}_k) \,|\, y_i = y_j \neq y_k\}$, where $\mathbf{x}_j$ is the nearest hit of $\mathbf{x}_i$, and $\mathbf{x}_k$ is the nearest miss of $\mathbf{x}_i$. In practice, we take one hit and one miss to construct the triplet for each sample, then we can get N triplets for the training set. The RDCML model restricts that the distance of each sample to its nearest hit should be separated by a large margin with the distances to its nearest miss. So the optimization problem is formulated as:

$$\begin{aligned} \min_{\mathbf{M}, \xi} \quad & \frac{1}{2}\|\mathbf{M}\|_F^2 + C \sum\nolimits_{i,j,k} \xi_{ijk} \\ \text{s.t.} \quad & \left(\mathbf{x}_i - \mathbf{x}_k\right)^T \mathbf{M}\left(\mathbf{x}_i - \mathbf{x}_k\right) - \left(\mathbf{x}_i - \mathbf{x}_j\right)^T \mathbf{M}\left(\mathbf{x}_i - \mathbf{x}_j\right) \geq 1 - \xi_{ijk}, \\ & \xi_{ijk} \geq 0, \quad \forall i, j, k, \quad \mathbf{M} \succcurlyeq 0, \end{aligned} \qquad (2)$$

where ξ_{ijk} denotes the slack variable, $\|\cdot\|_F^2$ denotes the Frobenius norm and C is the coefficient of the loss term. Here we introduce a matrix $\mathbf{X}_{ijk}$, where $\mathbf{X}_{ijk} = \left(\mathbf{x}_i - \mathbf{x}_k\right)\left(\mathbf{x}_i - \mathbf{x}_k\right)^T - \left(\mathbf{x}_i - \mathbf{x}_j\right)\left(\mathbf{x}_i - \mathbf{x}_j\right)^T$.

The RDCML model defined above is convex and can be solved using the standard SDP solvers. However, the high complexity of the general-purpose interior-point SDP solver makes it only suitable for small scale problems. In order to reduce the computational complexity, in the following we first analyze the Lagrange duality of the RDCML model, and then propose an algorithm by iterating between SVM training and PSD projection to learn the Mahalanobis distance metric.

By introducing the Lagrange multipliers λ, κ and $\mathbf{Y}$, we can derive the dual problem of Eq. (2) as follows:

$$\begin{aligned} \max_{\lambda, \mathbf{Y}} \quad & -\frac{1}{2}\left\|\sum\nolimits_{i,j,k} \lambda_{ijk} \mathbf{X}_{ijk} + \mathbf{Y}\right\|_F^2 + \sum\nolimits_{i,j,k} \lambda_{ijk} \\ \text{s.t.} \quad & 0 \leq \lambda_{ijk} \leq C, \quad \forall i, j, k, \quad \mathbf{Y} \succcurlyeq 0. \end{aligned} \qquad (3)$$

The KKT condition implies the following relationship between λ, $\mathbf{Y}$ and $\mathbf{M}$:

$$\mathbf{M} = \sum\nolimits_{i,j,k} \lambda_{ijk} \mathbf{X}_{ijk} + \mathbf{Y} \qquad (4)$$

The Lagrange duality allows us to solve the primal problem by solving its Lagrange dual problem, and the primal solution can then be obtained based on the KKT conditions. For the optimization in (2), we can first solve the dual problem in (3), and then obtain $\mathbf{M}$ using (4).

2.2 Optimization

Although the problem in (3) is also a semidefinite programming problem, fortunately it can be efficiently solved using the alternative optimization method. That is, we first fix $\mathbf{Y}$ and solve the sub-problem on λ. Then, by fixing λ, we solve the sub-problem on $\mathbf{Y}$. By alternatively updating $\mathbf{Y}$ and λ, we can obtain the optimal solution of the dual problem in (3). Given $\mathbf{Y}$, the sub-problem on λ is:

$$\max_{\lambda} \quad -\frac{1}{2} \sum_{i,j,k} \sum_{l,m,n} \lambda_{ijk} \lambda_{lmn} \langle \mathbf{X}_{ijk}, \mathbf{X}_{lmn} \rangle + \sum_{i,j,k} p_{ijk} \lambda_{ijk} \tag{5}$$
$$\text{s.t.} \quad 0 \le \lambda_{ijk} \le C$$

where $p_{ijk} = 1 - \langle \mathbf{X}_{ijk}, \mathbf{Y} \rangle$ and $\langle \mathbf{X}_{ijk}, \mathbf{Y} \rangle = \text{tr} \left(\mathbf{X}_{ijk}^T \mathbf{Y} \right)$ denotes the inner product of matrices $\mathbf{X}_{ijk}$ and $\mathbf{Y}$. Equation (3) is a quadratic programming (QP) problem, and can be efficiently solved by the SVM solvers such as LibSVM [8]. Given λ, the sub-problem on $\mathbf{Y}$ is:

$$\min_{\mathbf{Y}} \quad \| \mathbf{Y} - \mathbf{Y}_0 \|_F^2 \quad \text{s.t.} \quad \mathbf{Y} \succcurlyeq 0 \tag{6}$$

where $\mathbf{Y}_0 = -\sum_{i,j,k} \lambda_{ijk} \mathbf{X}_{ijk}$. We can use the eigenvalue decomposition to solve the problem (6). Through the eigenvalue decomposition of $\mathbf{Y}_0$, i.e., $\mathbf{Y}_0 = \mathbf{U} \mathbf{\Lambda} \mathbf{U}^T$ and $\mathbf{\Lambda}$ is the diagonal matrix of the eigenvalues, the solution to the sub-problem on $\mathbf{Y}$ can be explicitly expressed as $\mathbf{Y} = \mathbf{U} \mathbf{\Lambda}_+ \mathbf{U}^T$, where $\mathbf{\Lambda}_+ = \max(\mathbf{\Lambda}, 0)$.

Algorithm 1. The Modified SVM Algorithm of RDCML

Input: $\mathcal{T} = \{(\mathbf{x}_i, \mathbf{x}_j, \mathbf{x}_k) :$ the class label of $\mathbf{x}_i$ and $\mathbf{x}_j$ are the same, while the class label of $\mathbf{x}_i$ and $\mathbf{x}_k$ are different$\}$
Output: $\mathbf{M}$
1: Initialize $\mathbf{Y}^{(0)}$, $t \leftarrow 0$
2: **repeat**
3:    Update $\lambda^{(t+1)}$ by solving the sub-problem (5) using an SVM solver
4:    Update $\mathbf{Y}_0^{(t+1)} = -\sum_{i,j,k} \lambda_{ijk}^{(t+1)} \mathbf{X}_{ijk}$
5:    Update $\mathbf{Y}^{(t+1)} = \mathbf{U} \mathbf{\Lambda}_+ \mathbf{U}^T$, where $\mathbf{Y}_0^{(t+1)} = \mathbf{U} \mathbf{\Lambda} \mathbf{U}^T$, and $\mathbf{\Lambda}_+ = \max(\mathbf{\Lambda}, 0)$
6:    $t \leftarrow t + 1$
7: **until** the difference between the objective functions of (2) and its dual problem (3) lower than a threshold t
8: $\mathbf{M} = \sum_{i,j,k} \lambda_{ijk} \mathbf{X}_{ijk} + \mathbf{Y}$
9: **return** $\mathbf{M}$;

Table 1. Statistics of the UCI datasets used in our experiments

Dataset	# of training samples	# of test samples	Feature dimension	# of classes
Breast Tissue	96	10	9	6
Cardiotocography	1,914	212	21	10
ILPD	525	58	10	2
Letter	16,000	4,000	16	26
Parkinsons	176	19	22	2
Statlog Segmentation	2,079	231	19	7
Sonar	188	20	60	2
SPECTF Heart	80	187	44	2

3   Experimental Results

The proposed RDCML method is evaluated by kNN classification ($k = 1$) using two kinds of datasets: the UCI datasets and the handwritten digit datasets. We compare the proposed RDCML method with several state-of-the-art metric learning methods, including four pairwise constrained methods, i.e., NCA [7], ITML [1], MCML [3] and LDML [4], and one relative distance constrained method, i.e., LMNN [2]. Moreover, we also compare RDCML with the kNN classifier ($k = 1$) with the Euclidean distance. We set the coefficient parameter $C = 1$ in the RDCML model.

For each dataset, if there is no defined partition of the training set and the test set, we evaluate the performance of each method by 10-fold cross validation. The classification error rate is obtained by averaging the 10 runs. We implement RDCML based on the popular SVM toolbox LibSVM [8]. The source codes of NCA, ITML, LDML, MCML, and LMNN are also online available, and we tuned their parameters to get the best results. All the experiments are run on a workstation with 4 Intel Core i7 CPUs and 16 GB RAM.

3.1   Results on the UCI Datasets

In this subsection, we use eight datasets from the UCI Machine Learning Repository [9] to evaluate the proposed method. On the SPECTF Heart and Letter datasets, the defined training set and test set are employed to run the competing methods. On the other datasets, we use 10-fold cross validation to test the metric learning models. Table 1 summarizes the basic information of the 8 UCI datasets.

We compare the RDCML with the other competing methods. Table 2 lists the classification error rates of the seven methods on the 8 UCI datasets. RDCML achieves the best performance on Breast Tissue, ILPD, Letter, and Statlog Segmentation datasets. In order to provide a fair comparison on the performances of these models on multiple datasets, we list the average ranks of these models [16]. For each dataset, we rank the classification error rates of these models in the ascending order, and calculate the average rank for each model. One can see from Table 2 that the proposed RDCML achieves the best average rank.

Table 2. Comparison of the classification error rates (%) and average ranks of different metric learning models on the UCI datasets

Dataset	Euclidean	NCA	ITML	MCML	LDML	LMNN	RDCML
Breast Tissue	**30.74**	34.90	37.75	**30.74**	47.00	34.40	**30.74**
ILPD	34.69	34.22	33.69	34.69	35.94	34.22	**32.78**
Letter	4.10	2.70	2.88	4.10	10.05	3.55	**2.58**
Statlog Segmentation	2.58	2.38	2.29	2.79	2.96	2.54	**2.08**
Cardiotocography	21.69	**16.06**	19.80	20.66	22.06	19.11	19.53
Parkinsons	**4.11**	7.89	6.22	12.84	7.25	5.16	4.74
SPECTF Heart	38.40	**27.27**	35.19	29.85	33.06	34.66	34.22
Sonar	14.33	12.93	14.76	24.19	22.76	**11.67**	14.57
Average Rank	4.125	2.875	4	4.75	6.125	3.25	**2.125**

Table 3. Statistics of the handwritten digit datasets used in our experiment

Dataset	# of training samples	# of test samples	Feature dimension	# of classes
PenDigits	7494	3498	10	16
Semeion	1436	159	10	256
USPS	7291	2007	10	256
MNIST	60000	10000	10	784

3.2   Results on the Handwritten Digits Datasets

Apart from the UCI datasets, we assess the RDCML model on four handwritten digit datasets: MNIST, PenDigits, Semeion, and USPS. On the PenDigits, USPS and MNIST datasets, we use the defined training/test partitions to train the models and calculate the classification error rates. On the Semeion datasets, we use 10-fold cross validation to evaluate the metric learning methods, and the error rates are obtained by averaging over the 10 runs. Table 3 summarizes the basic information of the four handwritten digit datasets.

As the dimensions of the samples from USPS, Semeion and MNIST are relatively high, we first represented them into the vector form and used Principle Component Analysis (PCA) to reduce the feature dimension to 100, and train the metrics in the PCA subspace. Since MCML requires very large memory space (over 30 GB) on the MNIST dataset and cannot be run in our PC, we do not report the experimental result of MCML on this dataset. One can see from Table 4 that the proposed RDCML method outperforms all the other methods on the four handwritten digit datasets. Table 4 also lists the average ranks of these models in the last row. Of course RDCML achieves the best average rank on the handwritten digit datasets. The result shows that RDCML is effective in handwritten digits classification.

Table 4. Comparison of the classification error rates (%) and average ranks of different metric learning models on the UCI datasets

Dataset	Euclidean	NCA	ITML	MCML	LDML	LMNN	RDCML
PenDigits	2.36	2.23	2.39	2.36	6.15	2.51	**2.23**
Semeion	8.53	5.67	5.61	11.13	6.10	5.41	**5.21**
USPS	5.18	5.58	6.23	5.18	11.88	5.42	**4.98**
MNIST	2.86	5.56	2.72	N/A	8.67	2.26	**2.17**
Average Rank	3.75	3.75	4.25	4.33	6.25	3.5	1

4   Conclusion

In this paper, we propose a metric learning model, i.e., Relative Distance Constrained Metric Learning (RDCML). For each triplet, RDCML restricts that the Mahalanobis distance between $\mathbf{x}_i$ and $\mathbf{x}_j$ should be smaller than that between $\mathbf{x}_i$ and $\mathbf{x}_k$, and adopts the Frobenius norm regularizer and the hinge loss. We further propose a modified SVM algorithm for solving RDCML. The experimental results show that RDCML achieves better or comparable classification accuracy when compared with the state-of-the-art metric learning methods.

RDCML sheds some new insights on the formulation and optimization of metric learning. In the future, we will investigate effective metric models for some specific learning tasks like face verification, ranking and retrieval, and extend the iterative SVM algorithm for solving these models.

Acknowledgement. This work was supported in part by the National Natural Science Foundation of China under Grant 61271093, Grant 61471146, and the Program of Ministry of Education for New Century Excellent Talents under Grant NCET-12-0150.

References

1. Davis, J., Kulis, B., Jain, P., Sra, S., Dhillon, I.: Information-theoretic metric learning. In: Proceedings of the 24th International Conference on Machine Learning, pp. 209–216 (2007)
2. Weinberger, K.Q., Saul, L.K.: Distance metric learning for large margin nearest neighbor classification. Journal of Machine Learning Research 10, 207–244 (2009)
3. Globerson, A., Roweis, S.: Metric learning by collapsing classes. In: Advances in Neural Information Processing Systems, pp. 451–458 (2005)
4. Guillaumin, M., Verbeek, J., Schmid, C.: Is that you? Metric learning approaches for face identification. In: Proceedings of IEEE International Conference on Computer Vision, pp. 498–505 (2009)
5. Wang, F., Zuo, W., Zhang, L., Meng, D., Zhang, D.: A kernel classification framework for metric learning, arXiv:1309.5823 (2013)
6. Li, X., Shen, C., Shi, Q., Dick, A., Hengel, A.: Non-sparse linear representations for visual tracking with online reservoir metric learning. In: Proceedings of IEEE International Conference on Computer Vision and Pattern Recognition, pp. 1760–1767 (2012)

7. Goldberger, J., Roweis, S., Hinton, G., Salakhutdinov, R.: Neighborhood components analysis. In: Advances in Neural Information Processing Systems, pp. 513–520 (2004)
8. Chang, C.C., Lin, C.J.: LIBSVM: a library for support vector machines. ACM Transactions on Intelligent Systems and Technology 2, 27:1–27:27 (2011)
9. Bache, K., Lichman, M.: UCI Machine Learning Repository (2013), http://archive.ics.uci.edu/ml
10. Shen, C., Kim, J., Wang, L., Hengel, A.: Positive Semidefinite metric learning using boosting-like algorithms. Journal of Machine Learning Research 13, 1007–1036 (2012)
11. Bellet, A., Habrard, A., Sebban, M.: A Survey on Metric Learning for Feature Vectors and Structured Data, arXiv:1306.6709 (2013)
12. Mensink, T., Verbeek, J., Perronnin, F., Csurka, G.: Metric Learning for Large Scale Image Classification: Generalizing to New Classes at Near-Zero Cost. In: Proceedings of the 12th European Conference on Computer Vision, pp. 488–501 (2012)
13. Xing, E.P., Ng, A.Y., Jordan, M.I., Russell, S.: Distance metric learning with application to clustering with side-information. In: Advances in Neural Information Processing Systems, pp. 505–512 (2002)
14. Tsuyoshi, K., Nozomi, N.: Metric learning for enzyme active-site search. Bioinformatics 26(21), 2698–2704 (2010)
15. Bi, J., Wu, D., Lu, L., Liu, M., Tao, Y., Wolf, M.: Adaboost on low-rank PSD matrics for metric learning. In: Proceedings of the 2011 IEEE International Conference on Computer Vision and Pattern Recognition, pp. 2617–2624 (2011)
16. Demsar, J.: Statistical comparisons of classifiers over multiple data sets. Journal of Machine Learning Research 7, 1–30 (2006)
17. Kedem, D., Tyree, S., Weinberger, K.Q., Sha, F., Lanckriet, G.: Nonlinear metric learning. In: Proceedings of Advances in Neural Information Processing Systems, pp. 2582–2590 (2012)
18. Parameswaran, S., Weinberger, K.Q.: Large Margin Multi-Task Metric Learning. In: Advances in Neural Information Processing Systems, pp. 1867–1875 (2010)
19. Shen, C., Kim, J., Wang, L.: A scalable dual approach to semidefinite metric learning. In: Proceedings of IEEE International Conference on Computer Vision and Pattern Recognition, pp. 2601–2608 (2011)
20. Shen, C., Kim, J., Liu, F., Wang, L., Hengel, A.: Efficient dual approach to distance metric learning. IEEE Transactions on Neural Network and Learning Systems 25(2), 394–406 (2014)
21. Liu, M., Vemuri, B.C.: A robust and efficient doubly regularized metric learning approach. In: Proceedings of 2012 European Conference on Computer Vision, pp. 646–659 (2012)
22. Schultz, M., Joachims, T.: Learning a distance metric from relative comparisons. Advances in Neural Information Processing Systems 16, 41–48 (2004)

An Efficient Webpage Classification Algorithm
Based on LSH

Junjun Liu, Haichun Sun, and Zhijun Ding[*]

The Key Laboratory of Embedded System and Service Computing, Ministry of Education,
Tongji University, Shanghai 200092, China
liujunjuntj@163.com, sunhaichun1985@126.com,
Zhijun_ding@outlook.com

Abstract. With the explosive growth of Internet information, it is more and
more important to fetch real-time and related information. And it puts forward
higher requirement on the speed of webpage classification which is one of
common methods to retrieve and manage information. To get a more efficient
classifier, this paper proposes a webpage classification method based on locality
sensitive hash function. In which, three innovative modules including building
feature dictionary, mapping feature vectors to fingerprints using Locality-
sensitive hashing, and extending webpage features are contained. The compare
results show that the proposed algorithm has better performance in lower time
than the naïve bayes one.

Keywords: Explosive growth, Webpage classification, Locality-sensitive hashing, Fingerprint, Extending webpage features.

1   Introduction

With the rapid development of network and communication technologies, the Internet
has connected tens of millions of networks and the information is undergoing a verit-
able explosion of growth. The Internet has become a huge library without catalog [1].
According to "The 31st times China Internet development state statistical report", the
webpage number has increased by 36.1 billion in one year. That is, there are 0.99
million new webpages appearing every day. How to quickly fetch the interesting information and mining implicit knowledge from the Internet have been widespread
concerned by experts and scholars.

Webpage classification can be divided into more specific ones: subject classification,
functional classification, sentiment classification, and other types of classification [2]. This
paper focuses on subject classification. I.e., judging whether a webpage is about "arts",
"sports" and "computer" or not. Most existing webpage classification algorithms are based
on common text classification ones like k-Nearest Neighbor algorithm (KNN) [3, 15, 16],
Support Vector Machine (SVM) [4, 17, 18] and naïve bayes (NB) [5,19]. However, their
performance cannot meet the efficiency requirement of webpage classification on the

---

[*] Corresponding author.

H. Wang et al. (Eds.): ICYCSEE 2015, CCIS 503, pp. 250–257, 2015.

Internet. KNN's time complexity is linear, but that of its classifying-phase is not. The SVM classification algorithm, its learning ability and computational complexity is independent of the dimension of the feature space, thereby cannot meet the requirements of real-time webpage classification. NB owns the best time performance among all classification algorithms. But its speed is still not well meeting the requirements of webpage classification.

Locality-sensitive hashing (LSH) [6] is a method of performing probabilistic dimension reduction of high-dimensional data. It is raised by Piotr Indyk and Rajeev Motwani to solve problems of neighbor search in main memory. LSH has been applied to several domains including Near-duplicate detection [7, 8], Hierarchical clustering [9], and Genome-wide association study [10]. Its basic idea is to hash the input items so that similar items are mapped to the same buckets with high probability. The best advantages of LSH are that it improves space utilization and reduces time complexity. This paper aims to achieve a low time complexity and high accuracy classification algorithm, which is more suitable for webpage classification on the Internet.

The remaining of the paper is organized as follow: Section 2 presents a webpage classification model based on LSH. Section 3 describes the algorithm in detail. Section 4 presents the comparative experiments between the proposed algorithm and the NB algorithm. And, Section 5 concludes this paper.

2   Webpage Classification Model Based on LSH

The proposed webpage classification model is presented in Fig. 1.

- **Preprocessing.**

1. Parse document into a word stream, remove stop words;
2. Calculate the weight of words; and,
3. Use words threshold to retain only significant words and obtain feature vector and weight vector.

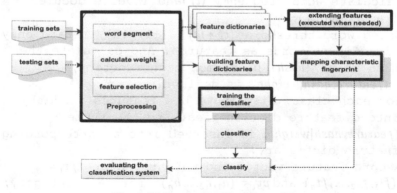

Fig. 1. Webpage classification model

Definition 1 (webpage keyword vector (kv) and keyword weight vector (pv)): A keyword vector is defined as a vector composited by higher frequency words. And a weight vector is composed by tf [11, 12] of the corresponding dimension. Respectively, they are shown as follows:

$$kv = (wd_1, wd_2, \dots, wd_n)$$
$$pv = (tf_1, tf_2, \dots, tf_n)$$

where wd_i represents a keyword.

Definition 2 (webpage (category) feature vector (fv)): A feature vector is defined as the intersection between kv and a feature dictionary. A fv is represented as follows:

$$fv = (ft_1, ft_2, \dots, ft_n)$$

Definition 3 (Union of feature vectors): The union of two feature vectors is defined as that appending words in the second vector to the first one and do not merger similar items. That is,

$$(ft_{11}, ft_{12}, \dots, ft_{1n_1}) U (ft_{21}, ft_{22}, \dots, ft_{2n_2})$$
$$= (ft_{11}, ft_{12}, \dots, ft_{1n_1}, ft_{21}, ft_{22}, \dots, ft_{2n_2})$$

3   Webpage Classification Algorithm Based on LSH

The webpage classification algorithm based on LSH (WCA_LSH) has four main parts including building feature dictionary presented in Section 3.1, training classifier phase presented in Section 3.2, Extending features presented in Section 3.3 and classify phase presented in Section 3.4.

3.1   Building Feature Dictionary

Pseudo-code for the algorithm of building feature dictionaries is as follows.

```
1) Get document;
2) Parse document and remove stop words;
3) Calculate word frequency (tf)and inverse document
   frequency(idf);
4) Using word threshold (tf * idf) [11, 12], retain only
   significant words as features. And calculate the
   hash value (hash) by common hash function,
   a (feature, hash|weight) is defined.
5) For each class, distribute the features randomly
   into d feature dictionaries. And the tuple
   (feature, hash|weight) is inserted into a corresponding
   feature dictionary.
6) Record feature vector and weight vector (fv_i =
   (ft_1, ft_2, ..., ft_n) and pv_i = (h_1, h_2, ...h_n)) for each category;
```

3.2   Training Classifier

For each category, we generate d fingerprints by mapping its feature vector to the d feature dictionaries. The method used to map characteristic fingerprint is one kind of LSH, named simhash. Simhash [13] was developed by Moses Charikar. This paper aims to obtain a better classification performance in a lower time complexity by using LSH.

Based on the algorithm of mapping characteristic fingerprint, we use feature dictionaries to train classifier. For each category, we repeat the following procedure.

```
1) Input feature dictionaries and fv and pv:
   fvᵢ = (ft₁,ft₂,…,ftₙ),  pvᵢ = (h₁,h₂,…hₙ);
2) For each feature dictionary, calculate the inter-
   section of the dictionary and fv. The same as pv.
3) Using the algorithm presented in [13] to map the
   corresponding fingerprint fᵢⱼ(i = 1,2,…,m,j = 1,2,…,d);
4) Record d fingerprints of every category.
```

At last, a category is presented by d fingerprints f_{11} , f_{12} ,..., f_{1d} . And, $f_{11}, f_{12},...,f_{1d}; f_{21}, f_{22},..., f_{2d};...;$ and $f_{m1}, f_{m2},..., f_{md}$ represent m categories.

3.3   Extending Features

Let $fv_j = (ft_{j1}, ft_{j2}, ..., ft_{jn_{ji}})$ and $pv_j = (h_{j1}, h_{j2}, ... h_{jn_{ji}})$ represent feature vector and weight vector of the category $i(i = 1,2, ..., m)$ under the feature dictionary j (the intersection of feature dictionary j and feature vector of category i). Where n_{ji} represents the size of fv_j. And $p_j = (t_{j1}, t_{j2,...,} t_{jk})$ and $q_j = (q_{j1}, q_{j2,...,} q_{jk})$ represent feature vector and weight vector of one webpage under the feature dictionary j. Where k represents the size of p_j. The procedure of extending features showed in Fig. 2 and the basic idea is as follows.

```
1) If 0 < k < min {nⱼᵢ, i = 1,2,…,m}, namely, when the fea-
   tures number of webpage is less than the minimum
   class feature numbers, we need extend its fea-
   tures.
2) Use union operation to extend feature vector. As-
   sume pⱼ is the feature vector which needs to be ex-
   tended And fvⱼ_cᵢ = (ftⱼ₁,ftⱼ₂,…,ftⱼₖ)U(tⱼ₁,tⱼ₂,…,tⱼₙⱼᵢ), store
   the size of fvⱼ_cᵢ into a variable k_cᵢ.
3) Calculate the weight vector of extended feature
   vector. (qⱼ₁,qⱼ₂,…,qⱼₖ)U(hⱼ₁/k_cᵢ, hⱼ₂/k_cᵢ, … hⱼₙⱼᵢ/k_cᵢ)。
```

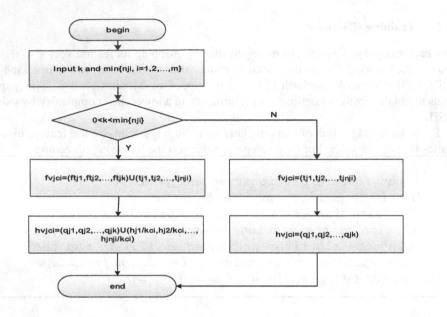

Fig. 2. Extending features

3.4   Classify Webpage

After a classifier is trained, we classify all documents in testing examples to evaluate the classification system. For each webpage, the classification process is shown in Fig. 3.

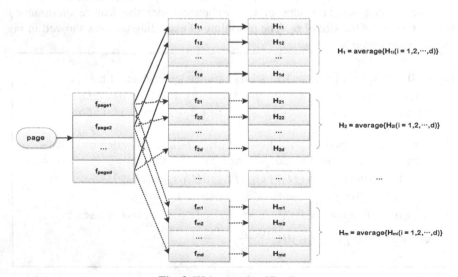

Fig. 3. Webpage classification

4   Experiment

In this paper, all the experiments are implemented in the same hardware and software environments. There are eleven sugon servers as datanodes and one as namenode. The hardware of each machine is 16-core 2.2GHz and 64G RAM. The programming language is Java. We use the corpus organized by Fudan university computer information and technology department International Database Center Natural Language Processing group. The comparative baseline is a distributed NB classification algorithm developed by Apache mahout [11, 12].

4.1   Experiment Prepare

In our experiments, C3, C11, C19, C31 and C39 represent Art, Space, Computer, Environment and Sports categories respectively.

4.2   Determining Initial Parameters

In this part, we determined initial parameters of our algorithm including the threshold for choosing features for every category and the count of feature dictionaries (d) by several comparative experiments. The results are shown in Figs. 4and 5.

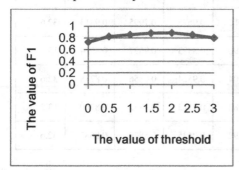 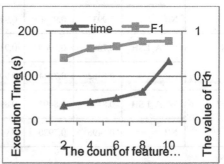

Fig. 4. The value of F1 under different threshold value

Fig. 5. The execution time and F1 value under different size of feature dictionaries

In NB, The same method is used to select features, namely $tf * idf$. So we use NB algorithm to determine the first parameters, threshold. From Fig. 4, we can see that the value of $F1$ of NB is the highest when threshold equals 2.0, so we choose 2.0 as the threshold in our paper.

Fig. 5 shows the relation among the value of d , execution time and $F1$ value in WCA_LSH algorithm. From Fig. 5, we can see that in the range of 1 to 10, the larger the d, the larger the $F1$ as well as longer the Algorithm's execution time. And the value of $F1$ is stabilizes with the increasing of d. We choose 8 as d's .In that case, the WCA_LSH classification can obtain higher accuracy in lower time.

4.3   Comparative Experiments

The comparative results are shown in Table 2. In Table 2, the column CAT is abbreviation of category and ALG is abbreviation of algorithm. P,R and $F1$ represent the precision, recall and $F1$ value respectively presented in Section 3.4. The column AVG represents the average value of P,R or $F1$ and AVG time represents the average time of execution time.

From Table 2, we can know that although the average precision and recall of WCA_LSH are slightly lower than the NB one, the former has better stability than the latter, and the classification time of the latter is 2.5 times than the former. What is more, for C3 and C11 two categories, the recall value of WCA_LSH is higher than NB one. In summary, the proposed classification algorithm can obtain good classification precision with obvious advantage in its efficiency.

Table 1. The comparative result between WCA_LSH and NB classification algorithms

Measure	CAT ＼ ALG	C3	C11	C19	C31	C39	AVG	AVG time (s)
P	WCA_LSH	0.9669	0.9109	0.8783	0.9352	0.7943	0.8971	13.2
	NB	0.9937	0.9846	0.9112	0.9675	0.7995	0.9313	32.6
R	WCA_LSH	0.7089	0.7960	0.9190	0.8883	0.9697	0.8564	13.2
	NB	0.6362	0.7944	0.9904	0.9524	0.9856	0.8718	32.6
$F1$	WCA_LSH	0.8180	0.8495	0.8982	0.9112	0.8732	0.8700	13.2
	NB	0.6496	0.7920	0.9492	0.9435	0.8829	0.8687	32.6

Acknowledgments. This work is partially supported by the National Natural Science Funds of P.R. China under Grants No. 61173042 and No. 90818023, HongKong, Macao and Taiwan Science and Technology Cooperation Program of China under Grant No.2013DFM10100.

References

1. Zhou, X.S., Li, S.: Modeling and Simulation of Webpage Automatic Classification. Computer Simulation 28(10), 121–124 (2011)
2. Qi, X., Davison, B.D.: Web Page Classification: Features and Algorithms. ACM Computing Surveys (CSUR) 41(2), 12 (2009)
3. Shi, K., Li, L., Liu, H.: An Improved KNN Text Classification Algorithm Based on Density. In: 2011 IEEE International Conference on Cloud Computing and Intelligence Systems (CCIS). IEEE (2011)

4. Liu, X.L., Ding, S.F., Zhu, H., Zhang, L.W.: Appropriateness in Applying SVMs to Text Classification. Computer Engineering and Science 32(6), 106–108 (2010)
5. Zhang, W., Gao, F.: An Improvement to Naive Bayes for Text Classification. Procedia Engineering 15, 2160–2164 (2011)
6. Gionis, A., Indyk, P., Motwani, R.: Similarity Search in High Dimensions Via Hashing. In: Proc. 25th VLDB, pp. 518–529 (1999)
7. Manku, G.S., Jain, A., Das Sarma, A.: Detecting Near-Duplicates for Web Crawling. In: Proceedings of the 16th International Conference on World Wide Web, Banff, Alberta, Canada, pp. 141–150 (2007)
8. Das, A.S., Datar, M., Garg, A., Rajaram, S.: Google News Personalization: Scalable Online Collaborative Filtering. In: Proceedings of the 16th International Conference on World Wide Web, pp. 271–280. ACM (2007)
9. Koga, H., Ishibashi, T., Watanabe, T.: Fast Agglomerative Hierarchical Clustering Algorithm Using Locality-Sensitive Hashing. Knowledge and Information Systems 12(1), 25–53 (2007)
10. Brinza, D., Schultz, M., Tesler, G., Bafna, V.: RAPID Detection of Gene–gene Interactions in Genome-wide Association Studies. Bioinformatics 26(22), 2856–2862 (2010)
11. Anil, R., Dunning, T., Friedman, E.: Mahout in Action. Manning (2011)
12. Mahout,
    http://mahout.apache.org/users/classification/bayesian.html
13. Charikar, M.S.: Similarity Estimation Techniques from Rounding Algorithms. In: Proceedings of the Thirty-fourth Annual ACM Symposium on Theory of Computing, pp. 380–388. ACM (2002)
14. Yang, Y.: An Evaluation of Statistical Approaches to Text Categorization. Information Retrieval 1(1-2), 69–90 (1999)
15. Steenwijk, M.D., Pouwels, P.J.W., Daams, M., van Dalen, J.W., Caan, M.W., Richard, E., Barkhof, F., Vrenken, H.: Accurate White Matter Lesion Segmentation by K Nearest Neighbor Classification with Tissue Type Priors (kNN-TTPs). NeuroImage: Clinical 3, 462–469 (2013)
16. Thilina, K.M., Choi, K.W., Saquib, N., Hossain, E.: Pattern Classification Techniques for Cooperative Spectrum Sensing in Cognitive Radio Networks: SVM and W-KNN approaches. In: 2012 IEEE Global Communications Conference (GLOBECOM), pp. 1260–1265. IEEE (2012)
17. Köknar-Tezel, S., Latecki, L.J.: Improving SVM Classification on Imbalanced Time Series Data Sets with Ghost Points. Knowledge and information systems 28(1), 1–23 (2011)
18. Dukart, J., Mueller, K., Barthel, H., Villringer, A., Sabri, O., Schroeter, M.L.: Meta-Analysis Based SVM Classification Enables Accurate Detection of Alzheimer's Disease Across Different Clinical Centers Using FDG-PET and MRI. Psychiatry Research: Neuroimaging 212(3), 230–236 (2013)
19. Rosen, G.L., Reichenberger, E.R., Rosenfeld, A.M.: NBC: The Naive Bayes Classification Tool Webserver for Taxonomic Classification of Metagenomic Reads. Bioinformatics 27(1), 127–129 (2011)

Semi-supervised Affinity Propagation Clustering Based on Subtractive Clustering for Large-Scale Data Sets

Qi Zhu, Huifu Zhang, and Quanqin Yang

School of Computer Science and Engineering, Hunan University of Science and Technology,
Xiangtan 411201, China

Abstract. In the face of a growing number of large-scale data sets, affinity propagation clustering algorithm to calculate the process required to build the similarity matrix, will bring huge storage and computation. Therefore, this paper proposes an improved affinity propagation clustering algorithm. First, add the subtraction clustering, using the density value of the data points to obtain the point of initial clusters. Then, calculate the similarity distance between the initial cluster points, and reference the idea of semi-supervised clustering, adding pairs restriction information, structure sparse similarity matrix. Finally, the cluster representative points conduct AP clustering until a suitable cluster division. Experimental results show that the algorithm allows the calculation is greatly reduced, the similarity matrix storage capacity is also reduced, and better than the original algorithm on the clustering effect and processing speed.

Keywords: subtractive clustering, initial cluster, affinity propagation clustering, semi-supervised clustering, large-scale data sets.

1   Introduces

Affinity propagation algorithm (AP)[1-3] is a clustering algorithm widespread concern in recent years, AP algorithm to each data point as a candidate class represents the point, avoiding the clustering results is limited to select the initial class representative points. But AP algorithm for iterative process prone to swings and algorithms need to store two great the similarity matrix, making storage complexity, which limits the ability of AP algorithm for processing large-scale data.

Semi-supervised clustering[4] combines the unsupervised learning and supervised learning characteristics, improve the quality of clustering, is one of the important research directions in the field of data mining, machine learning and pattern recognition.

Due to the limitations AP algorithm for processing large data sets, proposed an improved affinity propagation clustering algorithm. The subtraction clustering algorithm is introduced, drawing on ideas semi-supervised clustering, adding constraints, making the similarity matrix storage reduction, improve processing speed, shorten the operation time.

H. Wang et al. (Eds.): ICYCSEE 2015, CCIS 503, pp. 258–265, 2015.

2   Related Work

Reference [5] for some of their own data sets with complex structures, nuclear methods and affinity propagation algorithm is combined ,making the original cluster space is mapped into a high dimensional feature space, taking into account the spatial attribute data, using the data in the original spatial distribution manifold to constrain and adjust kernel function. Reference [6] The basic idea is to test data sets were randomly divided into two parts, and then SAP algorithm were used to obtain the corresponding class represents a collection point in each section, the last two classes will be combined into a new set of representative point data sets and then run once SAP algorithm. Algorithm for the data size is halved due to the PSAP, and therefore the amount of time spent computing time of approximately one eighth of the original. Reference [7] First, use the information to adjust the data set pairwise constraints distance matrix, calculated the distance manifold, resulting similarity measure based on manifold distance, and finally use the new similarity matrix conduct AP clustering algorithm. The objective function of the algorithm article adds penalty factor. Reference[8] characterization data distribution from the start, integrated global and local distribution of data, design of a data variable similarity metric calculation method effectively reflect the actual distribution of data clustering.

3   Affinity Propagation Clustering

AP algorithm is a clustering algorithm based on neighbor information dissemination. The similarity $s(i, j) = -\left\| x_i - x_j \right\|^2$ between data points composed of the similarity matrix , $S_{N \times N}$ as input, $S(i, i)$ is the same value P. generally, P is the intermediate values of dual points similarity. The P bigger ,the larger the number of clusters. AP algorithm delivery two types message: Between any two data points x_i and x_j , the responsibility $r(i, j)$ represents the point x_j as a representative of a class representative points x_i suitable level. the availability $a(i, j)$ represents the point x_i select point x_j as a representative of a class representative points suitable level. Core algorithm is that these two messages constantly update process, update the following formula:

$$r(i, j) \leftarrow s(i, j) - \max_{j' \neq j} (a(i, j') + s(i, j'))$$

(1)

$$\text{if } i \neq j, \ a(i, j) \leftarrow \min_{i \neq j} \{0, r(j, j) + \sum_{i' \neq i, i' \neq j} \max(0, r(i', j))\}$$

(2)

$$a(j, j) \leftarrow \sum_{i' \neq i, i' \neq j} \max(0, r(i', j))$$

(3)

Introducing damping k, eliminate oscillations may occur.

$$r_{new}(i, k) \leftarrow \lambda \times r_{old}(i, k) + (1-\lambda) \times r(i, k) \tag{4}$$

$$a_{new}(i, k) \leftarrow \lambda \times a_{old}(i, k) + (1-\lambda) \times a(i, k) \tag{5}$$

Among them, λ ($0 < \lambda < 1$) the higher the value (default is 0.5). Determine the point i cluster representative points:

$$k = \arg \max_k \{a(i, k) + r(i, k)\} \tag{6}$$

Among them, if $i=k$, then point i itself is the cluster center; If $i \neq k$, then point k is the point i cluster representative.

4   Semi-supervised Clustering

Semi-supervised clustering is to extract part of the unknown sample as a priori knowledge, commonly used a priori knowledge representation as pairwise constraints.

(1)If x_i and x_j belong to the same cluster, that is Must-link constraints $(x_i, x_j) \in M \Rightarrow s(i, j) = 0 \& s(j, i) = 0$

(2)If x_i and x_j belong to different clusters, that is Cannot-link constraints, $(x_i, x_j) \in C \Rightarrow s(i, j) = -\infty \& s(j, i) = -\infty$

5   Subtractive Clustering

Subtraction clustering algorithm [9] proposed by Chiu, consider a collection of N data points in an M-dimensional space. The algorithm process is as follows [10]:

Step1: Each data point is the candidate cluster centers, density index definition data point x_i is:

$$D_i = \sum_{j=1}^{n} \exp\left(-\frac{\|x_i - x_j\|^2}{\left(r_a/2\right)^2}\right) \tag{7}$$

With the highest density index data points x_{ci} as the first cluster center, D_{ci} is the density index. $r_a = \frac{1}{2} \min_k \left\{\max_i \{\|x_i - x_k\|\}\right\}$.

Step2: other data point density index is defined as:

$$D_i = D_i - D_{ci} \exp\left(-\frac{\|x_i - x_{c1}\|^2}{\left(r_b/2\right)^2}\right) \quad i \neq c_i$$

(8)

To avoid a similar distance in the cluster, and generally $r_b = (1.2\sim1.5)r_a$,

$$\text{Step3:} \quad \frac{D_{c_{k+1}}}{D_{c_1}} < \delta$$

(9)

Repeat the above repair process, it is determined whether or satisfies equation (9), when the exit was established, if true, then go to Step2. Nikhil R Pal and Debrub Chakraborty[11] found in the experiment $\delta = 0.5$ is to get the maximum number of clusters limit C_{max} .the r_a, r_b bigger, the fewer the number of clusters generated, conversely, the more the number of clusters generated [12].

6   Semi-supervised Affinity Propagation Clustering Based on Subtractive Clustering

Improved algorithm based on the above ideas is called Semi-supervised Affinity Propagation Clustering Based on Subtractive Clustering (SAP-SC), the algorithm idea is as follows:

1. Using subtractive clustering in data sets, according to the density value, obtain initial cluster centers in turn;

2. Calculate the similarity distance between the initial cluster centers;

3. Add the semi-supervised clustering Must-link constraints and Cannot-link constraints;

4. The cluster representative points conduct AP clustering, calculate $r(i, j)$ and $a(i, j)$, continually update information for each sample point.

5. Eliminate the shock until the clustering results are stable, to achieve convergence of purpose; determine the number of classes meets the requirements, and if not, then change the p value, repeat iterative process.

SAP-SC algorithm is described as follows:

Step1. Using the formula (7) calculate the candidate cluster centers x_i data point density index;

Step2.Using the formula (8)   calculate the density of other data points;

Step3. Reaching restrictions $\frac{D_{c_{k+1}}}{D_{c_1}} < \delta$, stop counting the number of initial cluster centers;

Step4. Take the number of initial cluster center is $C_i < \sqrt{N}$, wherein, C_i is the initial cluster center, N is the number of data points;

Step5.Initialize p, order $a(i, j)=0$ and $r(i, j)=0$, $\forall x_i$, $x_j \in X$ the initial cluster center , calculate the Euclidean distance between any two data points

$$D_{ij} = \left(\left\| x_i - x_j \right\|_2^2 \right)^{\frac{1}{2}}$$
;

Step6.Add a pair restriction information adjust the distance matrix;

Step7.Using the AP clustering algorithm, with the calculation formula(1)-(3)for the iteration;

Step8.According to equations(4)(5)eliminate oscillation to achieve the purpose of convergence; Analyzing whether the number of classes to meet the requirements. If not, change the value of p , repeat the iterative process until the number of clusters meet the requirement, then final output the clustering result.

7   Simulation

This experiment in memory 2GB, processor AMD Turion (tm) II Dual-Core Mobile M520 2.30GHz, 32-bit operating system windows7, using matlab programming.UCI data sets used in the experiment to verify the performance of the algorithm, the test data sets as shown in Table 1.

Table 1. The basic situation UCI datasets

datasets	number of samples	number of attributes	class number
Vowel	990	10	11
Yeast	1484	8	10
Image Segmentation	2310	19	7
Wave-form	5000	21	3
Landsat Satellite	6435	36	7

Experiments using 10-fold cross-validation method, each drawn from the original data set 90% as the training data set, the remaining 10% as the test data set. In determining the number of pairs of constraints, taking the average cross-validation results of 10 times, it represents semi-supervised clustering algorithm in a fixed number of pairwise constraints on clustering performance of a data set.

7.1   Evaluation

F-measure of precision and recall composition used for clustering evaluation. Precision and recall the definition of a class and its classification as follows:

$\text{Prec}(i, C_j) = \dfrac{N_{ik}}{N_k}$, $\text{Rec}(i, C_j) = \dfrac{N_{ik}}{N_k}$ Wherein, N_{ik} is the number of categories i in the cluster k ; N_k is the number of k clusters of all the objects ; N_i is the number of i in the classification of all objects. F-measure is defined as:

$$F-\text{measure}(i,C_j)=\frac{2\text{Prec}(i, C_j)\text{Rec}(i, C_j)}{(\text{Prec}(i, C_j)+\text{Rec}(i, C_j))}$$ The F-measure by each category i of F-measure

weighted average: $F=\dfrac{\sum_i [|i| \times F(i)]}{\sum_i |i|}$ Wherein, $|i|$ is the number i of the classification of

all objects.

Entropy is a measure of the purity of cluster indicators, for each cluster C_k , the Entropy is $E_k = -\sum_t p_{tk} \log(p_{tk})$ .Wherein, p_{tk} is the probability of cluster k sample belongs to class t. Entropy is a weighted average across all clusters Entropy of $E = -\sum_k (\frac{N_k}{N} \times E_j)$. Wherein: N represents the total number of data sets of samples, N_k represents the number of the number of samples in cluster k.

7.2   Results and Analysis

Experiments with AP、SAP and SAP-SC three algorithms on five data sets were tested, Figure (a), (b)were given three algorithms random experiment 20 times the average F-measure and Entropy indicators, table 2 shows three algorithms the number of iterations and average running time. By the subtractive clustering method the parameter values can be obtained, selecting the optimal parameters as follows: $r_b = 0.6$, $r_a = 0.5$, $\delta = 0.5$, the maximum number of iterations is 100. The results shown below:

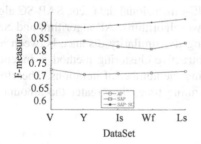

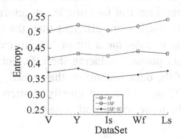

Fig. 1. (a) F-measure indicators with three algorithms on five data sets

Fig. 2. (b) Entropy indicators with three algorithms on five data sets

V:Vowel Y:Yeast Is:Image Segmentation Wf: Wave-form   Ls:Landsat Satellite

Table 2. Three algorithms iterations and average running time

data sets	AP		SAP		SAP-SC	
	number of iterations	average running time (s)	number of iterations	average running time (s)	number of iterations	average running time (s)
Vowel	33	78.24	38	108.43	23	61.74
Yeast	38	122.43	29	162.13	34	92.44
Image Segmentation	47	210.92	44	301.27	37	150.22
Waveform	59	428.85	53	448.51	46	278.65
Landsat Satellite	67	533.23	59	553.35	48	313.23

From the above results the following conclusions can be drawn:

a)From Figure (a), (b), with the same number of constraints for the different algorithms of F-measure Index and Entropy Index. Compared with the original AP algorithm, the performance of SAP-SC clustering algorithm has been greatly improved, the clustering effect is also better than SAP algorithm. This is because this paper based on semi-supervised algorithm metrics way is better reflect the law of data sets, making the clustering structure of the data set is more straightforward. In some data sets data structures have more overlap, the Euclidean distance measure cannot be separated, are not well calculated similarity distance. The subtraction clustering can find density of data points as potential cluster center, in the case of large volumes of data can be quickly and effectively classified.

b) Known from Table 2, a small amount of data, three algorithms consumed almost the same time. But for complex structures, multi-dimensional data sets, SAP-SC algorithm is much more efficient than the other two algorithms, AP algorithm and SAP algorithms consume a lot of time in calculating the similarity distance between any two data points , reducing the efficiency. Subtractive clustering method is independent of the data dimension, thus greatly reducing the number of iterations. Therefore, SAP-SC algorithm can greatly shorten the running time, the greater the amount of data, the more obvious advantages.

8   Summary

From the simulation experiments, we verify that the operating efficiency of the algorithm, improved clustering algorithm on the speed has improved greatly. Since subtractive clustering in determining the number of clusters to consider parameters, and further research is needed on how the clustering process automatically remove outlier problem. Therefore, improve this problem is our next work to be done.

Acknowledgements. This research has been partially supported by the national natural science foundation of China (51175169) and the national science and technology support program (2012BAF02B01).

References

1. Demiriz, A., Benneit, K.P., Embrechts, M.J.: Semi-supervised clustering using genetic algorithm. In: Proc of Intelligent Engineering systems through Artificial Neural Networks, pp. 809–814 (1999)
2. Liu, X., Yin, M., Luo, J.: An Improved Affinity Propagation Clustering Algorithm for Large-scale Data Sets. In: 2013 Ninth International Conference on Natural Computation, pp. 894–899 (2013)
3. Zhang, X., Furtlehner, C., Germain-Renaud, C., Sebag, M.: Data Stream Clustering with Affinity Propagation. IEEE Transactions on Knowledge and Data Engineering 26(7), 1644–1656 (2014)
4. Frey, B.J., Dueck, D.: clustering by passing messages between data points. Science 315(5814), 972–976 (2007)
5. Fu, Y.-D., Lan, J.-L.: Kernel-based adaptation for affinity propagation clustering algorithm. Application Research of Computers 29(5), 1644–1647 (2012)
6. Li, X., Wang, L., Song, Y.: Parallel computation of semi-supervised clustering algorithm based on affinity propagation. Computer Engineering and Applications 47(7), 149–152 (2011)
7. Feng, X.-L., Yu, H.-T.: Semi-supervised affinity propagation clustering based on manifold distance. Computer Engineering and Applications 28(10), 3656–3658 (2011)
8. Jun., D., Suo-Ping, W., Fan-Lun, X.: Affinity Propagation Clustering Based on Variable-Similarity Measure. Journal of Electronics & Information Technology 32(3), 509–514 (2010)
9. Chiu, S.L.: Fuzzy model identification based on cluster estimation. Journal of Intelligent and Fuzzy Systems 2(3), 267–278 (1994)
10. Cai, W., Cheng, J.: Fuzzy Clustering Based on Subtractive Clustering. Lanzhou Jiao Tong University Learned Journal 30(6), 50–54 (2011)
11. Nikhil, R.P., Chakraborty, D.: Mountain and subtractive clustering method: Improvements and generalizations [J]. International Journal of Intelligent Systems 15(4), 329–341 (2000)
12. Dash, M., Huan, L., Scheuermann, P., TanK, L.: Fast hierarchical clustering and its validation. Data & Knowledge Engineering 44, 109–138 (2003)

Gene Coding Sequence Identification Using Kernel Fuzzy C-Mean Clustering and Takagi-Sugeno Fuzzy Model

Tianlei Zang, Kai Liao, Zhongmin Sun, Zhengyou He, and Qingquan Qian

School of Electrical Engineering, Southwest Jiaotong University,
610031, Sichuan Province, China
zangtianlei@126.com

Abstract. Sequence analysis technology under big data provides unprecedented opportunities for modern life science. A novel gene coding sequence identification method is proposed in this paper. Firstly, an improved short-time Fourier transform algorithm based on Morlet wavelet is applied to extract the power spectrum of DNA sequence. Then, threshold value determination method based on kernel fuzzy C-mean clustering is used to combine Signal to Noise Ratio (SNR) data of exon and intron into a sequence, classify the sequence into two types, calculate the weighted sum of two SNR clustering centers obtained and the discrimination threshold value. Finally, exon interval endpoint identification algorithm based on Takagi-Sugeno fuzzy identification model is presented to train Takagi-Sugeno model, optimize model parameters with Levenberg-Marquardt least square method, complete model and determine fuzzy rule. To verify the effectiveness of the proposed method, example tests are conducted on typical gene sequence sample data.

Keywords: gene identification, power spectrum analysis, threshold value determination, kernel fuzzy C-mean clustering, Takagi-Sugeno fuzzy identification.

1 Introduction

With the completion of the Human Genome Project (HGP), obtaining effective biological information from numerous DNA sequences with physical or mathematical methods has important theoretical significance and practical value for biology, medical science and pharmacy [1]. Thus, gene coding sequence identification is considered as a hotspot in the field of bioinformatics under big data [2-5].

How to identify coded sequence (i.e. exon) in a given DNA sequence is a problem as well as the most fundamental and primary issue in bioinformatics. The common method of gene identification is statistics. It generally requires regarding DNA sequences with known coded sequences information as training dataset so as to determine parameters in the model [6]. However, the accuracy of gene identification will decrease greatly under the condition of inadequate gene information. Therefore, the key of gene identification is widely attached to the use of signal processing and analysis method to discover gene coded sequences. Its basic steps are as follows. To begin with, map symbol sequences of four nucleotides A, T, G and C into the

H. Wang et al. (Eds.): ICYCSEE 2015, CCIS 503, pp. 266–275, 2015.

corresponding numerical value sequences according to certain rules. Besides, conduct discrete Fourier transform to indicator sequences respectively so as to obtain the square power spectrum of each sequence, and add them together to obtain the power spectrum of the whole DNA sequences [7].

Research shows that, for the N DNA sequence, the power spectrum curve of exon sequence has a high spectrum peak at frequency N/3, but intron does not have a similar peak. This statistical phenomenon is called as 3-cycle of basic group which is generally considered as important feature information that can be used to identify gene coded sequence [8].

For DNA sequence identification, there are problems to be studied. Firstly, explore some rapid calculation method of power spectrum and SNR for long DNA sequence. Secondly, study threshold value determination method of gene identification for DNA sequence of certain gene type. Thirdly, consider factors such as the influence of random noise of DNA sequence and study methods for determining two endpoints of gene exon interval accurately. The proposed method in this paper attempts to provide solutions for problems above.

This paper is organized as follows. In section 2, extracting power spectrum using improved short-term Fourier transform based on Morlet Wavelet is introduced. Section 3 presents threshold value determination based on kernel fuzzy C-mean clustering. In section 4, exon interval endpoint identification based on Takagi-Sugeno fuzzy model is described. Experimental results and analytical work are discussed in section 5. Finally, the conclusion is given in section 6. Figure 1 shows the structure of gene coding sequence identification method.

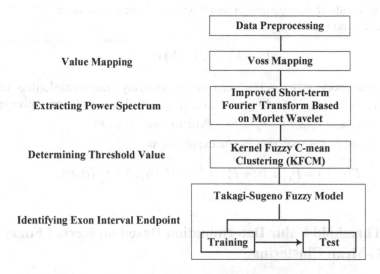

Fig. 1. The structure of gene coding sequence identification method

2   Extracting Power Spectrum Using Improved Short-Term Fourier Transform Based on Morlet Wavelet

According the 3-cycle feature of exon in combination with scale change feature of wavelet and sliding window feature of short-time Fourier transform, a fast algorithm of extracting gene sequence power spectrum using improved short-time Fourier transform based on Morlet wavelet is proposed.

Short-time Fourier transform form based on Gaussian window [9] is

$$\varphi_{STFT}(x,b,a) = e^{-\frac{(x-b)^2}{2}} e^{ja(x-b)}. \tag{1}$$

Where, a is scalability factor and b is shift factor. For the N DNA sequence, Morlet wavelet transform frequency can be fixed at $w_0 = (N/3)$.

Short-time Fourier transform function based on Morlet wavelet is

$$\varphi(x,b,a) = e^{-\frac{(x-b)^2}{2a}} e^{j\omega_0(x-b)}. \tag{2}$$

Then, the transformation formula of gene sequence is

$$U(a,b) = \sum u(x)\varphi(x,b,a) = \sum u(x)e^{-\frac{(x-b)^2}{2a}} e^{j\omega_0(x-b)}. \tag{3}$$

Where, a is scale change parameter and b is displacement parameter.

Power can be expressed as

$$P(a,b) = |U(a,b)|^2. \tag{4}$$

Four nucleotide types ATGC can be respectively transformed, thus obtaining $U_A(a,b)$, $U_B(a,b)$, $U_C(a,b)$ and $U_D(a,b)$ and further calculating power spectrum of each sequence- $P_A(a,b)$, $P_B(a,b)$, $P_C(a,b)$ and $P_C(a,b)$.

The overall power spectrum can be expressed as

$$P(a,b) = P_A(a,b) + P_T(a,b) + P_G(a,b) + P_C(a,b). \tag{5}$$

3   Threshold Value Determination Based on Kernel Fuzzy C-Mean Clustering

SNR data of DNA exon and intron sequence can be classified into two types by clustering analysis algorithm. The discrimination threshold value can be determined through calculates the weighted sum of clustering centers obtained. As kernel fuzzy C-mean clustering algorithm (KFCM) [10] well solves the dependence of traditional clustering algorithm on data distribution and has a high accuracy, threshold value determination method based on KFCM clustering is studied in this paper.

3.1   Kernel Fuzzy C-Mean Clustering Model

For the sample set $X=\{x_i|\ i=1, 2, \ldots, N\}$, value function of fuzzy C-mean clustering algorithm (FCM) can be expressed as

$$J_m(U,V) = \sum_{i=1}^{N}\sum_{j=1}^{C}\mu_{ij}^{m}\left\|x_i - v_j\right\|^2. \tag{6}$$

Where, μ_{ij} must meet

$$\forall i, \sum_{j=1}^{C}\mu_{ij} = 1; \forall i,j, \mu_{ij} \in [0,1]; \forall j, \sum_{i=1}^{N}\mu_{ij} > 0. \tag{7}$$

Where, C is the clustering number, N is the number of samples, $U = [\mu_{ij}]_{C\times N}$ is fuzzy C division matrix, μ_{ij} is the membership value of sample x_i corresponding to the jth clustering, $V=[v_j]$ is the set constituted by C clustering centers and m is index weight influencing the fuzzy degree of membership matrix. With lagrangian multiplier method, the objective function can be established as follows.

$$\bar{J}(U,V,\lambda) = J_m(U,V) + \sum_{i=1}^{N}\lambda_i(\sum_{j=1}^{C}\mu_{ij} - 1). \tag{8}$$

Non-linear mapping $\phi: x \rightarrow \phi(x)$ is introduced, and then sample distance in feature space can be defined as:

$$\left\|\phi(x_i) - \phi(v_j)\right\| = K(x_i,x_i) + K(v_j,v_j) - 2K(x_i,v_j). \tag{9}$$

Where, K is kernel function. The value function of KFCM is

$$J_\phi = \sum_{i=1}^{C}J_i = \sum_{i=1}^{C}\sum_{j=1}^{N}\mu_{ij}^{m}\left\|\phi(x_i) - \phi(v_j)\right\|^2. \tag{10}$$

This paper uses Gaussian kernel function

$$K(x,y) = \exp[-(x-y)^2/\sigma^2]. \tag{11}$$

Formula (9) is substituted with formula (11), $K(x,x)=1$, then formula (10) can be expressed as

$$J_\phi = 2\sum_{i=1}^{C}\sum_{j=1}^{N}\mu_{ij}^{m}[1-K(x_i,v_j)]. \tag{12}$$

Partial derivatives of J_ϕ about μ and v are calculated respectively. Updated formulas of new clustering center V and membership matrix U are obtained.

$$v_j = \frac{\sum_{i=1}^{N}\mu_{ij}^{m}K(x_i,v_j)x_i}{\sum_{i=1}^{N}\mu_{ij}^{m}K(x_i,v_j)}, \mu_{ij} = \frac{(1-K(x_i,v_j))^{-1/(m-1)}}{\sum_{j=1}^{C}(1-K(x_i,v_j))^{-1/(m-1)}}. \tag{13}$$

3.2 Discrimination Threshold Value Calculation

Clustering center V (two clustering center values are contained in this problem) is finally obtained in SNR data clustering of DNA sequence. The discrimination threshold value can be selected according to

$$R_0 = \sum_{i=1}^{2}w_i v_i. \tag{14}$$

Where, w_i is weight coefficient of two clustering centers, characterizing the degree of importance attached to exon and intron.

4 Exon Interval Endpoint Identification Based on Takagi-Sugeno Fuzzy Model

The difficulty of gene identification consists in accurate discrimination of two endpoints of gene exon interval. The method of fuzzy reasoning can be used to identify and correct two endpoints of exon and accurately locate DNA sequence coding area by training and test under a given threshold value. This paper studies exon interval endpoint identification model based on Takagi-Sugeno (T-S) model.

4.1 Takagi-Sugeno Fuzzy Model

Assume that there are N groups of data samples $(x_1^j,x_2^j,\cdots,x_m^j,y^j)$, $j=1,2,\cdots,N$ and take the first data sample $(x_1^1,x_2^1,\cdots,x_m^1,y^1)$ as the first clustering center $\rho = (\rho_1^1,\rho_2^1,\cdots,\rho_m^1)$, where $\rho_i^1 = x_i^1(i=1,2,\cdots,m)$. The corresponding fuzzy subset is A_i^1 and membership mean is ρ_i^1. Proper variance σ and clustering radius r are selected according to the object to be identified.

For the k th data sample $(x_1^k, x_2^k, \cdots, x_m^k)$, $k = 2, 3, \cdots, N$, assume that m clusters exist and clustering centers are respectively $\rho^1, \rho^2, \cdots, \rho^M$, the distance between this data sample and these m clustering centers is calculated with fuzzy likelihood function [11] (take the reciprocal of fuzzy likelihood function value $\sum_{l=1}^{m} \mu_{A_i^l}(x_i^k) \Big/ m$)

$$D_l = m \Big/ \sum_{l=1}^{m} \mu_{A_i^l}(x_i^k), l = 1, 2, \cdots, M .$$ (15)

Assume that $D_{\min}^k$ is the minimum distance between this sample and these clustering centers, i.e. the nearest cluster of the k th data sample. If $D_{\min}^k > r$, this data sample is considered as a new clustering center

$$\rho^{M+1} = (\rho_1^{M+1}, \rho_2^{M+1}, \cdots, \rho_m^{M+1}), \rho_j^{M+1} = x_j^k .$$ (16)

L clusters are obtained after N groups of sample data are considered, i.e. L fuzzy rules. Then, fuzzy model structure can be established. Antecedent parameters ρ and σ have been obtained in the clustering process. The conclusion part is input and output linear function. Conclusion parameter P is obtained with Levenberg-Marquardt least square method.

Based on the consideration of structure self-adaption, the concept of "confidence" of rules obtained is introduced. When the confidence of a rule is lower than certain threshold value $Q_{\min}$, obsolete rules are rejected dynamically. The "confidence" of the lth rule is defined as

$$Q(l) = (1 - \delta) \times \varepsilon \times \min_i \left(\frac{1}{S_h(A_i, A_l)} \right) + \delta \frac{n_l}{T} .$$ (17)

Where, ε is constant, $S_h(A_i, A_l)$ is likelihood function between fuzzy sets A_i and A_l, T is number of identification steps, n_l is the times that l is selected as the nearest cluster in T-step identification; $\delta \in (0,1)$ is weight, which is used to adjust their action degree in confidence.

4.2 Steps for Exon Interval Endpoint Identification

Steps for exon interval endpoint identification based on T-S model are as follows.

Step1. Determination of sample sequence length and sequence coding

Cut out coding sequence with a length L_A ($L_A = L_1 + L_2$) for each exon interval endpoint respectively as model input, where L_1 and L_2 are upstream and downstream length of interval endpoint respectively.

Each basic group in sample sequence is provided with a real number. The corresponding relationship is

$$A \rightarrow 1.5; C \rightarrow 0.5; G \rightarrow -0.5; T \rightarrow -1.5; others \rightarrow 0.$$

Cut out a sequence with a length $L_A = 40$ for each possible interval endpoint sample and encode it into L_A dimensional vector as input of T-S model.

Step2. Interval endpoint identification based on T-S model

① Training. For exon samples and intron samples, make their ideal output respectively as 0.95 and -0.95, use Levenberg-Marquardt least square method to calculate T-S model structure and antecedent and consequent parameters through iteration.

② Test. Take the sample to be identified as input, calculate one real number output value through T-S model, set up a proper threshold value and discriminate the type of this sample.

5   Example Test and Analysis

KFCM, FCM, mean square method, weighted mean method with standard deviation and empirical threshold value method are compared by using sample data of gene sequence of 100 persons and rats and 200 mammals [12]. The discrimination accuracy statistics of threshold value determination of data of two types of gene with five methods are shown in Table 1. The overall accuracy of threshold value discrimination is defined as $A = (S_n + S_p)/2$, where sensitivity $S_n = T_P/(T_P + F_N)$, specificity $S_p = T_N/(T_N + F_P)$, T_P is the number of exons correctly determined, T_N is the number of introns correctly determined, F_N is the number of introns wrongly determined, F_P is the number of exons wrongly determined.

Table 1. Discrimination accuracy statistics of five threshold value determination methods

Evaluation index	Method	Human and mouse	Mammal
	KFCM	3.4898	2.1440
	FCM	3.4997	2.1650
Threshold value R_0	Empirical threshold value	2.0000	2.0000
	Mean	4.3880	2.5710
	Weighted mean	5.6199	2.9016
	KFCM	0.7166	0.6412
	FCM	0.7143	0.6396
Total accuracy A	Empirical threshold value	0.6165	0.6366
	Mean method	0.6909	0.6339
	Weighted mean	0.6544	0.6354

According to table 1, threshold value determination method based on KFCM has better classification performance.

The performance of exon position identification algorithm based on T-S model is verified with human mitochondrial gene sequence NC_012920 [12]. At first, power spectrum of gene sequence is calculated by power spectrum algorithm. Then, the positions of exon are preliminarily determined by threshold value determination algorithm. Figure 2 shows the analysis result of power spectrum. The preliminary determined positions of exon endpoints are shown in Table 2.

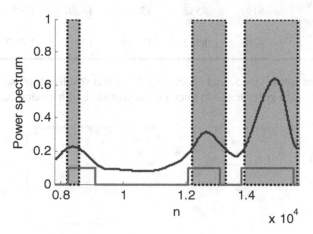

Fig. 2. Analysis result of power spectrum

In figure 2, the actual and preliminary determined intervals of exons are represented by solid and dashed boxes respectively. Endpoint sequence is tested with sliding window according to exon endpoint obtained and T-S fuzzy model established (50 points in forward and back direction respectively), as shown in Figure 3.

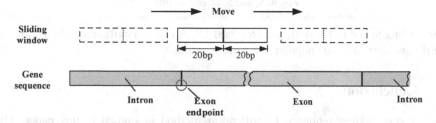

Fig. 3. Schematic diagram of sliding window test

100 groups of sequences obtained through sliding window are used as input of T-S fuzzy model. According to the sliding sequence identification result at endpoint 1, the identification result of sequence at exon endpoint 1 obtained through power spectrum is close to the feature of exon endpoint sequence (the identification result near this sequence section is closer to ideal output 0.95). Thus, the identification value of endpoint 1 is used as the start endpoint of this exon. According to the sliding sequence identification result at endpoint 2, the identification result of the 82nd sequence is closer to the feature of endpoint sequence. Therefore, the result is corrected as $8582+(82-50)\times5=8742$, and the correction method of other endpoints can be used in the same way. The corrected positions of exon endpoints are shown in Table 2.

Table 2. The preliminary determined and corrected positions of exon endpoints

	Endpoint 1	Endpoint 2	Endpoint 3	Endpoint 4	Endpoint 5	Endpoint 6
Preliminary determined positions	8197	8582	12199	13307	13933	15702
Corrected positions	8197	8742	12199	13217	13883	15652

Fig.4 shows the sensitivity S_n and specificity S_p of test results on 60 sequences selected from 100 human genes and 200 mice genes provided with random sampling.

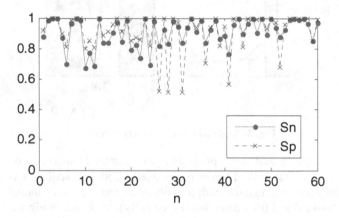

Fig. 4. S_n and S_p of test results

According to the test results, the proposed DNA exon identification algorithm can identify and correct exon endpoint to a certain extent.

6   Conclusion

A novel gene coding sequence identification method is studied in this paper. The improved short-time DFT algorithm based on Morlet wavelet can effectively reduces the amount of calculation and quickly extract the power spectrum of DNA sequence. Besides, the threshold value determined with KFCM has a better classification accuracy compared to FCM, mean square method, weighted mean method with standard deviation and empirical method. Moreover, Takagi-Sugeno fuzzy identification model can effectively identify and correct gene exon interval endpoint and has high accuracy. In this paper, threshold value is obtained by weighted sum of two clustering centers. Thus, the selection method of its weight coefficient is to be further studied. Meanwhile, it is necessary to deeply explore the identification of DNA exon sequences with repeated and consecutive segments.

References

1. Wang, Y.H., Zhang, C.T., Dong, P.X.: Recognizing shorter coding regions of human genes based on the statistics of stop codons. Biopolymers 63(3), 207–216 (2002)
2. Archana, Y., Manjula, P.N., Pavithra, K., et al.: Computational DNA sequence analysis. International Journal of Computer Trends and Technology 4(5), 1264–1268 (2013)
3. Kotlar, D., Lavner, Y.: Gene prediction by spectral rotation measure: a new method for identifying protein-coding regions. Genome Research 13(8), 1930–1937 (2013)
4. Randi, M., Zupan, J., Vikic-Topic, D., et al.: A novel unexpected use of a graphical representation of DNA: Graphical alignment of DNA sequences. Chemical Physics Letters 431(4-6), 375–379 (2006)
5. Liao, B., Ding, K.: Graphical approach to analyzing DNA sequences. Journal of Computational Chemistry 26(14), 1519–1523 (2005)
6. Som, A., Sahoo, S., Chakrabarti, J.: Coding DNA sequences: statistical distributions. Mathematical Biosciences 183(1), 49–61 (2003)
7. Yan, M., Lin, Z.S., Zhang, C.T.: A new fouirer transform approach for protein coding measure based on the format Z-curve. Bioinformatics 14(8), 685–690 (1998)
8. Yin, C., Yau, S.: Prediction of protein coding regions by the 3-base periodicity analysis of a DNA sequence. Journal of Theoretical Biology 247(1), 687–694 (2007)
9. Owens, F.J., Murphy, M.S.: A short-time Fourier transform. Signal Processing 14(1), 3–10 (1988)
10. Saritha, A.K., Ameera, P.M.: Image segmentation based on kernel fuzzy C means clustering using edge detection method on noisy images. International Journal of Advanced Research in Computer Engineering & Technology (IJARCET) 2(2), 399–406 (2013)
11. Takagi, T., Sugeno, M.: Fuzzy identification of systems and its application to modeling and control. IEEE Trans on Systems, Man and Cybernetics 15(1), 116–132 (1985)
12. Task A.: Gene identification problem and its algorithms. National Graduate Mathematical Contest in Modeling 2012, China (2012), http://gmcm.seu.edu.cn/_s2/00/47/c31a71/page.psp (accessed April 5, 2014)

Image Matching Using Mutual k-Nearest Neighbor Graph

Ting-ting Li[1], Bo Jiang[1], Zheng-zheng Tu[1], Bin Luo[1,2], and Jin Tang[1,2,*]

[1] School of Computer Science and Technology, Anhui University, Hefei 230601, China
[2] Key Lab of Industrial Image Processing and Analysis of Anhui Province, Hefei 230039, China
{litingtingad,ahhftang}@gmail.com

Abstract. Though weighted voting matching is one of most successful image matching methods, each candidate correspondence receives voting score from all other candidates, which can not apparently distinguish correct matches and incorrect matches using voting scores. In this paper, a new image matching method based on mutual k-nearest neighbor (k-nn) graph is proposed. Firstly, the mutual k-nn graph is constructed according to similarity between candidate correspondences. Then, each candidate only receives voting score from its mutual k nearest neighbors. Finally, based on voting scores, the matching correspondences are computed by a greedy ranking technique. Experimental results demonstrate the effectiveness of the proposed method.

Keywords: Image matching, mutual k-nn graph, spectral technique.

1 Introduction

Image matching is one of the important techniques in image processing, which has been widely used in many computer vision tasks including remote sensing, medical image analysis, 3D reconstruction and so on [1-5]. As an important aspect of image, image feature point plays an important role in analyzing images. Image feature point has a good adaptability for image deformation, occlusion and gray change. Hence, image feature point matching is widely used. Meanwhile, establishment of correct matching correspondences between image feature points is a fundamental problem in the field of computer vision, which is also a research hotspot. In fact, solving correct correspondence between feature points is a typical NP problem, and approximate methods are required. Generally, image feature point matching methods are mainly classified into two categories: First, through the similarity measurement between feature points to obtain the initial matches, and then use an algorithm to remove incorrect matches to get more precise matches [6]. Second, considering the geometrical relationship between feature points, combined with similarity measurement between feature points, matching results are obtained directly [7], [8].

Leordeanu and Hebert [7] proposed an effective method using a spectral relaxation technique. The main process is as follows. Firstly, a graph is constructed whose nodes

---

* Corresponding author.

H. Wang et al. (Eds.): ICYCSEE 2015, CCIS 503, pp. 276–283, 2015.

represent candidate correspondences and edges denote the affinities between candidate correspondences. Then, optimal matching between feature points is obtained by utilizing the principal eigenvector of affinity matrix. This method usually leads to better results [7], [9], [10]. However, when the number of candidate correspondences is large, the principal eigenvector computation is time-consuming [10]. Yuan et al [10] proposed a weighted voting method using a greedy technique. Each candidate correspondence not only gives its voting score to other candidate matches, but also receives voting scores from other candidate correspondences. Then, the optimal matches are computed by simple addition and ranking operation. This method significantly reduces the running time, but matching precision is difficult to maintain. Voting is an effective decision-making method, which has been widely used in image matching task [11], [12]. In this paper, we propose a new voting method by using the mutual k nearest neighbor graph model, in which each candidate match only receives voting from its mutual k nearest neighbors to make correct candidates get higher votes. Promising experimental results show the benefits of our method.

2   Mutual k Nearest Neighbor Graph

Assume that there are n vertices in the graph, for each vertex v_i, its k-nearest neighbors $v_1, v_2 \ldots v_k$ can be found through some kind of distance measurement. Vertex v_i connects to its k nearest neighbors with k directed edges. The direction of each edge is that v_i points to its k-nearest neighbors. In this way, a k-nn graph is constructed, where $V=\{v_1, v_2, \ldots, v_n\}$, $E=\{e_{ij} \mid v_i, v_j \in V, v_j \in KNN(v_i)\}$, $KNN(v_i)$ denotes the set of k nearest neighbors of vertex v_i. Mutual k-nn graph [13], [14] is a variant of the standard k-nn graph. In mutual k-nn graph, there is a connection between two vertices only if the rule of the neighborhood is fulfilled by both vertices, i.e., there is an edge between v_i and v_j if and only if $v_i \in KNN(v_j)$ and $v_j \in KNN(v_i)$. In contrast, in a k-nn graph, there is an edge between vertex v_i and v_j, if one of them belongs to the k-nearest neighbors of the other. Hence, the mutual k-nn graph is a sub-graph of the k-nn graph with the same data and the same value of k. The mutual k-nn graph is considered more restrictive. Mutual k-nn graph bi-directional mining vertical information is more than k-nn graph one-way mining vertical information.

3   Image Matching Based on Mutual k-nn Graph

In [10], each candidate match receives voting from all other candidate matches, however, incorrect candidate usually makes negative contribution to the whole voting result. To overcome this problem, we propose an image matching method which is based on mutual k-nn graph, called mutual k-nn matching (MKM). In MKM, each candidate only receives voting from its mutual k nearest neighbors. In this case, each correct matching correspondence usually receives voting from correct matches.

3.1   Mutual k-nn Graph Representation

In order to obtain mutual k nearest neighbors of each candidate, we need to represent matching correspondences as a mutual k-nn graph. The details are as follows. Let P and Q be two images to be matched. In image P, n key-points $(x_1, y_1)\ldots (x_n, y_n)$, indexed by $I=(1, 2, \ldots, n)$, are detected. In image Q, n' key-points $(x_{1'}, y_{1'})\ldots (x_{n'}, y_{n'})$, indexed by $I'=(1', 2', \ldots, n')$, are detected. Each pair of $c_i=(i, i')$, where $i \in I$ and $i' \in I'$, can be regarded as a candidate correspondence. In mutual k-nn graph, each candidate is taken as a vertex, i.e. $V= \{v_1, v_2 \ldots v_n \mid v_i=(i, i')\}$. There is an edge between v_i and v_j if each of them belongs to the k-nearest neighbors of the other. The measurement of near neighbor relationship is according to the similarity between candidate correspondences. The similarity S using geometric information is defined as follows:

$$S(c_i,c_j)=1-\left\|d_{ij}-d_{i'j'}\right\|\Big/\max_{u,v,u',v'}\left(\left\|d_{uv}-d_{u'v'}\right\|\right) \tag{1}$$

where $d_{ij}=\sqrt{(x_i-x_j)^2+(y_i-y_j)^2}\Big/\sqrt{\sigma_i^2+\sigma_j^2}, d_{i'j'}=\sqrt{(x_{i'}-x_{j'})^2+(y_{i'}-y_{j'})^2}\Big/\sqrt{\sigma_{i'}^2+\sigma_{j'}^2}.\sigma_i$ is the scale of feature point i in image P, (x_i, y_i) is the position of i in image P. $\sigma_{i'}$ is the scale of feature point i' in image Q, $(x_{i'}, y_{i'})$ is the position of i' in image Q. Note that similar definition is also presented in the work [15].

The similarity (Eq. (1)) is based on pair-wise geometric relationship between candidate matches. We can see that if both two candidates are correct, the geometric consistency is well preserved by them. Thus, in mutual k-nn graph, the higher similarity between two matches, the bigger probability of an edge exists between them. In contrast, incorrect matches happen in unstructured, random way and they fail to meet geometric consistency. Therefore, the similarities between incorrect matches and other candidates are low. The mutual k-nn graph bi-directionally verifies neighborhood for each pair of vertices. Hence, the number of mutual k nearest neighbors of incorrect matches is few. From above discussion, we can draw a conclusion that correct matches mainly receive high voting scores from correct matches, and vice-versa. The difference of voting score between correct and incorrect matches is obvious. Therefore, we can make use of voting score to distinguish correct and incorrect matches.

3.2   Image Matching Based on Mutual k-nn Graph

In section 3.1, how to represent matching correspondences as a graph is introduced. Then correct matches will be got according to voting score, which is similarity between two candidates. Each candidate receives and accumulates the voting scores given by its mutual k nearest neighbors and the final score is divided by the number of mutual k nearest neighbors. The overall algorithm can be summarized as below. Fig. 1 shows the voting score for each correspondence. Fig. 1(a) is initial matches, which consists of 100 correct correspondences (cyan lines) and 50 incorrect correspondences (red lines). In Fig. 1(b), every correspondence is marked by a round dot. Height of round dot in vertical direction represents for voting score for every correspondence. It is clear that correct matches get higher scores than incorrect ones.

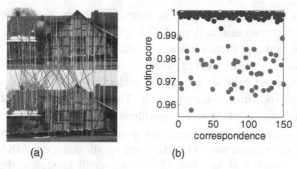

(a)   (b)

Fig. 1. An example of voting scores for 'house' image matching

---

Algorithm 1.   Image matching based on mutual k-nn graph (MKM)

1: Calculate similarity matrix S between candidate matches O according to Eq(1)

2: Build mutual k-nn graph using similarity S as

$G_{MKNN}=(V,E)$,   where $V=\{v_1,v_2,...,v_n$ $|v_i$ $=(i,i')\}$, $E=\{e_{ij}$ $|$ $v_j\in KNN(v_i) \wedge v_i\in KNN(v_j)\}$

3: Calculate voting scores for each c_i as: $score(c_i)=\sum_{j\neq i}\delta(c_i,c_j)*S(c_i,c_j)$

   where $\delta(c_i,c_j)=1$ if $(c_i,c_j)\in E$, 0 otherwise

4: Normalize: $score(c_i)=score(c_i)/degree(c_i)$

5: **While(** O is not empty) **do**

6:     Find the current correct match c^* in O as: $c^*=\arg\max_{c_i\in O} score(c_i)$

7:     Add match c^* to the correct correspondence set C and remove c^* from O

8:     Remove all potential candidates in conflict with $c^*=(i, i')$ from O. These are
       matches of the form (i, k) and (q, i') for one-to-one matching constraints.

9: **End while**

3.3   Computational Complexity

There are two main parts for the whole proposed method: mutual k-nn graph construction and voting, i.e., $O=O_{knn} + O_{voting}$. Mutual k-nn graph construction process can be performed in $O(kn\log n)$, where n is the number of candidate matches. Computing the score of each correspondence can be performed in $O(kn)$.

4   Experimental Results

In order to evaluate the performance of the proposed method, a comparison between MKM, SM [7], WVM [10] and RANSAC [16] has been performed. The computation of the affinity matrix for these comparison methods is $W(c_i, c_j)=S(c_i, c_j)$. The tested images are selected from Zurich Building Image Database (ZuBud) [17]. For each image, the SIFT [6] features are computed firstly, then they are matched based on nearest neighbor distance ratio [18] in order to obtain the candidate matches. The ground truth has been manually marked. For each image pair, 200 correct matches are randomly selected and incorrect matches are then generated. The results are presented with the average recall rate [18] and the average precision rate [18] because the added random incorrect matches are regenerated every time. The algorithms are repeatedly executed 100 times for each recall. Through the average precision versus recall curve, performances of different algorithms are compared.

Fig. 2 shows the results. Here, the initial matches contain 200 correct and 100 incorrect matches. We can see that (1) at the same recall value, our MKM always gains higher precision than WVM and SM. (2) As the recall increases, the precision decreases due to incorrect matches, and MKM maintains higher results than other methods. To further validate the robustness of MKM method, we evaluated our method in the case of larger number of incorrect matches, as shown in Fig. 3. The initial matches contain 200 correct and 150 incorrect matches. Here, we note that at the same recall value, MKM still obtains higher precision than WVM and SM, indicating the robustness of the proposed MKM method. This is because in mutual k-nn graph, no matter how many incorrect matches increase, each candidate only receives voting from its mutual k nearest neighbors. That is, increase of incorrect matches will not affect k nearest neighbors of correct matches significantly. Therefore, each correct match mainly receives voting from correct matches. Correct matches can get higher voting scores than incorrect matches. Fig.4 further shows the comparison results with RANSAC. Note that our MKM slightly outperform RANSAC. However, MKM is obviously faster than RANSAC. Fig.5 shows some matching results, in which initial matches containing 200 correct and 150 incorrect matches are show in Fig.5 (a). Fig.5 (b-d) shows matching results using WVM, SM, MKM, respectively, as they return 150 correspondences from initial matches. Note that our MKM generally performs better than SM and WVM.

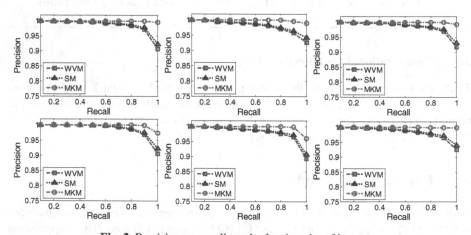

Fig. 2. Precision vs. recall results for six pairs of images

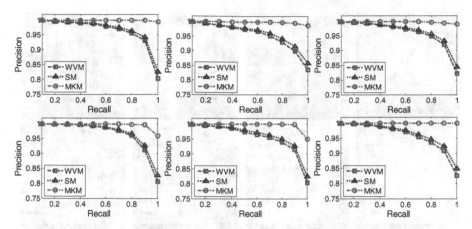

Fig. 3. Precision vs. recall results when larger number of incorrect matches exists

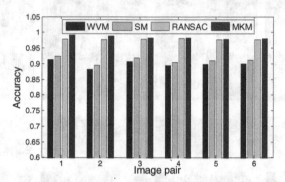

Fig. 4. Comparison results between different matching methods

5   Conclusions

A new image matching method based on mutual k-nn graph is proposed. Each candidate is treated not only as a candidate but also a voter. Our method takes the advantage of the fact that the incorrect matches generally have few mutual k nearest neighbors which happen in unstructured, random way. For each candidate, voting is summed from its mutual k nearest neighbors, and it reflects how other candidate correspondences support this candidate. The experiments show that the proposed algorithm obtains better matching results.

Acknowledgment. This work is supported by the National Natural Science Foundation of China (No. 61402002, 61472002); the Natural Science Foundation of Anhui Higher Education Institutions of China (No. KJ2014A015, KJ2013A007).

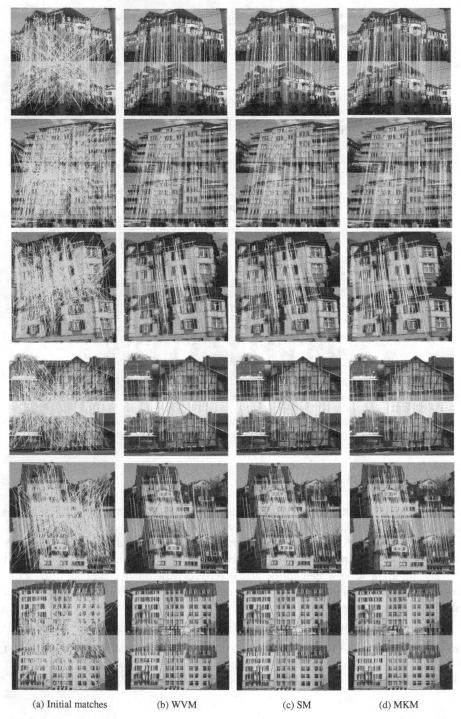

(a) Initial matches   (b) WVM   (c) SM   (d) MKM

Fig. 5. Matching results of different methods (incorrect matches are marked by red lines)

References

1. Huang, C.R., Chen, C.S., Chung, P.C.: Contrast context histogram—an efficient discriminating local descriptor for object recognition and image matching. Pattern Recognition 41, 3071–3077 (2008)
2. Misra, I., Manthira Moorthi, S., Dhar, D., et al.: An automatic satellite image registration technique based on Harris corner detection and random sample consensus outlier rejection model. In: 1st International Conference on Recent Advances in Information Technology, pp. 68–73. IEEE Computer Society, United States (2012)
3. Shin, D., Tjahjadi, T.: Clique descriptor of affine invariant regions for robust wide baseline image matching. Pattern Recognition 43, 3261–3272 (2010)
4. Zhang, G.P., Wang, Y.H.: Robust 3D face recognition based on resolution invariant features. Pattern Recognition Letter 32, 1009–1019 (2011)
5. Liu, Z.X., An, J.B., Meng, F.R.: A Robust Point Matching Algorithm for Image Registration. In: 4th International Conference on Machine Vision. SPIE, United States (2012)
6. Lowe, D.G.: Distinctive Image Features from Scale-invariant Keypoints. Int. J. Comput. Vision 60, 91–110 (2004)
7. Leordeanu, M., Hebert, M.: A spectral technique for correspondence problems using pairwise constraints. In: 10th IEEE International Conference on Computer Vision, pp. 1482–1489. IEEE Press (2005)
8. Caelli, T., Kosinov, S.: An eigenspace projection clustering method for inexact graph matching. IEEE Trans. Pattern. Anal. Mach. Intell. 26, 515–519 (2004)
9. Tao, D.C., Li, X.L., Wu, X.D., et al.: Geometric mean for subspace selection. IEEE Trans. Pattern Anal. Mach. Intell. 31, 260–274 (2009)
10. Yuan, Y., Pan, Y.W., Wang, K.Q., et al.: Efficient image matching using weighted voting. Pattern Recogn. Lett. 33, 471–475 (2012)
11. Lipman, Y., Funkhouser, T.: Möbius voting for surface correspondence. ACM Trans. Graphics 28, 1–12 (2009)
12. Kin-Chung Au, O., Tai, C.-L., Cohen-Or, D., et al.: Electors voting for fast automatic shape correspondence. Computer Graphics Forum 29, 645–654 (2010)
13. Berton, L., De Andrade Lopes, A.: Informativity-Based Graph: Exploring Mutual kNN and Labeled Vertices for Semi-Supervised Learning. In: 4th International Conference on Computational Aspects of Social Networks, pp. 14–19. IEEE Computer Society, United States (2012)
14. Ozaki, K., Shimbo, M., Komachi, M., et al.: Using the Mutual k-Nearest Neighbor Graphs for Semi-supervised Classification of Natural Language Data. In: 15th Conference on Computational Natural Language Learning, pp. 154–162. Association for Computational Linguistics (ACL), United States (2011)
15. Carneiro, G., Jepson, A.D.: Flexible Spatial Configuration of Local Image Features. IEEE Trans. Pattern Anal. Mach. Intell. 29, 2089–2104 (2007)
16. Fischler, M.A., Bolles, R.C.: Random sample consensus: a paradigm for model fitting with applications to image analysis and automated cartography. Communications of the ACM 24, 381–395 (1981)
17. Shao, H., Svoboda, T., Van Gool, L.: Zubud-zurich buildings database for image based recognition. Technical Report 260, Computer Vision Laboratory, Swiss Federal Institute of Technology (2003)
18. Mikolajczyk, K., Schmid, C.: A performance evaluation of local descriptors. IEEE Trans. Pattern Anal. Mach. Intell. 27, 1615–1630 (2005)

A New Speculative Execution Algorithm Based on C4.5 Decision Tree for Hadoop

Yuanzhen Li[1], Qun Yang[1], Shangqi Lai[2], and Bohan Li[1]

[1] Nanjing University of Aeronautics and Astronautics, NanJing, China
{liyuanzhen,qun.yang,bhli}@nuaa.edu.cn
[2] The University of Hong Kong, HongKong, China
aquas@hku.hk

Abstract. As a distributed computing platform, Hadoop provides an effective way to handle big data. In Hadoop, the completion time of job will be delayed by a straggler. Although the definitive cause of the straggler is hard to detect, speculative execution is usually used for dealing with this problem, by simply backing up those stragglers on alternative nodes. In this paper, we design a new Speculative Execution algorithm based on C4.5 Decision Tree, SECDT, for Hadoop. In SECDT, we speculate completion time of stragglers and also of backup tasks, based on a kind of decision tree method: C4.5 decision tree. After we speculate the completion time, we compare the completion time of stragglers and of the backup tasks, calculating their differential value, and selecting the straggler with the maximum differential value to start the backup task. Experiment result shows that the SECDT can predict execution time more accurately than other speculative execution methods, hence reduce the job completion time.

Keywords: speculative execution, C4.5 decision tree, Hadoop.

1 Introduction

With the development of big data, Hadoop has become the next-generation Internet information processing platform [1], [2]. In Hadoop [3], jobs are decomposed into a series of tasks for paralleling running. This makes job execution time depend on slow-running tasks ‒ namely stragglers. Tasks may be slow for various reasons, such as hardware errors or software bugs, but the definitive causes may be hard to detect. Hadoop doesn't try to fix slow-running tasks, but tries to detect slow-running tasks in order to launch a backup for them. This is termed *speculative execution* of tasks. Although speculative execution mechanism is an effective way to solve the above problem, there are some unnecessary startup of backups as a result of inaccurately calculating completion time of task.

C4.5 decision tree algorithm is an extension of ID3 algorithm. It can be used for classification. It has several advantages: fast learning, high accurate classification, simple shape model and better robustness for noise data. In this paper, C4.5 decision tree is used to accurately classify historical information of successful *map/reduce* tasks in each node to construct decision trees. Besides, it is

H. Wang et al. (Eds.): ICYCSEE 2015, CCIS 503, pp. 284–291, 2015.
© Springer-Verlag Berlin Heidelberg 2015

convenient for the stragglers and backup tasks to quickly and accurately find a set of similar *map/reduce* tasks by traversal decision trees. These trees are used as the basics for the prediction of completion time of stragglers and of backup tasks. It is an efficient and feasible way for accurately prediction of completion time in speculative execution.

In this paper, in order to accurately predict the completion time of stragglers and of backup tasks, and reasonable select backup nodes, we propose a new speculative execution algorithm based on C4.5 decision tree (SECDT). SECDT establishes four decision trees — two weight decision trees (WDT_{map} and WDT_{reduce}) and two time decision trees (TDT_{map} and TDT_{reduce}) to classify historical information of successful *map/reduce* tasks. These decision trees are based on the weights of each phase of *map/reduce* tasks and the execution time of *map/reduce* tasks. The prediction of completion time of stragglers and of backup tasks are implemented by traversal WDT and TDT. SECDT further classifies work nodes into *map* quick nodes and *reduce* quick nodes. Thus, SECDT can start backup tasks for *map* stragglers on *map* quick nodes and start backup tasks for *reduce* stragglers on *reduce* quick nodes.

The rest of this paper is organized as follows. In section 2, we will introduce on related work. In section 3, we design a new speculative execution model, and then propose our algorithm, SECDT. In section 4, we describe and analyze our experiment. Finally, section 5 makes conclusions and illustrates our future work.

2   Related Work

Hadoop is an open source implementation of distributed storage and computing system. It reflects the absolute advantage in high-volume data processing [4]. In this framework, failure [5] is the normal event rather than the exception. In view of the undetermined failure for straggler, speculative execution is utilized to support fault tolerance in Hadoop. In the past few years, a lot of efficient speculative execution algorithms([6], [7], [8]) have been proposed to reduce the execution time of job sets.

Longest Approximate Time to End (LATE) [9], as a classical MapReduce scheduling algorithm, is widely used in speculative execution. There are three principles in LATE: prioritizing tasks to speculate, choosing fast nodes to run on and capping straggler tasks to prevent thrashing. It selects the tasks with longest remaining time and start up the backup tasks. In case any of the backup tasks finishes earlier than the original task, the original task will be terminated. However, because the progress of tasks is computed on the basis of fixed-weight of phases of *map/reduce* tasks, the performance of the job is poor.

To solve the above problem, many researchers proposed kinds of methods. In [10], a History-based Auto-Tuning (HAT) algorithm is proposed. It tunes the weights of phases of a *map/reduce* task according to the value of history tasks and uses the accurate weights of phases of the last historical task to calculate the progress of current tasks. Although the algorithm uses history tasks predict execution time, it just reserves weights of phases of the last task. In [11], a Self-Adaptive MapReduce scheduling algorithm (SAMR) is proposed. It predicts progress of tasks dynamically. SAMR reads historical information to adjust time weight of each stage of *map* and *reduce* tasks. However, it does not consider that different types of jobs have different weights of phases of *map/reduce* task. Besides, it falls short of solving a crucial problem that only hardware heterogeneity is considered. To overcome these defects, [12] proposes an Enhanced Self-Adaptive MapReduce scheduling algorithm (ESAMR) to improve the speculative re-execution.

It can be concluded that inaccurately predicting execution time are common among the previous research work, which causes the poor performance. In this paper, we propose and implement SECDT: a new Speculative Execution strategy based on C4.5 Decision Tree. We predict execution time of stragglers and of backup tasks accurately, to determine whether backup tasks to be started. Experiments show that it can effectively decrease the completion time of job set.

3   SECDT: A New Speculative Execution Algorithm Based on C4.5 Decision Tree

3.1   Model of SECDT

(1) Tasks in Hadoop are classified into two categories: *map* and *reduce*. The execution process of a *map* task comprises two phases and the execution process of a *reduce* task comprises three phases as shown in Table 1.

Table 1. phases of *map* and *reduce* task

map task		reduce task		
map	recoder	shuffle	sort	reduce
WM_1	WM_2	WR_1	WR_2	WR_3
$P(M_1)$	$P(M_2)$	$P(R_1)$	$P(R_2)$	$P(R_3)$

In Table 1, WM_1 is the weight of map phase and WM_2 is the weight of recorder phase in *map* task. Similarly, WR_1, WR_2 and WR_3 are respectively weights of shuffle, sort and reduce phase in *reduce* task. WM_1, WM_2, WR_1, WR_2 and WR_3 are between 0 and 1. Furthermore, the sum of WM_1 and WM_2 is 1 and the sum of WR_1, WR_2 and WR_3 is 1. In addition, the progress score of each phase is defined in Table 1. For example, $P(R_1)$ indicates the progress score of shuffle phase, which is calculated by $H(R_1)/sum(R_1)$. $H(R_1)$ is the number of key/value pairs that have been handled in R_1 phase and $sum(R_1)$ is the sum of key/value pairs that needed to be handled in R_1 phase. So are $P(M_1)$, $P(M_2)$, $P(R_2)$ and $P(R_3)$.

(2) The progress score of a task is calculated by equation (1), the sum of schedule of all phases in the task T.

$$prog_T = \begin{cases} \sum_{i=1}^{2} WM_i * P(M_i), & T \text{ is map task} \\ \sum_{i=1}^{3} WR_i * P(R_i), & T \text{ is reduce task} \end{cases} \tag{1}$$

(3) If a backup is allowed to started, it must need the following constraint (a):

$$estEndTime_1 > estEndTime_2$$

where $estEndTime_1$ is the predicted completion time of a straggler and $estEndTime_2$ is the predicted completion time of a backup.

3.2   SECDT Algorithm Design

SECDT is performed in four steps, including, 1) constructing WDTs and TDTs; 2) predicting the completion time of straggler; 3)in the meantime, choosing a quick node and predicting the completion time of backup, 4) judging whether it has the right to be speculative. The following introduces these four steps in details.

1) Constructing WDTs and TDTs by C4.5 Decision Tree algorithm

In Hadoop cluster, CPU speed(cpuSpeed), load, free memory(M), literacy rate of disk(disk), network transmission speed(network) and the amount of inputting data(inputData) play an important role in task execution. We firstly extract all of the above attribute information corresponding to completed tasks. But, cpuSpeed, M, disk, network and load are only reserved at NodeManager(NM), while tasks run on containers. Besides, container cannot communicate with NM directly.

To solve the above problem, *sigar* module, a portable interface for gathering system information, is introduced to get attribute information. We modify the heartbeat message which can convey these attribute values to containers. When a task is successful, the attribute values, weights of phase and execution time of task will be written in log files of HDFS. Each node can read their log files to construct decision trees; so tasks can extract all attribute information for the next step. Take *map* tasks for instance, respectively taking the weight of phases and execution time of tasks as target, we utilize C4.5 algorithm to construct a weight decision tree of *map* (WDT_{map}) and a time decision tree of *map* (TDT_{map}) in each node. So are WDT_{reduce} and TDT_{reduce} for *reduce* tasks. The WDTs and TDTs provide traversal search operation for the following stragglers and backup tasks.

2) Predicting the completion time of stragglers

In the step 2, decision trees are established successfully. Next, we predict the completion time of stragglers. Take a straggler γ for example, it is a *reduce* and runs on sort phase. Given the heartbeat message transfer mechanism modified in step 1), SECDT gets the corresponding attribute information of γ.

SECDT scheduler traverses the WDT_{reduce} from root, layer by layer until searching a leaf node. It can acquire a similar sub-set of tasks with the same weights of phases. According to the leaf node, the progress score of straggler γ is calculated by equation (1). The completion time of straggler γ, $estEndTime_1(\gamma)$, is easily calculated by equation (2) and (3).

$$estEndTime_1(\gamma) = estRunTime_\gamma + AttemptStartTime_\gamma \qquad (2)$$

$$estRunTime_\gamma = (currTime - AttempStartTime_\gamma)/prog_\gamma \qquad (3)$$

where $currTime$ is the current time, $AttemptStartTime_\gamma$ denotes the start time of task γ. But, we cannot avoid that some individual target instances do not exist, so that there is no feasible leaf node with matching. For this situation, the completion time of stragglers are calculated by the Hadoop default algorithm.

3) Predicting completion time of backup task

How to select the work node for backup task is equally important as how to select backup services. If the backup task is allocated at a node with poor performance, it will add workload to that node and waste system resources. Consequently, how to detect quick nodes are extremely vital. In SECDT, we use the success rate of $map/reduce$ tasks on a node to represent the performance of $map/reduce$ task of the node. The nodes with the highest success rate of $map/reduce$ task are $map/reduce$ quick nodes. Given a node Ω with M map tasks and R $reduce$ tasks. The success rate of $map/reduce$ task in Ω, denoted by MSR_Ω and RSR_Ω, are calculated by equation (4).

$$\begin{cases} MSR_\Omega = \frac{Map_{success}}{M} \\ RSR_\Omega = \frac{Reduce_{success}}{R} \end{cases} \qquad (4)$$

where $Map_{success}$ is the number of successful map tasks and $Reduce_{success}$ is the number of successful $reduce$ tasks in Ω. If the backup is a $reduce$ task, we will select $reduce$ quick node with the highest RSR as the work node.

After determined work node, we traverses the $TDT_{map/reduce}$ from root, layer by layer until searching a leaf node. If the leaf node only owns one historical task, the execution time of the historical task is the execution time of backup. If the leaf node owns a group of similar tasks, the execution time of backup, $runTime_2$, is the average time of all tasks of the leaf node. So the completion time, $estEndTime_2$, of backup is calculated by $currTime + runTime_2$.

4) Judging and Start up

Hadoop chooses a task for it from one of three categories. First, any failed tasks are given highest priority. Second, non-running tasks are considered. Last, stragglers are taken into consideration. In this step, we will discuss which tasks to be started.

After step 2) and 3), the completion time of stragglers and of backup tasks have been accurately computed. For these stragglers with meeting constraint (a), SECDT joins them into the queue Que_{task}. When Hadoop chooses stragglers to start up, it also selects the straggler with the maximum differential value between $estEndTime_1$ and $estEndTime_2$ in Que_{task}.

4   Experiments and Analysis

4.1   Experimental Environment

In this section, the experiments are performed to evaluate the proposed approach: SECDT. The experimental subjects are data-intensive job set and compute-intensive job set. We compare SECDT with the other two existing classical speculative execution mechanism: Hadoop default speculative execution (DEFAULT) and LATE. Table 2 provides configuration information of 16 virtual nodes used for experiment. We use two well-known Hadoop applications, *wordcount* and *PI* calculated by Monte Carlo method, as our benchmarks to evaluate performances of SECDT.

<table>
<tr><td colspan="4">Table 2. node configuration</td><td colspan="4">Table 3. PI jobs in a job set</td></tr>
<tr><td>Class</td><td>RAM</td><td>CPU</td><td>Numbers</td><td>Job</td><td>Numbers</td><td>Job</td><td>Numbers</td></tr>
<tr><td>1</td><td>1GB</td><td>2</td><td>4</td><td>50*50</td><td>5</td><td>100*100</td><td>5</td></tr>
<tr><td>2</td><td>2GB</td><td>2</td><td>4</td><td>200*200</td><td>5</td><td>300*300</td><td>5</td></tr>
<tr><td>3</td><td>3GB</td><td>2</td><td>4</td><td></td><td></td><td></td><td></td></tr>
<tr><td>4</td><td>4GB</td><td>2</td><td>4</td><td></td><td></td><td></td><td></td></tr>
</table>

4.2   Results and Analysis

The performances are evaluated with the completion time of job set and variation trend of a single job completion time. For *wordcount* applications, the input files of *wordcount* is continuous 7 days news data, one day data as one job, downloading from *datatang* site [13]. *PI* is calculated by Monte Carlo method based on the casting points. A *PI* job is expressed by $M_1 * M_2$, which M_1 is the number of *map* task and M_2 is the casting points in each *map* task. Table 3 provides a job set information.

Firstly, we compare the whole completion time of job set. In Fig. 1, on average, SECDT finishes the job set 13 percent faster than LATE and 30 percent faster than DEFAULT. In Fig. 2, SECDT finishes the job set 18 percent faster than LATE and 33 percent faster than DEFAULT. We ascribe the poorer results of DEFAULT and LATE into the reason that LATE adopts the static fixed-weight based method to predict the execution time of stragglers, while DEFAULT uses the current progress rate. Compared with SECDT, the completion time of DEFAULT and of LATE are inaccurately and hazy.

In order to observe the variation trends of job completion time, we respectively select the seventh day news job (Job A) and a 300*300 job (Job B) as the next research target, from the above job set of *wordcount* and *PI*. We note their results during being executed in 20 times. In the experiment, as the number of job increases, the SECET finds it easier to construct blameless WDTs and TDTs to predict the completion time of stragglers and of backup tasks.

From Fig. 3 and Fig. 4, the job completion time of DEFAULT and of LATE are fluctuated irregularly. But, in SECDT, we can see that, there is a downtrend

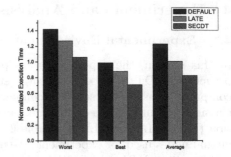

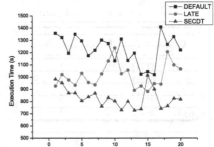

Fig. 1. completion time of *wordcount* job set

Fig. 2. completion time of *PI* job set

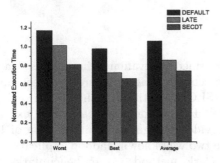

Fig. 3. trend of Job A completion time   **Fig. 4.** trend of Job B completion time

of execution time firstly, and then trends to steady in a certain number of times. In addition, in Fig. 3, at the 11 time, there is an obvious recovery time, called as turning point. As time goes and task increases, the amount of task data in log files is more and more. If historical information is excessive, the overhead for constructing decision trees may increase a large cost of speculative execution, which reduces our algorithms performance. Thus, to avoid large accumulation of historical information, we add a time threshold, default 3 hour (it can be set by user), to clear historical information. So it is the reason why turning point exists. As Fig. 4 shows, there is a turning point at the 15 time.

Generally speaking, SECDT completes a job set at least 13 percent faster than LATE and 30 percent faster than DEFAULT. No matter whether of the completion time of job set or single job, the experimental results further support the argument that SECDT is a feasible and efficient speculative execution algorithm both in theory and practice.

5 Conclusion

Traditional speculative execution cannot predict execution time accurately, because the completion time of stragglers and of backup tasks are respectively calculated based on current progress rate and the average time of all successful tasks. To address the problem, we propose and implement SECDT: a new speculative execution algorithm based on C4.5 decision tree. To get attribute values corresponding to current operating task, we modify the heartbeat message transfer mechanism. Furthermore, SECDT uses the weights of phases and execution time of success *map/reduce* tasks in log files to construct WDTs and TDTs; and then, it classifies cluster nodes into *map* quick nodes and *reduce* quick nodes. We presume to assign backups, against *reduce* stragglers, to *reduce* quick nodes, and vice versa. Afterwards, stragglers get a series of current attribute information of work node and themselves, to accurately predict execution time by traversal WDT. In the meantime, backup tasks get current attribute information of this node to gain the execution time by traversal TDT. Last, we compare the completion time between stragglers and backup tasks and select a straggler with the maximum differential value to start backup task.

Acknowledgements. This paper is financed and supported by the National Natural Science Fund(41301407) and Jiangsu Provincial Natural Science Fund(BK20130819).

References

1. Shvachko, K., Kuang, H., et al.: The hadoop distributed file system. In: 2010 IEEE 26th Symposium on Mass Storage Systems and Technologies (MSST), pp. 1–10 (2010)
2. Bhandarkar, M.: Hadoop: a view from the trenches. In: Proceedings of the 19th ACM SIGKDD International Conference on Knowledge Discovery and Data Mining, pp. 1138–1138 (2013)
3. Apache (2014), http://hadoop.apache.org/docs/r2.2.0/
4. Hsu, I.: Multilayer context cloud framework for mobile Web 2.0: a proposed infrastructure. International Journal of Communication Systems 26(5), 610–625 (2013)
5. Kim, Y.P., Hong, C.H., Yoo, C.: Performance impact of JobTracker failure in Hadoop. International Journal of Communication Systems (2014)
6. Wang, B.Y., Pu, X.Y.: Study of an improved hadoop speculative execution algorithm. Applied Mechanics and Materials 513, 2281–2284 (2014)
7. Hayashi, A., Grossman, M., et al.: Speculative execution of parallel programs with precise exception semantics on gpus. LCPC (2013)
8. Xu, H., Lau, W.C.: Optimization for Speculative Execution of Multiple Jobs in a MapReduce-like Cluster. arXiv preprint arXiv (2014)
9. Chen, Q., Liu, C., Xiao, Z.: Improving mapreduce performance using smart speculative execution strategy (2013)
10. Chen, Q., Guo, M., et al.: HAT: history-based auto-tuning MapReduce in heterogeneous environments. The Journal of Supercomputing 64(3), 1038–1054 (2013)
11. Chen, Q., Zhang, D., et al.: Samr: A self-adaptive mapreduce scheduling algorithm in heterogeneous environment. In: 2010 IEEE 10th International Conference on Computer and Information Technology (CIT), pp. 2736–2743. IEEE (2010)
12. Sun, X.: An enhanced self-adaptive MapReduce scheduling algorithm. University of Nebraska (2012)
13. China news summary of November 2013 (2013), http://www.datatang.com/data/45718

An Improved Frag-Shells Algorithm for Data Cube Construction Based on Irrelevance of Data Dispersion

Dong Li[1], Zhipeng Gao[1], Xuesong Qiu[1], Ran He[2], Yuwen Hao[3],
and Jingchen Zheng[3,*]

[1]State Key Laboratory of Networking and Switching Technology,
Beijing University of Posts and Telecommunications
{saiy,gaozhipeng,xsqiu}@bupt.edu.cn
[2]Space Star Technology Co., Ltd. China
cnmoon@qq.com
[3]General Hospital of Chinese People's Armed Police Forces, Beijing, China
{how_yuwen,saiyep}@163.com

Abstract. On-Line Analytical Processing (OLAP) is based on pre-computation of data cubes, which greatly reduces the response time and improves the performance of OLAP. Frag-Shells algorithm is a common method of pre-computation. However, it relies too much on the data dispersion that it performs poorly, when confronts large amount of highly disperse data. As the amount of data grows fast nowadays, the efficiency of data cube construction is increasingly becoming a significant bottleneck. In addition, with the popularity of cloud computing and big data, MapReduce framework proposed by Google is playing an increasingly prominent role in parallel processing. It is an intuitive idea that MapReduce framework can be used to enhance the efficiency of parallel data cube construction. In this paper, by improving the Frag-Shells algorithm based on the irrelevance of data dispersion, and taking advantages of the high parallelism of MapReduce framework, we propose an improved Frag-Shells algorithm based on MapReduce framework. The simulation results prove that the proposed algorithm greatly enhances the efficiency of cube construction.

Keywords: OLAP, MapReduce, Data cube construction, Frag-Shells, Data dispersion.

1 Introduction

On-Line Analytical Processing (OLAP) is a hot topic in the field of data warehouse and decision support systems. OLAP analysis is a multi-dimensional data analysis based method. Most OLAP analysis need cube pre-computation before data analysis. However, the challenge is that if all of data cubes should be calculated in advance, the storage space required may *explode*, especially when the cube contains a lot of dimensions. So pre-computing all the data cubes is unrealistic. If the amount and the

---

* Corresponding author.

H. Wang et al. (Eds.): ICYCSEE 2015, CCIS 503, pp. 292–299, 2015.

scale of data cubes are too large, a more rational choice is partially pre-computation that only the cubes may be useful are pre-computed.

Frag-Shells algorithm is a solution to this issue. The algorithm can deal with very high dimension database, and quickly calculate on-line small local cubes. Since the cube shell is pre-computed, the data for a query is accessible from the cube shell directly, without the need to calculate the bottom data. Thus a lot of time can be saved. And required storage space is small.

When the dimension of each fragment is less than 5, with the increasing of data amount, the storage space shows a linear growth. So compared with other cube construction, if the *queried dimension* is less than 4, the Frag-Shells algorithm is highly recommended. But the original Frag-Shells algorithm has the defect that it excessively depends on the dispersion of data. Therefore, we propose an improved Frag-Shells algorithm which is irrelevance to data dispersion and only need to traverse the original data for one time to calculate cube shell, thus the proposed algorithm would outperform the original one.

The MapReduce framework is proposed by Google for parallel computing on large data sets. As Frag-Shells algorithm is of a certain parallelism it can be improved and applied in the MapReduce framework, from which the improved algorithm will benefit much.

2   Related Work

MapReduce framework is the model used for massive data processing and parallel computing. It is based on two steps, namely Map and Reduce. In the Map function, the entire job is divided into some smaller sub-problems to be assigned to the different working nodes. The input value of map function will be assembled into intermediate key/value pairs, which will be handed over to the Reduce function. The Reduce function collects and merges all the pairs generated by the Map function, and somehow combine to form the output.

In parallel computing data cube Reference [1] verifies the feasibility of the data cube construction based on MapReduce framework by comparing the performance of experiments in the physical cluster and virtual cluster. Reference [2], introduces the use of MapReduce framework to calculate data cube by three methods, and compare the performance of them. The two works only verify the feasibility of calculating the cube in MapReduce framework, but did not specify how to create a simulation model. Reference [3] proposed a method to build quotient cube with MapReduce framework. Reference [4] presents the full data cube computing applications in MapReduce. Works [3], [4] only present two explicit examples to tell the advantages of parallel computing data cube, but they did not consider the influence of the data dispersion to the simulation results.

In this paper, we propose the improved Frag-Shells algorithm, and apply it to the MapReduce framework. The details of the improvement are specified in chapter 3, together with Frag-Shells simulation models based on MapReduce framework. In chapter 4, the feasibility and performance of our algorithm are to be verified through tense simulation. And this paper is concluded in chapter 5.

3   Implementation

3.1   Frag-Shells Algorithm

The basic idea of the Frag-Shells algorithm is as follows: The multi-dimensional data set is divided into *n-dimensional* disjoint dimension fragments, each of which will be converted into the inverted index form. Then compute the shell fragment cubes, keeping the inverted index associated with the original cube. Using pre-computed shell fragment cubes, we can dynamically assemble and calculate the data cube needed. This can be effectively done through the collection of intersection operation on the inverted index. We will show how to form the inverted index as follows.

Table 1. Dataset

TID	A	B	C	D	E	F	G
1	a_1	b_1	c_1	d_2	e_1	f_1	g_2
2	a_1	b_2	c_2	d_3	e_2	f_1	g_2
3	a_1	b_2	c_2	d_1	e_2	f_1	g_3
4	a_2	b_2	c_1	d_1	e_3	f_1	g_3
5	a_2	b_3	c_1	d_2	e_1	f_1	g_1
6	a_3	b_2	c_1	d_3	e_2	f_1	g_2
7	a_3	b_3	c_1	d_2	e_2	f_1	g_1
8	a_3	b_3	c_2	d_2	e_1	f_1	g_2

For each dimension of each attribute value table 2, we list all tuple identifiers (*TID*) which have the same attribute value. For example, attribute a1 is in tuples of 1, 2 and 3, and then the *TID* list contains three values, i.e., 1, 2, 3. Now each dimension attribute in the original database is scanned to calculate the corresponding *TID* list, as is shown in the following table 2.

Table 2. TID list of each attributes

Attributes	TID list	size	Attributes	TID list	size	Attributes	TID list	size
a_1	{1,2,3}	3	c_1	{1,5,6,7}	4	e_2	{2,3,6,7}	4
a_2	{4,5}	2	c_2	{2,3,8}	3	e_3	{4}	1
a_3	{6,7,8}	3	d_1	{3,4}	2	f_1	{all}	8
b_1	{1}	1	d_2	{1,5,7,8}	4	g_1	{5,7}	2
b_2	{2,3,4,6}	4	d_3	{2,6}	2	g_2	{1,2,6,8}	4
b_3	{5,7,8}	3	e_1	{1,5,8}	3	g_3	{3,4}	2

Table 3 retains all the information of the original database which is a lossless storage. The next step is to construct a shell fragment. According to predefined dimension separating plan, we will separate all of dimension into different collections. The separate plan could be designed according to the query requirements in specific project or the distribution statistics on query dimension after a period of time. It's beneficial to put the dimensions, frequently queried together with each other, into the same group. Then in every fragment, calculate the full data cube, and retain the inverted index. In the above example, we take the separating plan of $\{(A, B, C), (D, E), (F, G)\}$ as our grouping criteria, according to the value of each attribute cube intersect the table TID list depth-first order to calculate local full data cube.

For 1-dimensional cube is already calculated in table 2. To get 2-dimensional cube AB, we need all the attribute values of cube A and B in table 1, and intersect them to get the TID list of every pairs, i.e. 2-dimensional cube AB. Similarly we can calculate the cube AC and BC. The 3-dimensional cube ABC is getting by intersection of cube C and cube AB, which is already calculated.

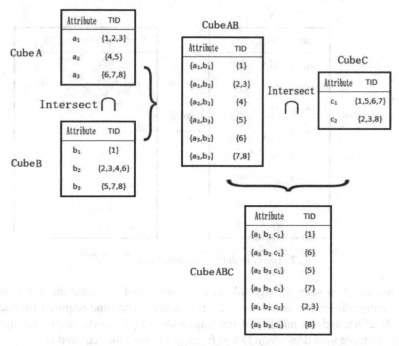

Fig. 1. Cube ABC

Fragments (D, E) and (F, G) are calculated in the same way. So far, we have achieved the Frag-Shells algorithm for computing the data cube.

We can see that, in the process of calculating multidimensional cubes by doing set intersection of fragment-dimensional cubes, it needs two dimensional of K properties values to make the intersection of a collection. Considering the TID number of inverted index is determined by data dispersion K, which is $|TID| = N/K$, N is the number of all multidimensional data. So every intersection needs $K \cdot K \cdot \left(\frac{N}{K}\right) = N \cdot K$ times calculation. Therefore, the cost time of the original algorithm depends on data dispersion. And in real life, for specific analysis needs, each dimension contains a lot of specific values. Therefore, the data dispersion K would be much bigger. This will slow down the processing speed of the original algorithm, so it should be improved.

3.2   The Improved Frag-Shells Algorithm

The original Frag-Shells algorithm repeatedly scans the data in table 1 in the calculation of the each dimensions of the multidimensional cube. Such as computing cube AB and AC, it needs to scan TID list of every attribute values of cube A twice. This increases the times of read operation and greatly slows down the algorithm.

To improve the original algorithm we still take the dimension separate plan. Only in the process of calculating the full cube of each fragment, the idea of **MultiWay** array aggregation is adopted. While scanning each dimension, we calculate all the combination of all the dimensions in each fragment. For example, for fragment $\{1, a_1, b_1, c_1\}$, we will generate the following $2^3=8$ kinds of the corresponding list of values:

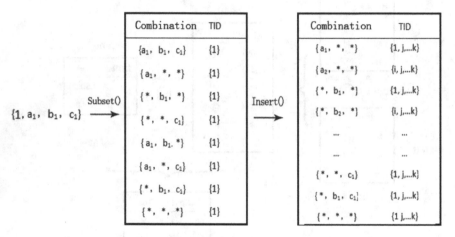

Fig. 2. The combination of dimensions

The algorithm will first scan through all the data tuples and calculate the combined value of the entire dimension attributes in each fragment. The time required for each fragment is $O(2^F n)$, and the number of the fragments is D/F, so the improved algorithm in a given tuple total dimension D and fragment F, the time required is

$$O(\ f(n)\ ) = O(\frac{D}{F}2^F n) = O(n) \tag{1}$$

The time required of the improved algorithm has a linear relationship with data amount, and it has no related with data disperse degree K. Because when calculating a tuple data, it only needs to calculate all combinations of attributes within the fragment, which has no effect on other data.

3.3   Improved Frag-Shells Algorithm Model Based on MapReduce Framework

Map Function. The main target of Map function is to separate all dimensions into a number of collections, according to predefined dimension separate plan. It also takes record of the TID information. Such as tuple $(1, a_1, b_1, c_1, d_2, e_1, f_1, g_2)$ will be processed by Map function into $(1, " a_1, b_1, c_1; 1")$, $(2," d_2, e_1; 1")$ and $(3," f_1, g_2; 1")$, written in the context and handed over to Reduce processing.

Reduce Function. Reduce function accepts parameter context key-value pairs from the Map function, and assigns them to different Reduce nodes according to the value of fragment number. In each Reduce function, the fully data cube of the fragment is calculated.

4   Experimental Results

4.1   Simulation Settings

In our simulation, we use experimental machine equipped with Hadoop framework as a server. On the server, there are four virtual Ubuntu systems, one as the master Hadoop node, the other three as slave nodes. Configuration of each node is a single-core processor 2.4GHz, 1G RAM.

4.2   The Experimental Results

We designed three groups of simulations. We compare the time cost in different amounts, different data dispersion degrees and different segments of data cube computation by the original Frag-Shell algorithm and the improvement algorithm based on MapReduce framework. We get the following results. OFS represents the original Frag-Shells algorithm, and IFS represents the improved Frag-Shells algorithm.

1. Given the data dispersion degree $K = 20$, fragment $F = 3$, the different amounts of data N for simulation results:
2. Given the data amount $N = 200000$, fragment $F = 3$, the different data dispersion degree K the simulation results are as shown in figure 4.

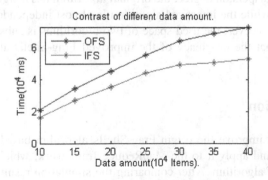

Fig. 3. Contrast of different data amount

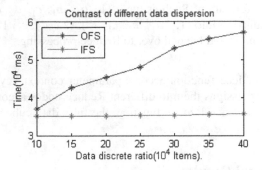

Fig. 4. Contrast of different data dispersion

3. Given the data amount $N = 200000$, data dispersion $K = 20$, and the different fragment of data F. Simulation results are as shown in figure 5.

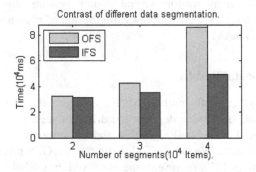

Fig. 5. Contrast of different data fragment

Therefore, through the comparison of simulation results, we can see that whether it is different amount of data, different data dispersion, or different fragment method, the improved Frag-Shells algorithm is always better than the original one. Especially on different data dispersion degree, the original algorithm has a significant increase in processing time, while the improved algorithm is almost independent of the data dispersion degree. However, the used space of two algorithms is substantially the same. So we can see that the efficiency of the improved Frag-Shells algorithm is greatly improved.

5 Conclusion

In this paper, we improve the origin Frag-Shells algorithm based on irrelevance of data dispersion, and apply it to the MapReduce framework, which greatly improves the parallelism of algorithm. After comparing the simulation results, we can see both the obvious advantages in efficiency of improved algorithm and the high performance it benefits from parallel computing in MapReduce framework.

But this research is still in the basic stage, we still have more questions need to be studied. In this paper, we just thought of one of algorithms to calculate the data cube and apply it to MapReduce framework. In data warehousing, data mining area, given the current background of big data, there are many more studies need to be carried out based on our current basis.

Acknowledgements. This paper is supported by NSFC (61272515, 61372108, 61121061), Beijing Higher Education Young Elite Teacher Project (YETP0474), the Open Research Fund of The Academy of Satellite Application (2014_CXJJ-DSJ_01) and NDRC High Tech Dept. ([2013]-2140).

References

1. Arres, B., Kabbachi, N., Boussaid, O.: Building OLAP cubes on a Cloud Computing environment with MapReduce. In: International Conference on Computer Systems and Applications, pp. 1–5. IEEE Press, Ifrane (2013)
2. Abelló, A., Ferrarons, J., Romero, O.: Building cubes with MapReduce. In: Proceedings of the ACM 14th International Workshop on Data Warehousing and OLAP, pp. 17–24. ACM Press, Glasgow (2011)
3. Zhang, J., Zhang, J.M.: Building quotient cube with MapReduce In hadoop. In: Advanced Materials Research, China, pp. 1031–1035 (2013)
4. Sergey, K., Yury, K.: Applying map-reduce paradigm for parallel closed cube computation. In: First International Conference on Advances in Databases, Knowledge, and Data Applications, pp. 62–67. IEEE Press, Gosier (2009)
5. Wang, Y., Song, A., Luo, J.: A mapreducemerge-based data cube construction method. In: 9th International Conference on Grid and Cooperative Computing, pp. 1–6. IEEE Press, Nanjing (2010)
6. Brezany, P., Zhang, Y., Janciak, I., et al.: An Elastic OLAP Cloud Platform. In: IEEE Ninth International Conference on Dependable, Autonomic and Secure Computing (DASC), pp. 356–363. IEEE Press, Sydney (2011)
7. Dehne, F., Zaboli, H.: Parallel real-time OLAP on multi-core processors. In: Proceedings of the 2012 12th IEEE/ACM International Symposium on Cluster, Cloud and Grid Computing, pp. 588–594. IEEE Press, Ottawa (2012)
8. Lakshmanan, L.V.S., Pei, J., Han, J.: Quotient cube: how to summarize the semantics of a data cube. In: Proceedings of the 28th International Conference on Very Large Data Bases, pp. 778–789. Morgan Kaufmann Press, Hongkong (2002)
9. Dean, J., Ghemawat, S.: MapReduce: Simplified Data Processing on Large Clusters. Communications of the ACM 51(1), 107–113 (2008)
10. Dehne, F., Eavis, T., Hambrusch, S., Rau-Chaplin, A.: Parallelizing the data cube. Distributed and Parallel Databases 11(2), 181–201 (2002)

An Availability Evaluation Method of Web Services Using Improved Grey Correlation Analysis with Entropy Difference and Weight

Zhanbo He, Huiqiang Wang, Junyu Lin,Guangsheng Feng, Hongwu Lv,
and Yibing Hu

College of Computer Science and Technology
Harbin Engineering University, Harbin, China
hzbhzb316@163.com

Abstract. Web services is one of the basic network services, whose availability evaluation is of great significance to the promotion of users' experience. This paper focuses on the problem of availability evaluation of Web services and proposes a method for availability evaluation of Web services using improved grey correlation analysis with entropy difference and weight (EWGCA).This method is based on grey correlation analysis, and use entropy difference to illustrate the changes of availability, set weight to quantize availability requirements of different operations or transactions in services. Through simulation experiment in high load scenarios for Web services, the experiment result shows that our method can realize hierarchical description and overall evaluation for availability of Web services accurately in the case of smaller test sample volumes or uncertain data even in the field of big data.

Keywords: Web services, availability evaluation, big data, grey correlation analysis, entropy difference, weight.

1   Introduction

Availability is a key indicator of computer network services. Availability evaluation of Web services is a method to investigate, analyze and evaluate computer network services, which aims at ensuring availability characteristics of Web services in all kinds of real stress conditions.

At present, it still has not formed a authoritative theoretical framework or evaluation standard for comprehensive availability evaluation of Web services. Existing researches generally can be divided into two ideas: one idea generally builds availability models based on queuing theory or test data collected previously to complete availability forecast. For example, it proposes a calculation model of Web services availability in [1] through introducing Markov process and queuing theory in the prediction model, but this method is limited to prediction. [2] proposes LUCS model to predict atomic Web services availability based on collecting past data, which can only apply to availability prediction before using Web services. The authors in [3] design a Web availability model of multi-level supply chain services according to the queuing

H. Wang et al. (Eds.): ICYCSEE 2015, CCIS 503, pp. 300–308, 2015.
© Springer-Verlag Berlin Heidelberg 2015

theory to evaluate availability, but this method is only applicable to the services of supply chain mode. The other idea introduces   metrics to quantify evaluate availability indicators. [4] involves a availability test model, however, it does not distinguish the difference between efficiency indicators and cost indicators. [5] proposes a availability measurement evaluation method, but whether the metric values can be applied to select services remains to be verified.

This paper focuses on availability evaluation of Web services in high load scenarios, according to theory of grey system and grey relational sequence model [6],which is founded by Chinese professor Deng Julong, and in the case of smaller test sample volumes or uncertain data, we can realize hierarchical description and overall evaluation using improved grey correlation analysis with entropy difference and weight(EWGCA).

This paper is organized as follows: Section 2 introduces availability evaluation algorithm of Web services using EWGCA. Section 3 realizes hierarchical description and overall evaluation for availability of Web services accurately. Section 4 makes a conclusion.

2   Improvement of Grey Correlation Analysis with Entropy Difference and Weight (EWGCA)

In this paper, according to the maximum and minimum values of each operation or transaction on evaluation indicators in Web services, we propose a method of EWGCA.

2.1   Establishment and Standardization of Interval Numbers Decision Matrix

This paper takes the interval numbers to establish decision matrix to achieve accurate evaluation [7].

We suppose S is the services operations set involved in a Web service, S = {S_1, S_2,..., S_n}, A is services performance evaluation indicator set, A = {A_1, A_2,..., A_m} (m, n are all positive integers).Use Web services performance evaluation tool to test and record services situation, then attribute values interval numbers $[x_{ij}^L, x_{ij}^U]$ of services operation S_i on performance evaluation indicator A_j, in which x_{ij}^L represents the minimum value of services operation S_i on performance evaluation indicator A_j, x_{ij}^U represents the maximum value. Thus, we can get the decision matrix of services operations set S on services performance evaluation indicator set A.

$$X = \begin{vmatrix} [x_{11}^L, x_{11}^U] & [x_{12}^L, x_{12}^U] & \cdots & [x_{1m}^L, x_{1m}^U] \\ [x_{21}^L, x_{21}^U] & [x_{22}^L, x_{22}^U] & \cdots & [x_{2m}^L, x_{2m}^U] \\ \cdots & \cdots & \cdots & \cdots \\ [x_{n1}^L, x_{n1}^U] & [x_{n2}^L, x_{n2}^U] & \cdots & [x_{nm}^L, x_{nm}^U] \end{vmatrix}$$

Because the indicators in indicator set have different dimensions, decision matrix need to be standardized when making decisions, we use the method of interval numbers decision matrix in[8] to realize standardization.

If A is a efficiency indicator, we use following formula ,

$$[r_{ij}^{L} = \frac{x_{ij}^{L}}{\sum_{i=1}^{n} x_{ij}^{U}}, r_{ij}^{U} = \frac{x_{ij}^{U}}{\sum_{i=1}^{n} x_{ij}^{L}}], i = 1,2,...,n; j \in I_1 \tag{1}$$

If A is a cost indicator, we use following formula,

$$[r_{ij}^{L} = \frac{\frac{1}{x_{ij}^{U}}}{\sum_{i=1}^{n} \frac{1}{x_{ij}^{L}}}, r_{ij}^{U} = \frac{\frac{1}{x_{ij}^{L}}}{\sum_{i=1}^{n} \frac{1}{x_{ij}^{U}}}], i = 1,2,...,n; j \in I_2 \tag{2}$$

Thus, $r_{ij}^{L}, r_{ij}^{U} \in [0,1]$ ($i = 1,2,...,n$; $j = 1,2,...,m$), standardized interval numbers decision matrix $R = [r_{ij}^{L}, r_{ij}^{U}]_{n \times m}$.

2.2   Confirmation of Optimal Subordinate Degree of Services

Then we improve the method in [7] to construct optimal vector G and suboptimal vector B.

Definition 1: We suppose $r_{ij}^{+} = \max\{r_{ij} \mid 1 \le i \le n\}$ ($j = 1,2,...,m$), and corresponding decision value is [r_{ij}^{L+}, r_{ij}^{U+}], we call G = {$g_1, g_2,..., g_m$} = {[r_{i1}^{L+}, r_{i1}^{U+}],[r_{i2}^{L+}, r_{i2}^{U+}],..., [r_{im}^{L+}, r_{im}^{U+}]}^{T} system optimal vector.

Definition 2: We suppose $r_{ij}^{-} = \min\{r_{ij} \mid 1 \le i \le n\}$ ($j = 1,2,...,m$), and corresponding decision value is [r_{ij}^{L-}, r_{ij}^{U-}], we call B={$b_1, b_2,...,b_m$}={[r_{i1}^{L-}, r_{i1}^{U-}], [r_{i2}^{L-}, r_{i2}^{U-}],...,[r_{im}^{L-}, r_{im}^{U-}]}^{T} system suboptimal vector.

We can obtain weight vector after determining each indicator weight: w = (w$_1$, w$_2$,..., w$_m$), w$_i$ > 0, ($i = 1,2,...,m$), $\sum_{i=1}^{m} w_i = 1$. According to grey correlation analysis[6], for a row vector R$_j$ =(r$_{j1}$, r$_{j2}$,...,r$_{jm}$)T in decision matrix $R = [r_{ij}^{L}, r_{ij}^{U}]_{n \times m}$, the correlation coefficient between R$_j$ and optimal vector G or suboptimal vector B is illustrated in following formulas.

$$\xi_{R_j,G} (k) = \frac{\min_j \min_k |r_{jk} - g_k| + \rho \max_j \max_k |r_{jk} - g_k|}{|r_{jk} - g_k| + \rho \max_j \max_k |r_{jk} - g_k|} \tag{3}$$

$$\xi_{R_j,B} (k) = \frac{\min_j \min_k |r_{jk} - b_k| + \rho \max_j \max_k |r_{jk} - b_k|}{|r_{jk} - b_k| + \rho \max_j \max_k |r_{jk} - b_k|} \tag{4}$$

$|r_{jk} - g_k|$ is the euclidean distance between(r_{jk}^{L}, r_{jk}^{U})and(g_k^{L}, g_k^{U}), ρ is resolution coefficient, $\rho \in [0,1]$, we generally set ρ =0.5. Then the weighted correlation coefficient between R_j and optimal vector G or suboptimal vector B is $\gamma(R_j, G) = \sum_{k=1}^{m} w_k \xi_{R_j G} (k)$

and $\gamma(R_j, B) = \sum_{k=1}^{m} w_k \xi_{R_j B} (k)$.

The bigger $\gamma(R_j, G)$ is, the correlation degree between the evaluated services and system optimal vector G is bigger, and services availability is better. The bigger $\gamma(R_j, B)$ is, the correlation degree between the evaluated services and system suboptimal vector B is bigger, and services availability is worse. To obtain system optimal solution vector: $u=(u_1, u_2,...,u_n)$, according to least square method 'distance square sum minimum', we build Lagrange function[7].

$$F(u_j) = \sum_{j=1}^{m} [(1-u_j)\gamma(R_j, G)]^2 + [u_j \gamma(R_j, B)]^2 \tag{5}$$

According to (5), we set $\dfrac{\partial F(u_j)}{\partial u_j} = 0$, then we can get

$u_j = \left\{1 + [\gamma(R_j, B)/\gamma(R_j, G)]^2\right\}^{-1}$ $(j=1,2,\cdots,n)$, u_j represents the optimal subordinate degree of the evaluated operation S_j.

2.3 Implementation of Hierarchical Description and Overall Evaluation

Now, we propose 'available entropy' to evaluate availability of Web services. The smaller the entropy value is, the better for availability of Web services. For a certain services performance evaluation indicator A_j, its entropy is defined as $H = -\log_2^{A_j}$. We use the entropy difference method in [9] to describe availability changes before and after the evaluation.

Definition 3: Availability probability of the services is defined as: $P(u_j | u_j \geq \delta) = \dfrac{u_j}{u_{max}}$, $(j = 1, 2,..., n)$, among which, u_j represents subordinate degrees of services operation S_j to system optimal vector, u_{max} is the system maximum subordinate degree, δ is the threshold of system subordinate degree, whose value is 0.1. If $u_j < \delta$, it means the services is not available.

Definition 4: We suppose P_1 is the availability probability of the services operation S_j before evaluation(before applying high load) and P_2 is the availability probability of the services operation S_j after evaluation(after applying high load), we use entropy difference $\Delta P = -\log_2(P_2/P_1)$ to illustrate the availability changes of services operation S_j. If $\Delta P \leq \delta$ (δ is an availability threshold, we generally set $\delta \leq 0.9$)

it means system can provide normal Web services. If $\Delta P > \delta$, it means system can't provide normal services.

Relationship between ΔP and availability level is showed in table 1.

Table 1. Hierarchical description

ΔP	<0	[0,1)	[1, 3)	≥ 3
Availability Level	good	nor-mal	bad	not available

Definition 5: We suppose H_1 is the available entropy (weighted sum of available entropy in each service) of services before evaluation (before applying high load), H_2 is the available entropy after evaluation(after applying high load),

so $\Delta H = H_2 - H_1 = \sum_{i=1}^{n} w_i' P_i^{(2)} - \sum_{i=1}^{n} w_i' P_i^{(1)} = \sum_{i=1}^{n} w_i' \Delta P_i$, (i=1,2,...,n) , w_i' is the weight

of services operation S_j relative to the whole Web services, ΔH is the overall evaluation result of availability of Web services.

3   Availability Evaluation of Web Services Using EWGCA

We can obtain test sample values in 3.1 through a simulation test in high load scenarios, and then use the test data to implement the hierarchical description and overall evaluation in 3.2.

3.1   Simulation Experiment in High Load Scenarios for Web Services

Load test simulation experiment is conducted by WebTours booking services.

We use LoadRunner11.00, a performance testing tool. The experiment environment uses a computer with 64 - bit OS serving as Web server and test machine at the same time, Intel (R) Core (TM) 2 Duo CPU, 4G of memory.

Simulation process: simulate 10 Vusers concurrent access, begin with two Vusers, then adding 2 Vusers every 15 seconds, and   10 Vusers keep connection and sustained concurrent access in 10 minutes, then reduce at the same speed, 2 Vusers every 15 seconds.

The operations include: login.pl, Search Flights, reservations.pl, reservations.pl_2, reservations.pl_3, Itinerary and SignOff. This paper will define login.pl as transaction login, define Search Flights, reservations.pl, reservations.pl_2 and reservations.pl_3 as transaction find_confirm_flight, Itinerary as transaction Itinerary, SignOff as transaction SignOff. As a result, the Web services operation (transaction) set is S = {login, find_confirm_flight, itinerary, signoff}.Considering the evaluation factors of Web services' availability, we suppose services performance evaluation indicator set A = { Average Transaction Response Time, Web Page Average Download Time}.

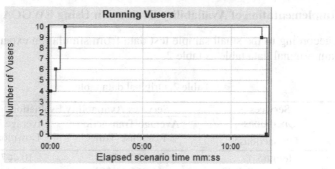

Fig. 1. Actual running vuser curve

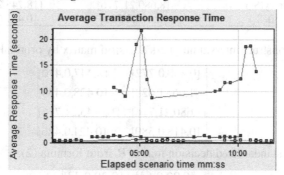

Fig. 2. Average transaction response time curve

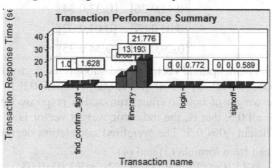

Fig. 3. Summary of transaction response time

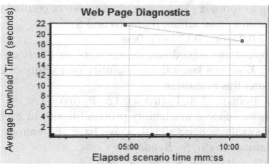

Fig. 4. Average download delay time curve

3.2   Implementation of Availability Evaluation Using EWGCA

Firstly, according to the small sample test data from simulation experiment in 3.1, we can obtain original data table as table 2.

Table 2. Original data table

Services Operations (Transactions)	Services Availability Evaluation Indicator	
	Average Transaction Response Time	Average Web Page Download Delay
login(S_1)	[0.430,0.772]	[0.447,0.470]
find_confirm_flight(S_2)	[1.065,1.628]	[0.428,0.434]
itinerary(S_3)	[8.680,21.776]	[18.745,21.776]
signoff(S_4)	[0.418,0.589]	[0.421,0.426]

Then we can construct interval numbers decision matrix by original data.

$$X = \begin{vmatrix} [0.430,0.772] & [0.447,0.470] \\ [1.065,1.628] & [0.428,0.434] \\ [8.680,21.776] & [18.745,21.776] \\ [0.418,0.589] & [0.421,0.426] \end{vmatrix}$$

We can obtain standardized decision matrix R from formula (2).

$$R = \begin{vmatrix} [0.22,0.64] & [0.30,0.33] \\ [0.11,0.26] & [0.33,0.34] \\ [0.01,0.03] & [0.01,0.02] \\ [0.29,0.65] & [0.34,0.35] \end{vmatrix}$$

Next step, we can obtain optimal vector G and suboptimal vector B.

G={ [0.29,0.65] , [0.34,0.35] }, B={ [0.01,0.03] , [0.01,0.02] }

Now, we set the weight of both average transaction response time and web page download delay are all 0.5, that is, the indicator weight vector is w = {0.5, 0.5}. We set resolution coefficient $\rho = 0.5$. The weighted correlation degree of each service vector can be obtained from formula (3) and (4).

r(R1,G)=0.88   r(R1,B)=0.12, r(R2,G)=0.97   r(R2,B)=0.03
r(R3,G)=0.02   r(R3,B)=0.98, r(R4,G)=0.99   r(R4,B)=0.01

Then, according to formula (5), we can obtain system optimal subordinate degree of the services operation (transaction) S_1, S_2, S_3 and S_4 .

$u_1 = 0.85$, $u_2 = 0.94$, $u_3 = 0.12$, $u_4 = 0.98$.

According to the optimal subordinate degree u_j of each service operation (transaction), by definition 3, we can know the availability probability of each service operation (transaction) after this evaluation.

$P_2 (u_1) = 0.87$, $P_2 (u_2) = 0.96$, $P_2 (u_3) = 0.12$, $P_2 (u_4) = 1$.

By definition 4, we suppose availability probabilities P_1 of each service operation (transaction) before evaluation are all 0.9, so we have got 'entropy difference' of services operation (transaction) S_1, S_2, S_3 and S_4 .

$\Delta P_1 = 0.04, \Delta P_2 = -0.09, \Delta P_3 = 2.94, \Delta P_4 = -0.14.$

According to the corresponding relationship between ΔP and availability level, hierarchical description of this Web services is showed in table 3.

Table 3. Availability hierarchical description

Services Operations (Transactions)	ΔP	Availability Hierarchical Description
login(S_1)	0.04	normal
find_confirm_fligh(S_2)	-0.09	good
itinerary(S_3)	2.94	bad
signoff(S_4)	-0.14	good

Finally, by definition 5, supposing the weights that each service operation (transaction) S_1, S_2, S_3 and S_4 relative to Web services are all 0.25, we can calculate the availability of the whole Web services.

$$\Delta H = H_2 - H_1 = \sum_{i=1}^{n} w_i' P_i^{(2)} - \sum_{i=1}^{n} w_i' P_i^{(1)} = \sum_{i=1}^{n} w_i' \Delta P_i = 0.6875.$$

As a result, the overall evaluation result of the Web services is 'normal'.

4   Conclusions

This paper focuses on availability evaluation of Web services and offers an method (EWGCA). Compared with other methods proposed in the papers before , this method of EWGCA will only need the maximum and minimum values on evaluation indicator of the services. Eventually, the accuracy and credibility of this evaluation result have both improved by 10% at least. In view of the work in this paper, it still should be applied to more real environment, which need further research and more experiments.

Acknowledgements. This research is supported by the National Natural Science Foundation of China (61370212), the Research Fund for the Doctoral Program of Higher Education of China (20122304130002), the Natural Science Foundation of Heilongjiang Province (ZD 201102) and the Fundamental Research Fund for the Central Universities (HEUCFZ1213, HEUCF100601).

References

1. Duan, Y., Huang, Y.: Research on Availability Prediction Model of Web services. In: 2011 International Conference on Computer Science and Services System (CSSS), pp. 1590–1594. IEEE (2011)
2. Silic, M., Delac, G., Krka, I., Srbljic, S.: Scalable and Accurate Prediction of Availability of Atomic Web Services. IEEE Transactions on Services Computing PP(99), 1, doi:10.1109/TSC.2013.3

3. Ma, S., Jiang, M., Li, L., Hu, M.: Availability Analysis and Evaluation of Multi-layer Supply Chain Services Based on Web. In: Second International Symposium on Intelligent Information Technology Application, IITA 2008, December 20-22, vol. 2, pp. 186–190 (2008)

4. Deng, F., Chen, H., Jiang, L.: A Usability Evaluation Model Based on the Degree of Usability for Web services. In: e-Business and Information System Security (EBISS), May 22-23, pp. 1–4 (2010)

5. Shao, L., Zhao, J., Xie, T., Zhang, L., Xie, B., Mei, H.: User-Perceived Services Availability: A Metric and an Estimation Approach. In: IEEE International Conference on Web Services, ICWS 2009, July 6-10, pp. 647–654 (2009)

6. Ju-Long, D.: Grey system theory. Huazhong University of Science and Technology Press, Wuhan (1990)

7. Guosheng, Z., Huiqiang, W., Jian, W.: Reseach on evaluation of network survivability situation based on grey correlation analysis. Journal of Chinese Computer Systems 27(10), 1861–1864 (2006)

8. Zhi-Ping, F., Xian-Bin, G., Quan, Z.: Methods of normalizing the decision matrix for multiple attribute decision making problems with intervals. Journal of Northeastern University 20(3), 326–329 (1999)

9. Yi-Rong, Z., Ming, X., et al.: A study on the evaluation technology of the attack effect of computer networks. Journal of National University of Defense Technology 24(5), 24–28 (2002)

Equal Radial Force Structure Pressure Sensor Data Analysis and Finite Element Analysis

Lian-Dong Lin[1,*], Cheng-Jun Qiu[1], Xiang Yu[2], and Peng Zhou[1]

[1] Heilongjiang Province Key Lab of Senior-Education for Electronic Engineering,
Heilongjiang University, Harbin 150080, China
[2] Heilongjiang Institute of Technology, Heilongjiang University, Harbin 150080, China
lld_web@hotmail.com

Abstract. As big data is very important today, we creative a force sensor with the AT-cut quartz crystal resonator and analyze the experimental data. Quartz crystal resonator has the characteristic that the resonance frequency changes by the external force, which has high precision, fast-speed response. Also it has the superior feature in the temperature and frequency stability. But it also has weakness, because of quartz crystal resonator has low degree of mechanical characteristic and weak to stress concentration by bending that the quartz crystal resonator had been hardly applied to the force measurement. The objective of this study is to construct the sensor mechanism that safely maintains the quartz crystal resonator for the external force with flat structure. We using finite element multiphysics simulation software designed and implemented an innovative structure-equal radial force structure, According to the measured data, applied load equivalent radial force structure between size and the frequency of the quartz monitor chip has a good linear relationship. The proposed force sensor is flat, small, and sensitive. It can be applied to several usages such as medical treatment and contact force detection of human.

Keywords: FEM, force sensor, equal radial force structure.

Introduction

Pressure sensors are widely used in military and industrial fields, such as aviation, aerospace, food, metrology, robotics. The pressure sensor which based on the principle of resonant has high accuracy, the advantages of digital output, without A / D conversion, in the digital signal processing field, began to gradually replace the analog signal output of the pressure sensor. Quartz material has good stability, which has a piezoelectric effect can be directly used to excite resonance sensor. It is difficult to measure the bio-signal when large load is impressed. In such cases, the force sensor is required not only for high sensitive measurement but also wide range of measurement.

Conventional force sensors used for medical treatment are diaphragm type sensors, strain gauge type sensors [1] and piezoelectric vibration type sensors [2]. However,

---

* Corresponding author.

H. Wang et al. (Eds.): ICYCSEE 2015, CCIS 503, pp. 309–319, 2015.

up to now, there is no sensor that satisfies high sensitivity, wide measurement range, high-speed response, small size, and high durability, etc.Especially, it is difficult to measure the bio-signal when large load is impressed. In such cases, the force sensor is required not only for high sensitive measurement but also wide range of measurement. Therefore, we paid attention to the quartz crystal that is one of the piezoelectric elements which provides us function of self-sensing. The quartz crystal generates a charge which is proportional to the applied external force, and has high sensitivity, excellent temperature stability, and frequency stability. Therefore, the quartz crystals have been used for various sensors such as gas sensors [3], temperature sensors[4], and DNA sensors [5]. Needless to say, it has been researched and developed as a force sensor. As an example of a commercially available force sensor, there are piezo-electric force sensors produced by the Kistler corporation. Sensing principle of these force sensors is based on detection of the generated charge by the external force. However, the force sensors of this principle have disadvantage in static force measurement since they are easy to receive the electric drift and noise.

1   Quartz Crystal Resonators

Resonant pressure sensor composed by mechanical conversion element, resonant sensing element and support element composition. It working principle is shown in fig1. Mechanical conversion element converting an external pressure to AT-cut quartz crystal sensor which make it turn into thickness-shear vibration mode. By applying an external force to the QCR, its resonant frequency changes proportional to the external force. The block diagram of quartz crystal resonators we designed is shown in Fig.1.

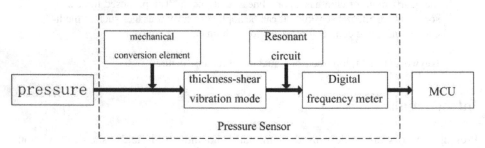

Fig. 1. Block diagram of quartz crystal resonators

The quartz crystal resonator (QCR) has a characteristic that the resonance frequency changes by an external force [6-7], and it can reduce influence of the electric drift by detecting the resonance frequency change. The output of a QCR is easily treated in a form of high-speed digital signal. Therefore, we used this characteristic as the sensing principle for the force sensor. However, the QCR is weak to the stress concentration by bending, and it is necessary to design a retention mechanism of the resonator. We developed a miniaturized force sensor with the QCR which is supported by a thin and stable retention mechanism. We evaluated the characteristic of this mechanism by the numerical analysis, and investigated the basic operation characteristic of this sensor.

1.1 Piezoelectric Equations

The following expression defined the equation of the surface charge of the piezoelectric material with external force:

$$q_i = d_{ij}\sigma_j \qquad (1)$$

qi--Charge density (C/cm2);Qi--The total amount of charge (C);σj--stress in j direction (N/cm2); dij-- Piezoelectric constant(C/N),(i=1,2,3, j=1,2,3,4,5,6).

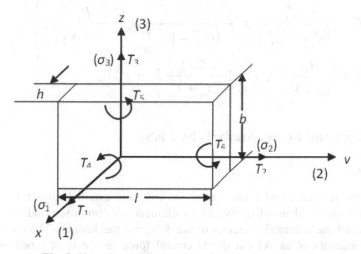

Fig. 2. Piezoelectric element coordinate system of notation

Fig.2 shows Piezoelectric element coordinate system of notation. Electric field pointing to the crystal axis positive direction is positive, otherwise it is negative. The equations (2) define crystal surface charge density in any state generated by force:

$$
\begin{cases}
q_1 = d_{11}\sigma_1 + d_{12}\sigma_2 + d_{13}\sigma_3 + d_{14}\sigma_4 + d_{15}\sigma_5 + d_{16}\sigma_6 \\
q_2 = d_{21}\sigma_1 + d_{22}\sigma_2 + d_{23}\sigma_3 + d_{24}\sigma_4 + d_{25}\sigma_5 + d_{26}\sigma_6 \\
q_3 = d_{31}\sigma_1 + d_{32}\sigma_2 + d_{33}\sigma_3 + d_{34}\sigma_4 + d_{35}\sigma_5 + d_{36}\sigma_6
\end{cases}
\qquad (2)
$$

q1、q2、q3 is the charge density that perpendicular to the x-axis, y-axis, and z-axis;σ1、σ2、σ3 is Stress in the direction of the corresponding; σ4、σ5、σ6 is Shear stress perpendicular in the direction of the corresponding; dij (i=1, 2, 3, j=1, 2, 3, 4, 5, 6) is Piezoelectric constant.

1.2 The Piezoelectric Vibrator Radial Vibration Mathematical Modeling

Using cylindrical coordinates to analysis thin wafer piezoelectric vibrator. We can get Piezoelectric equations which is shown in the below.

$$
\left.\begin{aligned}
x_r &= s_{11}^E X_r + s_{12}^E X_\theta + d_{31} E_z \\
x_\theta &= s_{12}^E X_r + s_{11}^E X_\theta + d_{31} E_z \\
D_z &= d_{31} X_r + d_{31} X_\theta + \varepsilon_{33}^X E_z
\end{aligned}\right\}
\tag{3}
$$

$$
\left.\begin{aligned}
X_r &= \frac{s_{11}^E}{\left(s_{11}^E\right)^2 - \left(s_{12}^E\right)^2} x_r - \frac{s_{12}^E}{\left(s_{11}^E\right)^2 - \left(s_{12}^E\right)^2} x_\theta - \frac{d_{31}}{s_{11}^E + s_{12}^E} E_z \\
X_\theta &= \frac{s_{11}^E}{\left(s_{11}^E\right)^2 - \left(s_{12}^E\right)^2} x_\theta - \frac{s_{12}^E}{\left(s_{11}^E\right)^2 - \left(s_{12}^E\right)^2} x_r - \frac{d_{31}}{s_{11}^E + s_{12}^E} E_z \\
D_3 &= \frac{d_{31}}{s_{11}^E + s_{12}^E}\left(x_r + x_\theta\right) - \frac{2 d_{31}^2}{s_{11}^E + s_{12}^E} E_z + \varepsilon_{33}^X E_z
\end{aligned}\right\}
\tag{4}
$$

2   Design and Stress Analysis by FEM

2.1   Design

To sustain the resonator and avoid destruction from the external impact, tension and bend, a retention mechanism of the sensor element was proposed and constructed. Figure.3 shows the schematic diagram of the designed mechanism. The conventional sustaining structure of an AT-cut quartz crystal force sensor is designed under the condition that a force will be applied from the direction which is in parallel to the electrical axis of the resonator [8, 10-11]. However, it is difficult to make the force sensor thin by these sustaining structures. In order to have a thinner and more compact sensor, we designed a retention mechanism which allows a force applied from the direction perpendicular to the electrical axis. As shown in Fig.3, the vertical compression force from the upper part is converted into the compression force of the quartz crystal in horizontal direction by using conical platform structure. Also, this structure the bending stress given to the QCR can be decreased and the bending moment is almost counterbalanced.

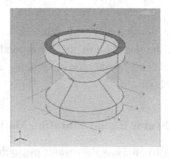

Fig. 3. Illustration of piezoelectric effect

We performed FEM analysis of the retention mechanism by using COMSOL software (COMSOL. Inc). This QCR structure like the shape of the "cone Taiwan", the quartz crystal resonator is fixed at the center of the upper and lower and quartz crystal resonator has a circular hole for electrode.

2.2   Stress Analysis by FEM

Figures.4 show the results of the FEM analysis, where the perpendicular compression stress of 4.3 MPa was impressed from the upper part of this sensor. Only the board springs and the QCR were considered as an analytical object, and the exterior parts are substituted by supposing proper boundary conditions.

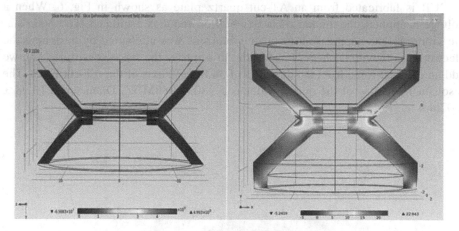

Fig. 4. Longitudinal section pressure

Only the board springs and the QCR were considered as an analytical object, and the exterior parts are substituted by supposing proper boundary conditions.Figure.4 shows the distribution of stress in x axis, where the tensile stress is assumed to be positive.

Figure.5 shows structure transverse section performance analysis. Fig.5 (a)shows transverse section of the total displacement. Fig.5 (b) shows transverse plane strain.

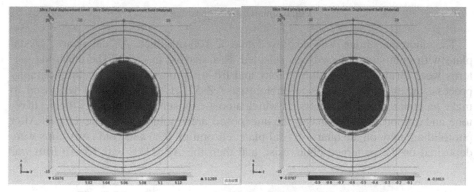

Fig. 5. Structure transverse section performance analysis diagram

From simulation we can confirmed that the QCR is impressed with the uniform compression stress in x axis. The bending stress in Fig. 4(b) is smaller than conventional construction,and the bending was hardly generated. The amount of the deformation of this sensor is very little. We confirmed high stiffness of this sensor based on the FEM analysis. Further, the proposed sensor structure is robust from unexpected external disturbances.

2.3   Design and Fabrication of Quartz Crystal Resonator

- Stress analysis by FEM

A QCR is fabricated from an AT-cut quartz plate as shown in Fig. 6. When a voltage is applied, the thickness-shear vibration mode is oscillated. To obtain stable vibration, the suppression of other vibrations was attempted by designing the electrodes in the circle shape. In addition, to improve the stress sensitivity[12], we add an external force from the direction of 34.8° of x axis of the AT-cut QCR. The resonant frequency of the quartz wafer is 5.980~5.988MHZ, Diameter is 13.98 ± 0.02mm.

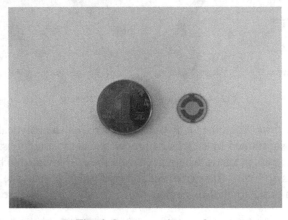

Fig. 6. Quartz monitor wafer

The micromachining was employed and a 100-μm-thick AT-cut quartz crystal plate with both-side polished was selected as a starting material. The electrode patterns were formed by photolithography and lift-off process. In the photolithography process, an image reversal type photoresist AZ-5214E (Clariant Co.) was used to make an undercut profile pattern, in which a post-exposure bake reverses the positive-tone and a flood exposure makes the unexposed areas soluble in the developer. After photolithography on the quartz crystal plate, chromium (Cr) and gold (Au) films were deposited on the patterned photoresist, and then the AZ5214E with Cr-Au film was

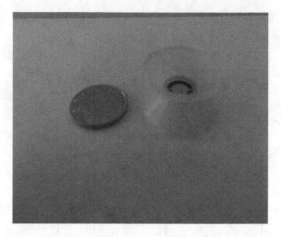

Fig. 7. Equal radial force structure nested for quartz monitor wafer

removed by the stripper. The process mentioned above was performed on both sides of the AT-cut quartz crystal substrate. Finally, fig.7 shows equal radial force structure nested for quartz monitor wafer.

- Characterization of quartz crystal resonator

The resonant frequency of quartz monitor wafer was calculated from following equation;

$$f = \frac{\sqrt{c/\rho}}{2h} \tag{5}$$

where h is the thickness, c is the elastic coefficient, and ρ is the density. The Quality factor of the fabricated QCR was calculated from following equation:

$$Q = \frac{2\pi f \cdot L}{R} \tag{6}$$

where R and L are the equivalent electrical resistance and inductance detected from the analyzer, respectively.

- Oscillation circuit

To obtain the sustained oscillation at the resonant frequency, it is necessary to amplify the vibration signal using an oscillation circuit. Figure 8 shows a three-point oscillator circuit by inductance for the QCR. The oscillation circuit that consists of a feedback quartz crystal and a transistors is connected with the oscillation circuit.

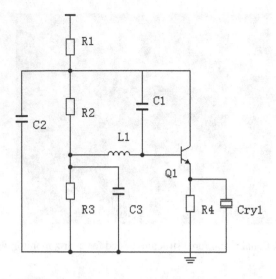

Fig. 8. Quartz crystal resonator oscillation circuit

Fig.9 show the sustained oscillation at 6MHZ frequencies is detected by Frequency Counter and oscilloscope.

Fig. 9. Steady amplitude waveforms

3   Performance Testing

Figure 10 shows the sensor and oscillation circuit. Dimensions of sensor devices is $\Phi = 15mm \times 40mm$ .The voltage of 5.0 Vp-p was impressed to the oscillation circuit and the QCR was oscillated. The oscillation is confirmed with the oscilloscope (TDS 210, Tektronix Ltd.), while the frequency change is detected using a frequency counter (SF842A,ShengFeng Tech.).

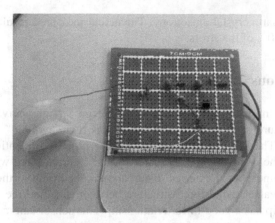

Fig. 10. The sensor and oscillation circuit

Fig11 shows the frequency shift of the force sensor, when the loading stress is impressed up to 100 kPa. It was observed that the resonance frequency increased proportional to the loading force, and the basic force sensing capability was investigated.

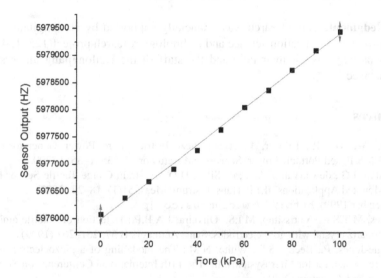

Fig. 11. Performance of the force sensor with the quartz crystal resonator
(Temperature: 22.0°C, Humidity: 35%, Input: 5 V)

As the results, after linear fits we get linear expressions between the resonant frequency of the sensor and the pressure:

$$f = 34.01364p + 5975996 \qquad (7)$$

Sensitivity of the sensor is 34 Hz/kPa and the nonlinearity error is calculated to be 0.01%. The measured frequency of the device is slightly lower than simulation results, that because of quartz crystal anisotropy. The actual measured results and simulation results has about 10% of the errors.

4   Conclusions

We proposed the miniaturized force sensor base on an innovative structure-equal radial force structure in this paper. The sensor has two "pyramid" type structure to bearing pressure. The FEM analysis shows that the proposed retention mechanism is safely maintains the quartz crystal resonator for the external force with flat structure. It is robust from unexpected external disturbances. According to the measured data, applied load equivalent radial force structure between size and the frequency of the quartz monitor chip has a good linear relationship.This sensor can measure static load and we confirmed effectiveness of our sensor. In experiments, the maximum load was set 100kPa, and the output sensitivity of the sensor is 34 Hz/kPa. the nonlinearity error is calculated to be 0.01%.Relation of the temperature and the sensor output is linear, and the temperature compensation can be easily performed. The proposed force sensor is flat, small, and sensitive. It can be applied to several usages such as medical treatment and contact force detection of human.

Acknowledgment. This research was financially supported by Heilongjiang Provincial Department of Education science and technology research project(12531542). We thank all participants for their help and the staff of the Heilongjiang university for their assistance.

References

1. Choi, W., Xia, R., Brewer, J., et al.: Piezoelectric Micro Power Generator(PMPG): A MEMS-Based Portable Power. Sensors and Actuators 74, 3551–3551 (2005)
2. da Silva, J.G., de Carvalho, A.A., da Silva, D.D.: A Strain Gauge Tactile Sensor for Finger-Mounted Applications. IEEE Trans. Instrum. Meas. 51(1), 18–22 (2002)
3. (December 1999), http://www.olympus.co.jp
4. Gomes, M.T.S.R., Verissimo, M.I.S., Oliveira, J.A.B.P.: Detection of volatile amines using a quartz crystal with gold electrodes. Sens. Actuators B 57, 261–267 (1999)
5. Glynne-Jones, P., Beeby, S.P., White, N.M.: The modelling of a piezoelectric vibration powered generator for Microsystems. In: The 11th International Conference on Solid-State Sensors and Actuators (2001)
6. Griva, G., Ilas, C., Eastham, J.F., et al.: High performance sensorless control of induction motor drives for industry applications. In: Proceedings of the Power Conversion Conference, Nagaoka, Japan, vol. 2, pp. 535–539 (1997)
7. Albert, W.C.: Force sensing using quartz crystal flexure resonators. In: 38th Annual Symposium on Frequency Control, pp. 233–239. IEEE (1984)
8. Itoh, H., Horiuchi, N., Nakazawa, M.: An analysts of the longitudinal mode quartz tactile sensor based on the Mason equivalent circuit. In: 50th Proceedings of the 1996 IEEE International Frequency Control Symposium, pp. 572–576. IEEE (1996)
9. Dumlet, B., Bourquin, R., Shibanova, N.: Frequency-output force sensor using a multimode doubly rotated quartz resonator. Sens Actuators A 48, 109–116 (1995)

10. Dong, Y.G., Wang, J.S., Feng, G.P., Wang, X.H.: Self-Temperature-Testing of the Quartz Resonant Force Sensor. IEEE Trans. Instrum. Meas. 48(6), 1038–1040 (1999)
11. Narumi, K., Fukuda, T., Arai, F.: Miniaturization of a wide range load sensor using AT-cut quartz crystal resonator. In: International Symposium on Micro-Nano Mechatronics and Human Science, MHS 2009, pp. 477–482. IEEE (2009)
12. Wang, Z., Zhu, H., Dong, Y., Feng, G.: A thickness-shear quartz force sensor with dual-mode temperature compensation. IEEE Sens. J. 3(4), 490–497 (2003)
13. Mitchell, J.P., Beer, J., Yancy, A., et al.: The Theory of Piezoelectric Shells and Plates. Journal of the National Medical Association 100(5), 572-4 (2008)
14. Comsol, A.B.: COMSOL multiphysics user's guide. Version (September, 2005)
15. Transistor technological SPECIAL, Electric circuit parts use manual The first collection, CQ Publishing Co. (2005)

Knowledge Acquisition from Forestry Machinery Patent Based on the Algorithm for Closed Weighted Pattern Mining

Huiling Yu[1,2], Jie Guo[2], Dongyan Shi[3,*], Guangsheng Chen[2], and Shanshan Cui[2]

[1] Postdoctoral Research Station of Mechanical Engineering,
Harbin Engineering University, Harbin, China
yhl2016@163.com
[2] College of Information and Computer Engineering,
Northeast Forestry University, Harbin, China
{yhl2016,gj2618,kjc_chen,css1989}@163.com
[3] College of Mechanical and Electrical Engineering,
Harbin Engineering University, Harbin, China
920537172@qq.com

Abstract. The application of big data mining can create over a trillion dollars value. Patents contain a great deal of new technologies and new methods which have unique value in the product innovation. In order to improve the effectiveness of big data mining and aid the innovation of products of forestry machinery, the algorithm for closed weighted pattern mining is applied to acquire the function knowledge in the patents of forestry machinery. Compared with the other algorithms for mining patterns, the algorithm is more suitable for the characteristics of patent data. It not only takes into account the importance of different items to reduce the search space effectively, but also avoids achieving excessive uninteresting patterns below the premise that assures quality. The extensive performance study shows that the patterns which are mined by the closed weighted pattern algorithm are more representative and the acquired knowledge has more realistic application significance.

Keywords: Forestry machinery patent, Knowledge acquisition, Closed frequent pattern, Weighted frequent pattern.

1 Introduction

Forestry issues have become one of common concerns in the international community. To satisfy the demand of forestry production, we need to improve the performance of forestry machinery about efficiency, energy consumption, and safety. Patents as knowledge carriers integrate technical information, economic information and commercial information which have unique value [1].Acquiring the function kowledge lies in the patents can assist the innovation of products effectively as well as reducing

---

* Corresponding author.

H. Wang et al. (Eds.): ICYCSEE 2015, CCIS 503, pp. 320–325, 2015.
© Springer-Verlag Berlin Heidelberg 2015

the funds of research and development so that the enterprise can launch the new products which are low-cost and high-quality for a short time.

The amount of patent literature is huge and the structure of it is complex. The traditional technique of conventional data processing has been unable to cope with the problems from big data. In order to solve these problems, we need to break through the traditional methods and propose new methodological reformation according to the characteristics of big data. While the researches about acquiring the knowledge from patents are relatively less. The methods which have been applied include text mining [2], natural language processing [3], the combination of OLAP and K-MEANS and so on. With the wide application of data mining technology in various fields, Hui Zhang [4] put forward a modified algorithm of association rule. Even though the new association rules algorithm solves the problem of Apriori algorithm to a certain extent, but it does not take into account the importance of each data item, easily produce too many uninteresting rules.

Forestry machinery patents contain quite many data items and mining the too short pattern doesn't have a practical use of meaning. Based on the above characteristics, the algorithm of closed weighted frequent pattern mining is applied to mining frequent patterns from forestry machinery patent so then obtains the functional knowledge. By means of giving data items different weights to distinguish their degree of importance, not only can reduce the search space, but also can discovery more interesting patterns. At the same time, the use of closed pattern constraint could ensure the same results of pattern mining under the condition of reducing the number of patterns and avoid generating short patterns. The experimental results shows that the quality of knowledge which is acquired from patents based on the algorithm of closed weighted frequent pattern mining is better than the existing methods.

2   Concepts and Definitions

2.1   Data Sources and Pretreatment

The statement of published patent has a fixed format which includes: title, abstract, claims, specification and drawings. We use the OCR software to recognize the image file of patents, the ICTCLAS segmentation system to decompose the content of specification and finally get the feature components. Due to the different expressing habits, there are great deals of feature components which have different definition but have the same meaning, such as "fruit bag" and "basket." In order to guarantee the accuracy of data we establish the thesaurus about the field of forestry machinery to take the place of synonyms in the feature components.

2.2   Establishment of Weighted Model

Let W = {w1, w2...wm} correspond to the set of items I = {i1, i2...im}. Each wm in W represents the importance of each im in I.

Definition 1 Weight of a Pattern. For a pattern P = {p1, p2...pn}, the weight of P is formally defined as follows.

$$W(P) = \frac{\sum_{i=1}^{n} wi}{|P|}$$

(1)

Definition 2 Weighted Support. The weighted support of a pattern P is generated by multiplying the support of P and the weight of P.

$$Wsup(P) = sup(P) \times W(P)$$

(2)

Definition 3 Weighted Frequent Pattern. If the weighted support of a pattern P is not less than the minimum weighted support, the pattern is weighted frequent.

Definition 4 Maximum Weight and Maximum Weighted Support. Even though a pattern is not frequent, its super pattern would probably be frequent that the mining of weighted frequent pattern does not satisfy the anti-monotone. Set the maximum weight as follows.

$$Wmax = \max\{w1, w2, \ldots, wm\}$$

(3)

Let the maximum weighted support as follows.

$$Wmsup(P) = sup(P) \times Wmax$$

(4)

3   Closed Weighted Frequent Pattern Mining

Unil Yun [5]. proposed the algorithm of closed weighted frequent pattern mining which is modified by the frequent pattern growth method FP-growth. The algorithm mainly consists of two phases: (1) Sort the frequent items by the weight value ascending order and then construct a weighted FP tree; (2) Mine FP tree and traversal it from bottom to up to get the frequent patterns. Illustrate the algorithm in detail as follows.

Table 1. Construct a weighted FP-tree

Construct a weighted FP-tree
Input: A database: TDB, A minimum weighted support threshold: Wminsup, A weight set: W, A maximum weight value: Wmax.
Output: A weighted FP-tree
1. Scan database once; Calculate support and weight of each item. 2. Compute the maximum weighted support of each item; Sort the items whose maximum weighted supports are not less than the minimum weighted support threshold according to the weight value ascending order. 3. Remove the weighted infrequent items in each transaction and sort the other items according to the sequence in step 2. 4. Insert the transactions which are disposed by step 3 into the FP-tree.

Table 2. Mine FP-tree from bottom to up

Mine FP-tree from bottom to up

WCFp_growth(FP-tree, CWpattern, α)

Input: A weighted FP-tree, A hash table store closed weighted frequent patterns: CWpattern, A condition prefix: α.

Output: Closed weighted frequent patterns

1. If FP-tree contains a single path D then {
2. Get all the combinations β by the node in path D and then put the weighted frequent pattern β ∪ α into temporary hash table tpattern
3. For each pattern P in tpattern {
4. If P not meet the closed check: closed_Subcheck(P, CWpattern)){
5. CWpattern=CWpattern ∪ P } } }
6. Else FP-tree contains multiple paths then{
7. For each κ in the header{
8. Let X=α ∪ κ
9. If sup(X)* Wmax≥Wminsup {
10. If X not meet the closed check: closed_Subcheck(X, CWpattern)){
11. If sup(X)* W(X)≥Wminsup { CWpattern=CWpattern ∪ X } }
12. Construct the condition tree of X: FP-treeX
13. If FP-treeX ≠Φ then { WCFp_growth(FP-treeX, CWpattern, X) } } } }

4   Experimental Results and Analysis

We evaluate the mining results on the performance of the algorithm of closed weighted frequent pattern and compare it with the algorithm PCAR which has been used to acquire knowledge from patents. The two hundred patents for test are derived from the shear class in forestry machinery. Through the pretreatment we get a set of feature components from each patent which contains 1408 data items and the maximum length record is 73. A random sample of 40, 80, 120, 160, 200 patents are selected to test.

Fig.1 is the operation results of closed weighted frequent pattern mining algorithm (WCloset) under different minimum weighted support and different amount of data: (1) The operation time decreases with the weighted support increase. The reason is that the higher support the items have, the more items are cut according to the weight of items in primary stage and the less running time in later stage. (2) The running time rises as well as the increase of data quantity. The results validate the stability of the algorithm operation. Compared with the other algorithm PCAR, the quantity of patterns which is mined by WCloset is lesser so that it avoids acquiring unimportant or boring patterns. The results are shown in Fig.2. From Fig.2.we can see that when set

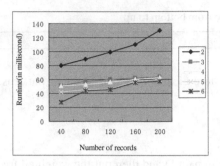
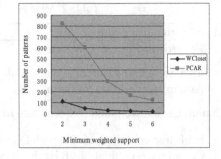

Fig. 1. WCloset operation result   **Fig. 2.** Algorithm comparisons

minimum weighted support 3, WCloset mined 45 frequent patterns, while PCAR mined 602 frequent patterns. Select 6 frequent patterns from each set of patterns. The results are shown as follows.

After mining patterns from patents, we need to identify the patterns and mark their functions. For example, the first combination of components in Table 3 could realize the shear function. The third combination not only can shear branches, but also can transform the length of hand shank to cut the higher branches. The forth combination could shear and collect object at the same time which could be used to pick fruit. We can know that there are some useful combinations in Table 4, however it also has a lot of useless and shorter combinations. Such as the first combination it doesn't have the functional significance. When designers carry out their research, we can provide the functional knowledge for them to assist them obtain the solution quickly. The functional knowledge could also be applied to patent classification. Compared with other classification methods, the method based on the functional knowledge classifies patents more exact and reliable.

Table 3. Six frequent patterns from Wcloset

ID	Component set
1	Scissors; Spring; Hand shank; Cramp ring
2	Spring; Fruits bag; Upright; Bracing wire; Cramp ring
3	Scissors; Hand shank; Expansion link; Cramp ring
4	Scissors; Hand protectors; Hand shank; Spring; Cramp ring; Fruits bag
5	Scissors; Spring; Hand shank; Fruits bag; Pin roll; Electrical machinery
6	Scissors; Hinge; Spring; Hand shank; Cramp ring

Table 4. Six frequent patterns from PCAR

ID	Component set
1	Hand shank; Spring
2	Spring; Fruits bag
3	Scissors; Fruits bag
4	Hand protectors; Hand shank; Spring
5	Scissors; Spring; Hand shank; Fruits bag
6	Scissors; Spring; Hand shank; Fruits bag; Electrical machinery

5   Conclusion

Knowledge acquisition has been widely studied at home and abroad, but the number of researches to acquire knowledge from patents is lesser. According to characteristics of forestry machinery patent data, integrating the closed pattern mining with weighted pattern mining acquires hidden knowledge. The experimental analysis proved the robustness and effectiveness of the algorithm for patents and ensured access to gain the valuable knowledge.

Acknowledgment. Foundation: Supported by the Fundamental Research Funds for the Central Universities(DL12EB01-02, DL12CB05) and Heilongjiang Postdoctoral Fund(Grant No. LBH-Z11277) and Natrual Science Foundation for Returness of Heilongjiang Province of China(LC2011C25).

References

1. Kang, B., Motohashi, K.: The role of essential patents as knowledge input for future R&D. World Patent Information 05, 1–9 (2014)
2. Seol, H., Lee, S., Kim, C.: Identifying new business areas using patent information: A DEA and text mining approach. Expert Systems with Applications 38, 2933–2941 (2011)
3. Wang, W.M., Cheung, C.F.: A Semantic-based Intellectual Property Management System (SIPMS) for supporting patent analysis. Engineering Applications of Artificial Intelligence 24, 1510–1520 (2011)
4. Zhang, H., Qiu, Q., Feng, P., Wang, Z.: An automated method for acquiring design knowledge from produce patents. Journal of Harbin Engineering University 30, 785–791 (2009) (in Chinese)
5. Yun, U.: Mining lossless closed frequent patterns with weight constraints. Knowledge Based Systems 20, 86–97 (2007)

Information Propagation with Retweet Probability on Online Social Network

Xing Tang, Yining Quan, Qiguang Miao[*], Ruihong Hou, and Kai Deng

School of Computer Science and Technology, Xidian University, Xi'an, Shaanxi, China
qgmiao@126.com

Abstract. The rapid development of online social network has attracted a lot of research attention. On online social network, people can discuss their ideas, express their interests and opinions, all of which are demonstrated by information propagation. So how to model the information propagation cascade accurately has become a hot topic. In this paper, we firstly incorporate the retweet probability into the traditional propagation models. To find the accurate retweet probability, we introduce the logistic regression model for every user based on the extracted features. With the crawled real dataset, simulation is conducted on the real online social network and moreover some novel results have been obtained. The homogenous retweet probability in the original model has underestimated the speed of information propagation, despite the scale of information propagation is almost at the same level. Besides, the initial information poster is really important for a certain propagation, which enables us to make effective strategies to prevent epidemics of rumor on social network.

Keywords: retweet probability, online social network, infectious model, diffusion model, logistic regression.

1   Introduction

With the advent of the Internet and mobile platform, increasing people join the online social network and leave their digital footprints. Online Social Network Service (SNS) has attracted much attention because of its rapid growing users. SNS websites provide a platform for people to build and keep social relationships on the web. With their relationships, users can express their views on hot events occurred in the real life, follow other users according to their interests and express their own opinions in the virtual world. Nowadays, the popularity of these websites and the maturity of big data processing tools enable researchers on social science to collect sufficient social data to verify traditional theories.

Information is the indicative of many social behaviors. People in the online social network mostly communicate with each other by means of posting information, retweeting and commenting. All of these actions consist the information cascade on the online social network. So it is the key to give an accurate model of information propagation which helps us understand user traces better in the social data.

---

[*] Corresponding author.

H. Wang et al. (Eds.): ICYCSEE 2015, CCIS 503, pp. 326–333, 2015.

To model the information diffusion, several traditional infectious disease models have been widely adopted in information propagation due to the deep analogy between the spread of information and epidemical disease [1] . Some works have already done to enlarge the applications of these models into the social network [2, 3]. Besides the disease model, some diffusion models in the sociology are also adopted [1]. While treating the infectious probability and diffusion probability as a constant or some certain distribution, these models fail to make use of the additional social features which can greatly affect the information propagation.

Retweet probability has been investigated a lot. Using many models, researchers have worked on several ways to study the retweet behavior. Without the additional information, Saito et.al [4] built an expectation-maximization model to find how to compute the retweet probability on the whole social network. While as the method for gathering more information about social network develops, Bongwon Suh et al. [5] applied the statistical tools to discover the correlation between social features and retweet behavior. Based on the discoveries, increasing researches focused on how to make use of different social features and meanwhile apply various learning model to determine the retweet probability [6-8].

In this paper, we investigate how individual probability affects the information propagation on online social network. Firstly, aiming to simulate the individual probability, we compute the retweet probability with logistic regression model and compare it with the global retweet probability, which is mostly considered in previous works. Moreover, analogous to the infectious probability and diffusion probability, the retweet probability is associated with every connection which is treated as directed edge, and both of them are input into the information propagation model. Then several experiments are conducted with propagation model on the SINA Weibo, an online social network in China. Lastly, we compare the results and give some explanations for these results.

2   Dataset Description

SINA Weibo is currently the one of the most popular online social networks in China. We crawled and collected a part of user profiles and tweet information in nearly one and a half months with our data mining platform which includes a cluster consisting of 8 computers. The specific data format is shown in Table 1. Totally, we gather 21,246 users and their corresponding tweets starting from user's registration time to Nov.1st, 2013.

Table 1. User data format

Parameters	id	name	tweet		verified	location
Description	user id	nick name	tweet list		verified status	user cities
Parameters	status_counts	fansNum	followsNum	follows		
Description	tweet count	fan count	follows count	follows list		

Retweeting is the action of reposting someone else's tweet inside your own tweet timelines. Usually users will receive his follows tweet in his own tweet timelines, under different circumstance, they can use *Retweet* button to retweet a certain tweet to show its interest or enlarge its influence. In our dataset, if one tweet retweet other's tweet, one of its field named rid will be set the retweeting tweet's id, the data format is as shown in Table 2.

Table 2. Weibo data format

Parameters	mid	text	pid	Rid
Description	tweet id	tweet content	original tweet id	retweet id

3   Retweet Probability Computation

In traditional information propagation model, the propagating rate is the probability determine how likely the virus or information will flow between people in infected state and susceptible state. In the online social network, when one of user follows post a tweet, this tweet will insert into user's timeline.

According to [9], the features between users are linearly related to the retweeting behavior. To obtain the probability, we introduce the logistic regression model here to compute the retweet probability. The logistic regression [10] is defined as follows:

$$p_i(retweet) = \frac{1}{1+\exp(-w^T x_i)} . \tag{1}$$

Then the features should be extracted from our dataset. Aiming to find how the general tweet flow through the whole social network, the social user information and structure information are considered without the particular tweet information. The features are as follows:

User information:

Gender: it is represented by 0/1 for female or male.

Verified: a verified user is the one who has real identity both online and offline, thus 0/1 for verified or not verified.

Tweet number: The tweet number is the count of tweets posted by user in a certain period. Because most of the values are greater than 1, the normalization is used here.

Fans number, Follow number: the count of follows and fans.

Location: In our dataset, the location of a user is a string which can be divided to province, city, district, so we can use string matching here to compute the similarity.

Structure information:

Co-follow number: From the structural view, due to our dataset is a directed network, the common follow can show the common interest between two users.

Mutual follow: In our dataset, mutual follow is considered as an important attribute. In the structural view of point, mutual follow is the bidirectional edge, while most of other edges is directed. 0/1 is used here for representing whether they are mutual follows or not.

Based on these extracted features, the Weka toolkit is used to train the LR model. To be specific, suppose user A has his follows, the follows who have never be retweeted by A is labeled as 0, and the one who has been retweeted once will be labeled as 1. Then according to the features between the 1 set and 0 set, the LR model can be trained.

Moreover, as to every user, several follows are kept as test sample to quantify the result, thus the problem being modeled as a binary classification problem. For every particular user, we split its follows into train and test set randomly, which is set as the input of the model. From the Weka output, the numerical result values can be averaged over the whole network users.

As we know, most of the previous works are devoted into a global logistic regression model. To compare individual result and global result, Table.3 shows the average measurements including precision, recall and F-value respectively on both individual and global model.

Table 3. Results on both individual and global model

	Precision (%)	Recall (%)	F values (%)
Individual	72.27	71.04	71.65
Global	65.26	69.86	65.41

From the F-value, it is clearly noticed that individual retweet probability learning is better than the global model learning.

After the model is trained, the retweet probability between users can be obtained. Together with the structure information, this information will be input into the model in next section.

4   Information Propagate Model

After determining the directed connection retweet probability, we should proceed to deal with tweet propagation model.

Firstly, diffusion model is considered here. An independent cascade model starts with an initial set of active nodes N. When node u becomes active for the first time in time step t, its provided with one chance p to activate each of its currently inactive neighbor v. In the propagation process, $p_{u,v}$ is selected independently as a constant probability. In our dataset, user u posts a tweet, and his followers has a particular chance to retweet the tweet. In our model $IC-p$ uses the retweet probability to simulate the propagating process.

On the other hand, with the similarity between virus flow and tweet propagation, the virus transmission model is also adopted here. In virus transmission model, when user has contact with each other, the virus has a certain probability flows through each people. In our Weibo dataset, the virus can be represented by the tweet.

In virus model [11, 12], there are usually three states for individual, S is the susceptible people, I is the infected ones and R is the recovered people. Compared with these, users in our dataset can be one of following states. S is the user state when the tweet has never been retweeted, I is the state which once the user retweet the tweet and the user often switch from S to I. Besides, the virus model deals with the undirected social network, while in Weibo dataset the connection is directed like: $A \rightarrow B$, user B is unable to ask user A to follow him, and moreover, any tweet posted by user B will infect A's timelines, that is the information flow is in the reverse direction.

Based on the discussed state, the Weibo dataset can be treated as SIS model. Users can retweet a tweet many times. It is easily noticed that in these steps, the infected probability p is chose from certain probability distribution or just set as a constant, which is irrelevant to the user attributes. However, as to online social network, the retweet probability is actually related to the user attributes. In the case of that, the infected probability here is modified as the retweet probability in our work. On the other hand, because tweet has its own content which is lacked in virus model, we only treat the personal information and structural information here to simplify the model.

In conclusion, the propagation algorithm for $SIS - p$ can be as follows:

Algorithm 1. $SIS - p$ propagation model

1: for u in $network.nodes$:

2:　$\text{Pr}_{network.edges[u]} = LR(u, tweets[u])$

3: end for

3: randomly select a user to be added into $set(I)$

4: while step $< maxStep$:

5:　for u in $set(I)$:

6:　　for user v in $u.fans$:

7:　　　retweet with the probability $\text{Pr}_{v \rightarrow u}$

8:　　　if retweeted:

9:　　　　user v is added into $set(I)$

10:　　end if

11:　end for

12:　remove u from $set(I)$

13:　end for

14:　$count(set(I))$ is added into retweet number

15: end while

16: return retweet number

---

5   Experiment and Discussion

To find how retweet probability affects the propagation model, we conduct several experiments. SIS model and IC model with constant probability and the models with retweet probability learned from the logistic regression beforehand are respectively simulated.

In this experiment, we initially randomly choose a user as a poster, then simulate the propagation process in the network. 100 users are selected in the initial phase and the average retweet number is calculated. The results of IC and SIS model are demonstrated respectively in Fig.1 and Fig.2

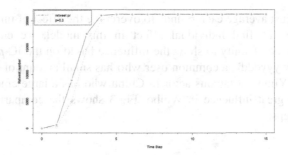

Fig. 1. IC and IC-p model

The result shows that the constant retweet probability really underestimates the speed of information propagation in real social network, 4 steps for red one and 6 steps for green one. However, the scales of both models are almost the same. This is easy to explain in the real circumstance. As to constant probability, the value of p is chosen at random which cannot reflect the real situation, moreover, user treats his every fans as homogenous users. Retweet or not is considered as a constant action for his users. Introducing retweet probability can solve this problem, which enables user retweet based on their individual situation. In fact, from the learned probability, there are nearly more than half retweet probability exceeds 0.3 which we randomly set here. In IC model, every user has opportunity to be infected in every step which is not in the real circumstance.

On the other hand, both the models are unable to propagate the whole network, which indicates in real network there exist many users never retweet. Observing our dataset really verifies this point.

From the algorithm 1, users in SIS model can be divided into two separated state sets: S and I. so the whole population will converge to two sets from which users will switch states. From Fig.2, the retweet probability really accelerates the converge process. Moreover, the SIS-p model shows more great amplitude compared with SIS model, indicating that the retweet probability has more effects to divide the population into the S and I state set, which is well accord with reality, for some users are more influential than others.

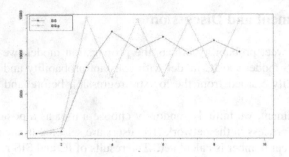

Fig. 2. SIS and SIS-p model

Above all, we just average each rounds to overcome the effect of initial user on the final steady state. To find individual affect in this model, we chose two users represent different community to show the influence based on the IC-p model. One's nickname is "wwwyyyddd", a common user who has small amount of fans in Weibo, another is Zhang Yimou, a famous actor in China who has a large amount of fans in Weibo and has a great influence in Weibo. Fig.3 shows the comparison of retweet number in very steps.

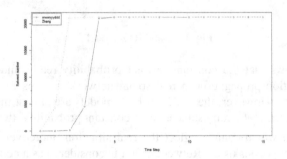

Fig. 3. IC-p on different posters

Clearly, the curve of Zhang goes to the steady state very fast in nearly 2 steps, while the red one stays small in three steps until it happened that one of its tweet goes wildly.

6   Summary

In this paper, we devote our work to discuss how individual retweet probability affects the information propagation on real online social network. Based on the logistic regression model, the retweet probability for every user is calculated with a certain accuracy. Then the simulation result shows that the retweet probability really influence the information propagation and results give more details about the differences between high influence-level user and common user which sheds light on the future study of influence propagation and rumor control.

However, information propagation model has its own shortage. In our next work, we will focus on adopting a more accurate model framework. Moreover, learning individual retweet probability is suffered from the problem of sparse, so how to model more accurate retweet probability with context factor and find the solution to overcome the sparse problem will be investigated further.

Acknowledgement. The work was jointly supported by the National Natural Science Foundations of China under grant No. 61472302，61272280，41271447, and 61272195; The Program for New Century Excellent Talents in University under grant No. NCET-12-0919; The Fundamental Research Funds for the Central Universities under grant No. K5051203020，K5051303016，K5051303018, BDY081422, and K50513100006; Natural Science Foundation of Shaanxi Province, under grant No.2014JM8310；The Creative Project of the Science and Technology State of xi'an under grant No. CXY1341(6).

References

1. Guille, A., et al.: Information Diffusion in Online Social Networks: A Survey. SIGMOD Record 42(2), 17 (2013)
2. Xu, B., Liu, L.: Information diffusion through online social networks (2010)
3. Yan, Q., et al.: Information Propagation in Online Social Network Based on Human Dynamics. Abstract and Applied Analysis 2013, 6 (2013)
4. Saito, K., Ohara, K., Yamagishi, Y., Kimura, M., Motoda, H.: Learning diffusion probability based on node attributes in social networks. In: Kryszkiewicz, M., Rybinski, H., Skowron, A., Raś, Z.W. (eds.) ISMIS 2011. LNCS, vol. 6804, pp. 153–162. Springer, Heidelberg (2011)
5. Suh, B., et al.: Want to be Retweeted? Large Scale Analytics on Factors Impacting Retweet in Twitter Network (2010)
6. Petrovic, S.V.S.A., Osborne, M., Lavrenko, V.: Rt to win! predicting message propagation in twitter. In: 5th ICWSM (2011)
7. Luo, Z., et al.: Who Will Retweet Me?: Finding Retweeters in Twitter. In: SIGIR 2013. ACM, New York (2013)
8. Hong, L., Dan, O., Davison, B.D.: Predicting Popular Messages in Twitter. In: WWW 2011. ACM, New York (2011)
9. Yang, Z., Rong, L.U., Qing, Y.: Predicting Retweeting in Microblogs. Journal of Chinese Information Processing 26(4), 109–114 (2012)
10. Hosmer, D.W., Lemeshow, S., Sturdivant, R.X.: Introduction to the logistic regression model. Wiley Online Library (2000)
11. Pastor-Satorras, R., Vespignani, A.: Epidemic dynamics and endemic states in complex networks. Phys. Rev. E. 63, 066117 (2001)
12. Pastor-Satorras, R., Vespignani, A.: Epidemic Spreading in Scale-Free Networks. Phys. Rev. Lett. 86, 3200–3203 (2001)

A Data Stream Subspace Clustering Algorithm

Xiang Yu[1], Xiandong Xu[1], and Liandong Lin[2]

[1]Department of Computer Science and Technology, Heilongjiang Institute of Technology,
Harbin, China
yuxpointfly@163.com
[2] Heilongjiang Province Key Lab of Senior- Education for Electronic Engineering,
Heilongjiang University, Harbin, China
1267013@qq.com

Abstract. The main aim of data stream subspace clustering is to find clusters in subspace in rational time accurately. The existing data stream subspace clustering algorithms are greatly influenced by parameters. Due to the flaws of traditional data stream subspace clustering algorithms, we propose SCRP, a new data stream subspace clustering algorithm. SCRP has the advantages of fast clustering and being insensitive to outliers. When data stream changes, the changes will be recorded by the data structure named Region-tree, and the corresponding statistics information will be updated. Further SCRP can regulate clustering results in time when data stream changes. According to the experiments on real datasets and synthetic datasets, SCRP is superior to the existing data stream subspace clustering algorithms on both clustering precision and clustering speed, and it has good scalability to the number of clusters and dimensions.

Keywords: data mining, data stream, subspace clustering, feature selection, dimension reduction.

1 Introduction

Recently, researches on data stream mining are motivated by more and more applications on continuous stream data, such as customer click streams, multimedia data, etc. By contrast with traditional data sets, data stream consists of a series of dynamic data objects which are massive and unordered[1,2,3]. Since not all data objects in data stream are maintained, it is necessary to be fault tolerant when clustering data stream. With the level of data collection technology increases and more and more characteristics data stream have, it is difficult to cluster data stream in high-dimensional space or cluster data stream in subspace effectively and efficiently. Clustering data stream in subspace needs to scan data sets several times, so it is almost impossible to process data stream with traditional subspace clustering methods.

HPStream[4] is a classic subspace clustering algorithm for high-dimensional data stream, Since it has several advantages such as fast clustering, high clustering precision, etc., HPStream is always used as the comparison to other algorithms on clustering speed and clustering precision. After data are partitioned into clusters, HPStream

H. Wang et al. (Eds.): ICYCSEE 2015, CCIS 503, pp. 334–343, 2015.

chooses clustering dimension with heuristic strategy. However, HPStream needs to determine the number of average dimensionality as parameter, which is hard to determine generally[5][6]. And there exists an obvious high-dimensional clustering problem. In 2007, Sun proposes GSCDS, a data stream subspace clustering algorithm based on grid which can cluster high-dimensional data stream[7]. In certain clustering subspace, GSCDS has high clustering speed and precision. When partitioning data space, the top-down partition method is adopted and then the subspace is related to clusters. GSCDS can identify clusters in different subspace with arbitrary shapes and the parameter of subspace need not to be predefined, and it has low computing complexity which is related to dimensionality of data space d, the number of grid cells m and the loop number of region partition, that is Q_{dmn}). However, when clusters in different subspace, GSCDS needs to run several times from top to down. In 2007, Park proposes a data stream subspace clustering algorithm based on grid[8][9], and it is based on, cell tree, a clustering algorithm on all dimensions. Park partitions data space into grid cells at different level and the grid cells are stored in a tree structure called sibling tree. In sibling tree, the clusters in each subspace can be easily found. There are three phases in this algorithm. In the first phase, a sibling tree is constructed by partition data space into grid cells at different levels. With data stream flows, the new coming data object falls into the corresponding grid cell according to its dimensional value, and the statistics of the grid cell is updated. In the second phase, dense grid cells are further partitioned. In the third phase, sparse grid cells are merged. The algorithm can cluster in all subspaces at different levels, but it needs to determine several parameters during the construction of sibling tree, and the parameters will further influence the final clustering results.

This paper proposes a new clustering algorithm SCRP(Subspace Clustering based on Region Partition), which can minimize the parameter influence on clustering and adopts down–top strategy. When partitioning regions, the distribution situations of data points on each dimension are considered. When clustering, the dense regions on each dimension overlap and then subspace clusters appear. Since the partition regions are formed on the basis of grid cells, SCRP has the advantages of grid clustering algorithms which can fast cluster and are not sensitive to outliers[10,11,12]. SCRP can find subspace clusters effectively and adjust the subspace cluster information according to the changes of data stream. When data stream changes, the region information will also change which will further influence the subspace clusters.

2   Definition

2.1   Concepts of SCRP

Dense grid Cell: The grid cells whose dense support exceed the dense threshold ρ.
Dense Region: The region consists of connected dense grid cells on each dimension.
Partition Regions: The region sets consist of dense regions and sparse regions which are formed by merging dense grid cells and sparse grid cells.

Region-tree : region–tree is a limited set formed of $k(k \leq 2^d)$ subspace nodes. Region-tree has only one root node, and the number of sons involved in each node will not exceed d, where d is the dimensionality of data space.

2.2   Principle of Region Partition in SCRP

Suppose D is a set consists of n d-dimensional data points, $D = \{x_1, x_2, ..., x_n\}$, $D \subseteq \mathbf{R}^d$. Data point x is denoted as $x_i = \{x_{i1}, x_{i2}, ..., x_{ij}, ..., x_{id}\}$, where $j = 1, 2, ..., d$, and x_{ij} is the value of x_i on dimension j. When data stream flows, the slipping window is defined by time and the statistics information of each corresponding node in region-tree is updated according to the data in slipping window, such as the average value of data points in a region, etc. Then cluster in subspace according to requirements.

First, SCRP finds all dense regions on each dimension, and then merge connected dense regions and sparse regions to form partition regions. The set $R^1 = \{R_1^1, R_2^1, ..., R_d^1\}$ involves dense regions on each dimension, where R_i^1 denotes the set of dense regions on dimension i. For example, in 3-dimensional subspace, C_{ac} and C_{bc} are the clusters which are overlapped by dense region R_c^1 with R_a^1 in subspace ac and R_c^1 with R_b^1 in subspace bc, as shown in Figure 1.

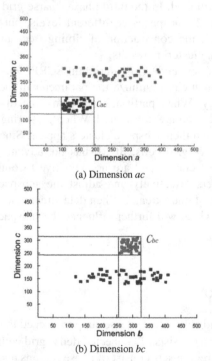

(a) Dimension ac

(b) Dimension bc

Fig. 1. Two dimensional subspace clusters in three-dimensional space

Project the data in $D = \{x_1, x_2, ..., x_n\}$ on dimension i, the result is shown in figure 2. According to the distribution situation of data projection, when region partition on dimension i completes, the result of region partition is kept in R_i^1. Similarly, the result of ordered partition regions on other dimensions is kept in the corresponding sets. Finally, the dense regions on each dimension are kept in $R^1 = \{R_1^1, R_2^1, ..., R_d^1\}$.

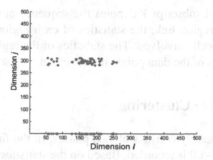

Fig. 2. Data projection on the ith dimension

2.3   SCRP Clustering

In SCRP, when clustering, the similarities of overlapped regions in subspace are compared with the specified threshold. If the similarity exceeds the threshold, the overlapped region is identified as a cluster of the subspace. The region similarity is denoted as $\text{sim}(R_i^1, R_j^1) = |R_i^1 \cap R_j^1|$, where $R_i^1, R_j^1 \in R^1$. In SCRP, a tree structure called region-tree is used to preserve subspace clusters. In region-tree, we can find clusters in all subspaces, and the sequence of dimensions is not important. Region-tree is constructed according to the predefined order $d_1 \rightarrow d_2 \rightarrow ... \rightarrow d_d$. Figure 3 shows a

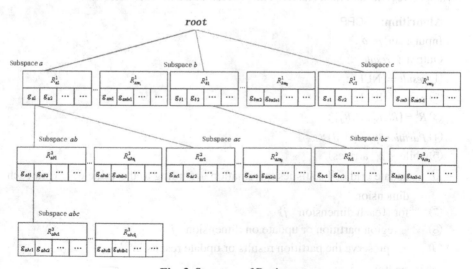

Fig. 3. Structure of Region-tree

region-tree in 3-dimensional data space, and a, b, c denotes three dimensions in data space respectively. In each data space, different regions and the grid cells involved in the regions are preserved in the region link structure. R_X^i denotes the region link in i-dimensional data space which records the basic information of each region in the data space, where subscript X denotes the region sequence number and the dimensions involved. The region in region-tree involves at least one grid cell. In grid cell g_Y, the statistics is recorded, and subscript Y denotes the sequence number of grid cells and the of the subspace. In region link, the statistics of each region can be computed by the statistics of the grid cells involved. The statistics of the region will continually be updated with the changes of the data points in the region.

3   SCRP Subspace Clustering

First, SCRP clustering mainly consists of two phases. In the first phase, the statistics information of each grid cell is recorded. Based on the statistics information, the partition regions are formed on each dimension. With the changes of data stream, the partition results and the corresponding statistics of regions is updated. When the density of grid cells changes, the regions formed by connected grid cells also change, and the partition regions need to be renewed.

The update process of region partition is as follows:

Sparse Region Split: When the density of sparse region exceeds the threshold, then split the sparse region into grid cells and reunite the grid cells into regions.

Dense Region Split: When the density of dense region falls under the threshold, split the dense region into grid cells and reunite the grid cells into regions.

In the second phase, construct region-tree, and record the information of each re-gion and the grid cells involved. Based on the information, compute the similarity of overlapped regions and then cluster. The process of SCRP is as follows:

Algorithm: SCRP.
Input : sp, ρ, δ.
Output : $result$.
① $result$ = NULL;
② $root$ = NULL;
③ $R^1 = \{R_1^1, R_2^1, ..., R_d^1\}$;
④ $PartitionSpace(\delta, d)$;
⑤ while (x_i arrives)
⑥   update grid cell statistics information according to the value of x_i on each dimension.
⑦   for (each dimension j)
⑧   region partition or update on dimension j
⑨   preserve the partition results or update results in R_j^1
⑩   endfor

⑪endwhile

⑫ $root = Construct\_RT(R^1)$;

⑬ $result = SP\_Clustering(root, sp)$;

⑭return $result$;

In the algorithm, sp denotes the clustering subspace, ρ, δ denotes the dense threshold and the partition parameter of grid cell respectively, $result$ preserves the clustering result, R^1 preserves the partition regions on each dimension, and $root$ denotes the root of region-tree.

SCRP first partitions d-dimensional data space with parameter δ, and updates the statistics information of grid cells according to the arrival of data points. When clustering in subspace, the connected grid cells on each dimension whose densities exceed threshold ρ are merged, further the partition regions are formed and preserved in R^1, then construct region-tree and cluster in the subspace sp.

4   Experiments Analysis

In order to analyze the clustering speed, clustering precision and scales, KDD-CUP99, a real-world data set of MIT, together with a synthetic data set containing 500000 60-dimensional data objects generated by a data generator are used. In the synthetic data set, the domain of data point on each dimension is [0,500], and each data point is involved in one cluster or exists as an outlier. In data space, the outliers in the synthetic data set are less than 5 percent of the total data points and are uniformly distributed.

4.1   Quality Evaluation

Commonly, when clustering data stream at different time or in different subspace, the results will be different, and it is hard to evaluate the clustering quality when simulating data stream with the data in synthetic data set. Since CLIQUE is a clustering algorithm with high precision, its clustering result is always used as the measure to evaluate the clustering quality of other algorithms.

In the experiments, the clustering result of CLIQUE is used as the precision measure when compare the clustering quality between SCRP and HPStream. CLIQUE runs several times with different partition granularities, and the best clustering result is used as the precision measure. When simulate data stream changing, CLIQUE needs to run several times to guarantee the clustering precision. In order to test the validity of the algorithm, the synthetic data set is divided into two parts to simulate the wave change of data stream.

In this experiment, parameter δ is set $\delta = 20$ in SCRP, when the new coming data points increase to 1000, the region density and the region-tree need to be updated because of the density decay. The parameter of HPStream is set $l = 20$, where l denotes the average dimensionality. The comparison results of clustering precision are shown in figure 4 and figure 5. As can be seen, the clustering quality of SCRP is superior to HPStream.

When data stream changes, the historical information decays to reduce the influence on SCRP clustering. Other than HPStream, SCRP preserves historical information but reduces its influences gradually. According to the changes of data stream, SCRP updates the statistics information of regions and gets accurate clustering results. When wave changes coming, the stability of SCRP is superior to HPStream. Compared with HPStream, SCRP does not need to divide data into clusters and provide the average dimensionality in advance, in addition, SCRP does not have too much requirements on parameters.

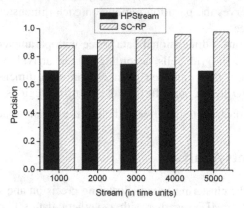

KDD-CUP99, stream speed 200/s

Fig. 4. Comparison of precision

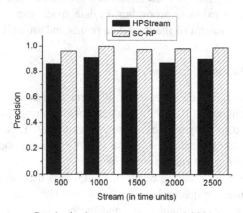

Synthetic data set, stream speed 200/s

Fig. 5. Comparison of precision

4.2   Runtime Evaluation

In the experiment, the parameter of SCRP is $\delta = 20$, when the new coming data points increase to 1000, the region density and the region-tree need to be updated because of the density decay. The average dimensionality is set $l = 20$. As shown in figure 6, compared with HPStream, SCRP needs less time, and it becomes more obvious when the region-tree is formed and tends to stable.

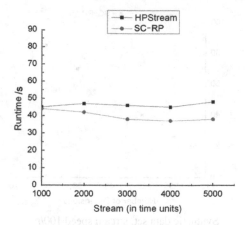

Synthetic data set, stream speed 200/s

Fig. 6. Comparison of runtime

4.3   Scalability Evaluation

In the experiment, the parameter of SCRP is $\delta = 20$, when the new coming data points increase to 1000, the region density and the region-tree need to be updated because of the density decay. The experiment aims to test the scalability of SCRP when the data dimension and the number of clusters change. As shown in figure 7, SCRP has good scalability to the number of clusters, and its runtime is almost linear to the number of clusters in subspace, as the number of clusters increases, the runtime also increases.

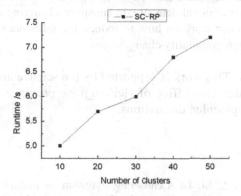

Synthetic data set, stream speed 100/s

Fig. 7. Counts of clusters in subspace

SCRP is different to CLIQUE, its performance can hardly be influenced by the dimensionality, and it has good scalability to the changes of dimensionality. As shown in figure 8, the runtime of SCRP is almost linear to the dimensionality, as the dimensionality increases, the runtime also increases.

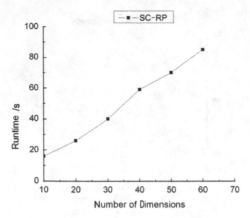

Synthetic data set, stream speed 100/s

Fig. 8. Change of dimensions

5   Conclusion

In this paper, a region-tree structure is designed to preserve the changes of data stream, and SCRP, a data stream subspace clustering algorithm based on region partition is proposed. Compared to the current data stream subspace clustering algorithms, SCRP can cluster in subspaces at all levels and costs less time. In addition, SCRP can hardly be influenced by outliers and it can record and adjust the clustering results according to the changes of data stream in region-tree. The experiments on real data set and synthetic data set show that SCRP has good effectiveness and applicability, and it is superior to traditional data stream subspace clustering algorithms. In the future, we will extend our work on how to reduce the influences on clustering quality when the grid partition granularity changes.

Acknowledgments. This work is supported by the science and technology project 12531542 of the education office of Heilongjiang province. We thank professor Qi haoliang for enlightening discussions.

References

1. Ling, C., Lingjun, Z., Li, T.: A clustering algorithm for multiple data streams based on spectral component similarity. Information Sciences 183(1), 35–47 (2012)
2. Weiguo, L., Jia, O.: Clustering algorithm for high dimensional data stream over sliding windows. In: Proc of the 10th Int. Conf. on Trust, Security and Privacy in Computing and Communications, pp. 1537–1542. IEEE, Piscataway (2011)
3. Halkidi, M., Koutsopoulos, I.: Online Clustering of distributed streaming data using belief propagation techniques. In: Proc of the 12th Int. Conf. on Mobile Data Management, pp. 216–225. IEEE, Piscataway (2011)

4. Aggarwal, C., Han, J., Wang, J., et al.: A framework for clustering evolving data streams. In: Proc of the 29th Int. Conf. on VLDB, pp. 81–92. Morgan Kaufmann (2003)
5. Parsons, L., Haque, E., Huan, L.: Subspace clustering for high dimensional data: A review. ACM SIGKDD Explorations Newsletter 6(1), 90–105 (2004)
6. Yihong, L., Yan, H.: Mining data streams using clustering. In: Proc of the 4th Int. Conf. on Machine Learning and Cybernetics, pp. 2079–2083. IEEE, Piscataway (2083)
7. Yufen, S.: Research on clustering algorithm based on grid. Huazhong University of Science and Technology, Wuhan (2006)
8. Park, N.H., Lee, W.S.: Cell tree: An adaptive synopsis structure for clustering multi-dimensional on-line data stream. Data & Knowledge Engineering 4(3), 1–22 (2007)
9. Park, N.H., Lee, W.S.: Grid-based subspace clustering over data streams. In: Proc of the ACM Conf. on Information and Knowledge Management, pp. 801–810. ACM, New York (2007)
10. Yanwei, Y., Qin, W., Jun, K., et al.: An on-line density-based clustering algorithm for spatial data stream. Acta Automatica Sinica 38(6), 1051–1059 (2012)
11. Dutta, B.R., Angelov, P.: Evolving local means method for clustering of streaming data. In: Proc of the 2012 Int. Conf. on World Congress on Computational Intelligence, pp. 1–8. IEEE, Piscataway (2012)
12. Lingjuan, L., Xiong, L.: An improved online stream data clustering algorithm. In: Proc of the 2nd Int. Conf. on Business Computing and Global Informatization, pp. 526–529. IEEE, Piscataway (2012)

Fine-Grained Access Control for Big Data
Based on CP-ABE in Cloud Computing

Qi Yuan[1,2,*], Chunguang Ma[1], and Junyu Lin[1]

[1] Harbin Engineering University, Harbin City 150001, China
[2] Qiqihar University, Qiqihar City 161006, China
foreveryuanqi@126.com

Abstract. In Cloud Computing, the application software and the databases are moved to large centralized data centers, where the management of the data and services may not be fully trustworthy. This unique paradigm brings many new security challenges, which have not been well solved. Data access control is an effective way to ensure the big data security in the cloud. In this paper, we study the problem of fine-grained data access control in cloud computing. Based on CP-ABE scheme,we propose a novel access control policy to achieve fine-grainedness and implement the operation of user revocation effectively. The analysis results indicate that our scheme ensures the data security in cloud computing and reduces the cost of the data owner significantly.

Keywords: big data, Fine-grained, Access control, Cloud computing, CP-ABE.

1   Introduction

With the development of networking technology and the increasing need of computing resources, many organizations are apt to outsource their storage and computing need. We commonly refers to this kind of new economic and computing model as cloud computing. In the future, more and more businesses will obtain the services from cloud computing technologies.

Benefits of cloud computing are clear while cloud computing also poses kinds of risk[1].If the security problems in cloud computing are not well solved, it will impede the fast growth of cloud computing. In fact, the confidentiality of data has become the biggest hurdle in cloud computing. In order to share data, users publish these data on Cloud Servers, meanwhile, they require fine-grained data access control according to permission of data consumer under the distributed cloud computing access control architecture [2].

Extensive works have been done on data access control in the past decade years, and various research such as [3][4] went into implementing fine-grained access control effectively. Traditionally, researchers usually assume that the data are stored in a fully trusted server which acts as an omnipotent reference monitor[5] responsible for confirming and administering the access control policy. However, this assumption

---

* Corresponding author.

H. Wang et al. (Eds.): ICYCSEE 2015, CCIS 503, pp. 344–352, 2015.
© Springer-Verlag Berlin Heidelberg 2015

does not hold in cloud computing any more, because cloud servers can obtain the outsourced data and leak these data to an unauthorized user for high benefit. In order to help the data owner achieve fine-grained access control of data stored in an untrusted cloud server, a reliable solution is that the data are encrypted by using certain encryption algorithm and decryption key is disclosed only to authorized users. Unauthorized users, even if Cloud Server, cannot decrypt the data for lack of the decryption key. Existing access control schemes either introduce high complexity on key management or fail to revoke the user effectively. Attribute-based access control schemes [4][6] have solved the above-mentioned problem effectively by providing fine-grained data access control and preventing collusion attack. S.Roy et.al.[7] constructed a Ciphertext Policy Attribute-Based Encryption (CP-ABE) scheme have been a promising technique [8-10] for access control of encrypted data. However, CP-ABE scheme can't solve effectively the attribute revocation problem of data access control in cloud computing.

In this paper, we propose a novel fine-grained data access control scheme based on CP-ABE for cloud computing which achieves fine-grainedness and data confidentiality of data access control in cloud computing. Our scheme is more flexible because it does not need the trusted third party.The data owner can authorize most of computation assignments to the reliable authorized users and implement user revocation effectively in our scheme.

2   Model and Preliminary

2.1   System Model

Our system model consists of three types of entities: Data Owner, Data Consumer(User) and Third Party Authority, which is in accord with [11]. Cloud Server is considered to be honest but curious. We assume that each entity initializes a public/private key pair, and other entities can easily obtain its public key when necessary.

2.2   System Framework

In this section, we define the framework based on CP-ABE,which contains the following algorithms.

Setup: Let G_0 be a bilinear group of prime order P and g be a generator of G_0. Furthermore, let $e : G_0 \times G_0 \rightarrow G_1$ denote a bilinear map and function $H : Z_p^* \rightarrow G_0$ map any attribute to an element in G_0,where $H(i) = g^i$. We choose three random exponents $\alpha, b, \gamma \in Z_p$.The parameter d denotes the number of attributes of every private key.We choose a d-degree polynomial $v(x)$ at random, and moreover, its initial value is set to be $v(0) = b$.The master key MK is (b, γ, g^a) and the public parameters PK are $PK = (G_0, g, h = g^b, h^\gamma, f = g^{\frac{1}{\gamma}}, e(h, g)^a, \{g^{v(0)}, g^{v(1)}, ..., g^{v(d)}\})$.

Encryption (PK, M, T): Let T be a tree that represent an access structure. Define a random polynomial q_x for each node x of T in the top-down manner starting from the root node R. For each non-root node x, $p_x(0) = p_{parent(x)}(index(x))$ .where $parent(x)$ represents x's parent and $index(x)$ is x's unique index given by its parent. For the root node R, $p_R(0) = s$,then output the ciphertext

$$CT = (T, C1 = Me(h, g)^{as}, C2 = (h^\gamma)^s, \forall y \in Y1 : C1_y = g^{q_y(0)}, C2_y = H(i)^{q_y(0)},$$

$$\forall y \in Y2 : C3_y = h^{q_y(0)+u_y}, C4_y = (V(i))^{u_y}, C5_y = g^{u_y}, i = F(att(y))).$$

Note $Y1$ and $Y2$ be a set of the leaf nodes in T with positive attributes and negative attributes respectively. For each node $y \in Y2$, u_y is randomly chosen from Z_p. We define a function $V(i): Z_p \to G_0$ as $V(i) = g^{\upsilon(i)}$. Function $F: \{0,1\}^* \to Z_p^*$ maps each attribute to an unique integer in Z_p^*.

KeyGen(MK, S): Choose a random $r \in Z_p$ and a random $r_j \in Z_p$ for each attribute $j \in S$, compute the key as

$$SK = (D = g^{(a+r)\gamma}, D1 = g^r, \forall j \in S : D1_j = h^r \cdot H(j)^{r_j}, D2_j = g^{r_j}, D3_j = (V(j))^r).$$

Delegate (SK, S^*): This algorithm employs a secret key SK , which is for a set S of attributes, and another set S^* that $S^* \subset S$. Choose a random r^* and $r_k^*, \forall k \in S$, and then create a new secret key as $SK^* = (D^* = D \cdot f^{r^*}, D1^* = g^r \cdot g^{r^*},$

$$\forall k \in S^* : D1_k^* = D1_k \cdot h^{r^*} \cdot H(k)^{r_k^*}, D2_k^* = D2_k \cdot g^{r_k^*}, D3_k^* = D3_k \cdot (V(k))^{r^*}).$$

Decrypt(CT, SK): Firstly, compute $DecryptNode(CT, SK, x)$ for leaf nodes. If the node x is a leaf node, then let $i = att(x)$. On the one hand, if x corresponds to a positive attribute and $i \in S$, then $DecryptNode(CT, SK, x) = e(D1_i, C1_x)/e(D2_i, C2_x)$ $= e(h, g)^{rq_x(0)}$,If x corresponds to a positive attribute and $i \notin S$,we define $DecryptNode(CT, SK, x) = \perp$ .On the other hand, if x corresponds to a negative attribute and $i = att(x) \notin S$,then we set $S' = S \cup i$ and compute Lagrangian coefficients $\{\sigma_j\}_{j \in S'}$. Then $DecryptNode(CT, SK, x) = e(D1, C3_x)/e(D5_x, \prod_{j \in S}(D3_j)^{\sigma_j}) \cdot e(D1, D4_x)^{\sigma_i}$ $= e(h, g)^{rq_x(0)}$,If the node x corresponds to a negative attribute and $i \in S$,then we define $DecryptNode(CT, SK, x) = \perp$. It aggregates these pairing results in the bottom-up manner using the polynomial interpolation technique. Finally, it may recover the blind factor $e(h, g)^{as}$ and output the message M .

3   Our Scheme

In this section, we describe the construction of our scheme.

Definition and Notation: We assume that attributes of the user in our system can be divided into two categories. (i) static attributes which remain unchanged for a long time, and (ii) dynamic attributes which can be updated, if necessary. The file owner assigns an access tree for its own data file which can be depicted by Boolean formula including AND, OR, NOT and threshold. Analogously, we divide this access tree into two distinct parts, as shown in Figure.1. More precisely, the tree $T1$ contains the static attributes and the tree $T2$ includes the dynamic attributes, respectively. The static attributes in $T1$ are file information and a dummy. We assign the root node of $T1$ to be an AND gate and the other child nodes can be any threshold gate.

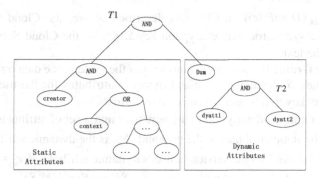

Fig. 1. An Access Tree Combines Static Attributes and DynamicAttributes

The Cloud Server maintains a list CUL which records the user's IDs and the version of user's dynamic attributes. The server also keeps a ciphertext history list CHL which stores different version of ciphertexts as well as the relevant versions. For the purpose of user grant/revocation, the data owner maintains a list UL with all users' IDs and a list DAL which contains the last updated dynamic attributes and relevant versions.

System Setup: This algorithm outputs the system public parameter PK and the system master key MK by calling the algorithm CP-ABE Setup. The data owner signs each element of PK and sends PK along with these signatures, i.e. $(PK, \delta_{O,PK})$ to the Cloud Server, and then the data owner chooses the dynamic attribute set SD and sets the version of the dynamic attribute set Ver_{SD} .Finally, the data owner stores (Ver_{SD}, SD) in the DAL list.

New File Creation: To reduce the cost of encryption/decryption, we combine CP-ABE system with symmetric key cryptosystem. Before uploading a file to the Cloud Server, the data owner processes the data file as follows.

1)Select a unique ID for the data file.

2)Select a symmetric encryption scheme E ,and produce a symmetric encryption key SSK ,then encrypt the data file using SSK ,i.e. $E_{SSK}(DataFile)$.

3) Define two access trees T_1 and T_2 for the data file and encrypt SSK using CP-ABE. Firstly, the data owner calls $Encryption(SSK, PK, T1)$ to generate the ciphertext CT .From Encryption(PK, M, T) presented in section 2, we see that $CT = (CT_{Dum}, CT^*)$, and then the data owner re-encrypts CT_{Dum} according to the dynamic attribute's access

tree $T2$ by calling the function $Encryption(CT_{Dum}, PK, T2)$ to generate the ciphertext CT'_{Dum}. Then, we denote CT by (CT'_{Dum}, CT^*).

4) Send the tuple $(ID, Ver_c, CT, E_{SSK}(DateFile), \delta_{o,(ID,Ver_c,CT,E_{SSK}(DataFile))})$ to the Cloud Server. Obviously, $Ver_c = Ver_{SD}$.

On receiving the tuple, the Cloud Server will verify the signature $\delta_{o,(ID,Ver_c,CT,E_{SSK}(DataFile))}$ .If correct, the Cloud Servers will store the ciphertext $(Ver_c, CT, E_{SSK}(DataFile))$ in CHL. As described above, the Cloud Server has no idea about the symmetric data encryption key SSK. So the Cloud Server cannot decrypt the ciphertext.

New User Grant: If a new user wants to join the system, the data owner defines a set of attributes S including static and dynamic attributes for the user. The corresponding secret key to this user is generated as follows.

1) Assign a unique identity ω to the new user and a set of attributes $S = \{S_s, S_d\}$,where S_s is the static attributes of the user and S_d is the dynamic attribute set picked from the set SD and set the version of user's dynamic attribute $Ver_\omega = Ver_{SD}$.

2) Generate a secret key SK for ω, i.e., $SK \leftarrow Key(MK, S)$ .According to the construction of KeyGen, SK can be denoted by (SK_s, SK_d).

3) Encrypt the tuple $(\omega, S, Ver_\omega, SK, \delta_{O,(\omega,S,Ver_\omega,SK)})$ with user ω's public key k, i.e. $C = (\omega, S, Ver_\omega, SK, \delta_{O,(\omega,S,Ver_\omega,SK)})_k$

4) Store the IDs of the user ω in UL.

5) Send the tuple $(\omega, Ver_\omega, C, \delta_{O,(\omega,Ver_\omega,C)})$ to the Cloud Server.

Once the Cloud Server receives the tuple $(\omega, Ver_\omega, C, \delta_{O,(\omega,Ver_\omega,C)})$, then,

1) Verify $\delta_{O,(\omega,Ver_\omega,C)}$ and proceed if correct.

2) Store (ω, Ver_ω) in the system user list CUL.

3) Forward C to the user .

When the user receives C, firstly, it will decrypt C with its private key, and verifies the signature $\delta_{O,(\omega,S,Ver_\omega,SK)}$ .If the signature is correct, the user accepts (S, Ver_ω, SK) as the attribute set, the version of dynamic attribute and the secret key.

User Revocation: Whenever a user ω is revoked, we have two methods to remove the user.

The first method: when the data owner wants to remove a user ω, it updates the dynamic attribute set SD as well as Ver_{SD} .Then it re-encrypts the data file and updates user's secret keys except the removed user ω .The data owner acts as follow:

1) Remove ω from the user list UL.

2) Update the dynamic attribute list DAL and change Ver_{SD} and the dynamic attribute set SD.

3) Design a new tree $T2$, the dynamic attribute set SD satisfied $T2$, Re-encrypt CT'_{Dum} to CT'_{Dum_new} with a new $T2$.

4) Send the tuple $(\omega, Ver_c, CT'_{Dum\_new}, \delta_{(\omega, Ver_c, CT'_{Dum\_new})})$ to the Cloud Server, where Ver_c is equal to Ver_{SD}.

Once the Cloud Server receives the information from the data owner, it will verify the signature $\delta_{(\omega, Ver_c, CT'_{Dum\_new})}$. If correct, the Cloud Servers will remove the user ω from CUL and update the old ciphertext in CHL $CT = \{CT'_{Dum}, CT^*\}$ with $CT' = \{CT'_{Dum\_new}, CT^*\}$. Whenever a user except the user ω requires a file, the date owner will update its secret key.

Second method: we can revoke a user easily by the attaching negative attributes. We use a non-monotonic access structure to express negative constraints in the access policy. The basic idea for revoking user is to associate the negative constraints with access policy of ciphertext which contains the ID of the revoked user by using non-monotonic access structure. Whenever a data owner want to revoke a user ω, it can attach a leaf node $(NOT\ \omega)$ to the tree $T1$. In addition, this scheme also allows us to revoke a set of users who have the same attributes. Augmenting the original access formula T with (NOT ("attribute A" AND "attribute B")), we can effectively revoke all the users with attributes A and B from the system.

File Access: In this algorithm, the Cloud Server responds user request on the data file access, and contacts the data owner to update the user secret key if necessary. As user's secret keys need not to be updated in the second method, then we update the secret key aiming at the first method.

First, the user sends a data file access request $REQ = (u, Ver_u, ID)$, where u is the identity of the user, Ver_u is the version of its dynamic attribute set and ID is the number of the request file.

On receiving the access request, the Cloud Server acts as following:

1) Check the request if the user is a valid system user according to CUL.

2) If true, check whether Ver_u is equal to Ver_{SD}.

3) If true, the user need not update the secret key, and then sends the ciphertext which it requested..

4) If not true, the user needs updating the secret key and then sends u the identity of the user to the data owner.

When the data owner receives the data.

1) Check whether the user u is a valid user in UL.

2) If true, assign a new dynamic attribute set S'_d for the user u and create a new key $SK'_d = Key(MK, S'_d)$.

3) Encrypt the tuple (S'_d, Ver'_u, SK'_d) with the user ω's public key k, i.e. $CP = (S'_d, Ver'_u, SK'_d)_k$, where Ver'_u is equal to Ver_{SD}.

4) Send the tuple $(u, Ver'_u, CP, \delta_{(u, Ver'_u, CP)})$ to the Cloud Server.

Once receiving the tuple, the Cloud Server replaces the old user's dynamic attribute version Ver_u with new version Ver'_u in CUL, and send the tuple $(u, Ver_u, CP, \delta_{(u,Ver_u,CP)}, CT)$ to the user u. The user decrypts the tuple with its static secret key and the new dynamic secret keys.

User Delegation: As described above, the algorithm would bring about a huge computation overhead for the data owner to grant user and might require the data owner to be always online to provide secret key update service for users. To solve this problem, we propose a user delegation operation. For the sake of clarity, we divide the whole users into several categories. Every user in the same category has the same dynamic attributes. The data owner chooses a local reliable authority for every category and local authority has the right to grant/revoke the user. We assume that the local authority of category i has a set of attributes S_i and the secret key SK_i.

When a new user v wants to join the system, the data owner assigns a category to the user v (assume in category i). Then the data owner authorizes the local authority i to implement User Grant operation. The local authority i acts as the data owner choosing a meaningful set of attributes $\bar{S}_i \subset S_i$ for this user v and calls the CP-ABE function $Delegate(SK, \bar{S}_i)$ to generate the secret key for this user, and then sends the tuple (v, Ver_v) to the Cloud Server.

When a user has been revoked, the data owner updates the dynamic attributes and the secret key components for the user's local authority. If the user whose dynamic attributes and secret key components are not updated wants to access a file, it can ask its local authority to update the dynamic attributes and the secret key components. Consequently, the data owner does not need to handle a heavy computation and can be always on online to provide user grant/revocation.

4   Efficiency of Our Scheme

4.1   Security Analysis

Now we analyze the security of the proposed scheme.

Fine-grained access control: In our algorithm, the data owner defines and executes expressive access structure for each data file. More specifically, the access structure of each file is described as a logic formula over data file information.

Preventing collusion: CP-ABEscheme shows it can resist the collusion attack. On the one hand, we do not disclose any secret key components to the Cloud Server in our scheme. So the Cloud Server has no useful information for a user to access the unauthorized files. The Cloud Server can't collude with users. On the other hand, each user only has its own secret keys, which can not collude together to access the file that it have no rights to access. This is guaranteed by CP-ABE scheme.

Keeping data file confidentiality: In our scheme, data files are encrypted by symmetric encryption scheme (such as AES) with the symmetric key SSK and the symmetric key SSK is directly encrypted using CP-ABE. As showed above, the Cloud Server can't decrypt ciphertext to obtain SSK and the data file. As symmetric encryption scheme and CP-ABE are secure so far, our scheme can keep data file confidentiality for each operation doesn't leak any information about the data file.

4.2   Perfomance Analysis

We analyze and compare the computation complexity of the proposed scheme with the scheme[4] in Table.1. The parameter $|N|$ is the number of users, which is almost similar to$|L|$.

Table 1. Scheme complexity comparison

Operation	Complexity in scheme[5]	Complexity in our scheme
File Creation	$\mathcal{O}(|I|)$	$\mathcal{O}(|L|)$
User Grant	$\mathcal{O}(|L|)$	$\mathcal{O}(|I|)$
User Revocation	$\mathcal{O}(|N|)$	$\mathcal{O}(|L|)$
File Access	$\mathcal{O}(\max(|L|,N))$	$\mathcal{O}(|I|)$

The main computation in System Setup operation is the same as function Setup.
The computation overhead of New File Creation includes two parts: one part is generated by encrypting the data file using the symmetric key SSK and the other part is produced by encrypting the SSK using CP-ABE. The computation overhead of the first part lies on the size of the underlying data file that is necessary for any cryptographic algorithm. The computation overhead of the second step is $\mathcal{O}(|L|)$ multiplication operations over G_0 and a pair operation from $G_0 \rightarrow G_1$, where L represents the attribute set of leaf nodes in the access tree. The operation New User Grant is performed interactively by the Cloud Server , the data owner and the user. Here, we just count the complexity of the data owner. The computation mainly overhead consists of the generation of the user secret key as well as encryption of the user secret key using the user's public key. The first part takes $\mathcal{O}(|I|)$ multiplication operations on G_0 ,where I denotes the attribute set of user. The second step takes much less times compared with the first part. Finally, we consider the computation overhead of the data owner in the User Revocation and File Access operation. In first method of User Revocation, the overhead computation is caused by generating new ciphertext components, which accounts for $\mathcal{O}(|L|)$ multiplication operations over G_0 . File Access operation that handles the new secret key components for the dynamic attribute takes $\mathcal{O}(|I|)$ multiplication operation over G_0 .

From Table 1, the complexity of our scheme is same as the scheme[4]. However, we can delegate the User Revocation operation and File Access operation to a reliable user, which can reduced the cost of the data owner significantly.

5 Conclusion

Due to sharing the physical resources in the untrusted cloud servers , It has also brought users a great number of security challenges about cloud computing. access control for big data is an effective approach to ensure data security in the cloud. In this paper, we presented a new access control scheme based on CP-ABE. Specifically,we adopt the idea that data file is encrypted by a symmetric key that is later encrypted using the CP-ABE algorithm. To introduce revocation facility into our first scheme, we define both static and dynamic attributes for each user that enable more fine-grained access control and attach negative attribute to access tree. In order to release the data owner from heavy computation task, we introduce the operation User Delegation to deal with new user grant and user revocation.The analysis results indicate that the proposed scheme can realize use revocation and reduce the cost of the data owner efficiently.

Acknowledgements. This research is supported by a grant from National Natural Science Foundation of China (No. 61170241, 61472097).This paper is funded by the International Exchange Program of Harbin Engineering University for Innovation-oriented Talents Cultivation.

References

1. Feng, D.G., Zhang, M., Zhang, Y., et al.: Study on Cloud Computing Security. Journal of Software 22(1), 71–83 (2011)
2. Almutairi, A., Sarfraz, M., Basalamah, S., et al.: A distributed access control architecture for cloud computing. IEEE Software 29(2), 36–44 (2012)
3. Wang, X.M., Fu, H., Zhang, L.: Research Progress on Attribute-Based Access Control. Chinese Journal of Electronics 38(7), 1660–1667 (2010)
4. Yu, S., Wang, C., Ren, K., et al.: Achieving Secure, Scalable, and Fine-grained Data Access Control in Cloud Computing. In: INFOCOM 2010 Proceedings IEEE, San Diego, CA, March 14-19, pp. 1–9 (2010)
5. Anderson, J.P.: Computer security technology planning study. Air Force Electronic Systems Division, Report ESD-TR-73-51 (1972)
6. Hur, J., Noh, D.: Attribute-Based Access Control with Efficient Revocation in Data Outsourcing Systems. IEEE Transactions on Parallel and Distributed Systems 22(7), 1214–1221 (2011)
7. Roy, S., Chuah, M.: Secure data retrieval based on ciphertext policy attribute-based encryption (CP-ABE) system for the DTNs. Technical Report (2009)
8. Zhiguo, W., Jun'e, L., Deng, R.H.: HASBE: A Hierarchical Attribute-Based Solution for Flexible and Scalable Access Control in cloud Computing. IEEE Transactions on Information Forensics and Security 7(2), 743–754 (2012)
9. Bianchi, G., Capossele, A., Petrioli, C.: AGREE:exploiting energy harvesting to support data-centric access control in WSNs. AD HOC NETWORKS 11(8), 2625–2636 (2013)
10. Kan, Y., Xiaohua, J.: Expressive, Efficient, and Revocable Data Access Control for Multi-Authority Cloud Storage. IEEE Transactions on Parallel and Distributed Systems 25(7), 1735–1744 (2014)
11. Wang, Q., Wang, C., Li, J., Ren, K., Lou, W.: Enabling public verifiability and data dynamics for storage security in cloud computing. In: Backes, M., Ning, P. (eds.) ESORICS 2009. LNCS, vol. 5789, pp. 355–370. Springer, Heidelberg (2009)

The New Attribute-Based Generalized Signcryption Scheme[*]

Yiliang Han[1,2], Yincheng Bai[1], Dingyi Fang[2], and Xiaoyuan Yang[1]

[1] Key Lab. of Cryptology and Information Security, Department of Electronic Technology;
Engineering University of CAPF, Xi'an 710086, China
[2] College of Information Science and Technology
Northwest University, Xi'an 710127, China
yilianghan@hotmail.com

Abstract. An attribute-based generalized signcryption scheme based on bilinear pairing has been proposed. By changing attributes, encryption-only mode, signature-only mode, and signcryption mode can be switch adaptively. It shows that the scheme achieves the semantic security under the decisional bilinear Diffie-Hellman assumption and achieves the unforgeability under the computational Diffie-Hellman assumption. It is more efficient than traditional way and can be used to secure the big data in networks.

Keywords: Attribute-based Encryption, Generalized signcryption, Diffie-Hellman assumption, Random oracle model, Big data.

1 Introduction

With the development of information technology, big data [1] in network faces enormous risks. Traditional public key cryptography encrypts the individual messages with the public key of the recipient, which damages the users' privacy as well as reduces the efficiency. It fails to perform the operation if the recipient will not be able to get the key. Therefore, the traditional public key cryptography has been unable to meet the needs of big data in networks. Confidentiality and authentication of the communication are two important security features. Zheng [2] proposed the new cryptographic primitive of signcryption in 1997, which encrypts and the signs the message within a logical step while keeping the lower cost than traditional way. But the signcryption scheme will lose its advantage, when we only need the encryption or the signature operation. Switching between encryption, signature and signcryption inevitably consumes a lot of resources. The concept of generalized signcryption was proposed by Han [3] in 2006, which proposed the flexibility to choose a module encryption, signature or signcryption. It can be applied to the broader communications environment.

[*] This work is partially supported by Natural Science Foundation of China (61103231, 61272492, 61462408, 61103230), the Project funded by China Postdoctoral Science Foundation (2014M562445).

H. Wang et al. (Eds.): ICYCSEE 2015, CCIS 503, pp. 353–360, 2015.

Identity-based cryptography [4] simplifies the certificate management problems of traditional public-key cryptosystem, which can regard some users' information (such as age, height, etc.) as a public key. Boneh et al [5] proposed first identity-based encryption scheme in 2001 based on bilinear. Sahai and Waters [6] had taken into account the problems of users' privacy, and proposed an encryption scheme based on fuzzy identity. It describes the identity of a user is multiple attributes instead in this scheme, which is the prototype of ABE (attribute-based cryptography) scheme. Goyal [7] in 2006 divided attribute-based encryption into key-based strategy (Key Policy ABE, KP-ABE) and ciphertext-based strategy (Ciphertext Policy ABE, CP-ABE) encryption schemes. Then Hemanta [8] proposed signature scheme based on the property in 2008. Gagne [9] first proposed signcryption (Attribute Based Signcryption, ABSC) scheme based on attributes. In 2012, Zhang [10] proposed an attribute-based signcryption scheme with dynamic threshold. Liu and others [11] proposed the signcryption scheme based on attributes in the vector space in 2013. Since the privacy of signature attributes and other issues are not well resolved, more improved ABSC schemes [12] have been constantly proposed.

An attribute-based generalized signcryption scheme for big data is proposed. It inherits the features of ABE and generalized signcryption simultaneously. Compared with other scheme [13], it achieves the confidentiality and further broaden the application environment. In random oracle model, it is indistinguishable against adaptive chosen ciphertext attacks and unforgeable against chosen message attacks.

2   Preliminaries

2.1   Bilinear Pairing

Let g be generators of the group G and G_1, which prime number is p. If the map $e: G \times G \to G_1$ is called bilinear pairing, then need to meet the following properties:

- Bilinearity. If any $g_1, g_2 \in G, x, y \in {}_R Z_q{}^*$, it has $\langle g_1{}^x, g_2{}^y \rangle = \langle g_1, g_2 \rangle^{xy}$.
- Non-degeneracy. For any $g_1, g_2 \in G$, it has $\langle g_1, g_2 \rangle \neq 1$.
- Computability. For any $g_1, g_2 \in G$, an efficient algorithm is able to evaluate $\langle g_1, g_2 \rangle \in G_1$.

2.2   Computational Assumption

Definition 1. (Decision Bilinear Diffie-Hellman Problem, DBDH.) If there is no attacker is able to win the following game by at least the advantage of ε, the hypothesis ε – DBDH is founded. Challenger B randomly selects parameters $a, b, c, z \in Z_p$ and $\eta \in \{0\ 1\}$. If $\eta = 1$ then B outputs the parameters $(g, g^a, g^b, g^c, e(g,g)^{abc})$; if $\eta = 0$, the output is $(g,\ g^a, g^b, g^c, e(g,g)^z)$. The guess of η by attacker A is η'. If $|\Pr[A(g, g^a, g^b, g^c, e(g,g)^{abc}) = 1] - \Pr[A(g, g^a, g^b, g^c, e(g,g)^z) = 1] | \geq \varepsilon$, then the advantage of solving the problem of DBDH by A is at least ε.

Definition 2. (Computational Diffie-Hellman Problem, CDH.) If there is no attacker can win the following game by at least the advantage of ε, the hypothesis $\varepsilon - \mathrm{CDH}$ is founded. Challenger B randomly selects parameters $a, b \in Z_p$ and outputs (g, g^a, g^b) and attackers try to get the result $g^{ab} \in G$. If $\Pr[A(g, g^a, g^b) = g^{ab}] \ge \varepsilon$, then the advantage of solving the problem of DBDH by A is at least ε.

3   Attribute-Based Generalized Signcryption Scheme

3.1   Initialization

Parameter Generation. PKG (Public Key Generator) randomly selects $s, r \in Z_q{}^*$. The default attribute set $M = \{M_1, M_2, \cdots, M_{r-1}\}$ composed by r-1 attributes. Selects bilinear pairing $e : G \times G \to G_1$, selects $g, g_2 \in G, k \in Z_q{}^*$ uniformly and calculates $g_1 = g^k$, $E = e(g_1, g_2)$. Lagrange coefficient is $\Theta_{x,z}(y) = \prod_{z \in Z, z \ne x}(y - z / x - z)$. Randomly selects the set of attributes $\Omega \in Z_q{}^*$. $H : \{0,1\}^* \to G$ and $H_1 : \{0,1\}^* \times \{G\}^* \to G$ are hash functions. Encryption and decryption algorithms are respectively $E_\theta(\cdot)$ and $D_\theta(\cdot)$. The master key of system is k, the public parameters is $params = (s, r, q, G, G_1, e, g, g_1, g_2, E, H, H_1, E_\theta(\cdot), D_\theta(\cdot))$.

Key Generation. Private key generator randomly selects the r-1 degree polynomial $l(n)$ and satisfies $l(0)=a$. The set of attributes for a user ID is $\gamma_{ID} \in \Omega$, and generates attribute set $\gamma_{ID}' = \gamma_{ID} \cup M$. Randomly selects $t_n \in Z_q{}^*$. Computing users' private key $R_n = (r_{no}, r_{n1}) = (g_2^{l(n)} H(i)^{t_n}, g^{t_n})$.

3.2   Signcryption and Unsigncryption

Signcryption. The assertion supported by this scheme is $T_{\theta, \gamma^*}(\cdot)$, which γ^* is the set of attributes, θ for the threshold, in which $1 \le \theta \le r$. If the attribute set γ' includes at least θ elements in the set of attributes γ^*, then γ' satisfies the assertion $T_{\theta, \gamma^*}(\cdot)$. The encryption assertion of the scheme is $T_{\theta', \gamma_2^*}(\cdot)$ and the signature assertion is $T_{\theta, \gamma_1^*}(\cdot)$, which $|\gamma_1^*| = v_1$ and $|\gamma_2^*| = v_2$. The set of attributes for the sender is $\gamma_A = \{n_1, n_2, \cdots, n_{vA}\}$ and satisfies $T_{\theta, \gamma_1^*}(\gamma_A) = 1$. The subset of attributes for A is $\gamma_A' \subseteq \gamma_1^* \cap \gamma_A$. Default attributes in M contributes the default set of attributes $M_2' = \{n_{\theta'+1}, n_{\theta'+2}, \cdots, n_r\} \subset M$. Randomly selects $\eta \in Z_q{}^*$, calculates $\sigma_0 = g^\eta$ and $\{\sigma_n = H(n)^\eta\}_{n \in \gamma_2^* \cup M_2'}$. Randomly selects $v_1 + r - \theta$ random number $t_n' \in Z_q{}^*$ and calculates $\sigma_0' = [\prod_{n \in \gamma_A' \cup M_1} r_{n,0}^{\Theta_{n,\eta}(0)}][\prod_{n \in \gamma_1^* \cup M_1} H(n)^{t_n'}] \cdot H_1(m, \{\sigma_n\}_{n \in \gamma_2^* \cup M_2'})^\eta$, $\theta = E^\eta$ and $c = E_\theta(m) \oplus g^{\sigma_0'}$. If it is the encryption operation, the output ciphertext is

$\sigma = \{\sigma_0\,,\ \sigma_0',\{\sigma_n\}_{n\in\gamma_2*\cup M_2'},0,c\}$. Otherwise calculates $\{\sigma_n' = r_{n,1}^{\Theta_{n,\eta}(0)}\,g^{t_n'}\}_{n\in\gamma_A'\cup M_1'}$, $\{\sigma_n' = g^{t_n'}\}_{n\in\gamma_1*\cup M_1'}$, and outputs the ciphertext $\sigma = \{\sigma_0\,,\ \sigma_0',\{\sigma_n\}_{n\in\gamma_2*\cup M_2'},$ $\{\sigma_n'\}_{n\in\gamma_A*\cup M_1'},c\}$. When the above signcryption procedure is a separate signature process, then the sender private key is an empty set, so $c = E_\theta(m)\oplus g^{\sigma_0'} = E_0(m)$ $\oplus g^0 = m$.

UnSigncryption. The set of attributes for receiver B is $\gamma_B = \{n_1,n_2,\cdots,n_{vB}\}$, and satisfies $T_{\theta,\gamma_2*}(\gamma_B)=1$. If the signers of this ciphertext satisfies the assertion $T_{\theta,\gamma_1*}(\cdot)$, then continues to meet, otherwise terminates the operation. Calculates $H_1(m,\{\sigma_n\}_{n\in\gamma_2*\cup M_2'})$, if $E = e(g,\sigma_0')/[\prod_{n\in\gamma_1*\cup M_1'}e(H(n),\sigma_n')]e(H_1(m,\{\sigma_n\}_{n\in\gamma_2*\cup M_2'}),\sigma_0)(1)$ holds then continues the process, otherwise terminates the operation. B selects the set of attribute $\gamma_B' \subseteq \gamma_2*\cap\gamma_B$ .Calculates $\{r_{n0}' = r_{n0}\}_{n\in\gamma_B'\cup M_2'},\{r_{n0}' = H(n)\}_{n\in\gamma_2*/\gamma_B'},$ $\{r_{n1}' = r_{n1}\}_{n\in\gamma_B'\cup M_2'},$ $\{r_{n1}' = g\}_{n\in\gamma_2*/\gamma_B'}, \theta = \prod_{n\in\gamma_2*\cup M_2'}(e(r_{n0}',\sigma_0)/e(\sigma_n,r_{n1}'))^{\Delta_{n\eta}(0)}$ and $m = D_\theta(c)\oplus g^{\sigma_0'}$.

3.3 Adaptive Mode

Signature. If the private key for receiver is an empty set, then the encryption process is masked and $c = E_\theta(m)\oplus g^{\sigma_0'} = E_0(m)\oplus g^0 = m$.

Encryption. If $\{\sigma_n'\}_{n\in\gamma_A*\cup M_1'} = 0$, then the operation is encryption, and calculates $\{r_{n0}' = r_{n0}\}_{n\in\gamma_B'\cup M_2'}$, $\{r_{n0}' = H(n)\}_{n\in\gamma_2*/\gamma_B'}$, $\{r_{n1}' = r_{n1}\}_{n\in\gamma_B'\cup M_2'}$, $\{r_{n1}' = g\}_{n\in\gamma_2*/\gamma_B'}$, $\theta = \prod_{n\in\gamma_2*\cup M_2'}(e(r_{n0}',\sigma_0)/e(\sigma_n,r_{n1}'))^{\Delta_{n\eta}(0)}$ and $m = D_\theta(c)\oplus g^{\sigma_0'}$.

4   Security Analysis

4.1   Correctness

$$e(g,\sigma_0')/[\prod_{n\in\gamma_1*\cup M_1'}e(H(n),\sigma_n')]e(H_1(m,\{\sigma_n\}_{n\in\gamma_2*\cup M_2'}),\sigma_0)$$

$$= \frac{e(g,[\prod_{n\in\gamma_A'\cup M_1'}(g_2^{l(n)}H(n)^{t_n})^{\Theta_{n,\eta}(0)}][\prod_{n\in\gamma_1*\cup M_1'}H(n)^{t_n}]\cdot H_1(m,\{\sigma_n\}_{n\in\gamma_2*\cup M_2'})^\eta)}{[\prod_{n\in\gamma_1*\cup M_1'}e(H(n),g^{t_n\Theta_{n,\eta}(0)}g^{t_n'})\prod_{n\in\gamma_1*/M_A'}e(H(n),g^{t_n'})](H_1(m,\{\sigma_n\}_{n\in\gamma_2*\cup M_2'}),g^\eta)}$$

$$= \prod_{n\in\gamma_A'\cup M_1'}(g,g_2^{l(n)})^{\Theta_{n,\eta}(0)} = E$$

$$\theta = \prod_{n \in \gamma_2^* \cup M_2'} (e(r_{n0}', \sigma_0) / e(\sigma_n, r_{n1}'))^{\Delta_{nj}(0)}$$

$$= \prod_{n \in \gamma_2^* \cup M_2'} (\frac{e(g_2^{l(n)} H(n)^{t_n}, g^{\eta})}{e(H(n)^{\eta}, g^{t_n})})^{\Delta_{nj}(0)} \prod_{n \in \gamma_2^* / M_2'} (\frac{e(H(n), g^{\eta})}{e(H(n)^{\eta}, g)})^{\Delta_{nj}(0)}$$

$$= \prod_{n \in \gamma_2^* \cup M_2'} (g_2^{l(n)}, g^{\eta})^{\Delta_{nj}(0)} = E^{\eta}$$

4.2 Security Proofs

Theorem 1. If a DBDH problem is on the establishment of the group G, then the attacker can't win with non-negligible advantage. And the scheme has the security of selection attributes set.

Proof. If the attacker A executes at most q_{H_n} times H_n queries, q_{SC} times key genera-tion queries, q_{η} times signcryption queries and q_{US} times unsigncryption queries, which $n=1$, with the ε advantage of a polynomial time attack definitions 3 game.

System Setting. Algorithm F calculates $g_1 = g^a$ and $g_2 = g^b$. The default set of attribute is $M = \{M_1, M_2, \cdots, M_{r-1}\}$, for which randomly selects system parameters $s, r \in Z_q^*$. The set of signature attributes for challenger A is γ_1^*, and the threshold is $1 \leq \theta \leq r$. The set of default attributes randomly selected is $|M_1'^*| = r - \theta$, $M_2'^* \subseteq M$ and $|M_2'^*| = r - \theta'$. Their signcryption and encryption assertions are respectively $T_{\theta, \gamma_1^*}(\cdot)$ and $T_{\theta', \gamma_2^*}(\cdot)$.

Random Oracle Query. H_1 Oracle. Query n, and if n is in the list K_1 which saved H_1, then returned corresponding results. Otherwise, if $n \in \gamma_2^* \cup M_2'^*$, then randomly selected $x_n \in Z_q^*$ and recorded $H_1(n) = g_1^{x_n}$ into the list K_1, otherwise randomly selected $x_n, y_n \in Z_q^*$ and recorded $H_1(n) = g_1^{-x_n} g^{y_n}$ into the list K_1.

H_2 Oracle. Randomly selected $h_1, h_2 \in [1, q_{H_2}]$. Queried n and if it is saved in the list K_2 of H_2, returns the corresponding result. Or if $n \neq h_1, h_2$, then randomly selected $x_n, z_n \in Z_q^*$, and recorded $H_2(u_n, \{\sigma_m\}_{m \in \Omega}) = g_1^{z_n} g^{x_n}$ into the list K_2 which saved H_1. If $n = h_1$, then randomly selected $x_{h_1} \in Z_q^*$ and recorded $H_2(u_n, \{\sigma_m\}_{m \in \Omega}) = g^{x_{h_1}}$ into the list K_2. If $n = h_2$, then randomly selected $x_{h_2} \in Z_q^*$ and recorded $H_2(u_n, \{\sigma_m\}_{m \in \Omega}) = g^{x_{h_2}}$ into the list K_2.

Key Query. Defines the set O, O', P, making $P = O' \cup \{0\}, |O'| = r - 1, O \subseteq O' \subseteq P$ and $O = (\gamma_A \cap \gamma_2^*) M_n'$. If the parameter γ_n makes the formula $|\gamma_A \cap \gamma_2^*| < \theta'$ estab-lished, then let $N_n = (g_1^{a_n} H_1(n)^{b_n}, g^{b_n})$, which $n \in O', a_n, b_n \in Z_q^*$, wherein is that met

$l(i)=a_n$, $l(0)=a$ is a polynomial of degree $r-1$ $l(i)$. For $n \in O'$, let $b_n = (\Theta_{0,z}(n)/y_n)b+b_n{}'$, $l(n) = \sum_{m \in O'} \Theta_{m,z}(n)l(m) + \Theta_{0,z}(n)l(0)$, $N_n = ((g_1^{-y_n} g^{x_n})^{t_n'}$ $g_2^{(\Theta_{0,z}(n)x_n/y_n) + \sum_{m \in O'} \Theta_{m,z}(n)l(m)}, g_2^{(\Theta_{0,z}(n)/y_n)} g^{t_n})_{n \in \gamma_n}$. Otherwise, the query fails.

Signcryption Query. If $|\gamma_A \cap \gamma_2{}^*| < \theta'$, then the ciphertext is generated based on the private key $R_{An} = (r_{no}, r_{n1})$ by γ_A is sent to A. Otherwise $r-\theta$ default attributes in M makes the set $M_1{}'$, and $r-\theta'$ default attributes in M makes the set $M_2{}'$. If $\gamma_A \cup M_1{}' = \{n_1, n_2, \cdots, n_r\}$, then randomly selected $t_n, \eta' \in Z_q{}^*$. If $\eta = (-b/Z_{n_r}) + \eta'$, calculated $\sigma_0 = g^\eta = g_2^{-1/Z_{n_r}} g^{\eta'}$, $H_2(m, \{\sigma_n\}_{n \in \gamma_2{}^* \cup M_2{}'}) = g_1^{z_{n_r}} g^{y_n}$, $\sigma_0{}' = g_2^a H_2(m, \{\sigma_n$ $\}_{n \in \gamma_2{}^* \cup M_2{}'})^\eta \prod_{n \in \gamma_1{}' \cup M_1{}'} H_1(n)^{t_n} = (g_1^{z_{n_r}} g^{y_n})^\eta \prod_{n \in \gamma_1{}' \cup M_1{}'} H_1(n)^{t_n} g_2^{-y_n/Z_{n_r}}$, $\{\sigma_n{}' = g^{t_n}$ $\}_{n \in \gamma_1{}^* \cup M_1{}'}$, $\{\sigma_n = g^{y_n \eta} = g_2^{-y_n/Z_{n_r}} g^{y_n \eta'}\}_{n \in \gamma_2{}^* \cup M_2{}'}$ and $c_n = E_\theta(m_n) \oplus g^{\sigma_0{}'}$, Which θ is randomly selected. Finally outputs the ciphertext $\sigma = \{\sigma_0, \sigma_0{}', \{\sigma_n\}_{n \in \gamma_2{}^* \cup M_2{}'}, \{\sigma_n{}' \}_{n \in \gamma_A{}^* \cup M_1{}'}, c_n\}$.

Unsigncryption Query. If $|\gamma_A \cap \gamma_2{}^*| < \theta'$, then generated a private key, and perform decryption operation and returned the plaintext m_n or $\perp$. Otherwise, returned an invalid signature. If the attacker has executed the signcryption query, then the process is terminated. Finally A selected the equal length plaintext m_0, m_1 and the set of default attributes $M_1{}'^*, M_2{}'^*$ of $\gamma_1{}^*, \gamma_2{}^*$, but no key query has been executed. Otherwise, the query is terminated.

Challenge. The set of attributes $\gamma_A{}'^* \in \gamma_1{}'^*$ consists of θ attributes. After queries, we can get the key $r_{n,0}$ and $r_{n,1}$ generated by $\gamma_A{}'^*$. Calculated $\sigma_0{}^* = g^{c\eta^*}$, $\{\sigma_n{}^* = H_1(n)^{c\eta^*} = g^{c y_n \eta^*}\}_{n \in \gamma_2{}^* \cup M_2{}'^*}$, $\sigma_0{}'^* = [\prod_{n \in \gamma_A{}'^* \cup M_1{}'^*} r_{n,0}^{\Theta_{n,\eta}(0)}][\prod_{n \in \gamma_1{}^* \cup M_1{}'^*} H_1(n)^{t_n'}]$, $H_2(m, \{\sigma_n{}^*\}_{n \in \gamma_2{}^* \cup M_2{}'^*})^{c\eta^*} = [\prod_{n \in \gamma_A{}'^* \cup M_1{}'^*} r_{n,0}^{\Theta_{n,\eta}(0)}][\prod_{n \in \gamma_1{}^* \cup M_1{}'^*} (g_1^{-y_n} g^{x_n})^{t_n}]g^{c y_{hy} \eta^*}$, $\theta = C^{\eta^*}, c^* = E_\theta(m) \oplus g^{\sigma_0{}'^*}, \{\sigma_n{}'^* = r_{n,1}^{\Theta_{n,\eta}(0)} g^{t_n}\}_{n \in \gamma_A{}'^* \cup M_1{}'^*}, \{\sigma_n{}'^* = g^{t_n}\}_{n \in \gamma_1{}^*/M_A{}'^*}$ and output $\sigma^* = \{\sigma_0{}^*, \sigma_0{}'^*, \{\sigma_n{}^*\}_{n \in \gamma_2{}^* \cup M_2{}'^*}, \{\sigma_n{}'^*\}_{n \in \gamma_A{}'^*/M_A{}'^*}, c^*\}$. If $H_2(m, \{\sigma_n{}^*\}_{n \in \gamma_2{}^* \cup M_2{}'^*}) \neq g^{y_{hy}}$, then the game ends. This process can query whether the σ^* is legal or not, but cannot do the unsigncryption query of σ^* or the key query of $\gamma_2{}^*$.

Guess. If A outputs $b'=b$, then $C = e(g, g)^{abc}$, and wins in the game. The advantage analysis is: the probability of A did not query the $\gamma_2{}^*$ is at least $1/q_{H_1}$, the probability of rejection of a valid ciphertext is at most $q_{US}/2^s$, the probability of challenge $\gamma_1{}^*, \gamma_2{}^*$ is at least $t_1 = 1/(2 \; q_{H_1})$, the probability of selecting $M_1{}'^*, M_2{}'^*$ is at least $t_2 = 1/(r-\theta \; r-1)$, the probability of $H_2(m, \{\sigma_n{}^*\}_{n \in \gamma_2{}^* \cup M_2{}'^*}) = g^{y_{hy}}$, $b \in \{0,1\}$ is at least $1/q_{H_2}^2$. Then the advantage is calculated as: $\varepsilon' = t_1 t_2 ((\varepsilon + 1)$

$/2 - 1/1)(1 - q_{US} / 2^S) / q_{H_1} q_{H_2}^2 (r - \theta'/r - 1)$.

Theorem 2. If a CDH problem is on the establishment of the group G, then the attacker cannot win with non-negligible advantage. And the scheme has the unforgeability of selection of attributes set under the selection of plaintext.

Proof. Algorithm B receives the CDH problem instance and its purpose is the calculation of g^{ab} based on (g^a, g^b). The process is similar with Theorem 1. F sets the appropriate system parameters and B answers the random oracle, key, singcryption and unsigncryption queries of the attacker A. If A did not select γ_1^*, γ_2^*, $M_1'^*$, $M_2'^*$ or $H_2(m, \{\sigma_n^*\}_{n \in \gamma_2^* \cup M_2'}) \neq g^{y_h}$, then the game is over, and

$$e(g, \sigma_0'^*) / [\prod_{n \in \gamma_1^* \cup M_1'} e(H_1(n), \sigma_n'^*)]e(H_2(m, \{\sigma_n^*\}_{n \in \gamma_2^* \cup M_2'}, \sigma_0^*))$$

$$= e(g, \sigma_0'^* / \prod_{n \in \gamma_1^* \cup M_1'} \sigma_n'^{*x_n} \sigma_0^{*y_h}) = e(g, g^{ab})$$

Therefore, the advantage is calculated as follows.

$$\varepsilon' = [t_1 t_2 \varepsilon (1 - q_{US} / 2^S)] / [q_{H_1} q_{H_2} (r - \theta'/r - 1)]$$

$$> (r - \theta)(r - \theta')(2\varepsilon - q_{US} / 2^{S-1}) / [q_{H_2} q_{H_1}^3 (r - 1)^{2r - \theta - \theta'}]$$

5   Performance

The communication and computation cost are two important factors needed to be considered in the analysis, which were designed by the cost generated by ciphertext length and signcryption and the unsigncryption operation. Table 1 of this section compares with features similar to a representative of the similar attribute-based signcryption scheme [14]. They compare in the calculation of the amount and length of ciphertext and computational signcryption ciphertext length, where e, prospectively denotes the exponent arithmetic and bilinear operation, $|G|$, $|G_1|$ denotes the mode of the group, l represents the number of attributes and n_m represents the message length.

Table 1. Performance comparison

Scheme	Signcrypt	Unsigncrypt	Ciphertext length
Scheme[13]	$e+8p$	$(6+2l)e$	$(5+2l)|G_1|+n_m$
Ours	$e+6p$	$2le$	$4|G_1|+n_m$
Ours	$e+8p$	$(l+2)e$	$4|G_1|+n_m$
Ours	$e+9p$	$(3l+2)e$	$4|G_1|+n_m$

The results shows that the scheme has its equivalent efficiency while selecting signcryption mode, but since the program [14] can only achieve a single signcryption function, our scheme has great advantages in selecting encryption or signature module in terms of computation or storage areas. Without affecting the function selection of this program, our scheme has great prospect of implementation.

6   Conclusions

Combined with the advantages of generalized signcryption and attribute-based cryptosystem, this paper proposes an attribute-based generalized signcryption scheme for big data in networks, and proves the program to be safe in the random oracle model.

This scheme can be adaptively selected in the cryptographic module in a confidentiality or authentication or both of them required during big data communication, and be able to function only when a particular need for another operation can be automatically shielded, taking into account both the flexibility to choose the functionality and efficiency improvement, so the scope is wider.

References

1. Wang, Y., Jing, X., Cheng, X.: Network big data: present and future. Chinese Journal of Computers 36(6), 1125–1138 (2013)
2. Zheng, Y.: Digital signcryption or how to achieve cost (Signature & encryption) << cost(Signature) + cost(Encryption). In: Kaliski Jr., B.S. (ed.) CRYPTO 1997. LNCS, vol. 1294, pp. 165–179. Springer, Heidelberg (1997)
3. Han, Y., Yang, X.: New ECDSA-Verifable generalized signcryption. Chinese Journal of Computers 11, 2003–2012 (2006)
4. Shamir, A.: Identity-based cryptosystems and signature schemes. In: Blakely, G.R., Chaum, D. (eds.) CRYPTO 1984. LNCS, vol. 196, pp. 47–53. Springer, Heidelberg (1985)
5. Boneh, D., Franklin, M.: Identity-based encryption from the weil pairing. In: Kilian, J. (ed.) CRYPTO 2001. LNCS, vol. 2139, pp. 213–229. Springer, Heidelberg (2001)
6. Sahai, A., Waters, B.: Fuzzy identity-based encryption. In: Cramer, R. (ed.) EUROCRYPT 2005. LNCS, vol. 3494, pp. 457–473. Springer, Heidelberg (2005)
7. Goyal, V., Pandey, O., Sahai, A., et al.: Attribute based encryption for fine-grained access control
8. Ma, J., Prabhakaran, M., Rosulek, M.: Attribute-based signatures: achieving attribute-privacy and collusion-resistance [R/OL]. Cryptology ePrint Archive, Report 2008/328, http://eprint.iacr.org/2008/328
9. Gagné, M., Narayan, S., Safavi-Naini, R.: Threshold attribute-based signcryption. In: Garay, J.A., De Prisco, R. (eds.) SCN 2010. LNCS, vol. 6280, pp. 154–171. Springer, Heidelberg (2010)
10. Zhang, G.Y., Fu, X.J., Ma, C.G.: A Dynamic Threshold Attributes-based Signcryption Scheme. Journal of Electronics & Information Technology 34(11), 2680–2686 (2012)
11. Liu, J., Wang, J.D., Zhuang, Y.: Attribute-Based Signcryption Scheme on Vector Space. Acta Electronica Sinica 41(4), 776–780 (2013)
12. Meng, X.Y., Chen, Z., Meng, X.Y.: Privacy-Preserving Decentralized Key-Policy Attribute-Based Signcryption in Cloud Computing Environments. Applied Mechanics and Materials 475, 1144–1149 (2014)
13. Li, J., Au, M.H., Susilo, W., et al.: Attribute-based signature and its applications. In: Proceedings of the 5th ACM Symposium on Information, Computer and Communications Security, pp. 60–69. ACM (2010)
14. Chen, S.Z., Wang, H.B.: Efficient Attributes-Based Signcryption Scheme. Journal of Information Engineering University 12(5), 526–531 (2011)

An Improved Fine-Grained Encryption Method
for Unstructured Big Data

Changli Zhou*, Chunguang Ma*, and Songtao Yang

Harbin Engineering University, Harbin City 150001, China
zhouchangli888@gmail.com, machunguang@hrbeu.edu.cn

Abstract. In the big data protecting technologies, most of the existing data protections adopt entire encryption that leads to the researches of lightweight encryption algorithms, without considering from the protected data itself. In our previous paper (FGEM), it finds that not all the parts of a data need protections, the entire data protection can be supplanted as long as the critical parts of the structured data are protected. Reducing unnecessary encryption makes great sense for raising efficiency in big data processing. In this paper, the improvement of FGEM makes it suitable to protect semi-structured and unstructured data efficiently. By storing semi-structured and unstructured datum in an improved tree structure, the improved FGEM for the datum is achieved by getting congener nodes. The experiments show the improved FGEM has short operating time and low memory consumption.

Keywords: Big data, encryption, unstructured data, semi-structured data, IoT.

1 Introduction

The Internet of things (IoT) makes great evolution and innovation of computing pattern, which accompanied with the smaller and smarter development of computers. The dream of Mark Weiser[1], who proposed the concept of ubiquitous computing and aimed to make computers provided invisible service, are coming true. The human-centered mobile devices in our daily life enable anyone to get information service practically anywhere and anytime. IoT gives rise to the research of wireless communication technologies, such as MANET [2], VANET [3] and Mobile Internet. So that these distributed mobile devices are more efficient to transmit massive data in the ubiquitous network, but still confronted with big data processing issues.

As the Internet of Things offer us a pervasive communication dimension, what if our private data is exposed in this transparent world. It will bring too much burden if we encrypt all the transmitted private data which is massive in distributed mobile networks. As a result, the distributed mobile networks need lightweight and efficient protection methods [4, 5]. In this paper, we improve the data structure we proposed in our previous work to make it suitable to fine-grained encryption of unstructured data for big data security in IoT.

---

* Corresponding author.

H. Wang et al. (Eds.): ICYCSEE 2015, CCIS 503, pp. 361–369, 2015.

2   Related Works

There are sensing layer, transport layer, processing layer and application layer in IoT[6-8], the sensor layer includes mobile devices, limited resources, but generating huge amount of data, some involves big data privacy. As mentioned in [9], big data protection method can be divided into two aspects. One is the design of lightweight protecting algorithm, such as proposing data encryption algorithm; the other is protecting the crucial parts, which achieves minimum protecting operations and optimal effects. In this paper, we still consider from the latter perspective.

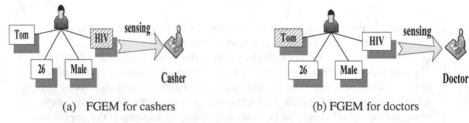

(a)   FGEM for cashers   (b) FGEM for doctors

Fig. 1. Fine-grained encryption method sketch

As a traditional and effective method, encryption has been widely studied. A lot of lightweight algorithms for the limited resources terminals have many achievements [10, 11]. But nearly all the proposed algorithms are from the perspective of protecting the entire parts of a data, without considering reducing unnecessary protections. Entire data encryption methods need to be improved for unstructured data.

The main idea of fine-grained encryption derived from multi-level encryption [12-14] and fine-grained access control technology [15-17] in the field of database. In our paper [9], we proposed the fine-grained encryption method due to the entire encryption brings too much operation of big data processing in IoT. In our research, it's easy to find that not all the parts of a piece of data need encryption in distributed environment. Such as a piece of patient's data contains name, sex, age, and disease, to different viewers, we only need to encrypt different parts, as shown in Fig.1. If the critical parts are protected according to different viewers, the others parts are meaningless.

818-89-9988	John	Male	24	He had tuberculosis and heart disease, tuberculosis has been cured. Allergic to pollen, cannot go out in the spring flower area,	The man is Italian immigrant, now Lives in Seventh Avenue of Brooklyn, New York. He and his full- time housewife have three kids, 2 girls and a boy. They usually go hiking on Saturday. He had tuberculosis and heart disease, tuberculosis has been cured. Allergic to pollen, cannot go out in the spring flower area, there have been record breathing difficulties, can't take powder drugs at the same time.

Fig. 2. Two kinds of unstructured data (semi-structured and unstructured data)

In our proposed method [9], FGEM can only encrypt structured data. But there is a lot of unstructured and semi-structured big data in IoT, as shown in Fig 2. Such unstructured datum should be concealed to different viewers.

3   Improved Fine-Grained Encryption Method

3.1   Network Model Description

We describe the network model in Fig.3 and the sensor layer is considered as the most foundational layer and the improved FGEM works on it. Through the Fig.3, we can see that the sensor and some mobile nodes generate datum and the sink nodes organize them, so the improved fine-grained encryption method runs on these sink nodes which have strong ability in computing and storage.

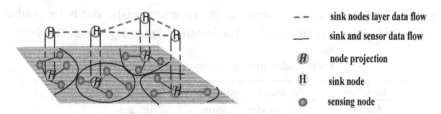

- - sink nodes layer data flow
___ sink and sensor data flow
Ⓗ node projection
Ⓗ sink node
● sensing node

Fig. 3. Network model of sensing layer

3.2   Fine-Grained Encryption Method

(1)   A Tree Structure

In this part, we propose a tree structure with pointers according to the graph structure of Object Exchange Mode (OEM) [18]. As shown in Fig.4, the graph structure contains object nodes and directed edges.

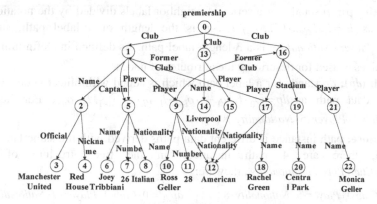

Fig. 4. OEM graph

The OEM can be represented by a digraph with a unique root node, as shown in Fig.4. The digraph is represented by a 3-tuples (r, V, E). The r stands for the root nodes, the V stands for the set of object nodes, and the E stands for the set of edges, the elements in set $E < o_i, l_n, oj >$ means object o_i points to object oj through label l_n. Label l_n represents the relationship between objects.

We remove the loop circuits in OEM to make it a tree structure, and use some objects with pointers which point some unstructured data or semi-structured data. That is the new tree structure. The tree structure is used for storing the data.

Definition 1: (O, E) represents a tree structure, O is a set of object nodes, E is a set of edges with labels, the neighbor objects can be presented by $< o_i, l_n, oj >$, that means object o_i points to object oj through label l_n.

Definition 2: Object node can be represented by 4-tuples $(object\text{-}id, label, type, value)$, $object\text{-}id$ is a node's unique identification, $label$ stands for an attribute that describes what kind of data item is stored in the object, $type$ type stands for attribute value type, and $value$ is attribute value. The identifiers' meanings are in Tab 1:

Table 1. Identifiers' meaning of the 4-tuples

Identifier	Identifier meaning	Value type
object-id	unique identification	integer
Label	attribute	char
Type	attribute value type	atomic, complex
Value	attribute value	char

There are two types of values of identifier type, they're atomic type, such as integer and char, and complex type, such as a pointer pointing to other objects. That is the key part for unstructured data encryption.

(2).   Key Concepts from Definitions

Label path (lp): lp stands for a series of neighbor labels divided by the notation ".", recorded as $lp = l_1 \cdot l_2 \cdot .... \cdot l_i \cdot .... \cdot l_{n-1} \cdot l_n$, n is the length of a label path, such as $lp = Club \cdot Player \cdot Nationality$ is a 3-length label path. As defined in Definition 2, the label of edges is used for describing the attributes of nodes.

Data path (dp): dp stands for a label path which is added with object nodes, we can record a data path as $dp = o_0 \cdot l_1 \cdot o_1 \cdot .... \cdot o_{i-1} \cdot l_i \cdot o_i \cdot .... \cdot o_{n-1} \cdot l_n \cdot o_n$. For instance, $dp = 0 \cdot Club \cdot 1 \cdot Player \cdot 5 \cdot Nationality \cdot 8$.

Path instance: path instance is different data paths derived with the same label path, In Fig.5, there are 4 path instances in the tree structure of label path $lp_0 = Club \cdot Player \cdot Nationality$, they are:

$dp_{00} = 0 \cdot Club \cdot 1 \cdot Player \cdot 5 \cdot Nationality \cdot 8$; $dp_{01} = 0 \cdot Club \cdot 1 \cdot Player \cdot 9 \cdot Nationality \cdot 12$;

$dp_{02} = 0 \cdot Hospital \cdot 17 \cdot Name \cdot 18 \cdot Disease \cdot 20$; $dp_{03} = 0 \cdot Hospital \cdot 17 \cdot Name \cdot 23 \cdot Disease \cdot 24$.

To construct such a data structure which is analogous as a thread tree, is actually change a two-dimensional table in a database into a one-dimensional structure, that is changing a nonlinear structure into a linear one, with the aim of implementing fine-grained encryption to unstructured data in distributed mobile networks.

Congener nodes: congener node is a set of last nodes in different path instances but with the same label path. As shown in Fig.5, the congener nodes set of label path lp_0 is a set of nodes, their identifier is {8, 12, 20, 24}.

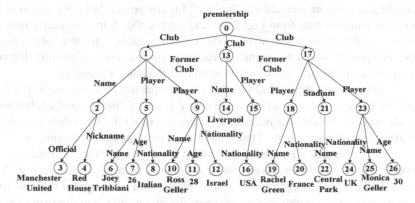

Fig. 5. Tree structure

Theorem 1: the congener nodes o_1 and o_2 in two data paths $< o_1, l_m, o_i >$ and $< o_2, l_n, oj >$, if $l_m = l_n$, nodes o_i and oj are also the congener nodes in the same set.

Theorem 1 is apparently right, we will not prove it. By Theorem 1, we can get any congener nodes set in the tree structure.

(3). The Improved Fine-Grained Encryption

In FGEM, if we want to encrypt all patients' bloodtype, it is converted to encrypt the congener nodes with the same label path $lp = Club \cdot Player \cdot Nationality$. So, searching for the congener nodes set means a lot. According to theorem 1, we can get every congener nodes set. For semi-structured data, as shown in Fig.2 (a), the last unstructured part is variable-length, it's difficult for the system to pre-allocated buffer, but still with labels such as medical history. For unstructured data, as shown in Fig.2 (b), there is no label, we cannot get the congener nodes set by label path, full text encryption is the only way. So we need to improve the tree structure.

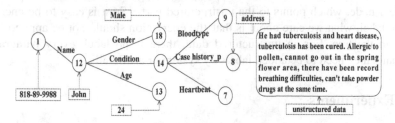

Fig. 6. Patient's information with unstructured part

For semi-structured data, the unstructured part is still with labels, but its attribute value type is complex as in Definition 2, i.e., the pointer. For the variable-length reason, system cannot pre-allocate appropriate size of storage space, adopting pointer can solve the problem. By the pointer, unstructured part can be stored in large space. But that means the pointer, actually the address pointing to the unstructured part of a data, will be encrypted following the label path when we searching for the congener nodes. We improve the tree structure and add a suffix "_ p" for any pointer label. As shown in Fig. 6, we pick a patient's data from Fig. 2 (a) and imitate Fig. 5 to construct a new tree structure. The label of the eighth object node is added a suffix "_p", that means the object node is a pointer. When traversing the tree for congener nodes, if meeting that notation of a pointer, it will encrypt the unstructured data the pointer points to.

For the unstructured data, it has no labels to any paragraph as shown in Fig.2. So we need to improve the tree structure. The key point of fine-grained encryption is label path, so it's necessary for unstructured data to extract the key labels or add confidentiality level of every part. That makes the fine-grained encryption work. As shown in Fig. 7, we pick a patient's information from Fig. 2 (b) and imitate Fig. 5 to construct a new tree structure. Every paragraph has its own main idea. In Fig. 2, we can easily find that two paragraphs are mainly about John's family and healthy condition. So we add labels to each paragraph even the article title, and construct the tree structure to store the data and then execute the fine-grained encryption. The labels can also be replaced by confidentiality levels to represent every part's importance when choosing encryption strength.

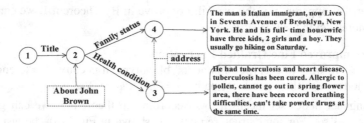

Fig. 7. Patient's information of unstructured data

We can find that the label is the most important notation to realize fine-grained encryption. For a semi-structured data, its unstructured part with label is not directly stored in the object node of a tree for its variable length reason. We add a pointer in the object node, which points to the unstructured part. So it is easy to be encrypted with the label, what we care about is that the encryption should not be applied on the pointer in the tree. For an unstructured data, there is no label, every paragraph is coordinate, they need to be added labels by system or artificially.

4   Experiments

To protect the critical part is the main idea of FGEM, so are the improved ones. The process of fine-grained encryption is mainly searching and encryption in the proposed tree structure. We take different amounts of data with the same label path " $lp = Hospital \cdot Name \cdot Condition \cdot Casehistory\_p$ " and" $lp = Title \cdot Economiccondition$ " as test objects and encrypt only one object node in Fig.6 and Fig.7 respectively. The data with that label path is stored in the tree structure. We take the unstructured part as the critical one of every data, which is encrypted, i.e., the label "Case history" and "Economic condition" lead to. Then we totally set 3 comparison groups: entire encryption with encryption algorithm IDEA, fine-grained encryption with sequential search and fine-grained encryption with index search. We find the improved FGEM with index search has better performances, the searching algorithm is much more considerable.

4.1   Operating Time

The FGEM considers from the perspective of data itself. It reduces unnecessary encryption overhead. So we don't consider raising efficiency of encryption algorithm. But we also realize that, raising efficiency of encryption algorithm and the simplified crypto-operation can both achieve the optimization technique, but we consider from the latter one. As shown in Fig. 8-9, we get some conclusions:

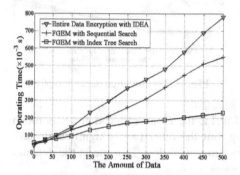
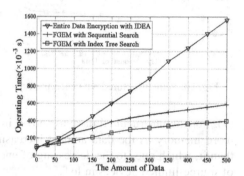

Fig. 8. Semi-structured data encryption   **Fig. 9.** Unstructured data encryption

(1). For the semi-structured data, as shown in Fig.8, entire data encryption costs more time as the data amount increasing and it shows that improved FGEM with sequential search costs shorter, that dues to time consumption of comparison one by one in the tree structure, But if we adopt an index for the tree, its consumption is not obviously increasing, because the index is significantly helpful to locate congener nodes storing the unstructured part of a data;

(2). For the unstructured data, as shown in Fig.9, entire data encryption cost much more time as the data amount increasing and it shows improved FGEM with sequential search costs much shorter, which dues to the labels comparison time is much shorter than the encryption cost. But if we only encrypt a part of the unstructured data, the time consumption lows as encrypting operation decreasing, the main direction of optimization is to reducing encrypted parts of unstructured data;

(3). For the two kinds of data, the tree structure with index has obvious advantages when the data amount is increasing. That meets the efficiency demand of big data processing in the IoT environment.

4.2   Memory Consumption

We discuss the consumption of internal memory when processing is carrying on, which is much less than external memory. As shown in Fig.9, we get some conclusions in the following:

(1). For the semi-structured data, as shown in Fig. 10, due to extra internal memory consumption for comparing, the one with sequential search has similar consumption as the entire encryption as data amount increasing. But the one with index needs less memory because it doesn't need to load all the parts into the memory.

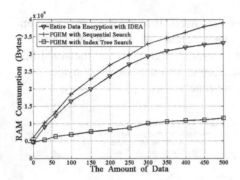
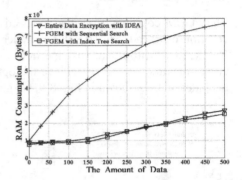

Fig. 10. Semi-structured data encryption   **Fig. 11.** Unstructured data encryption

(1). For the unstructured data, as shown in Fig.11, the entire encryption needs much more memory as the data amount increasing, but the other two is significantly less than it, which dues to the decreasing of encrypting operation. That meets the demand of source-limited devices in IoT.

5   Conclusions

In the paper, we have proposed an improved FGEM for unstructured data and semi-structured data. We have found that the fine-grained protection is a new kind of protecting idea in big data processing, reducing unnecessary encrypting operation is the key point. We believe that the idea is appropriate for the IoT environment. All the fine-grained encryption methods are just realizations of the idea, we will try to propose more protecting methods according to the idea in the following works.

Acknowledgements. This research is supported by a grant from National Natural Science Foundation of China (No. 61170241, 61472097), Specialized Research Fund for the Doctoral Program of Higher Education (No. 20132304110017), This paper is funded by the International Exchange Program of Harbin Engineering University for Innovation-oriented Talents Cultivation.

References

1. Weiser, M.: The computer for the 21st century. Scientific American 265(3), 94–104 (1991)
2. Wang, S.H.: An Exchange Framework for Intrusion Alarm Reduction in Mobile Ad-hoc Networks. Journal of Computers 8(7), 1648–1655 (2013)
3. Chen, W., Guha, R.K., Kwon, T.J., et al.: A survey and challenges in routing and data dissemination in vehicular ad hoc networks. Wireless Communications and Mobile Computing 11(7), 787–795 (2011)
4. Hui, Y., Wang, Y.: Design of Lightweight Encryption Algorithm Based on Optimized S-box. Communications Technology 5, 39 (2010)
5. Singh, S., Murshed, N., Kropatsch, W.G. (eds.): ICAPR 2001. LNCS, vol. 2013. Springer, Heidelberg (2001)
6. ITU-T. Recommendation Y. 2221, Requirements for support of ubiquitous sensor network (USN) applications and services in NGN environment. Geneva (2010)
7. Wolf, W.: Cyber-physical system. IEEE Computer 42(3), 88–89 (2009)
8. Fadlullah, Z.M., Fouda, M.M., Kato, N., et al.: Toward intelligent machine-to-machine communications in smart grid. IEEE Communications Magazine 49(4), 60–65 (2011)
9. Ma, C., Zhou, C., et al.: A Research of Fine-grained Encryption Method for IoT. Journal of Computational Information Systems 8(24), 10213–10222 (2012)
10. Matalgah, M.M., Magableh, A.M.: Simple encryption algorithm with improved performance in wireless communications. In: 2011 IEEE Radio and Wireless Symposium (RWS), pp. 215–218. IEEE (2011)
11. Varalakshmi, L.M., Florence, S.G.: An enhanced encryption algorithm for video based on multiple Huffman tables. Multimedia Tools and Applications, 1–13 (2013)
12. Yuping, Z., Xinghui, W.: Research and realization of multi-level encryption method for database. In: 2010 2nd International Conference on Advanced Computer Control (ICACC), vol. 3, pp. 1–4. IEEE (2010)
13. Zhou, X., Liu, J.: A novel efficient database encryption scheme. In: 2012 9th International Conference on Fuzzy Systems and Knowledge Discovery, FSKD 2012, pp. 1610–1614 (2012)
14. Sallam, A.I., El-Rabaie, E.S., Faragallah, O.S.: Encryption-based multilevel model for DBMS. Computers & Security 31(4), 437–446 (2012)
15. Kiviharju, M.: Towards pervasive cryptographic access control models. In: SECRYPT 2012 - Proceedings of the International Conference on Security and Cryptography, pp. 239–244 (2012)
16. Marian, H., Sascha, F., Michael, B., Thomas, M., Matthew, S.: Towards privacy-preserving access control with hidden policies, hidden credentials and hidden decisions. In: 2012 10th Annual International Conference on Privacy, Security and Trust, PST 2012, pp. 17–24 (2012)
17. Zhu, Y., Gail-Joon, A., Huang, D., Wang, S.: Towards temporal access control in cloud computing. In: Proceedings - IEEE INFOCOM, pp. 2576–2580 (2012)
18. Papakonstamfinous, Y., Wiom, J.: Object exchange across heterogeneous information sources. In: Proc. of Int. Conf. on Data Engineering (ICDE), pp. 251–260 (1995)

Research and Reflection on Teaching of C Programming Language Design

Hui Gao[1,2], Zhaowen Qiu[2,*], Di Wu[2], and Liyan Gao[3]

[1] Harbin Huade University, China
44117252@qq.com
[2] Institute of Information and Computer Engineering, Northeast Forestry University, China
qiuzw@nefu.edu.cn, 85312371@qq.com
[3] Harbin 47 middle school, China
chncsp@163.com

Abstract. C Language is a basic programming language and a compulsory foundation course for majors of science and engineering. Subjected to teaching periods, students have no time to do enough concrete integrated exercises. In teaching of C programming language design, both teachers and students have some problems. On the one hand teachers should change their teaching ideology, on the other hand, students should spontaneously study and improve their study motivation. In this way, students will improve their programming ability and apply what they learn comprehensively.

Keywords: C Programming Language Teaching, Teaching Ideology, Independent Learning, Practice.

1 Introduction

C language is a basic programming language, which contains the basic idea and conception of programming in its developing. In computer teaching of higher education, it is a required course of science and engineering major, and its contents are in sync with abroad. Objectively speaking, it will directly affect the teaching effect that how we teach or learn it. During these years' teaching in C language, I have thought a lot about how to improve teaching effect, and put these ideas into practices, and then they worked.

2 The Status and Problems of C Language Programming Teaching

Nowadays, the traditional C language programming teaching always means lectures and hands-on computer trainings in multimedia classroom in form. In content, they are always introductions and instructions, which follow three steps, conception, quick example and

---

* Corresponding author.

H. Wang et al. (Eds.): ICYCSEE 2015, CCIS 503, pp. 370–377, 2015.

guided practice. This three-step teaching method will be limited by teaching hours. While teaching students the designing ideas, methods, important conceptions and applications, teachers leave less time for students to take practice out of class[1].

In C language programming teaching, there exists some problems on both side, teachers and students.

2.1　Materials Arrangements Are Not Reasonable

Now, most materials selected as textbooks in C language programming teaching are unpractical and lack of interest. Students take it dull and show no interest. They would not like to study seriously, which affects the teaching effect and cause themselves troubles in future studies.

2.2　Outdated Teaching Methods Can Not Motivate Students

During classroom teaching, C language programming courses are always given in form of lectures, which cannot give full play to students' initiative. Heuristic teaching methods are not fully used, neither are advanced intuitive multimedia teaching means. Teachers do not know whether students prepare their lessons, or what they really need. All these, to some extent, dampen students' enthusiasm.

2.3　Teachers Are Not Fully Aware of the Differences between Various Majors, So They Cannot Arouse Students' Interest

Key points, to different majors, are supposed to be different. Teachers always ignore the differences, and give lectures to different majors with the same lesson plans. In this way, what students have learned can not meet the needs of practices in their majors, so that students can not take a right attitude to this course.

2.4　Students Lack of Awareness of Autonomous Learning, and Attach Little Importance to Practice Teaching

Most students just are all teacher-centered. They would not ask any question, or rather, they just do not know what to ask. After class, out of lack of diligence, they barely refer to relevant books or materials. Internet has provided us an open learning environment. But, instead of learning actively, students always indulge in games online. In this regard, teachers are responsible to guide students to learn the methods of autonomous learning[2]. C language is a practical computer language. However, there are many drawbacks in present practice course teaching. Many students do not understand the purpose of hands-on computer training, nor the contents, and do not pay attention in class. Some teachers irresponsible are not strict with students. Some others arrange curriculums unreasonably. These to some extent result in that students prefer games to exercises in practice course teaching. With difficulties in practical operations, students grow to dislike this course[4].

3   The Ways to Improve the Effect of C Language Programming Teaching

3.1   Textbooks Should be Helpful to Arouse Students' Initiative

Theory textbooks should meet the needs of students at different levels. And cases quoted ought to be practical, which students would get interested in when they read them. For example, as an application-oriented university, we should select those materials practical, full of cases, and finely explained. Such as Colleges and Universities C language Practical Course (3rd edition).

In addition, content arrangements of practice textbooks should be based on the students' professional interest. On the one hand, it should contain the key points. On the other hand, experimental cases should be relevant to professional practice. In this way, practice textbooks should be filled with integrated contents of training. After practice training, students will be able to do all exercises, and learn the approaches to design applications. Then they do improve a lot in practice abilities[1].

3.2   Change the Modes of Thoughts and the Methods of Teaching

To raise students' interest in C language and help them build positive thought patterns, teachers should give students a first impression about what programming is, by interpreting a comprehensive example program. And then teachers explain those statements in this program to give students a general idea about C language. During the whole teaching, teachers should focus on thoughts, methods and algorithms, which would help improve students' ability of analyzing and solving problems, and build correct thought patterns of programming design.

Take Guessing game program design for example. Rules are as follows. Player has seven opportunities to guess right the number generated by the program randomly. If he loses, the program will regenerate a new number. Player could prefer to go on or not. The program would tell true or false, lower or higher.

Step 1: Natural language analysis. The program "thinks out" a number between 1 and 100. When player guesses right, it shows "Right!", otherwise "Wrong!", and tells player the number he gives lower or higher. Each player has seven opportunities to guess right, and then the game is over. Player can guess numbers repeatedly, until he wants to stop.

First, on the whole, the basic function of the program is "Generate", "Guess" and "Quit or not". If not, "repeat again".

In this program, "Generate" and "Guess" are two sub-modules. In the overall design, we just need consider what they should do, but not how to do it.

Step 2: Flowchart Design.

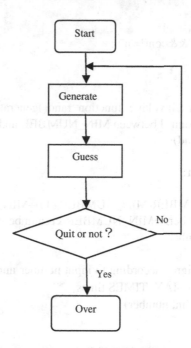

Step 3: Code design.

/*main function design, initialize the random seed, complete the framework design from top down。 */

```
#include    <stdlib.h>
#include    <stdio.h>
#include    <assert.h>
#include    <time.h>
#define MAX_NUMBER 100
#define MIN_NUMBER 1
#define MAX_TIMES 7
int MakeNumber(void);
void GuessNumber(int numbre);

main()
{
    int number;
    int cont;
    srand(time(NULL));
do{
        number=MakeNumber();
        GuessNumber(number);
        printf("Continue?(Y/N):");
        cont=getchar();
        while(getchar()!='\n')
```

```
                   {     ;    }
        }while(cont!='N'&&cont!='n') ;
                return 0;


}
```
/*sub-function design : invoking function rand()generate random number, after computing the value of control between MIN_NUMBER and MAX_NUMBER.*/
```
    int MakeNumber(void)
    {
            int number;
            num-
ber=(rand()%(MAX_NUMBER-MIN_NUMBER+1))+MIN_NUMBER;
            assert(number>=MIN_NUMBER&&number<=MAX_NUMBER);
            return number;
    }
```
/*the sub function design : according to input number number, tell right or wrong, larger or smaller,   within MAX_TIMES times。 */
```
    void GuessNumber (int number)
    {     int guess;
          int times=0;
          assert ( number>=MIN_NUMBER&&number<=MAX_NUMBER);

        do{
                times++;
                printf("Round %d:",times);
             scanf("%d",&guess);
             while(getchar()!='\n')
                {    ;    }
          if (guess > number)
                printf("Wrong! Too high!\n");
          else if (guess < number)
                printf("Wrong! Too low!\n");
          } while (guess != number&&times<MAX_TIMES);
            if(guess==number)
            printf("Congratulations!You 're so cool!\n");
            else
            printf("Mission failed after %d attempts.\n",MAX_TIMES);
    }
```
Step 4: Compile and Run. Along the line of structured programming design, we write three comparatively independent programs, complete the program design, and achieve what we expected. The interface runs as follows.

Figure 1 shows the computational result when player did not guess right within 7 times

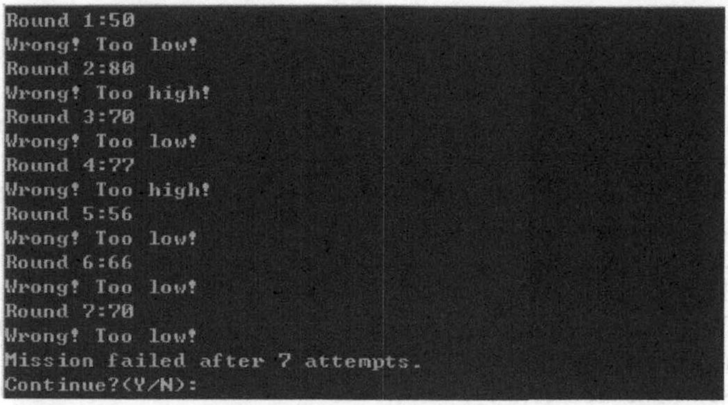

```
Round 1:50
Wrong! Too low!
Round 2:80
Wrong! Too high!
Round 3:70
Wrong! Too low!
Round 4:77
Wrong! Too high!
Round 5:56
Wrong! Too low!
Round 6:66
Wrong! Too low!
Round 7:70
Wrong! Too low!
Mission failed after 7 attempts.
Continue?(Y/N):
```

Fig. 1.

Figure 2 shows the computational result when player guessed right within 7 times.

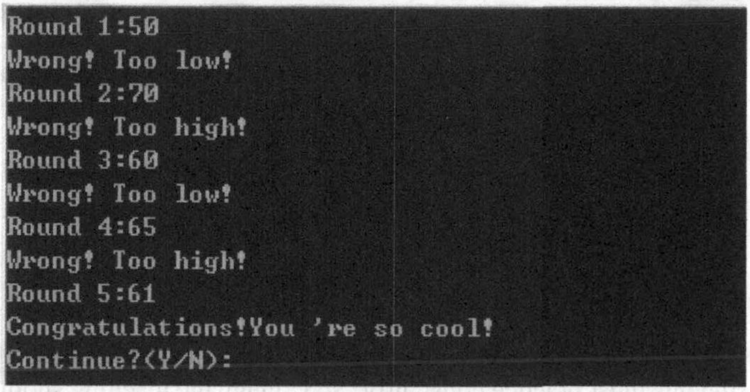

```
Round 1:50
Wrong! Too low!
Round 2:70
Wrong! Too high!
Round 3:60
Wrong! Too low!
Round 4:65
Wrong! Too high!
Round 5:61
Congratulations!You 're so cool!
Continue?(Y/N):
```

Fig. 2.

As to the algorithms involved in C language programming, teachers should selectively explain in detail and give students more practice trainings, meanwhile, the instructions in the relations between different algorithms. In addition, teachers could make use of multimedia during the teaching to enlarge the amount of information, such as demonstrating excellent software written in C language. By explaining these programs, teachers manage to teach students excellent programming thoughts and methods.

3.3   Arouse Interest of Different Majors

To arouse students' interest, teachers should first declare the importance and practicality of C language, make students understand it will help a lot in the future in their majors. Then, students will get interested in it.

C language is taught as a basic course. In the first lesson, teachers should begin with the function of C language, its evolution, its status in computer languages, the relation with vc++ and other relevant program design, the application in software design, the relation with embedded system and the application in it, etc. To illustrate the foundation status of C language in computer science, it is better to show students some programs written in C language, such as game scoring program, Monkeys Eat Peaches Problem, Han Xin Musters Soldiers Problem. The animation effects and graphical interfaces of these programs can effectively arouse students' interest.

In fact, researches from educational psychology show that interest comes from motivations, accompanies actions, and fulfills itself in results. For example, when teaching students of electronics, teachers could integrate the application designs from "Freescale" and "Bi Shen Cup" into his detailed explanation of features of C language programming. That means, through well-targeted learning, students get enlightments, knowledge and abilities. When they feel it useful and effective, interest and enthusiasm for this course will be aroused[3].

Therefore, in teaching, teachers should lead students to taste success and get the satisfaction of gains.

3.4   Encourage Students to Raise Questions and Arouse Students' Initiative to Study

In teaching, teacher should not tie students down to textbooks, but encourage them to claim freely different opinions on the algorithms and solutions, which can help students explore new ways to solve problems and develop creative ideas. Teachers should also give students more chances to question and find the best solution to problems actively. In this way, we can achieve the goal of encouraging students to study actively.

3.5   Give More Practice Training

C language programming is a practical course. In order to comprehend the theoretical knowledge, students must practice constantly. So C language teaching should be aimed at applying. Teachers should lay stress on the conformation of practical skills, pay more attention on the practice teaching, and arouse students' innovative consciousness. In teaching, teachers should commit themselves to a combined method of guidance and counseling, projection demonstration and computer operation. Teachers may improve the curriculum design to increase the proportion of practice, which can help students raise creative thoughts, improve the ability to analyze and solve problems independently, and the ability of programming[5].

4   Conclusion

In summary, in order to improve students' ability to apply the knowledge of C language comprehensively and help students form good programming habits, teachers should analyze the whole teaching process systematically, and implement teaching reform according to the actual conditions of different majors. Help students acquire the abilities of practicing, learning lessons and applying knowledge comprehensively. In this way, we do improve students' programming capabilities.

Acknowledgment. Foundation: C++ program design Key Course of Northeast Forestry University.

References

1. Xuebing, W.: The Innovative thoughts and Improvements in C Language Teaching. Hippophae (10), 178 (2010)
2. Haitao, B.: C Language Practice Constructions and Innovation for Independent. Institute Value Engineering (31), 272–273 (2012)
3. Yi, N., Eni, D., Yanlin, W.: How to Inspire Students' Interest in C Language Teaching. Journal of Dong guan University of Technology (01), 107–110 (2006)
4. Guixin, B., Hongbo, X.: Researches and Practices in C Language Teaching. Popular Science & Technology (06), 117 (2006)
5. Yuling, Z.: Warming up and Project driven in C Language Teaching. Vocational Education Research (04), 95–96 (2012)

A Peer Grading Tool for MOOCs on Programming

Zhaodong Wei and Wenjun Wu

State Key Laboratory of Software Development Environment
Beihang University, Beijing, China
{wzd,wwj}@nlsde.buaa.edu.cn

Abstract. In massive open online courses (MOOCs), peer grading will play an important role to promote MOOCs development. In this paper, we develop a peer grading tool for programming courses on MOOCs. It is capable of dealing with large and diverse student population, and providing them with targeted subjective assessment. This tool firstly partition the submissions into small chunks to reduce the task of reviewers and give us flexibility to scale the code review process. Next we use code normalization and chunks clustering to assign similar chunks to the same student for increasing reviewer efficiency. Besides, the tool use a random allocation strategy and workload classification to assure reviewers workload balance while every student can get diverse feedback. Finally our evaluation experiments on a number of students in school indicate that the tool has achieved a significant improvement over the peer grading on MOOCs.

Keywords: Peer grading. MOOCs, Code normalization, Review balance.

1   Introduction

As a MOOC student learns a course mainly by watching online course videos and complete homework assignments, it is essential to quickly provide them with accurate assessment results and high quality feedbacks on their homework submissions[1, 2]. Such an immediate evaluation of student homework also enables instructors to understand the progress of these students and adaptively design the sequent course materials for study. But the massive number of MOOC students in a MOOC course brings new challenges on homework assessment. It is impossible for an instructor with a few teach assistants to grade thousands and even tens of thousands of home submission within a short amount of time.

There are two scalable grading methods to generate assessment feedbacks on homework problems: (i) auto-grading based feedback[3, 4] and (ii) peer grading feedback[5]. For instance, in edX MOOC course "Introduction to Computer Science and Programming" offered by MITx, the instructor implemented an auto grader that executes every program submitted by students against a set of test cases and reports the failed test cases back to the students. Although the auto-grader can efficiently screen massive submissions and tell students whether their programs are correct, it has no way to reveal the coding errors related to the failed test cases and give the advices for students to improve their coding skills.

H. Wang et al. (Eds.): ICYCSEE 2015, CCIS 503, pp. 378–385, 2015.

The second method of peer grading[6-8] based feedback is being suggested as a potential solution to the problem[9]. In the course of "Human Computer Interaction" offered by Stanford on Coursera[10], it used a calibrated peer grading system. At the phase of calibration in the system, it requires students to correctly assess a training submission before they are allowed to grade other students submissions. Not like the auto-grading based feedback, the students can write their subjective evaluation of the code under review and directly declare which lines of the code has problems or how to correct them[11]. The major weakness of this approach is that every student has to review five other students' submissions, it is a large amount of work for the beginners which may reduce their enthusiasm or effect the accuracy of grading while peer review. In additions, the idea of partitioning code into small chunks was proposed in Caesar[12], a social code review tool designed for programming courses. Caesar also detects clusters of similar code to improve the efficiency of peer grading, which is shown useless by practice and is abandoned now. What's more, so many students who use Caesar while peer reviewing always had unfair treatment for unbalanced workload on grading other students' submissions.

In this paper, we present an improved peer grading tool based on original Caesar for programming courses on MOOCs. This tool enables the students on MOOCs to review others' submissions through partition and clustering, and generate high-quality feedbacks. Moreover, this tool uses workload classification to ensure the proportionality of all the students. Firstly, we introduce the workflow of the peer grading tool from partition to chunks assigning. Then we describe the design of the major components in this tool. Finally we present explanation evaluate the performance of the peer grading tool through several experiments.

2   Overview of the Tool

The goal of our peer grading tool is to divide the entire program submitted by students into multiple chunks that are more easily readable for peer reviewers who often lack enough skills and times for checking hundreds lines of programs. Figure 1 shows the workflow of the peer grading tool, which consists of three process steps: code partition, chunk clustering and job allocation.

The first step is to split the program files into code chunks, each of which represents a readable piece of code in the form of program functions. Students only need review several small pieces of source code instead of the entire code files. The partition step not only decreases the cost of review effectively but also gives us flexibility to deal with the small chunks rather than the whole submissions. When the number of chunks increases rapidly, it causes more workload for students to review them. The most obvious approach to increase the review efficiency is clustering, which attempts to assign similar code chunks to the reviewers. Thus, the second step is to group the similar chunks together so that the tool can allocate the similar code chunks to the reviewers with prior experiences. The third step is to allocate reviewing jobs to students and ensure workload balance among them. Because the chunks in cluster A may cost much more time than that in cluster B. So the workload balance becomes an essential part of the peer grading tool.

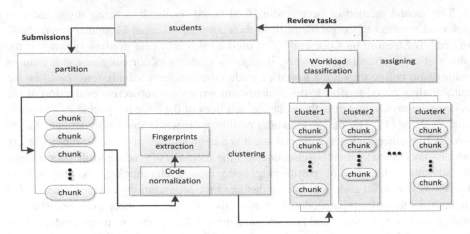

Fig. 1. Workflow of our peer grading tool

3   Partition and Clustering

As the first step in our peer grading tool, the partition of student submissions can directly affect the quality of the entire peer grading process. Note that it is also important to guarantee the content of chunk is readable and easily to understand. Based on the syntax parsing of source code submission, the tool separates files into pieces of functions so that the chunks can still be semantically meaningful to the reviewers. In most cases every function of a program has no invocation of other functions. It's a simple and effective strategy of partition. However, when a function calls other methods in the same file, it is difficult for reviewers to only check the chunk with the function. Fortunately, such cases rarely occur in the programming courses at the beginner level, which will be explained in Section 5.1. In order to cluster the chunks, the tool has to identity the characteristics of a chunk to calculate the distance between any pair of chunks. So we introduce two pre-processing steps before clustering including code normalization and fingerprints selection.

3.1   Code Normalization

The preprocessing in the Caesar workflow only involves removal of spaces and empty lines of source codes as well as simple character conversation. However, these simple operations can't guarantee the generation of precise fingerprints of source codes. For example, if we modify the names of some variables in a function, its fingerprints change a lot. Thus it seems significant that we must make the code more normalized. And we set three rules in code normalization. (1)Remove all the spaces, empty lines and the comments. (2) Delete all the language keywords, predefined identifiers and the name of library functions. (3) Replace the user defined identifiers with a special symbol.

3.2   Fingerprints Selection

After getting the normalized tokens, we calculate the hash value of all the k-gram substrings. The substrings are highly adjacent so that all parts of the tokens are taken into the comparison with that of other chunks. That means the token the length of which is n will generate n-k+1 substrings, and every fingerprint corresponds to the hash of every substring. The winnowing algorithm[13, 14] is used to choose a small subset of these fingerprints. It has greatly reduced calculation amount during fingerprints comparison without leaving out any substrings. In our work, we define the size of window as w, and select the hash values anyone of which is the lowest hash in the related window as the fingerprints of the chunk. As winnow algorithm does, we take all the substrings the size of which is k+w-1 into consideration. Finally, we get (n-k+1)-w+1 fingerprints for each chunk, and we can also record the position of each fingerprint in the chunk.

3.3   Hierarchical Clustering

Once we get the fingerprints of all the chunks, it is easy to calculate the distance between every two chunks, but we do not know the distribution of all the chunks because it is hard to get the common fingerprints that exist in most of the chunks, maybe they does not exist. So it is very appropriate to choose the hierarchical agglomerative clustering algorithm[15] to cluster the chunks. In our work, we use the Jaccard similarity coefficient to measure the similarity between the chunks or clusters. And the Jaccard

distance $J(C_i, C_j) = 1 - \frac{|F_{(C_i)} \cap F_{(C_j)}|}{|F_{(C_i)} \cup F_{(C_j)}|}$, $F(C_i)$ represents the number set of fingerprints of

chunk i, so the numerator above represents the number of set of fingerprints which exist in both of chunk i and chunk j. Likewise, the denominator below represents the number of set of fingerprints which exist in chunk i or chunk j. We regard each chunk as a leaf node, every time we choose the nearest two chunks as a new cluster according to the single-linkage approach and the new cluster equals a new node of the tree. After n-1 iterations, all the chunks become a big cluster, and in the meanwhile we get an integrated hierarchical tree. In the end, when we cut the tree in the specific depth to get the specific number of clusters whatever we want the chunks should be clustered into.

4   Assigning and Workload Balance

According to the context of peer grading on MOOCs, Cluster id, Submission diversity and Chunk quantity are seen as the main assigning strategies. All these strategies are designed to reduce review time for students during the peer grading, and ensure the quality and response time of reviews. First, we look for the student who has already been assigned with the chunks belonging to the same cluster. Practices have proven that students always worked efficiently when reviewing the similar contents. Second, to ensure that each submission can be reviewed by different students, we try to find the student who has already been assigned least chunks which belong to the same owner as the chunk being assigned does. Third, some students may be assigned with too many review tasks if we merely assign all the chunks which belong to the same cluster to one student. So we also control the total number of chunks assigned to each student.

Experimental data indicate that there is a high variety of reviewing time among individual students and variety of reviewing difficulties among code chunks. It is due to the different code structure in different functions. The statements that invoke the other methods or the nested loops are the dominant factors, while the functions only with assignment statements or simple logic are always easily to review. So the students who are assigned with the review tasks on the chunks have to exert more effort..

We have found the main factors that affect students reviewing, and designed a classification to identify chunk's difficulty to be reviewed. We record the time of every chunk review. So the system can automatically identify a chunk's difficulty by the time cost on it after peer grading. During the feature extraction, we get the number of method invocations and nested loops respectively through traversing the whole content of every chunk. The features are all integers, so it is easy to get the Euclidean distance among the chunks. Then we use K Nearest Neighbor classification to identify the review difficulty of a new chunk according to all the training data.

5   Data and Experiments

In our experiments, we run six programming assignments in our MOOC course of C Language Program Design to evaluate the performance of peer grading tool. Almost 200 students participated in the course and most of them submitted files on corresponding assignments.

5.1   Partition

Two preliminary studies[12] in Caesar's experiments showed that reviewers can still write a significant amount of meaningful feedback on the chunks they were given. Likewise, in our practice, the feedback given by students indicated that these functions are readable and understandable individually. We also ran a Readable test that we asked some other students who didn't participate in the course to randomly review 20 chunks in each assignment. And the feedbacks in the tests indicated that lots of chunks are difficult to understand to these new students. In Assignment 1 and 2, students always wrote only one function to solve a little problem, while students in other assignments preferred to design multiple methods.

5.2   Clustering

To evaluate the performance of clustering, we set a similarity threshold k as a termination criterion for hierarchical clustering. More accurate clustering results are produced when we decrease the value of k.

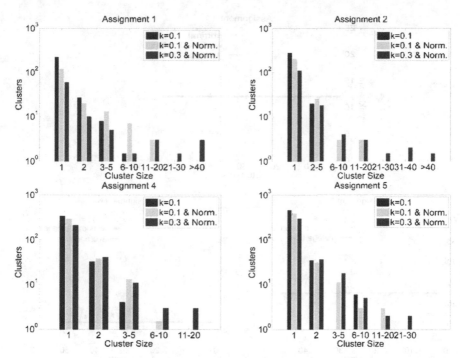

Fig. 2. Clustering performance. Bule bars show clustering result without code normolization; green and red bars are both in condition of code normolization with different k. The abscissa represents the quantity of chunks in cluster, and the ordinate represents the number of clusters.

Figure 2 shows the cluster size distributions of some assignments. Without code normalization, most chunks in Assignment 2 and 4 were alone and the biggest size of cluster did not exceed 3. We set two distance threshold k=0.3 and k=0.1 as the stop condition during clustering respectively, and the threshold k=0.3 has already request the high similarity (80%) between clusters. The distribution in Assignment 1,2 showed good clustering results, some clusters' size were even very close to the number of students when k equaled to 0.3. On the contrary, the cluster distributions in Assignment 4,5 were relatively unsatisfied, the maximal chunk size were approximately 20 only and the rest of chunks fall into very small clusters. This is due to the fact that the request in these two assignments were loosely specified, and multiple ideals of solving the problem exists which lead to various structures and logics in the functions.

5.3 Chunks Assigning

In the first stage of our research, we just tried to send the similar chunks to the same student to help faster review. Then we added the workload balance into chunks assigning in the second stage. We defined every chunk's difficulty level as easy, middle, or hard based on the time reviewer spent on it, the difficulty level together with the extracted features constituted the training data of workload KNN classification. At last, we get the best choice of k by using cross validation on training data.

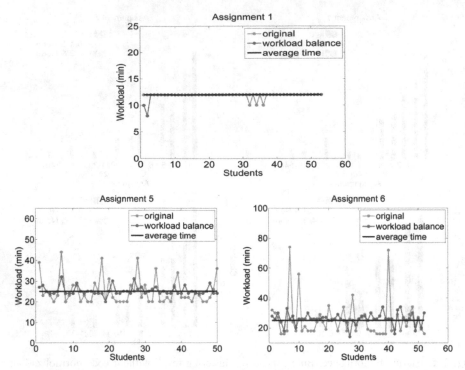

Fig. 3. workload balance performance. The abscissa represents student index, and the ordinate represents cumulative workload value. (the time cost during review).

Figure 3 shows the workload distribution throughout the process of peer grading. Assignment 5 and 6 show significant difference between two lines, which can be easily understood if a student was assigned all the hard chunks while another just needed to review easy functions. And the red lines indicate fair tasks among all the students by using the workload classification, their workloads moved up and down in a limited range. The data in experiments also shows quite a difference among the number of chunks assigned to every student. However, we can hardly find the difference between the red and blue lines in Assignment 1. All the chunks in this assignment do not involve any nested loop or invocation to another function. So these chunks were all classified to easy chunks and the workload classification seems meaningless in that case.

6   Conclusions and Future Work

Our paper presents improved peer grading tool for making simpler, more efficient, and more balanced peer review on MOOCs. We divide the submissions of student into small pieces of code, which makes the task of review easier and makes flexible operations on the chunks. Then we did chunk clustering to increase overall efficiencies of peer grading. We also design good strategies to help each student achieve abundant feedbacks while taking workload balance into consideration by using workload classification.

There also remain some issues to be addressed in future work. We have designed the peer grading tool for C/C++ Language programming courses only so far. It is very necessary to make the tool support most of the programming languages. Another issue is about the result of the peer grading. As the students are beginner programmers, their feedbacks may seems not very exact, and sometimes even incorrect. In addition, given the diverse background of MOOC students, there feedbacks on the same program can vary a lot. Thus, we should design some strategies or methods like probabilistic model to improve precision rate of peer grading.

References

1. Breslow, L., Pritchard, D.E., DeBoer, J., Stump, G.S., Ho, A.D., Seaton, D.: Studying learning in the worldwide classroom: Research into edX's first MOOC. Research & Practice in Assessment 8, 13–25 (2013)
2. Pappano, L.: The Year of the MOOC. The New York Times 2 (2012)
3. Cheang, B., Kurnia, A., Lim, A., Oon, W.-C.: On automated grading of programming assignments in an academic institution. Computers & Education 41, 121–131 (2003)
4. Ihantola, P., Ahoniemi, T., Karavirta, V., Seppälä, O.: Review of recent systems for automatic assessment of programming assignments. In: Proceedings of the 10th Koli Calling International Conference on Computing Education Research, pp. 86–93 (2010)
5. Ertmer, P.A., Richardson, J.C., Belland, B., Camin, D., Connolly, P., Coulthard, G., et al.: Using peer feedback to enhance the quality of student online postings: An exploratory study. Journal of Computer Mediated Communication 12, 412–433 (2007)
6. Kulkarni, C., Wei, K.P., Le, H., Chia, D., Papadopoulos, K., Cheng, J., et al.: Peer and self assessment in massive online classes. ACM Transactions on Computer-Human Interaction (TOCHI) 20, 33 (2013)
7. Shah, N.B., Bradley, J.K., Parekh, A., Wainwright, M., Ramchandran, K.: A Case for Ordinal Peer-evaluation in MOOCs. Presented at the NIPS Workshop on Data Driven Education, Harrah's Tahoe D (2013)
8. Liu, E.Z., Lin, S.S., Yuan, S.: Alternatives to Instructor Assessment: A Case Study of Comparing Self and Peer Assessment with Instructor Assessment under a Networked Innovative Assessment Procedures. International Journal of Instructional Media 29, 395–404 (2002)
9. Weld, D.S., Adar, E., Chilton, L., Hoffmann, R., Horvitz, E., Koch, M., et al.: Personalized online education—a crowdsourcing challenge. In: Workshops at the Twenty-Sixth AAAI Conference on Artificial Intelligence (2012)
10. Piech, C., Huang, J., Chen, Z., Do, C., Ng, A., Koller, D.: Tuned models of peer assessment in MOOCs, arXiv preprint arXiv:1307.2579 (2013)
11. Sadler, P.M., Good, E.: The impact of self-and peer-grading on student learning. Educational Assessment 11, 1–31 (2006)
12. Tang, M., Miller, R.C., Smith, A.C.: Caesar: A Social Code Review Tool for Programming Education (August 22, 2011)
13. Schleimer, S., Wilkerson, D.S., Aiken, A.: Winnowing: local algorithms for document fingerprinting. In: Proceedings of the 2003 ACM SIGMOD International Conference on Management of Data, pp. 76–85 (2003)
14. Sorokina, D., Gehrke, J., Warner, S., Ginsparg, P.: Plagiarism detection in arXiv. In: Sixth International Conference on Data Mining, ICDM 2006, pp. 1070–1075 (2006)
15. Johnson, S.C.: Hierarchical clustering schemes. Psychometrika 32, 241–254 (1967)

DIPP—An LLC Replacement Policy for On-chip Dynamic Heterogeneous Multi-core Architecture

Zhang Yang, Xing Zuocheng, and Ma Xiao

Science and technology on Parallel and distributed processing laboratory
National University of Defense Technology
ChangSha, China
zhangyang@nudt.edu.cn

Abstract. As the big data era is coming, it brings new challenges to the massive data processing. A combination of GPU and CPU on chip is the trend to release the pressure of large scale computing. We found that there are different memory access characteristics between GPU and CPU. The most important one is that the programs of GPU include a large number of threads, which lead to higher access frequency in cache than the CPU programs. Although the LRU policy favors the programs with high memory access frequency, the programs of GPU can't get the corresponding performance boost even more cache resources are provided. So LRU policy is not suitable for heterogeneous multi-core processor.

Based on the different characteristics of GPU and CPU programs on memory access, this paper proposes an LLC dynamic replacement policy--DIPP (Dynamic Insertion / Promotion Policy) for heterogeneous multi-core processors. The core idea of the replacement policy is to reduce the miss rate of the program and enhance the overall system performance by limiting the cache resources that GPU can acquire and reducing the thread interferences between programs.

Experiments compare the DIPP replacement policy with LRU and we conduct a classified discussion according to the program results of GPU. Friendly programs enhance 23.29% on the average performance (using arithmetic mean). Large working sets programs can improve 13.95%, compute-intensive programs enhance 9.66% and stream class programs improve 3.8%.

Keywords: Big data, Heterogeneous, Multicore, Replacement Policy, DIPP.

1   Introduction

The coming of the big data era has brought new challenges to the massive data processing. However, the traditional multi-core architecture can hardly meet the demands for large scale computing. Industrial community have integrated different kinds of cores on the same chip to constitute a heterogeneous multi-core processor architecture [1,2]. In the CPU-GPU heterogeneous systems, CPU is responsible for the execution of serial tasks and the logic control, while GPU has advantages on parallel computing. They can work together.

H. Wang et al. (Eds.): ICYCSEE 2015, CCIS 503, pp. 386–397, 2015.
© Springer-Verlag Berlin Heidelberg 2015

"Memory Wall" is already a very serious problem. It has become more acute in the heterogeneous CMP [3]. The performance of the on-chip memory system is an important factor in determining CMP's system performance. The goal of this paper is to effectively manage the shared LLC on heterogeneous CMP and optimize the overall performance when running multiple applications on heterogeneous CMP. The main work and innovation of this paper is as follows.

We compute the number of threads and memory access per thousand instructions in 20 kinds of GPU programs and 27 kinds of SPEC CPU2006 programs and find that GPU programs have higher access frequency than the CPU programs on average.

We presents a dynamic replacement policy DIPP in LLC for heterogeneous CMP based on different features between GPU and CPU programs. It changes the replacement policy dynamically to limit the cache resources GPU can acquire, reduce the thread interference in programs and achieve improvements on the overall performance.

2 Related Work

2.1 Dynamic Replacement Strategy

The traditional LRU can't meet the performance needs of the program with jitter. Most blocks do not hit in the transfer process, so cache space can't be fully utilized. LIP insertion strategy is to insert a new block onto LRU. It can not only put parts of the working set in cache so that they can be hit, but can protect cache free from jitter. BIP, promoted from LIP, can adapt to changes in the working set. DIP, proposed by Qureshi etc, is to select a replacement policy with lower miss rate between BIP and LRU [4].

LRU chain represents the new and recent characters of the cache block. Applications with far reuse distance and scanning program don't do well in LRU replacement policy. Most blocks inserted into LLC are not reused again. NRU is a kind of commonly used replacement policy which is similar to LRU. SRRIP extends NRU bit to two and sets the RRPV value of newly inserted blocks to a value close to the middle of the far reuse distance to reduce the impact of the scanning program. When the reuse distances of all the blocks are greater than the available cache resources, SRRIP will cause the jitter and there is no cache hits. BRRIP sets RRPV of the block to a far reuse distance in a low probability and saves parts of the working set to improve the hit rate. DRRIP uses group decision-making to choose a more suitable replacement policy for the current program between SRRIP and BRRIP [5].

2.2 Capacity Management

LRU divides shared cache implicitly based on demands. High demanded program can get more resources, but the applications with high demands may not be able to get a high performance matched with more cache resources. UCP is a kind of low-cost and real-time replacement policy. Each core is assigned a performance monitor separated from the shared cache. UCP is optimized to minimize the total miss rate in system and to divide the shared cache into ways for multiple programs [6]. When the shared cache between multi-cores are not properly managed, the processors with shared cache at the last stage is prone to the problem of low efficiency and poor equity. Cache division is a very efficient method for the isolation of threads with poor performances. Xie identified the performance problems when Thrashing programs exist and proposed a simple and effective cache managing method for this kind program: TC (Thrasher Caging) [7].

Xie collects the total memory access number and the total miss number of one program to LLC once every T cycles. The programs with large memory access number or with large cache miss number will cause great interference to other programs. TC isolates these programs to a fixed-size division.

By managing the insertion and improvement policy, PIPP can achieve pseudo-division of shared cache [8]. The threads with larger capacities are positioned closer to MRU for it has longer dwelling time and the threads with smaller capacities are inserted close to LRU so that it can be evicted in a short time. This method provides the dynamic insertion by not inserting a block into the head of the queue each time, not strictly dividing blocks and allowing cores stealing cache capacity to use all the available space.

3   The Program Features in Heterogeneous Multi-core Processors

A large number of cores are integrated in GPU, which use the SIMT execution mode. In SIMT, one single instruction can control multiple computing threads to fully exploit the parallelism of the program. The execution mode in SIMT contains a large number of threads, and thread switches in GPU can hide the memory access latency.

Access per kilo-Instruction (APKI) may reflect the frequency of the access to the memory by the program. Programs with higher APKI have more memory accesses. The difference in architecture between GPU and CPU leads to different access characteristics. We count up the number of threads and APKI of GPU and CPU programs.

Table 1. The number of threads on GPU and APKI

program	thread	APKI	program	thread	APKI
bfs	31264	66.11	convolution-sep	36864	72.41
fastwalsh	65536	100.35	histogram	8740	159.87
mergesort	65536	38.78	particlefilter	3704	43.42
volumerender	8192	4.91	quasirandom	1536	18.32
hotspot	131072	12.97	montecarlo	512	19.34
blackscholes	1920	135.72	mm	2112	45.37
convtex	147456	140.72	radixsortrhrust	1672	620.48
dwthaar1d	64	50.74	reduction	512	307.52
fdtd3d	3200	54.26	sad	3168	87.47
fft	4096	489.73	scalarprod	1024	176.62

Table 2. The number of threads on CPU and APKI

program	thread	APKI	program	thread	APKI
400.perlbench	1	6.18	401.bzip2	1	9.82
403.gcc	1	5.93	410.bwaves	1	0.04
416.gamess	1	0.04	429.mcf	1	67.24
433.milc	1	14.62	434.zeusmp	1	7.03
435.gromacs	1	3.69	437.leslie3d	1	40.91
444.namd	1	0.34	445.gobmk	1	4.08
450.soplex	1	36.01	453.povray	1	0.40
454.calculix	1	0.01	456.hmmer	1	3.53
458.sjeng	1	0.51	459.gemsFDT	1	21.07
462.libquantu	1	32.96	464.h264ref	1	2.05
465.tonto	1	2.01	470.lbm	1	37.70
471.omnetpp	1	31.89	473.astar	1	6.35
481.Wrf	1	6.99	482.sphinx3	1	14.18
998.specrand	1	0.01			

Comparing Table 1 and Table 2, we conclude that the GPU programs have higher access frequencies than CPU programs. To explore how much the performance promotion will be achieved by the GPU programs when more cache resources are provided, we use 4096 cache groups and enlarge the cache capacity by increasing the associativity from 1 way (256KB) to 32 ways (8M). The experiment result is shown in Figure 1.

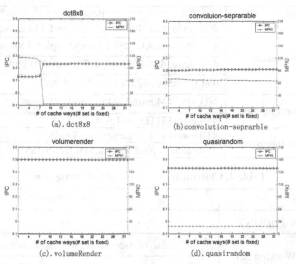

Fig. 1. The sensitive graph of cache capacity for GPU programs

We can conclude that GPU program requires a small amount of cache to improve the performance. When the cache capacity increases, its performance will not be significantly improved. Qureshi divides the programs into three categories according to the performance enhancement by the increase of cache capacities, those are low performance programs, performance saturated programs and high performance program.

We classify the simulation results of SPEC CPU2006 and NVIDIA SDK and Rodinia of GPU test set according to APKI and the utility speedup. Utility speedup is the decreased percentage of miss rate when allocated 32 ways compared with 1 way:

$$USpeedup = \frac{miss_{1way} - miss_{32way}}{miss_{1way}} \qquad (3\text{-}1)$$

SPEC CPU2006 programs are classified into three categories and the GPU programs are classified into four categories.

- CPU program: if APKI < α, it's compute-intensive class (C). If APKI > α, and USpeedup > β, it's cache-friendly class (A). Otherwise it is large working set (B).
- GPU program: if APKI < η, it's compute-intensive class (C). If APKI > η and USpeedup > θ, it's cache-friendly class (A). If APKI > η and USpeedup < θ, it's a large working class (B). If APKI > η and USpeedup = 0, it's a stream class (D).

Herein we set $\alpha = 5$, $\beta = 15\%$; $\eta = 5$, $\theta = 15\%$.

This paper uses MacSim heterogeneous multi-core simulator [9]. With the cache ways increased from 1 (256KB) to 32 (8MB), we measure APKI and USpeedup of SPEC CPU2006, NVIDIA SDK and Rodinia. Classification results are as follows:

Table 3. Classification results of CPU programs

Type		Benchmarks
A	Cache-Friendly	400.perlbench, 401.bzip2, 429.mcf,434.zeusmp
		435.gromacs,437.leslie3d, 450.soplex,456hmmer
		458.sjeng, 464.h264ref, 465.tonto, 470.lbm,
		471.omnetpp, 473.astar, 481.sphinx3
B	Large-Working Set	403.gcc, 444.namd, 445.gobmk, 453.povray
		459.gemsFDTD, 481.wrf
C	Compute-Intensive	410.bwaves, 416.gamess, 433.milc
		454.calculix ,462.libquantum, 988.specrand

Table 4. Classification results of GPU programs

Type		Benchmarks
A	Cache-Friendly	bfs
B	Large-Working Set	convolution-separable, fastwalshtransform
		histogram, mergesort, particlefilter, hotspot
C	Compute-Intensive	volumerender
D	Stream	quasirandom,montecarlo

4   DIPP-- LLC Dynamic Replacement Policy of Heterogeneous Multi-core Processors

In the heterogeneous multi-core processors using LRU, on-chip LCC resources are shared by simultaneously executed programs. However, LLC resources are not evenly distributed among them. Different program based on different number of cache accesses can get different sizes of cache resources. So GPU program will be favored because of the high frequency of memory access. Cache blocks of the GPU program will quickly evict the cache block of the CPU program. The unreasonable resource allocation results in inability to enhance the whole performance of the GPU program and reduce the performance of the CPU program.

LRU chain can reflect the reusability of cache on the data and instruction block. The chain's first location is called the MRU position, which has the highest reusability. The chain tail is called the LRU position, which has the lowest reusability. Once a cache block fails, a new block will be inserted into the MRU position and the block in the LRU position will be evicted. When the cache block hits, it will be promoted to the MRU position. The lifecycle of a cache block starts from the chain's first location, goes through all the position in LRU queue and ends by evicted from the queue. LRU can be refined into three parts: the eviction policy, the insertion policy and the promotion policy. The following figures are the LRU chain of DIPP with 8-way association.

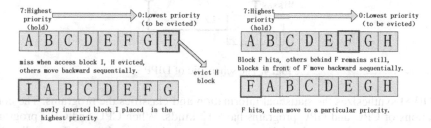

Fig. 2. Replacement policy of CPU program

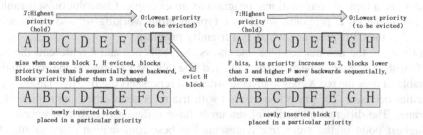

Fig. 3. Replacement policy of GPU program

The priority of every position is expressed by the use of a specific number. The priority of MRU is that the association number subtracts one. In this case it is 7. LRU's priority is 0. The priority of the block between MRU and LRU is in a descending order.

DIPP cache strictly limits the priority of cache blocks in the GPU program no more than 3. Whether inserted a new block or promoted when hit, the priority of the block is set 3. For the block in the CPU program, its block priority is 8, no matter it is hit or inserted. Every block is sequentially transferred from the highest priority position to its lowest priority position, and then evicted. Therefore, even if the GPU program continuously requests for the cache resources, the resource's upper limit to cache has been set. The cache block in the CPU program can be transferred sequentially from the position of MRU to LRU to gain more cache resources than the GPU program.

This replacement policy is just static. Although its design idea is simple, it can't make flexible adjustments based on certain features of a section of programs. DIPP can reduce the priority of GPU program in the LRU chain. OHAI (Online Heterogeneous Applications Identification) collects the failure request of L2 cache and simulates the LLC access on itself. It computes APKI and USpeedup of the CPU and GPU program every T cycles. DIPP can dynamically change the insertion strategy and promotion strategy of CPU and GPU based on the comprehensive information. DIPP circuit is composed by OHAI and the decision-maker, as shown below.

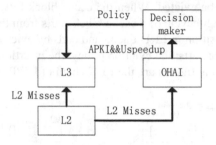

Fig. 4. The schematic of DIPP

OHAI synthesizes the statistical information and classifies the programs. The combinations of CPU and GPU programs have 12 kinds. When CPU and GPU programs are executed simultaneously, different program combinations have different distributions. We explore how to allocate resources reasonably to achieve better performance. We analyze a typical combinational program as an example. Convolution-separable is selected as the GPU program, which is a typical large working set and perlbench is selected as the CPU program, which is a friendly program.

In Figure 5, the non-marked blue line shows the perlbench and it represents the trend of IPC with increased cache capacities. The red line labeled x represents the convolution-separable. It uses the red X-axis in reverse order and represents the IPC trend as the cache capacities decrease. The green line marked with triangles shows the combination of the programs. The different allocation of resources have different overall performance and the highest point of the green line represents the best combination of programs. The highest point is in the case when X equals 25 in the forward order. That means we can get the optimal performance when perlbench has 25 ways. As the CPU program is at the upper end of the cache chain, we consider allocating one way less than the optimized way. Other combinations of programs can perform the same analysis. The design rules of the combined program constitute the program decider, which can make flexible adjusts to different kinds of CPU and GPU programs.

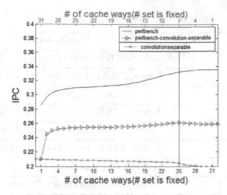

Fig. 5. Allocation of cache ways

The design rule is shown as the following graph,

Table 5. Design rules of decision maker

CPU	GPU	Design Rules
A	A	8-way assigned to GPU program,24-way assigned to CPU program
A	B	8-way assigned to GPU program,24-way assigned to CPU program
A	C	2-way assigned to GPU program,30-way assigned to CPU program
A	D	1-way assigned to GPU program,31-way assigned to CPU program
B	A	12-way assigned to GPU program,20-way assigned to CPU program
B	B	16-way assigned to GPU program,16-way assigned to CPU program
B	C	2-way assigned to GPU program,30-way assigned to CPU program
B	D	1-way assigned to GPU program,31-way assigned to CPU program
C	A	30-way assigned to GPU program,2-way assigned to CPU program
C	B	30-way assigned to GPU program,2-way assigned to CPU program
C	C	2-way assigned to GPU program,30-way assigned to CPU program
C	D	1-way assigned to GPU program,31-way assigned to CPU program

OHAI obtains the classified information every T cycles and classifies the CPU and GPU programs online. DIPP compares the classification condition of the CPU and GPU programs with the above table and chooses the best replacement policy dynamically to make the cache resources reasonable in the heterogeneous processors.

5   Experimental Results and Analysis

MacSim is a heterogeneous simulator whose input method is trace-driven and propulsion method is time-driven [9]. As a micro-architecture simulator, MacSim can simulate a complete architecture. The experimental setup is in the following table.

Table 6. The experimental configuration

CPU	1 core, 3GHZ, 4-wide
	8-way, 64KB L1 I/D (3-cycle) ,64B line
	8-way, 256KB L2(8-cycle) ,64B line
GPU	4 cores, 1.5GHZ, 1-wide
	8-way, 32KB, L1 D (2-cycle), 64B line
	8-way, 4KB, L1 I (2-cycle),64B line
L3 Cache	32-way, 8MB(4tiles,30-cycle)64B line
NOC	1 GHz, ring topology, 1 dimension
DRAM	2 Controllers, 8 banks, 2 channels
	1.6GHz, FR-FCFS

The result of the friendly program is shown below.

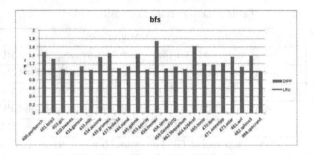

Fig. 6. Results of friendly program

As Figure 6 shows, bfs obtains a high performance improvement. The highest improved program is the program couple composed with 456.hmmer, and its IPC increases 74.27%. The worst performance improvement is the program couple composed with 433.milc, and its IPC only improves 3.95%. The average increase of IPC is 10.79%. We conclude that bfs is a cache-friendly program.

According to the design rules in Table 5, when the GPU-friendly programs are executed with the CPU-friendly program together, 24-way will be assigned to the CPU program and the remaining 24-way will be assigned to the GPU program. The improvement of the CPU-friendly program is remarkable when getting more capacities for the cache. The typical representatives are 400.perlbench, 401.bzip2, 434.zeusmp, 435.gromacs and 456.hmmer. In the heterogeneous multi-core processors using LRU, APKI of bfs is 66.91, while that is very low in typical CPU programs. In this case, the cache block of bfs will be evicted quickly from the cache blocks of CPU-friendly programs, which reduces the performance of the CPU program. DIPP avoids the interference between threads and improves the performance.

When the GPU-friendly programs are executed with the CPU large working set programs, the cache capacities acquired by the CPU large working class programs are reduced, compared with the CPU-friendly programs. The performance improvement is inferior to the above case. The typical programs are 403.gcc whose IPC improves 4.9% and 444.namd whose IPC increase is 12.55%. Though the GPU programs are friendly, its performance improvement is not obvious when it gets more cache capacities.

When the GPU-friendly programs and the compute-intensive programs are performed together, GPU programs get almost all of the resources, because the CPU programs are compute-intensive programs and they do not need lots of cache resources. Although the performance improvement of the GPU-friendly programs with more cache is less conspicuous than the CPU-friendly programs, it is better than the CPU compute-intensive programs. The typical representatives are 435.gromacs, whose IPC improved by 45.05% and 464.h264ref whose IPC improved by 61.94%.

In summary, when the GPU-friendly programs are executed with the CPU-friendly programs and with the CPU compute-intensive programs, it will obtain significant performance improvement.

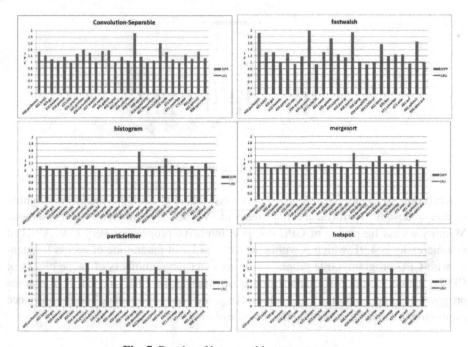

Fig. 7. Results of large working set programs

According to the results of Figure 7, these are typical CPU large working class programs. Such programs have also got a good performance. Convolution-separable is a large working class program. Performance improvement of it is very significant. Average increase of IPC is 10.52%. In this group the highest performance improvement is the program couple composed with 456.hmmer and its IPC improves by 91.11%. The worst is the program couple composed with 433.milc and its IPC drops by 1.91%.

When the GPU large working class programs are executed with the CPU-friendly programs, a small amount of cache capacities can guarantee the GPU large working class programs to achieve the peak performance and the CPU programs can obtain large increases with more cache capacities, thereby improving the overall performance. Representatives are 434.zeusmp and 456.hmmer.

When the GPU large working class programs are executed with the GPU large working class programs, 16 ways can't guarantee the performance improvement of the two programs. The performance can hardly improve. The representative is 462.libquantum. When the GPU large working class programs are executed with the CPU-intensive programs, most programs obtain a good performance improvement. When the cache capacities exceed the cache reuse distance, GPU programs will have good temporal locality and the performance improvement is significant.

In summary, when the GPU large working class programs carried out with the CPU-friendly programs and with the CPU compute-intensive programs, it will obtain significant performance improvements.

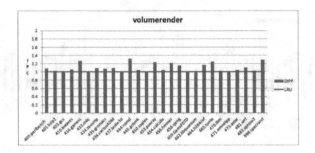

Fig. 8. Result of compute-intensive program

According to Figure 8, such program is a typical GPU compute-intensive program. It gets almost no performance improvement. We take volumerender for example. Memory access frequency of GPU compute-intensive program is very low. APKI of volumerender is 4.91, which is within the same order of magnitude of access frequency with CPU program. GPU program will not evict large numbers of cache blocks of CPU program. So the performances are close when using LRU and DIPP. But some programs also gets a promotion, such as 416.gamess and 444.namd. The performance promotion is caused by avoiding the thread isolation.

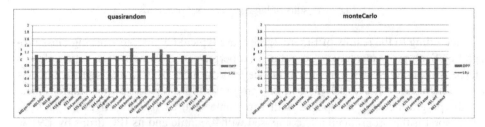

Fig. 9. Results of stream type programs

Figure 9 is a typical stream type program and the USpeedup of this type is 0. The enhancement of such programs is not conspicuous, especially monteCarlo, whose performance is hardly improved. Some programs' performances are even declined. We analysis montecarlo for example whose APKI is only 19.34, which is within the same order of the magnitude with most CPU programs. So even in the multi-core processors using LRU, cache blocks in the GPU programs will not evict the blocks of the CPU program massively. The same analysis can be performed to quasirandom whose performance improvement comes from the thread isolation.

6   Summary

Cache replacement policy is an important topic on the computer architecture to directly affect the performance of the overall system. GPU and CPU programs have different access characteristics. This paper has proposed a dynamic replacement policy DIPP in LLC for heterogeneous CMP, which can reduce the priority in the LRU chain of the cache block on the GPU, reduce the interference between threads and bring down the miss rate of the program. As a result, the performance of overall system is improved.

This paper selected SPEC CPU2006 program and 10 kinds of GPU programs to verify the performance of DIPP. According to the results, we can conclude that friendly program's average performance promote is 23.29% and that of large working class programs is 13.95%. The average enhancement for compute-intensive programs is 9.66% and that of stream type program is 3.8%. We can conclude that the new replacement policy is suitable for big data processing.

Reference

1. Fan, Z., Qiu, F., Kaufman, A., Yoakum-Stover, S.: GPU cluster for high performance computing. In: SC 2004: Proceedings of the 2004 ACM/IEEE Conference on Supercomputing. IEEE Computer Society, Washington, DC (2004)
2. Dally, W.J., Hanrahan, P., Erez, M., Knight, T.J., et al.: Merrimac: Supercomputing with Streams. In: SC 2003: Proceedings of Supercomputing Conference 2003, vol. 5270, pp. 35–42 (2003)
3. Hennessy, J.L., Patterson, D.A.: Computer Architecture: A Quantitative Approach, 4th edn. Morgan Kaufmann (2006)
4. Qureshi, M.K., Jaleel, A., Patt, Y.N., et al.: Adaptive insertion policies for high performance caching. ACM SIGARCH Computer Architecture News 35(2), 381–391 (2007)
5. Jaleel, A., Theobald, K.B., Steely Jr., S.C., Emer, J.: High performance cache replacement using re-reference interval prediction (RRIP). In: ISCA-32, pp. 60–71 (2010)
6. Qureshi, M.K., Patt, Y.N.: Utility-based cache partitioning: A low-overhead, high-performance, runtime mechanism to partition shared caches. In: MICRO-39, pp. 423–432 (2006)
7. Xie, Y., Loh, G.H.: Scalable shared-cache management by containing thrashing workloads. In: Patt, Y.N., Foglia, P., Duesterwald, E., Faraboschi, P., Martorell, X. (eds.) HiPEAC 2010. LNCS, vol. 5952, pp. 262–276. Springer, Heidelberg (2010)
8. Xie, Y., Loh, G.H.: PIPP: promotion/insertion pseudopartitioning of multi-core shared caches. In: ISCA-31, pp. 174–183 (2009)
9. MacSim, http://code.google.com/p/macsim/

Data Flow Analysis and Formal Method

Yanmei Li[*], Shaobin Huang, Junyu Lin, and Ya Li

College of Computer Science and Technology, Harbin Engineering University, Harbin, China
{liyanmei,huangshaobin,linjunyu,liya}@hrbeu.edu.cn

Abstract. Exceptions are those abnormal data flow which needs additional calculation to deal with. Exception analysis concerned abnormal flow contains a
lot of research content, such as exception analysis method, program verification. This article introduces another research direction of exception analysis
which based on formal method. The article analyses and summarizes those
research literatures referring exception analysis and exception handling logic
verification based on formal reasoning and model checking. In the article, we
provide an overview of the relationship and difference between traditional ideas
and formal method concerning program exception analysis. In the end of the article, we make some ideas about exception analysis based on formal semantic
study of procedure calls. Exception handling is seen as a special semantic effect
of procedures calls.

Keywords: Exception; Propagation, Formal method, Model Checking.

1 Introduction

Exception handling indicates those program codes which check the pre-condition,
then do something to avoid a system crash when the system is going to entrap into an
unexpected situation or the error states. However, using exceptions properly and fully
is not easy for most programmers. For an individual program, because the exception
codes usually mixed with the normal parts, which make it difficult to understand their
behavior. Therefore, research of exceptional analysis concludes analyzing and verifying the adequacy, completeness and consistency constraint of exception handling
logic. In addition, the research also contains the time and space efficiency analysis of
exception handling.

With the works of Dijkstra and Hoare in program verification and the works of Scott
and Stratchey in program semantic aspects, formal method becomes an effective way to
access a credible system. Thus, that formal method becomes a mainly and core technology of exception analysis comes naturally. Thus, it is strongly recognized as a rational
direction. Formal methods include formal specification and verification. Formal verification is to verify the existing program P: whether its behavior referring exception dealing
with in this paper meets the formal specification pre-specified? In other words, that we
focus is the formalization and verification which study the adequacy, completeness and
consistency problems of exceptional analysis in theory level.

---

[*]Crossponding author.

H. Wang et al. (Eds.): ICYCSEE 2015, CCIS 503, pp. 398–406, 2015.

2   Research Overview

The main characteristic of the customary exception analysis is visual graphic representation. Those research are based on data flow analysis, control flow and control dependence analysis which belong to program structured analysis method. Some researchers also use pointer analysis which greatly improved the accuracy of exceptional flow analysis. In addition, some scholars apply post-pass feasibility pruning method and Data Reach algorithm which further enhance precision of exception analysis. Some scholars are dependent on the type analysis, such as class hierarchy analysis and constraint set solving process. In fact, the above analytical method of constraint set and so on has introduced formal reasoning[1].

Different framework makes that methods and technology of exception analysis are more and more diversified. However, there are many problems in the current research. First of all, the core method of the exception analysis is single. Most of the exception analysis eventually are reduced to exception chain analysis. They use the traditional method, such as control flow analysis, control dependence and data flow analysis. However, the above method doesn't apply to the object-oriented program and components program. Therefore, an alternative research method occurred. That is exception analysis and verification based formal method. The comparison between them is shown as the figure 1.

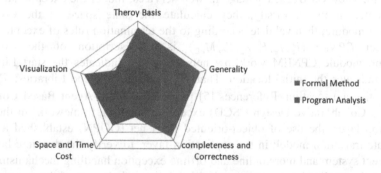

Fig. 1. Comparisons between program analysis and formal method

In the field of exception analysis, the advantage of a formal method is more obvious relative the program analysis. Although with program analysis exception analysis seems more intuitive and visualization, it's time and space cost is considerable large. Most case, propagation modeling is still not clear to program analysis. Formal method is more generality and has profound theoretical basis theoretical basis. High complexity, correctness and completeness is another advantage of a formal method.

3   Analysis Based on Formal Method

The common goal of exception analysis and formal methods is to ensure the stability and credibility of the programs. Their meeting points produce a new research direction which is exceptional analysis based on formal methods. It referred to the issues of program validation including exception mechanism. The core problem of research is the formalization to find fully abstract domain for program exception, such that complete formal reasoning and model checking of the exception.

3.1   Formal Modeling of Program Exception

On the respect of formalization of program behavior especially including exception, KO use language PPSL based on regular expression and predicate to formalize the exception. In which, regular expression indicates object name, and predicate indicates the permissible operating. E.g., read, predicate parameters include object names and the state variable group concerning programs' run. KO extend PPSL language and used a context-free grammar with state variable PE-Grammars [2]. They considered the object access of system resources, the sequence of related operations, synchronization and competition of inter-process and other security attributes. In references [3], with abnormal pattern state machine the author describes a runtime exception in which control flow containing runtime exception and applies an iterative equation to access control flow sequence of runtime exception.

Since the particularity of service-oriented software, exception handling mechanism is slightly different. In references [4], it established an exception handling model CPN (Colored Petri Nets) which concluding the normal control flow and exception handling logic. They via BPEL language in web services, model the exceptional in service oriented software. Firstly, they calculate the state space of the exception processing module; then validate according to the termination rules of exception handling. Let $CPN_E = (P_e, T_e, F_e, \Sigma_e, C_e, M_{e0})$ is a description of the exception processing module CPNEM workflow nets. That P_{ex} indicates the start Libraries; $P_{es} = \emptyset$ indicates the initial location. That P_{ef} indicates the end Libraries; $P_{ef} = \emptyset$ indicates the end location. References [5] proposed a Multi-Agent Based Computer Supported Collaborative Design CSCD) exception handling framework. In the communication layer, the use of object-oriented Petri net (OOPN) established a multi-agent state transition model; in the application layer, to verify the dynamic behavior of the agent system and more online and offline exception handling mechanism of the model; in the application layer, to verify the dynamic behavior of the agent system and more online and offline exception handling mechanism of the model [6].

In addition, Qiu et al define the PPL language based on CSP which formalize the description of the operational semantics of exception handling process. In the execution, a peer maybe in one of six states and N is the normal state. For the purpose of facilitating a two-step commitment, C and F are used to denote complete and finish states, respectively. The other states are special for CEH (coordinated exception handling), where E means the peer or its partners encounter some exceptions, EH (exception handling) means the peer is handling an exception, and EF means that the peer fails to handle an exception [7]. References [8] through the constituent elements of exception processing and exception processing modes are formally described, the use of exception processing exception processing model composed of elements linked to form a complete exception handling model.

3.2   Exception Handling Logic Verification

Formal methods advocator introduced of CSP, B method, Petri net to formalize exception handling. Representative work of exception handling logic analysis and verification: Hamadi [9] verified adaptive recovery network (self-adapting recovery net, SARN) T, such as its consistency constraints, reachability, liveness and bound that assured those exception handling behavior. In the literature [10], the author proposed exception flow analysis and verification base B method and specification language Alloy. Fig.5 is a schematic representation of the proposed approach to verifying CA action-based software systems. The system fault model and exceptional activity can be specified. This model identifies the exceptions that can be raised in each CA action and how they are handled. To verify a CA action-based design, it is necessary to translate it into a specification language with operatives.

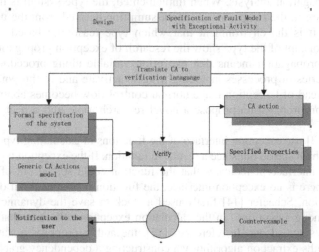

Fig. 2. State graph

Brito [11] proposed exception propagation formal analysis and verification methods, based B method and (communicating sequential processes TCSP); By this method he do the formal verification with his fault-tolerant software architecture. Pereira [12] based coordinated atomic action theory, proposed a model with the architecture of exception handling coordination mechanisms. This method is used to guide the formal specification of concurrent fault-tolerant systems. The model provides a set of predefined CSP processes and channels to specify, and coordinate exception handling of those components. He used the FDR (failure divergence refinement) to verify and coordination atomic action and concurrent security attributes associated exception handling. However, CSP based events process has certain defects in modeling capability, data abstraction, and B method has limitations when describing the concurrency control system.

Modeling process of exception handling based formal reasoning is long and complex, and thus not suitable for dynamic analysis. Using model checking tries to modeling and analysis exceptions have just begun. Conversion of exceptional state to migration system only relates to state diagram minimal subset that meet the requirements. In addition, concurrent exceptions are often modeled as a state when model checking for concurrent systems, ignoring a lot of useful object details.

4   Exception Prorogation Analysis

With the popularity of object-oriented concepts, types become a basic concept for exception handling. This makes that the exception handling mechanism is closed to the procedure call mechanism. It adds exceptions and their handling to the public framework of parameter matching and process calling. Type system is a formal tool for the study of language, which constructs a type system for existing language. It establishes a formal model, and on this model to study the language features such as exception propagation analysis. When implemented, the type system is the exception type information in the program and type assumption derived from the use of exception variable. It is the environment that which type reasoning based. This section based on the concept of the type show the research of exception propagation analysis.

Exception propagation means that exception variable along procedure call chain experienced series of processes including trigger, activate and re-thrown, caught that trigger and spread of Exception make normal control flow becomes abnormal control flow. Exception propagation propose a novel research topic and aspect in exception analysis

Robillard [13] use exception interface of the functions in calculating type's collection of exception which spread out since a recursive function. If the exception type is declared in exceptional interface, it indicates that the function could spread out of this type of exception; if there is no exception interface, the function may spread out of any certain type of exception. Schaefer [14] firstly used a stack to save the dynamic parent-child relationship among subroutines in the calculation exception collection that those recursive subroutines spread out. In references [15], the author proposes a static exception propagation paths extraction algorithm via constructing a dependency graph of the software associated with the exception types. His algorithm can analyze exception propagation hops, exception hierarchy and exception propagation boundaries.

5   Exception Analysis with Semantic of Procedure Call

Exception is based on procedure calls, and exception propagates throughout the procedures via arguments passing and type conversions of exception variable. General sense, the nature of exception propagation is the research referring procedure calls and its formal semantic study. We propose that exception analysis with formal semantic reasoning of procedure calls.

5.1   Extracting Semantics of Exception Handling Calling

Exception handling module can be seen as a special procedure call In order to ensure the correctness of the exception handling calls, procedure calls must be based on their pre-and post-conditions, and runs according to a predetermined specification. Via deviating the procedure calling semantic, Tit can generate the abstracted specification which can be used to describe the effect of the procedure calls If you want to ensure that there is no defect in exception analysis, the process may determine the semantics of exception handling according exception variables' including formal parameters state changes.

Let an Exception handling procedure call EH-Proc (al). For the list of parameters pl = (IN x : Type, OUT y : Type), where pl_1 reference parameter x , pl_2 referenced parameter y. For parameters type of IN, those formal parameter is assigned and initialized through a procedure call. The formal parameter of the value type and referring type as pl_i , we use the symbol as pl_2 ? indicate the initiation value of pl_i . In the list of parameters, the constraint can be formal description initially as:

$pl_T = pl_T\ ?\wedge pl_T \in T$

Let the precondition is Q, and Q_{proc} is the initial condition. We can get:

$$Q_{proc} \equiv \bigwedge_{i \in S_{vr}} (pl_i = pl_i\ ?\wedge pl_i \in TYPE_i) \wedge \bigwedge_{i \in S_{res}} (pl_i \in TYPE_i) \wedge Q \tag{1}$$

Where, S_{vr} indicates the true parameter, and S_{res} indicates the formal parameter. So that we can apply the predicate transformation of strongest post-condition to define the procedures' semantics. Let procedure is proc which including procedure S. According Q_{proc} , the post-condition is the form as: $R_{proc} = sp(Q_{proc}, S)$ and $R_{proc} = R_{proc}(E\,/\,v)$.

Where, $R_{proc}(E\,/\,v)$ indicate that the local variable v is replaced by expression E, and we can eliminate those local variable. If $R_{proc} \Rightarrow v = E$ and E is an expression, then there is $R_{proc} = R_{proc}(E\,/\,v)$. That $R_{proc}(E\,/\,v)$ indicates replace v with the expression E .

According (1), we can get the initial precondition of the module Div :

$Q_{Div} \equiv (\#v > l \wedge v = v? \wedge v \in seq\,Z \wedge w = w? \wedge w \in seq\,Z \wedge div \in seq\,Z)$

Applying the predicate transformation of strongest post-condition, then it get

$R_{Div} \equiv sp(Q_{Div}, S)$

$\equiv \#v > l \wedge v = v? \wedge w = w? \wedge v \in seq\,Z \wedge w \in seq\,Z \wedge div \in seq\,Z$

$\wedge \forall j \in [0..i-1](div[j] = v?[j] + w?[j]) \wedge i = l$

According $R_{proc} = R_{proc}(E\,/\,v)$, it can eliminate the local variable i :

$R_{Div} \equiv sp(Q_{Div}, S)$

$\equiv \#v > l \wedge v = v? \wedge w = w? \wedge v \in seq\,Z \wedge w \in seq\,Z \wedge$

$div \in seq\,Z \wedge \forall j \in [0..l](div[j] = v?[j] + w?[j])$

In this section, we can derivation the post-condition according to the precondition of the process. The post-condition is the process performed to achieve the effect, and it is also the semantics of the procedure calls.

5.2   Parameter Substitution and Precondition Checking

In the derivation of the post-condition, it is necessary to provide a method to substitute the true parameter into the formal parameter. Let R_{proc} is the post-condition of process proc. After substituted true parameter into the formal parameter, then the post-condition is expressed as R'_{proc}. The following is the substitution rules concerned about the precondition and post-condition:

If $B \Rightarrow Q'_{proc}$ is true, the value of the true parameter satisfied the precondition before the procedure calling. Otherwise, the procedure calling itself exit flaw. If it is established, then it satisfied the   precondition, otherwise the program exit flaw.

Through the above analysis process, we can get the algorithm to compute the strongest post-condition. Suppose that Q is the precondition, $proc(al)$ is a procedure call to procedure $proc(pl)$, where al is the true parameter list referring about formal parameter list of pl . Furthermore, there is $Q \equiv Q_1 \vee Q_2 \vee ... \vee Q_n$, and

$$sp(Q, proc(al)) = sp(Q_1, proc(al)) \vee ... \vee sp(Q_n, proc(al)), i \in [1...n] \ .$$

We can follow the algorithm steps as:

1. Precondition Checking: Computing whether the program status meets the preconditions before procedure calls. Let initial condition is the form as $B = B_1 \vee ... \vee B_n$
   。 For every B_i, computing $B_i \wedge \neg Q'_{proc}$ and according its outcome to determine. If it doesn't satisfy, there is an error and you don't need to continue the following steps.
2. Compute R'_{proc} : we can compute R'_{proc}. Then we may extract the semantics of exception handling.
3. Reduce R'_{proc} : we need to reduce R'_{proc} via eliminate the referring parameter with the formula of $R_{proc} = R_{proc}(E/v)$.
4. For the purpose of reduction of $sp(Q_i, proc(al))$, we use the expression of E replace v of Q_i, and invariable v is the true parameter concerned referring parameter list.   Where, predicate including invariable v in Q_i, there is $Q_i \Rightarrow v = E$.
5. Based on the above steps, we get $sp(Q_i, proc(al)) = Q_i \wedge R'_{proc}$.

According to the above algorithm, let us to review the previous program examples. We can compute $sp(Q_i, div(a,b,d))$ and $Q \equiv (a = a? \wedge b = b? \wedge a, b, d \in seq \ Z)$ .Firstly, Q can insure that the precondition of Q'_{div} is satisfied by $\#a > l$ before procedure calls. Then

$R'_{Div} \equiv \#a > l \wedge a \in seq \ Z \wedge b \in seq \ Z \wedge d \in seq \ Z \wedge \forall j \in [0..l] \ (d[j] = a[j] / b[j])$

$sp(Q_i, Div(a,b,d)) = Q_i \wedge R'_{Div}$

$= \#a > l \wedge a,b,d \in seq \ Z \wedge d \in seq \ Z \wedge$

$\forall j \in [0..l] \ (d[j] = a[j] / b[j]) \wedge a = a? \wedge b = b?$

Above is a derivation of the process specification. By solving procedure call post-conditions, which can detect the two defects of pre-conditions of violation and repeated defines of reference type parameter.

6   Conclusion

This paper summarizes the two kinds of research for exception analysis. Program structure approach is reliable and accurate; however, it has fatal weaknesses in the theoretical foundation such as versatility and correctness, completeness. It makes it not be long-term development, and is neglected by more and more people. Practice has proved that formal description for exception handling can effectively increase and deepen understanding and awareness. Thus, it meets the goal of verification. Other program testing and structural analysis methods cannot reach referred to the reason, systematic of the formal method. However, exception handling model based on formal methods cannot be reliably compared to exceptional structural analysis method. Therefore, these two methods are incorporated into a unified framework, which can add the reliability using formal methods.

Referencese

1. Fu, C.: Improving Software Reliability Using Exception Analysis of Object Oriented Programs. ProQuest (2008)
2. Ko, C., Ruschitzka, M., Levitt, K.: Execution monitoring of security-critical programs in distributed systems: a specification-based approach. In: Proceedings of the 1997 IEEE Symposium on Security and Privacy, pp. 175–187 (1997)
3. Jin, D.: Research on the Effect of Runtime Exception in Software Static Testing. Journal of Computers 34, 1090–1110 (2011)
4. Jiang, C.: Verification of Terminability for Exception Processing in Service-oriented Software. Journal of Frontiers of Computer Science & Technology 06 (2012)
5. Tian, F., Li, R., He, B., Zhang, J.: Modeling and Analysis of Exception Handling Based on Multi-Agent in Computer Supported Collaborative Design. Journal of Xian Jiaotong University 38, 5 (2004)
6. Pereira, D.P., de Melo, A.C.V.: Formalization of an architectural model for exception handling coordination based on CA action concepts. Science of Computer Programming 75, 333–349 (2010)
7. Cai, C., et al.: Global-to-Local Approach to Rigorously Developing Distributed System with Exception Handling. Journal of Computer Science & Technology, 238–249 (2009)
8. Wu, Q.: Exception Handling Model Based on Colored Petri Net in Service-oriented Software. Computer Science 38 (2011)
9. Hamadi, R., Benatallah, B., Medjahed, B.: Self-adapting recovery nets for policy-driven exception handling in business processes. Distributed and Parallel Databases 23, 1–44 (2008)
10. Castor Filho, F., Romanovsky, A., Rubira, C.M.F.: Improving reliability of cooperative concurrent systems with exception flow analysis. Journal of Systems and Software 82, 874–890 (2009)

11. Brito, P.H.S., De Lemos, R., Rubira, C.M.F., Martins, E.: Architecting fault tolerance with exception handling: verification and validation. Journal of Computer Science and Technology 24, 212–237 (2009)
12. Brennan, P.T.: Observations on program-wide Ada exception propagation. In: Proceedings of the Conference on TRI-Ada 1993, New York, pp. 189–195 (1993)
13. Li, S., Tan, G.: JET: exception checking in the Java native interface. SIGPLAN Not 46, 345–358 (2011)
14. Schaefer, C.F., Bundy, G.N.: Static analysis of exception handling in Ada. Software: Practice and Experience 23, 1157–1174 (1993)
15. Qiu, X., Zhang, L., Lian, X.: Static analysis for java exception propagation structure. In: 2010 IEEE International Conference on Progress in Informatics and Computing (PIC), pp. 1040–1046 (2013)

10-Elements Linguistic Truth-Valued Intuitionistic Fuzzy First-Order Logic System

Yingxin Wang[1], Xin Wen[2], and Li Zou[2,3]

[1] Marine Engineering College, Dalian Maritime University, China
[2] School of Computer and Information Technology, Liaoning Normal University, China
[3] State Key Laboratory for Novel Software Technology, Nanjing University, China

Abstract. This paper presents 10-elements linguistic truth-valued intuitionistic fuzzy algebra and the properties based on the linguistic truth-valued implication algebra which is fit to express both comparable and incomparable information. This method can also deal with the uncertain problem which has both positive evidence and negative evidence at the same time.10-elements linguistic truth-valued intuitionistic fuzzy first-order logic system has been established in the intuitionistic fuzzy algebra.

Keywords: linguistic truth-value, 10-elements linguistic truth-valued intuitionistic fuzzy logic, first-order logic system.

1 Introduction

In 1965, L.A.Zadeh put forward the concept of intuitionistic fuzzy sets [1]. In the following decades, the theory of fuzzy sets has been constantly developed and improved, and it is used in many fields. Intuitionistic fuzzy sets as a form of promotion of fuzzy set is well solved the hesitation and uncertainty by an additional amount [2, 3].

Nowadays, there exists many alternative methods to lattice-valued first-order logic, such as Y.Xu proposed the definition of generalized text, generalized prenex normal form and generalized Skolem standard, which are the basis of resolution principle based on lattice-valued first-order logic system [4]. D.Meng discussed the algorithm of transforming any formula to a reducible form, proposed the resolution principle by using ultrafilter and proved the soundness theorem and quasi-completeness theorem of resolution principle [5,6]. Y.Xu and J.J.Lai, et al. proposed α –resolution-based automatic reasoning in $LF(X)$ and its reliability and completeness respectively [7,11]. X.X.He got the α -unit algorithm in $L_n P(X)$ [8]. H.L.Zheng, et al. constructed a knowledge representation model with ten element linguistic value credibility factors, and achieved reasoning with linguistic value credibility factors [9]. W.T.Xu proposed the resolution principle based on ideal by taking an ideal as a reduced level but not an element of the real domain [10]. In 2014, W.T.Xu expanded general clause set to the general situation and proposed α –generalized resolution principle in $LF(X)$ [12].

H. Wang et al. (Eds.): ICYCSEE 2015, CCIS 503, pp. 407–417, 2015.

Most of the research for the first-order logic system above is based on the linguistic truth-valued lattice. In order to establish an uncertainty automatic reasoning system in the framework of the non-classical logic which is fit to express both comparable and incomparable information , and also deal with the uncertain problem which has both positive and negative evidence at the same time, this paper presents 10-elements linguistic truth-valued intuitionistic fuzzy first-order logic system based on linguistic truth-valued lattice implication algebra and the basic theory and algorithm of automated reasoning. This paper is organized as follows: Section 2 reviews the concept of intuitionistic fuzzy sets and its properties. Section 3 presents 10-elements linguistic truth-valued intuitionistic fuzzy algebra. Section 4 establishes 10-elements linguistic truth-valued intuitionistic fuzzy first-order logic system. We conclude in Section 5.

2   Preliminaries

Definition 1[2]. Intuitionistic fuzzy set is defined as follows:

$$A = \{(x, \mu_A(x), v_A(x)) \mid x \in U\},$$

where U is a discourse, $\mu_A(x): U \rightarrow [0,1]$ and $v_A(x): U \rightarrow [0,1]$ are the membership degree and nonmembership degree of the object $x \in U$ belonging to $A \subseteq U$ which satisfied with $0 \leq \mu_A(x) + v_A(x) \leq 1$ for any $x \in U$.

In the intuitionistic fuzzy set A, $\pi_A(x) = 1 - \mu_A(x) - v_A(x) (\forall x \in U)$ is called the degree of indeterminacy of x to A. In Zadeh's fuzzy set, if $\mu_A(x)$ is the membership degree of x to A, then $1 - \mu_A(x)$ is non-membership degree,i.e., $\pi_A(x) = 1 - \mu_A(x) - v_A(x) = 0$. Hence, the intuitionistic fuzzy set is an extension of fuzzy set.

For any intuitionistic fuzzy set $A = \{(x, \mu_A(x), v_A(x)) \mid x \in U\}$ and $B = \{(x, \mu_B(x), v_B(x)) \mid x \in U\}$, the operations of union ($\cup$) , joint ($\cap$) and complement (') are defined as follows :

$$A \cup B = \{(x, \max(\mu_A(x), \mu_B(x)), \min(v_A(x), v_B(x))) \mid x \in U\},$$
$$A \cap B = \{(x, \min(\mu_A(x), \mu_B(x)), \max(v_A(x), v_B(x))) \mid x \in U\},$$
$$A' = \{(x, v_A(x), \mu_A(x)) \mid x \in U\}.$$

All the intuitionistic fuzzy sets of U are denoted as $IFS(U)$, and the intuitionistic fuzzy sets have the following order relations : $\forall A, B \in IFS(U), A \leq B$ if and only if $\forall x \in U$, $\mu_A(x) \leq \mu_B(x)$ and $v_A(x) \geq v_B(x)$, naturally, $A = B$ if and only if $A \leq B$ and $B \leq A$.

3 10-Elements Linguistic Truth-Valued Intuitionistic Fuzzy Algebra

3.1 Linguistic Truth-Valued Intuitionistic Fuzzy Algebra Based on 10-Elements Linguistic Truth-Valued Lattice Implication Algebra

Definition 2. In the 10-element linguistic truth-valued lattice implication algebra $\mathcal{L}$ $_{v(5\times2)}=\{A,B,C,D,E,J,K,M,N,S\}$, for arbitrary $(h_i,t),(h_j,f)\in \mathcal{L}_{v(5\times2)}$, $((h_i,t),(h_j,f))$ is called an linguistic truth-valued intuitionistic fuzzy set if $((h_i,t),(h_j,f))$ satisfies $(h_i,t)'\geq(h_j,f)$, where the operation "'" is the negation of $\mathcal{L}_{v(5\times2)}$.

In the 10-element linguistic truth-valued lattice implication algebra $_{v(5\times2)}$ (Fig. 1), for (B,J), according to $x'=f^{-1}((f(x))')$ we get $B'=K$ in the Fig. 1, $K\geq J$, hence (B,J) is a linguistic truth-valued intuitionistic fuzzy set. Similarly, we can get that $(A,J),(B,K),(C,S),(D,M),(E,N),(C,K),(D,S),(E,M),(C,J),(D,K),\quad(E,S),(D,J),$ $(E,K),(E,J)$ are linguistic truth-valued intuitionistic fuzzy sets.

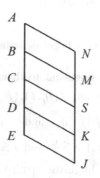

Fig. 1. The structure chart of $\mathcal{L}_{v(5\times2)}$

Theorem 1. For arbitrary $(h_i,t),(h_j,f)\in \mathcal{L}_{v(5\times2)}$, $((h_i,t),(h_j,f))$ is a linguistic truth-valued intuitionistic fuzzy set if and only if $i\leq j$.

Proof : For arbitrary $(h_i,t)\in \mathcal{L}_{v(5\times2)}$, $(h_i,t)'=(h_i,f)$. Hence, $(h_i,t)'\geq(h_j,f)$ if and only if $(h_i,f)\geq(h_j,f)$. In $\mathcal{L}_{v(5\times2)}$, $(h_i,f)\geq(h_j,f)$ if and only if $i\leq j$.

Corollary 1. For any $(h_i, t) \in \mathcal{L}_{v(5 \times 2)}$, the number of (h_i, t) which can compose the linguistic truth-valued intuitionistic fuzzy set with (h_j, f) is $5 - i + 1$. Hence in $\mathcal{L}_{v(5 \times 2)}$, there are

$$\sum_{i=1}^{5} (5 - i + 1) = \frac{5 \times 6}{2} = 15 .$$

We can get that the number of linguistic truth-valued intuitionistic fuzzy sets is 15, the structure chart of 10-element linguistic truth-valued intuitionistic fuzzy lattice is shown in the Fig. 2.

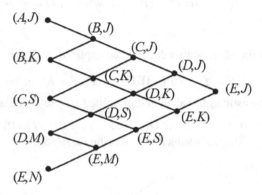

Fig. 2. The structure chart of 10-element linguistic truth-valued intuitionistic fuzzy lattice

Definition 3. In the 10-element linguistic truth-valued intuitionistic fuzzy lattice $LI_{10} = (VI_{10}, \cup, \cap)$ (Fig.2), for any $((h_i, t), (h_j, f)), ((h_k, t), (h_l, f)) \in LI_{10}$, $((h_i, t), (h_j, f)) \leq ((h_k, t), (h_l, f))$ if and only if $(h_i, t) \leq (h_k, t)$ and $(h_j, f) \geq (h_l, f)$, i.e. if and only if $i \leq k$ and $j \leq l$.

Definition 4. In the 10-element linguistic truth-valued intuitionistic fuzzy lattice $LI_{10} = (VI_{10}, \cup, \cap)$ (Fig.2), for any $((h_i, t), (h_j, f)), ((h_k, t), (h_l, f)) \in LI_{10}$, $((h_i, t), (h_j, f)) \leq ((h_k, t), (h_l, f))$ if and only if $i \leq k$ and $j \leq l$, then:

1) $((h_i, t), (h_j, f)) \to ((h_k, t), (h_l, f)) = ((h_{\min(5, 5-i+k, 5-j+l)}, t), (h_{\min(5, 5-i+l)}, f))$;

2) $((h_i, t), (h_j, f)) \cup ((h_k, t), (h_l, f)) = ((h_{\max(i,k)}, t), (h_{\max(j,l)}, f))$;

3) $((h_i, t), (h_j, f)) \cap ((h_k, t), (h_l, f)) = ((h_{\min(i,k)}, t), (h_{\min(j,l)}, f))$;

4) $((h_i, t), (h_j, f))' = ((h_{5-j+1}, t), (h_{5-i+1}, f))$.

Corollary 2. The linguistic truth-valued intuitionistic fuzzy lattice $LI_{10} = (VI_{10}, \cup, \cap)$ is a bounded distributive lattice (as shown in Fig.2), where

$((h_5,t),(h_5,f))=(A,J)$ and $((h_1,t),(h_1,f))=(E,N)$ are the maximum and minimum element of LI_{10} respectively.

In the following, for any $a,b \in LI_{10}$, the supremum of a and b is expressed in $\mathrm{Sup}\{a,b\}$, denoted by $a \cup b$; the infimum of a and b is expressed in $\mathrm{Inf}\{a,b\}$,denoted by $a \cap b$.

$$x_1 \cup x_2 \cup \dots \text{denoted as } \sum_i x_i ,$$

$$x_1 \cap x_2 \cap \dots \text{denoted as } \prod_i x_i .$$

The operations of 10-element linguistic truth-valued intuitionistic fuzzy lattice define as follows. (Table 1, 2, 3, 4)

Table 1. 10-element linguistic truth-valued intuitionistic fuzzy lattice implication operation

→	(A,J)	(B,K)	(C,S)	(D,M)	(E,N)	(B,J)	(C,K)	(D,S)	(E,M)	(C,J)	(D,K)	(E,S)	(D,J)	(E,K)	(E,J)
(A,J)	(A,J)	(B,K)	(C,S)	(D,M)	(E,N)	(B,J)	(C,K)	(D,S)	(E,M)	(C,J)	(D,K)	(E,S)	(D,J)	(E,K)	(E,J)
(B,K)	(A,J)	(A,J)	(B,K)	(C,S)	(D,M)	(B,J)	(B,J)	(C,K)	(D,S)	(B,J)	(C,J)	(D,K)	(C,J)	(D,J)	(D,J)
(C,S)	(A,J)	(A,J)	(A,J)	(B,K)	(C,S)	(A,J)	(A,J)	(B,J)	(C,K)	(A,J)	(B,J)	(C,J)	(B,J)	(C,J)	(C,J)
(D,M)	(A,J)	(A,J)	(A,J)	(A,J)	(B,K)	(A,J)	(A,J)	(A,J)	(B,J)	(A,J)	(A,J)	(B,J)	(A,J)	(B,J)	(B,J)
(E,N)	(A,J)	(A,J)	(A,J)	(A,J)	(A,J)	(A,J)	(A,J)	(A,J)	(A,J)	(A,J)	(A,J)	(A,J)	(A,J)	(A,J)	(A,J)
(B,J)	(A,J)	(B,J)	(C,K)	(D,S)	(E,M)	(A,J)	(B,J)	(C,K)	(D,S)	(B,J)	(C,J)	(D,K)	(C,J)	(D,J)	(D,J)
(C,K)	(A,J)	(A,J)	(B,J)	(C,K)	(D,S)	(A,J)	(A,J)	(B,J)	(C,K)	(A,J)	(B,J)	(C,J)	(B,J)	(C,J)	(C,J)
(D,S)	(A,J)	(A,J)	(A,J)	(B,J)	(C,K)	(A,J)	(A,J)	(A,J)	(B,J)	(A,J)	(A,J)	(B,J)	(A,J)	(B,J)	(B,J)
(E,M)	(A,J)	(A,J)	(A,J)	(A,J)	(B,J)	(A,J)	(A,J)	(A,J)	(A,J)	(A,J)	(A,J)	(A,J)	(A,J)	(A,J)	(A,J)
(C,J)	(A,J)	(B,J)	(C,J)	(D,K)	(E,S)	(A,J)	(B,J)	(C,J)	(D,K)	(A,J)	(B,J)	(C,J)	(B,J)	(C,J)	(C,J)
(D,K)	(A,J)	(A,J)	(B,J)	(C,J)	(D,K)	(A,J)	(A,J)	(B,J)	(C,J)	(A,J)	(A,J)	(B,J)	(A,J)	(B,J)	(B,J)
(E,S)	(A,J)	(A,J)	(A,J)	(B,J)	(C,J)	(A,J)	(A,J)	(A,J)	(B,J)	(A,J)	(A,J)	(A,J)	(A,J)	(A,J)	(A,J)
(D,J)	(A,J)	(B,J)	(C,J)	(D,J)	(E,K)	(A,J)	(B,J)	(C,J)	(D,J)	(A,J)	(B,J)	(C,J)	(A,J)	(B,J)	(B,J)
(E,K)	(A,J)	(A,J)	(B,J)	(C,J)	(D,J)	(A,J)	(A,J)	(B,J)	(C,J)	(A,J)	(A,J)	(B,J)	(A,J)	(A,J)	(A,J)
(E,J)	(A,J)	(B,J)	(C,J)	(D,J)	(E,J)	(A,J)	(B,J)	(C,J)	(D,J)	(A,J)	(B,J)	(C,J)	(A,J)	(B,J)	(A,J)

Table 2. 10-element linguistic truth-valued intuitionistic fuzzy lattice disjunction operation

∨	(A,J)	(B,K)	(C,S)	(D,M)	(E,N)	(B,J)	(C,K)	(D,S)	(E,M)	(C,J)	(D,K)	(E,S)	(D,J)	(E,K)	(E,J)
(A,J)	(A,J)	(A,J)	(A,J)	(A,J)	(A,J)	(A,J)	(A,J)	(A,J)	(A,J)	(A,J)	(A,J)	(A,J)	(A,J)	(A,J)	(A,J)
(B,K)	(A,J)	(B,K)	(B,K)	(B,K)	(B,K)	(B,J)	(B,K)	(B,K)	(B,K)	(B,J)	(B,K)	(B,K)	(B,J)	(B,K)	(B,J)
(C,S)	(A,J)	(B,K)	(C,S)	(C,S)	(C,S)	(B,J)	(C,K)	(C,S)	(C,S)	(C,J)	(C,K)	(C,S)	(C,J)	(C,K)	(C,J)
(D,M)	(A,J)	(B,K)	(C,S)	(D,M)	(D,M)	(B,J)	(C,K)	(D,S)	(D,M)	(C,J)	(D,K)	(D,S)	(D,J)	(D,K)	(D,J)
(E,N)	(A,J)	(B,K)	(C,S)	(D,M)	(E,N)	(B,J)	(C,K)	(D,S)	(E,M)	(C,J)	(D,K)	(E,S)	(D,J)	(E,K)	(E,J)
(B,J)	(A,J)	(B,J)	(B,J)	(B,J)	(B,J)	(B,J)	(B,J)	(B,J)	(B,J)	(B,J)	(B,J)	(B,J)	(B,J)	(B,J)	(B,J)
(C,K)	(A,J)	(B,K)	(C,K)	(C,K)	(C,K)	(B,J)	(C,K)	(C,K)	(C,K)	(C,J)	(C,K)	(C,K)	(C,J)	(C,K)	(C,J)
(D,S)	(A,J)	(B,K)	(C,S)	(D,S)	(D,S)	(B,J)	(C,K)	(D,S)	(D,S)	(C,J)	(D,K)	(D,S)	(D,J)	(D,K)	(D,J)
(E,M)	(A,J)	(B,K)	(C,S)	(D,M)	(E,M)	(B,J)	(C,K)	(D,S)	(E,M)	(C,J)	(D,K)	(E,S)	(D,J)	(E,K)	(E,J)

Table 2. (*Continued*)

(C,J)	(A,J)	(B,J)	(C,J)	(C,J)	(C,J)	(B,J)	(C,J)	(C,J)	(C,J)	(C,J)	(C,J)	(C,J)	(C,J)	(C,J)	(C,J)
(D,K)	(A,J)	(B,K)	(C,K)	(D,K)	(D,K)	(B,J)	(C,K)	(D,K)	(D,K)	(C,J)	(D,K)	(D,K)	(D,J)	(D,K)	(D,J)
(E,S)	(A,J)	(B,K)	(C,S)	(D,S)	(E,S)	(B,J)	(C,K)	(D,S)	(E,S)	(C,J)	(D,K)	(E,S)	(D,J)	(E,K)	(E,J)
(D,J)	(A,J)	(B,J)	(C,J)	(D,J)	(D,J)	(B,J)	(C,J)	(D,J)	(D,J)	(C,J)	(D,J)	(D,J)	(D,J)	(D,J)	(D,J)
(E,K)	(A,J)	(B,K)	(C,K)	(D,K)	(E,K)	(B,J)	(C,K)	(D,K)	(E,K)	(C,J)	(D,K)	(E,K)	(D,J)	(E,K)	(E,J)
(E,J)	(A,J)	(B,J)	(C,J)	(D,J)	(E,J)	(B,J)	(C,J)	(D,J)	(E,J)	(C,J)	(D,J)	(E,J)	(D,J)	(E,J)	(E,J)

Table 3. 10-element linguistic truth-valued intuitionistic fuzzy lattice conjunction operation

∧	(A,J)	(B,K)	(C,S)	(D,M)	(E,N)	(B,J)	(C,K)	(D,S)	(E,M)	(C,J)	(D,K)	(E,S)	(D,J)	(E,K)	(E,J)
(A,J)	(A,J)	(B,K)	(C,S)	(D,M)	(E,N)	(B,J)	(C,K)	(D,S)	(E,M)	(C,J)	(D,K)	(E,S)	(D,J)	(E,K)	(E,J)
(B,K)	(B,K)	(B,K)	(C,S)	(D,M)	(E,N)	(B,K)	(C,K)	(D,S)	(E,M)	(C,K)	(D,K)	(E,S)	(D,K)	(E,K)	(E,K)
(C,S)	(C,S)	(C,S)	(C,S)	(D,M)	(E,N)	(C,S)	(C,S)	(D,S)	(E,M)	(C,S)	(D,S)	(E,S)	(D,S)	(E,S)	(E,S)
(D,M)	(D,M)	(D,M)	(D,M)	(D,M)	(E,N)	(D,M)	(D,M)	(D,M)	(E,M)	(D,M)	(D,M)	(E,M)	(D,M)	(E,M)	(E,M)
(E,N)	(E,N)	(E,N)	(E,N)	(E,N)	(E,N)	(E,N)	(E,N)	(E,N)	(E,N)	(E,N)	(E,N)	(E,N)	(E,N)	(E,N)	(E,N)
(B,J)	(B,J)	(B,K)	(C,S)	(D,M)	(E,N)	(B,J)	(C,K)	(D,S)	(E,M)	(C,J)	(D,K)	(E,S)	(D,J)	(E,K)	(E,J)
(C,K)	(C,K)	(C,K)	(C,S)	(D,M)	(E,N)	(C,K)	(C,K)	(D,S)	(E,M)	(C,K)	(D,K)	(E,S)	(D,K)	(E,K)	(E,K)
(D,S)	(D,S)	(D,S)	(D,S)	(D,M)	(E,N)	(D,S)	(D,S)	(D,S)	(E,M)	(D,S)	(D,S)	(E,S)	(D,S)	(E,S)	(E,S)
(E,M)	(E,M)	(E,M)	(E,M)	(E,M)	(E,N)	(E,M)	(E,M)	(E,M)	(E,M)	(E,M)	(E,M)	(E,M)	(E,M)	(E,M)	(E,M)
(C,J)	(C,J)	(C,K)	(C,S)	(D,M)	(E,N)	(C,J)	(C,K)	(D,S)	(E,M)	(C,J)	(D,K)	(E,S)	(D,J)	(E,K)	(E,J)
(D,K)	(D,K)	(D,K)	(D,S)	(D,M)	(E,N)	(D,K)	(D,K)	(D,S)	(E,M)	(D,K)	(D,K)	(E,S)	(D,K)	(E,K)	(E,K)
(E,S)	(E,S)	(E,S)	(E,S)	(E,M)	(E,N)	(E,S)	(E,S)	(E,S)	(E,M)	(E,S)	(E,S)	(E,S)	(E,S)	(E,S)	(E,S)
(D,J)	(D,J)	(D,K)	(D,S)	(D,M)	(E,N)	(D,J)	(D,K)	(D,S)	(E,M)	(D,J)	(D,K)	(E,S)	(D,J)	(E,K)	(E,J)
(E,K)	(E,K)	(E,K)	(E,S)	(E,M)	(E,N)	(E,K)	(E,K)	(E,S)	(E,M)	(E,K)	(E,K)	(E,S)	(E,K)	(E,K)	(E,K)
(E,J)	(E,J)	(E,K)	(E,S)	(E,M)	(E,N)	(E,J)	(E,K)	(E,S)	(E,M)	(E,J)	(E,K)	(E,S)	(E,J)	(E,K)	(E,J)

Table 4. 10-element linguistic truth-valued intuitionistic fuzzy lattice inverse operation

'	(A,J)	(B,K)	(C,S)	(D,M)	(E,N)	(B,J)	(C,K)	(D,S)	(E,M)	(C,J)	(D,K)	(E,S)	(D,J)	(E,K)	(E,J)
	(E,N)	(D,M)	(C,S)	(B,K)	(A,J)	(E,M)	(D,S)	(C,K)	(B,J)	(E,S)	(D,K)	(C,J)	(E,K)	(D,J)	(E,J)

3.2   The Properties of 10-Elements Linguistic Truth-valued Intuitionistic Fuzzy Lattice LI_{10}

In the 10-element linguistic truth-valued intuitionistic fuzzy lattice $LI_{10} = (VI_{10}, \cup, \cap)$, from the last section, it is easy to get the following conclusion.

Theorem 2. For any $((h_i, t), (h_j, f)), ((h_k, t), (h_l, f)) \in LI_{10}$,

1. $((h_1, t), (h_1, f)) \rightarrow ((h_k, t), (h_l, f)) = ((h_5, t), (h_5, f))$;

2. $((h_5, t), (h_5, f)) \rightarrow ((h_k, t), (h_l, f)) = ((h_k, t), (h_l, f))$;

3. $(((h_i, t), (h_j, f)) \cap ((h_k, t), (h_l, f)))' = ((h_i, t), (h_j, f))' \cup ((h_k, t), (h_l, f))'$;

4. $(((h_i, t), (h_j, f)) \cup ((h_k, t), (h_l, f)))' = ((h_i, t), (h_j, f))' \cap ((h_k, t), (h_l, f))'$.

Proof : 1. It follows from Definition 4 (1) that

$$((h_1,t),(h_1,f)) \to ((h_k,t),(h_l,f))$$
$$= ((h_{\min(5,5-1+k,5-1+l)},t),(h_{\min(5,5-1+l)},f))$$
$$= ((h_5,t),(h_5,f)).$$

2.
$$((h_5,t),(h_5,f)) \to ((h_k,t),(h_l,f))$$
$$= ((h_{\min(5,5-5+k,5-5+l)},t),(h_{\min(5,5-5+l)},f))$$
$$= ((h_k,t),(h_l,f)).$$

3.
$$(((h_i,t),(h_j,f)) \cap ((h_k,t),(h_l,f)))'$$
$$= ((h_{\min(i,k)},t),(h_{\min(j,l)},f))'$$
$$= (h_{5-\min(j,l)+1},t),(h_{5-\min(i,k)+1},f)$$
$$= (h_{\max(5-j+1,5-l+1)},t),(h_{\max(5-i+1,5-k+1)},f)$$
$$= ((h_i,t),(h_j,f))' \cup ((h_k,t),(h_l,f))'.$$

4. It is similar to the proof of 3.

Corollary   3.   For   any $((h_i,t),(h_j,f))$ $\in LI_{10}$,
$((h_i,t),(h_j,f))'' = ((h_i,t),(h_j,f)).$

Proof :
$$((h_i,t),(h_j,f))''$$
$$= (((h_i,t),(h_j,f))')'$$
$$= ((h_{5-j+1},t),(h_{5-i+1},f))'$$
$$= ((h_{5-(5-i+1)+1},t),(h_{5-(5-j+1)+1},f))$$
$$= ((h_i,t),(h_j,f)).$$

Corollary 4. For any $((h_i,t),(h_j,f))$, $((h_k,t),(h_l,f))$, $((h_m,t),(h_s,f)) \in LI_{10}$,

1. $(((h_i,t),(h_j,f)) \cup ((h_k,t),(h_l,f))) \to ((h_m,t),(h_s,f))$
$= (((h_i,t),(h_j,f)) \to ((h_m,t),(h_s,f))) \cap (((h_k,t),(h_l,f)) \to ((h_m,t),(h_s,f)))$;

2. $((h_m,t),(h_s,f)) \to (((h_i,t),(h_j,f)) \cup ((h_k,t),(h_l,f)))$
$= (((h_m,t),(h_s,f)) \to ((h_i,t),(h_j,f))) \cup (((h_m,t),(h_s,f)) \to ((h_k,t),(h_l,f)))$;

3. $(((h_i,t),(h_j,f)) \cap ((h_k,t),(h_l,f))) \to ((h_m,t),(h_s,f))$
$= (((h_i,t),(h_j,f)) \to ((h_m,t),(h_s,f))) \cup (((h_k,t),(h_l,f)) \to ((h_m,t),(h_s,f)))$;

4. $((h_m,t),(h_s,f)) \to (((h_i,t),(h_j,f)) \cap ((h_k,t),(h_l,f)))$
$= (((h_m,t),(h_s,f)) \to ((h_i,t),(h_j,f))) \cap (((h_m,t),(h_s,f)) \to ((h_k,t),(h_l,f)))$.

Proof : $1.(((h_i,t),(h_j,f)) \cup ((h_k,t),(h_l,f))) \rightarrow ((h_m,t),(h_s,f))$

$= ((h_{\max(i,k)},t),(h_{\max(j,l)},f)) \rightarrow ((h_m,t),(h_s,f))$

$= ((h_{\min(5,5-\max(i,k)+m,5-\max(j,l)+s)},t),(h_{\min(5,5-\max(i,k)+s)},f))$

$= ((h_{\min(5,\min(5-i+m,5-k+m),\min(5-j+s,5-l+s))},t),(h_{\min(5,\min(5-i+s,5-k+s))},f))$

$= ((h_{\min(5,5-i+m,5-j+s)},t),(h_{\min(5,5-i+s)},f)) \cap ((h_{\min(5,5-k+m,5-l+s)},t),(h_{\min(5,5-k+s)},f))$

$= (((h_i,t),(h_j,f)) \rightarrow ((h_m,t),(h_s,f))) \cap (((h_k,t),(h_l,f)) \rightarrow ((h_m,t),(h_s,f)))$.

2.3.4   The Proof is Similar to the Proof of 1

Corollary 5. For any $((h_i,t),(h_j,f))$, $((h_k,t),(h_l,f))$, $((h_m,t),(h_s,f)) \in LI_{10}$,

$1.((h_i,t),(h_j,f)) \cup (((h_k,t),(h_l,f)) \cap ((h_m,t),(h_s,f)))$

$= (((h_i,t),(h_j,f)) \cup ((h_k,t),(h_l,f))) \cap (((h_i,t),(h_j,f)) \cup ((h_m,t),(h_s,f)))$

$2.((h_i,t),(h_j,f)) \cap (((h_k,t),(h_l,f)) \cup ((h_m,t),(h_s,f)))$

$= (((h_i,t),(h_j,f)) \cap ((h_k,t),(h_l,f))) \cup (((h_i,t),(h_j,f)) \cap ((h_m,t),(h_s,f)))$

Proof : $1.((h_i,t),(h_j,f)) \cup (((h_k,t),(h_l,f)) \cap ((h_m,t),(h_s,f)))$

$= ((h_i,t),(h_j,f)) \cup ((h_{\min(k,m)},t),(h_{\min(l,s)},f))$

$= ((h_{\max(i,\min(k,m))},t),(h_{\max(j,\min(l,s))},f))$

$= ((h_{\min(\max(i,k),\max(i.m))},t),(h_{\min(\max(j,l),\max(j,s))},f))$

$= ((h_{\max(i,k)},t),(h_{\max(j,l)},f)) \cap ((h_{\max(i,m)},t),(h_{\max(j,s)},f))$

$= (((h_i,t),(h_j,f)) \cup ((h_k,t),(h_l,f))) \cap (((h_i,t),(h_j,f)) \cup ((h_m,t),(h_s,f)))$

3.It is similar to the proof of 1.

4   Resolution Principle of 10-Elements Linguistic Truth-Valued Intuitionistic Fuzzy First-Order Logic System

4.1   10-Elements Linguistic Truth-Valued Intuitionistic Fuzzy First-Order Logic System

Definition 5. Let D be a nonempty domain. The n-place function which is defined in D^n, and the value belongs to LI_{10} is called n-place linguistic truth-valued intuitionistic fuzzy propositional function, where D^n stands for n-time cartesian product.

(Note that "n-place" stands for the order of predicate logic, when $n=1$, it is for first-order logic).

Definition 6 $(\forall x)$. The phrase "for any elements x" is said to be universal quantifier, denoted by $(\forall x)$. The phrase "there is an element x" is said to be existential quantifier, denoted by $(\exists x)$.

Let $G(x)$ be the linguistic truth-valued intuitionistic fuzzy first-order predicate, for any $x_0 \in D$, then $G(x_0)$ is said to be linguistic truth-valued intuitionistic fuzzy proposition. Hence, $(\forall x)G(x)$ means "For $x \in D$, there is $G(x)$..." Define the true-value of $(\forall x)G(x)$ as follows :

The true-value of $(\forall x)G(x)$ is $(A, J) \Leftrightarrow$ For any element $x \in D$, the true-value of $G(x)$ is (A, J).

The true-value of $(\forall x)G(x)$ is $(C, S) \Leftrightarrow$ There exists an element $x_0 \in D$ that the true-value of $G(x_0)$ is (C, S).

Similarly, $(\exists x)G(x)$ means that"There exists an element $x_0 \in D$ that $G(x_0)$ ...". Define the true-value of $(\exists x)G(x)$ as follows:

The true-value of $(\exists x)G(x)$ is $(A, J) \Leftrightarrow$ There exists an element $x_0 \in D$ that the true-value of $G(x_0)$ is (A, J).

The true-value of $(\exists x)G(x)$ is $(C, S) \Leftrightarrow$ For any element $x \in D$, the true-value of $G(x)$ is (C, S).

Definition 7. 10-element linguistic truth-valued intuitionistic fuzzy first- order language consists of the following:

(1)Individual constants: Denoted by $a, b, c, \ldots$. When the set D is given, it can be an element in D.

(2)Variables: Denoted by the lowercase letters $u, v, w, x, y, z, \ldots$. When the set D is given, any elements in D can be substituted in the variables.

(3)Function symbols: Denoted by the lowercase letters $f(x_1, x_2, \ldots, x_n)$, $g(y_1, y_2, \ldots, y_m)$, ... When the set D is given, it can be a mapping from D^n to D.

(4)Predicate symbols: Denoted by the capital letter $P, Q, R, \ldots$. When the set D is given, it can be ant predicates defined in D^n.

Definition 8. The linguistic truth-valued intuitionistic fuzzy lattice $LI_{10} = (VI_{10}, \cup, \cap)$ is a bounded distributive lattice. G is a formula of the first-order logic. If the true-values of the atoms in G belong to VI_{10}, then G is called a formula of 10-element linguistic truth-valued intuitionistic fuzzy first- order logic.

Definition 9. An interpretation of a formula G in 10-element linguistic truth-valued intuitionistic fuzzy first-order logic consists of a nonempty domain D, and an assignment of "values" to each constant, function symbol, and predicate symbol occurring in G as follows :

(1)To each constant, we assign an element in D.

(2)To each n-place function symbol, we assign a mapping from D^n to D.

  (Note that $D^n = \{(x_1,...,x_n) \mid x_1 \in D,...,x_n \in D\}$).

(3)To each n-place predicate symbol, we assign a mapping from D^n to VI_{10}.

5   Conclusion

Linguistic truth-valued intuitionistic fuzzy algebra based on linguistic truth-valued lattice implication algebra has become the basis of linguistic truth-valued intuitionistic fuzzy logic. Firstly, this paper establishes the 10-element linguistic truth-valued intuitionistic fuzzy algebra, and its related properties are obtained. On the basis of it, we set up the 10-element linguistic truth-valued intuitionistic fuzzy first-order logic, in which the value of proposition is no longer a simple number but the linguistic truth-valued which can express both comparable and incomparable information. This method can also deal with the uncertain problem which has both positive evidence and negative evidence at the same time. Information expressed in this way is more flexible and more closed to the thought of human. The further work is the reasoning method of this system, and applying the method into the field of decision making, risk analysis and expert system.

Acknowledgments. This work is partly supported by National Nature Science Foundation of China (Grant No.61105059,61175055,61173100), International Cooperation and Exchange of the National Natural Science Foundation of China (Grant No.61210306079), Sichuan Key Technology Research and Development Program (Grant No.2011FZ0051), Radio Administration Bureau of MIIT of China (Grant No.[2011]146), China Institution of Communications (Grant No.[2011]051), and Sichuan Key Laboratory of Intelligent Network Information Processing (Grant No.SGXZD1002-10),Liaoning Excellent Talents in University (LJQ2011116).

References

1. Zadeh, L.A.: Fuzzy sets. Information and Control. 8(3), 338–353 (1965)
2. Atanassov, K.: Intuionistic fuzzy sets. Fuzzy Sets and Systems. 20(1), 87–96 (1986)
3. Atanassov, K.: More on intuionistic fuzzy sets. Fuzzy Sets and Systems. 33(1), 37–45 (1989)
4. Xu, Y., Ruan, D., Qin, K.Y., Liu, J.: Lattice-valued logic. Springer (2003)
5. Meng, D., Xu, Y., Jia, H.D.: Resolution Principle Based on Six Lattice-Valued First-Order Logic L6F(X). In: Proceedings of 2005 IEEE Networking, Sensing and Control Conference, pp. 838–843 (2005)

6. Meng, D., Song, Z.M., Qin, K.Y.: Algorithm of transforming any formula in LF(X) into reducible form. Journal of Southwest Jiaotong University (Natural Science Edition) 38(4), 433–437 (2003)
7. Xu, Y., Liu, J., Ruan, D., Lee, T.T.: On the consistency of rule bases based on lattice-valued first-order logic LF(X). International Journal of Intelligent Systems 21, 399–424 (2006)
8. He, X.X., Xu, Y., Liu, J.: a-Input and a-unit resolution methods for generalized horn clause set. Journal of Donghua University (Eng. Ed.) 29(1), 66–70 (2012)
9. Zheng, H.L., Xu, B.Q., Zou, L.: An approach for knowledge representation based on ten-element lattice implication algebra. Computer Applications and Software. 30(1), 37–39 (2003)
10. Xu, W.T., Xu, Y.: Ideal resolution principle for lattice-valued first-order logic LF(X) based on lattice implication algebra. Journal of Shanghai Jiaotong University (Science) 17(2), 178–181 (2012)
11. Lai, J.J., Xu, Y.: Uncertainty reasoning of two models based on linguistic truth-valued lattice value first-order logic system Lv(n×2)F(X). Fuzzy Systems and Mathematics 27(4), 28–35 (2013)
12. Xu, W.T., Zhang, W.Q., Xu, Y., Zhang, D.X.: α-generalized resolution principle based on the lattice-valued first-order logic system. Journal of Xidian University 41(1), 168–173 (2014)

The Research of Heartbeat Detection Technique for Blade Server

Weiwei Jiang, Naikuo Chen, and Shihua Geng

Inspur Chaoyue Digital Control Electronics Co., Ltd. Jinan, 250104
jiangweiwei609@163.com

Abstract. The blade servers have been widely used in the telecommunications, financial and other big data processing fields as for the high efficiency, stability and autonomy. This study takes the hot redundancy design for dual-BMC management of blade servers as the research project, and puts forward a heartbeat detection program utilizing I2C for transmission of IPMI commands. And it's successfully applied to the blade servers to achieve a hot redundancy, monitor and management of master/slave BMC management module, which is more standardized, reliable, and easy to implement.

Keywords: BMC, Big data, Blade server, Heartbeat Detection, IPMI.

1 Introduction

The blade server is designed for special application industries and big data computing environment. Due to the high efficiency, stability and autonomy, it has been widely used in many big data processing fields, which involved in national security and people's livelihood such as telecommunications, financial and so on[1]. The BMC management module is the core of the blade server. Generally, the hot redundant dual-BMC management is acquired to the monitor, management, configure and diagnosis of blade server system. For the requirement of the cluster applications and distributed systems, the BMC heartbeat detection mechanism has also become one of the important mechanisms for the implementation of the serves. The aim of this study is to develop a platform of heartbeat detection.

This paper presents a heartbeat detection program utilizing I2C for transmission of IPMI commands. It has been successfully applied to the blade servers, and achieved the hot redundancy, monitoring and management between master and slave BMC management module, which is more reliable and easily.

2 Heartbeat Detection Technology

The heartbeat detection is a common technology for network fault detection in the servers to judge if the network connection is normal based on the messages (that is heartbeat packets) between the heartbeat detection software of server-side and client-side[2,3].

H. Wang et al. (Eds.): ICYCSEE 2015, CCIS 503, pp. 418–424, 2015.

The servers as the master node and the client as the peripheral node, the heartbeat detection is realized by the heartbeat packets sent by the client.

The principle is as follows, the peripheral nodes send heartbeat packets to the master node regularly, and the master nodes check if they have received the packets regularly. The period of the master nodes is longer than that of peripheral nodes, the master nodes will consider it is troubleshooting network connection and report to manager if they do not receive the packets from the responding nodes.

3   Hot Redundancy Hardware Architecture of Dual-BMC Management

The management module, which is the core part of the blade sever, mainly is the ARM and IPMI protocol [4]. It can get the health information from boards through I2C and GPIO execute relative commands. The server management board is equipped with one BMC module and the other units are equipped with independent BMC modules. Firstly, the BMC module of management board distributes independent IP addresses for the BMCs on other units by I2C, and interacts with BMC by IPMI Command or OEM Command through the network. It obtains information from each unit and make corresponding updates to the SDR periodically, or reset according to de command. The switching could be operated by GPIO to achieve remote control of VGA and USB when the long-range channel switching is detected.

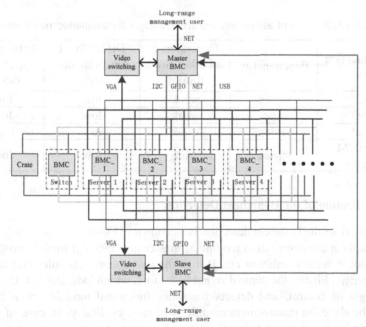

Fig. 1. Hot redundancy hardware architecture of dual-BMC management for blade server

4   Implementation of Heartbeat Detection Technology under Blade Sever Platform

4.1   Development Environment under Blade Sever Platform

Under the platform, the management software is developed by MDS tool. The MDS is one kind of development tools installed under Linux having good interactive interface. In the development environment of MDS, it will convert all the modules into the rom.ima file which needs download, including all the contents of the uboot, kernel and file system. The management software works when burning the rom.ima file into the flash chip.

4.2   Selection of Communication Mode

Initially, two communication modes of I2C and Ethernet are identified based on the above analysis of the characteristics. However, the structure of the whole management network for the blade sever is relatively complex, and many IP address resources are occupied. Both of the IP addresses for the master and slave management modules have to be clear and definite, when using the Ethernet mode, which depends on the stability of switch. The restrictions and dependent conditions are many, which is not conducive to improve the stability of the whole management system. Ultimately, the more simple, mature and stable communication mode of I2C is determined.

Table 1. Analysis of advantages and disadvantages for communication mode

Communication mode	Functionality	Dependence of other conditions	Difficulty of hardware design	Difficulty of software design
I^{2}C	strong	weak	low	low
GPIO	weak	weak	low	low
Serial	strong	weak	low	high
10M/100M Ethernet	strong	strong	low	low

4.3   Realization of the Heartbeat Detection

The flow chart of the heartbeat detection is as shown in Figure 2.

The heartbeat communication module of master management module program detects the slot is main whether or not. If it is, the management module enter the work mode of master. Firstly, the control signal is set to 1, which indicates that the module has the right of control, and detects the other functional modules; then check in whether the slave the management module is present cyclically. In case of yes, the master send heartbeat request signal to it and waits to receive the response of the slave one.

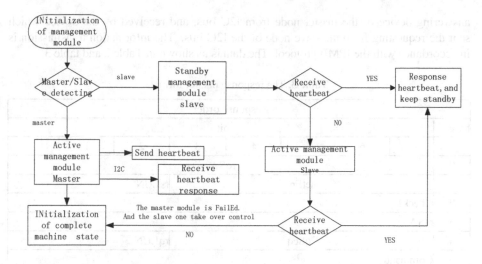

Fig. 2. Flow chart of the heartbeat detection

Table 2. The request signal packet format

byte	Request order							
	bit							
	7	6	5	4	3	2	1	0
0	rsSA							
1	netFn				rsLUN			
2	Checksum							
3	rqSA							
4	rqSeq				rqLUN			
5	Command							
6	Data							
7	Checksum							

Conversely, if the heartbeat communication module of the management module detect the slot is slave, the management module enter the work mode of slave. Firstly, the control signal is set to 0, which indicates that the module has not the right of control, and waiting for detecting of the other functional modules. Waiting for the heartbeat request signal sent by the master module. When the slave one receives the signal, responds to the master one. If it can not receive the request signal within10 consecutive cycles, the slave module considers the failure of master module, and the control signal of it is set to 1, and performs the switching of control right.

The transmission of IPMI command is based on the I2C bus, using the 'request- response' protocol. The transmission of request and response packets is through the I2C bus in "writing" mode. The request message is sent from the main node of I2C, and received by the slave node of I2C bus. Accordingly, the response is sent from the answering device of the master node from I2C bus, and received by the device which sent the requesting from the slave node of the I2C bus. The information transmission is in accordance with the IPMB protocol. The data is as shown in Table 2 and table 3.

Table 3. The response packet format

byte	Response order							
	bit							
	7	6	5	4	3	2	1	0
0	rqSA							
1	netFn					rsLUN		
2	Check1							
3	rsSA							
4	RqSeq					rqLUN		
5	Command							
6	Completion Code							
7-N	Data							
N+1	Check2							

5   Test and Validation

Apply this technique to the blade server (as shown is Figure 3), which is a typical big data processing device. By using RISC (Reduced Instruction Set Computer) construction , 32 bits Microcontroller takes on low- loss and low- consume (VDDCORE supply voltage is 1.8V), high capability, high executing speed (the unit is MIPS).

Fig. 3. The blade server

5.1   Slave Management Mode

The management module mode is forced to slave mode, and we send packets in accordance with the heartbeat request signal format via I2C analyzer. Then we can receive the response signal, and the data is in line with the definition of the content of the program, which denotes the slave model of the heartbeat communication module is normal. The debug message is as shown in Figure 4.

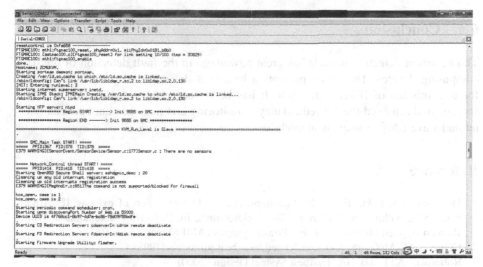

Fig. 4. Debug message of slave management mode

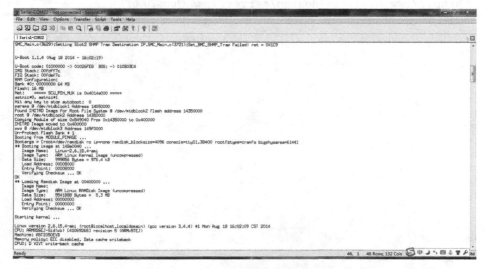

Fig. 5. Debug message of master management mode

5.2   Master Management Mode

The management module mode is forced to master mode, and we send packets in accordance with the heartbeat response signal format via I2C analyzer. Then we can see the debug messages printed in the debug interface (serial ports or SSH, as shown is Figure 5), which keeps according with the definition of the program, denoting the master model of the heartbeat communication module is normal.

6   Conclusion

The heartbeat detection module has great advantage in the fault detection of big data computing devices. This study presents a heartbeat detection program utilizing I2C for transmission of IPMI commands. It has been successfully applied to the blade servers, and achieved the hot redundancy, monitoring and management between master and slave BMC management module.

Reference

1. De Dominicis, C.M., Ferrar, P., Flammini, A., et al.: Integration of existing IEC61850-based SAS within new high-availability architectures. In: 2010 IEEE International Workshop on Applied Measurements For Power Systems (AMPS), vol. 12 (2010)
2. GGouda, M., McGuire, T.M.: Acclerated heartbeat protocols (1998)
3. Robertson, A.: Linux-HA Heartbeat System Design (2000)
4. IPMI Specification Second Generation v2.0.
   http://www.mtel.com/design/server/ipmi (June 2006)

The Monitoring with Video and Alarm System Based on Linux Router

Qiqing Huang, Yan Chen*, Honge Ren, and Zhaowen Qiu

Institute of Information and Computer Engineering, Northeast Forestry University, China
cheny0451@126.com

Abstract. As one of important safeguarded measures in security field, the video monitoring system have been applied in extensive ways such as building's security, information's acquisition and medical treatment. In current time, the intelligent home systems as well as the Internet of Things are forming a more and more mature industrial chain. With abundant of new technology applications, this industrial chain is developing towards to a more intelligent way. One of the most important applications is the video monitoring system developed by new technology. Deploy an appropriate design of video monitoring system in the several fields such as public security, traffic schedule, systems control and so on, is a major technological medium which can guarantee not wasting an amount of manpower resource as well as keeping the system in a good running status. This article based on the preponderance of the Linux system, especially the character of open source of it. Combine the Linux system and some particular routers; designing a real-time and efficient video monitoring system. It's functional and in the meantime it can discriminate abnormal information or the situation automatically and then raise the alarm. This system use matured B/S mode in software's framework, which is also lightweight class. In addition, this system manages the expenses in hardware cost more commendably.

Keywords: video monitoring, alarm, B/S framework.

1   Introduction

Since the conception of the Internet of Things technology was formally put forward by the International Telecommunication Union (ITU) in the year 2005, the technology of the Internet of Things has been acknowledged considered as the third revolution of the information industry following in the field of computers and internet with their high development. The internet of Things' industry started relatively early in our country, the accumulation of the technology has reached a relatively altitude---We can formulate the standard of the internet of Things along with the first echelon of countries and districts such as America and European Union [1].

The internet of Things uses Internet communication as its core, depend on such information's transducer and communication equipments as primary detector, radio frequency(RFID), infrared sensors, intellect IC card, GPS system, wireless communication

---

*Corresponding author.

H. Wang et al. (Eds.): ICYCSEE 2015, CCIS 503, pp. 425–432, 2015.
© Springer-Verlag Berlin Heidelberg 2015

device etc., using the agreement appointed, it can realize monitor and manage the objects that have been switched in the certain Internet. As one of the most typical as well as the most mature application of the internet of Things, the video monitoring system has been deployed in numerous fields and which can be proved that it really makes a great importance. This kind of the monitoring system is a great example of the organic combination which combine the optical sensor (camera) and the technology of radio communication.

Nowadays the major part of professional video monitoring system is mostly based on complex hard structure and C/S scheme, which also means a high expense on both using and maintaining, what's worse it is also suggest itself is not a beat choice for every situation, and what's more, great majority of them do not possess the function that identify the exception information which it collects and then set an alarm. So this kind of video monitoring system loses some usages in which   such as household and somewhere not enough in man power, both of these places actually like "nobody guarded" lacing of man power. Therefore this video monitoring system has disadvantages in spreading to more extensive throng. So because of that problems, developing a brand new video monitoring system which is both costs less and not seemed so complex, is exactly meets the need of numerous consumers.

2   The Design Scheme of the System

According to the design of the scheme, and combing with the detailed demand analysis. This new system is designed to aim at the object like household consumer, small-sized enterprise or organization. And this system also can involve with collecting and monitoring the information of some dangerous circumstances (example like isolated ward, storage room used for irradiation material). These kinds of using circumstance are not allowed people for monitoring and managing in a long time, which demands this system need to cohesive the promptly transmission of visual frequency used Internet communication and automatically judge the potential signal using algorithm, and in the meantime, these information which produced by this system should be sent to the end of sink timely. The design of the product must be market-oriented, so it needs a sufficient consideration in the problem of controlling the expense on costs. We make a research on this suitable design, and here is the final scheme:

(1) Optical camera: choose the TOSHIBA 720p supported the MJPG form with a USB interface as transducer part and link to the router;

(2) Employ the HUAWEI HG255D wireless router, which has 64M RAM. Insert the suitable Open Wrt system based on Linux kernel into the router [2];

(3) Use the wireless way to transfer the video between optical camera and the Internet, which means it won't be affected by the arrangements of traditional wires, and also can use the wireless router sufficiently;

(4) Server software: choose Tomcat. In the same time compile identification algorithm by Java;

(5) Data base: use MySQL. MySQL is suited in this system development environment, and also can associative the visual Navicat to realize the management of data;

(6) The final presentation of monitoring employs the framework of B/S form. The consumers can examine the information directly on the browser of the computers; there is no need to add extra expense of software resource;

(7) The alarm signal can regulate the mobile equipment according to the demand. For example we can compile Android App or access the service number of the OTT application in Eclipse.

The termination of the monitoring require relatively less compared to other kinds of system. Here is our recommended collocation table of the system:

(1) CPU: dikaryon ZHIQIANG CPU, 3600MHZ master frequency, 800MHZ bus line frequency, 2M second-class register;

(2) Hard disc: capacity bigger than 75GB, 1000rpm revolution speed, 8000KB register, SCSI interface form;

(3) Internal storage: 1G volume, 667 MHZ master frequency, DDR3 form;

We suggest choosing the old obsolete machine as the server, which is a win-win choice for reusing the old obsolete machine. In the same time the terminal machine can examine the video signal timely presented on browser directly using the server machine.

3   The Implementation Process of the System

This system adopts the router which integrated with TCP/IP communication agreement and has been implemented directly by stable open code Linux operation system. This scheme lets out the complex matching of the compatibility, just only need to compile the code according to the different demands of the clients and then could finish the transmission of the whole video information.

Meanwhile, the Exception information identification algorithm can be implemented in JSP with server code, so we choose Java as the programming language, Video Surveillance Program used in routers compiled by C language through the GCC compiler, select MySQL as the relational database management system.

Here are the steps which are included of the implementation of the system functions:

(1) Inject the OpenWrt system into the HUAWEI HG255D wireless route, using the demand "make dep" to generate the dependent relationship between the core programs. And use the demand "make zImage" to generate core mapping file. Relying on the demands of "make modules" and "make modules install", we can generate the module of the system which could be loaded.

(2) Choose OpenWrt firmware editions and the router hardware correspond to the hardware. Using the demand "make menuconfig", we can dispose the core, choosing the document system of YAFFS, which supports the NFS enablement. Start the USB device supported module. Then using the SECURECRT to judge whether the camera supports the connection of relevant platform and correctly install the relevant drive program of the camera or not.

The USB camera belongs to one kind of video equipment, so we need to start the module named Video4Linux [3]. Using the document library offered by Video4Linux (V4L) to compile the application program, the library provides the structure, variable names and functions, all of which have been defined. By this way, it's easy to realize the communication among related equipments, so that it can achieve the acquisition towards the video data. For example the header file "include / Linux / videodev.h" provided by V4L, in which defines some common structures and symbols used in the driver of the

VideoaLinux device; the "drivers / media / video / videodev.c" file provides some Kernel interface functions.

The video capture process requires plenty of data structures, here are some brief introductions of several commonly used data structures.

(1) struct video_capability: to store performance parameters such as device name , type, the maximum and minimum supported resolution, the information source of the camera et cetera.

(2) struct video_mbuf: Using the mmap mapping to the buffered frame information of the memory, such as size (frame size), offsets (deflection relatived to the base address of each frame), frames (the maximum supported frame amount).

(3) struct video_mmap: For the acquisition device memory mapping.

(4)struct video_picture: Contains the image acquisition frame property equipment.

(5) struct video_channel: Properties of the signal source.

(6) struct video_window: Contains the equipment acquisition window information.

(7) struct video_mmap: equipment used to capture the memory map.

(8) struct video_picture: contains the image frame properties collected from the device .

(9) struct video_channel: properties of the signal source .

(10) struct video_window: contains the information about equipment acquisition window.

Here is the process of video capture:

Step 1.Open the camera and call the system

"functionvfd=open("dev/v4l/video 1",O_RDWR)"to on / off the Camera. vfd is the file descriptor, if the open operation is successful,it will return"vfd", else it returns "-1". Call the video close function "close(vfd)" to turn off the video.

Step 2. Read the information call

the function "ioctl (vfd, VIDIOCGAP, & vfd_cap)" to read the information regarding the camera from the "struct video_capture", it can realize the control of I / O device. Then call the function "ioctl (vfd, VIDIOCGPICT, & vfd_pic)" to get the information of camera's "struct video_picture", such as image brightness , color, contrast, etc. , of which the VIDIOCGAP and VIDIOCGPICT are the control command parameters provided by ioctl.

Step 3. Set the parameters

Firstly, give the variable assignments , then call the function "ioctl (vfd, VIDIOCSPICT, & vfd_pic)" to change the color information of image; Call the function "ioctl (vfd, VIDIOCSWIN, & vfd_win)" to set the relevant parameters of the window.

Step 4. Read image

Directly calling the function method read () or calling the function mmap () to mapping the internal storage, both of which can collect the image. The function read () reads directly through the device number of image information from the kernel buffer. If the call read( ) doesn't completed when the data reading completed, the process hangs, the program enters the wait state, until finishing reading the related data then it returns the error message. the function mmap() is mapping the device files inside to the system memory, each process can access this file directly, and share the memory. It does not require any copies of data, which improves the access speed towards I / O device.

In addition, mmap () is nonblocking mode, it can avoid the phenomenon of wait indefinitely which due to the reason can't reading the data. So using mmap () method reads the image data can ensure the real-time of image.

Video Server, Using the interfaces of v4l2. The structure of which can reference to mjpeg-streamer:

```
struct vdIn {
                int fd;
                char *videodevice;
                char *status;
                char *pictName;
                struct v4l2_capability cap;
                struct v4l2_format fmt;
                struct v4l2_buffer buf;
                struct v4l2_requestbuffers rb;
                void *mem[NB_BUFFER];
                unsigned char *tmpbuffer;
                unsigned char *framebuffer;
                int isstreaming;
                int grabmethod;
                int width;
                int height;
                int fps;
                int formatIn;
                int formatOut;
                int framesizeIn;
                int signalquit;
                int toggleAvi;
                int getPict;
                int rawFrameCapture;
                /* raw frame capture */
                unsigned int fileCounter;
                /* raw frame stream capture */
                unsigned int rfsFramesWritten;
                unsigned int rfsBytesWritten;
```

```
                         /* raw stream capture */
                         FILE *captureFile;
                         unsigned int framesWritten;
                         unsigned int bytesWritten;
                         int framecount;
                         int recordstart;
                         int recordtime;
            };
```

Then write this structure into the drive to initialize the camera. This operation is completed by command ioctl involved include VIDIOC_QUERYCAP, VIDIOC_S_FMT, VIDIOC_S_PARM, VIDIOC_REQBUFS and VIDIOC_QUERYBUF, and using the mmap to mapping the memory.

Finally, we obtain the picture by ioct command, the command involved include VIDIOC_QBUF and VIDIOC_DQBUF.Then write the obtained data into the file, it turns out to be the pictures, the pictures continuously transmitted over a network and it turns to be the video.

The system requires load the module of camera dynamically, which, means the system can control the working of camera by disposal the blackout. The drive programs need to provide the implementation of fundamental in and out operating interface function such as open, read, write, and close etc. The memory mapping function is defined in struct file operations. Relying on the function ioctl, we can control the input and output interfaces [4].

Use WLAN transfer the video signal to the terminal end of the monitoring. If the distraction of the signal is relatively too interferential to come to pass real time, we can consider two methods to handle it: increasing the magnifications of the signal rely or choosing the wired transfer way gates through the port which linked to the terminal end of the monitoring to have a communication.

Choose the Tomcat as the server; compile the JSP page which can receive the data as well as show presentation on the browser in the Java circumstance. This system combine the real need, using the connection pool between the jdbc and JSP page to realize the control of the systematic authority.

Compile the objective-distinguish algorithm which based on dynamic video signal in the integrated environment of Eclipse. It contains following basic circumstances: call a smoke shell alarm by judging the ambiguity of the video signal in a smoking circumstance; judging the fire's status in the certain situation and call a fire alarm by identifying the brightness alteration information of the partial environment as well as matching the morphology of the flames. To meet the consumer's demands, this system can also set the function which can guard against the burglary: according to the classic motion-dot-graph algorithm, it can judge the exotic people's break-in in its silence status and then report to the police. There are plenties of these algorithms in the build-in tool box of the Matlab, and these algorithms have been used frequently. We can export the algorithm document which in a form of m to Java or directly connect these algorithms to the server.

According to the demand of customers, compile the Android APP. Using the socket way to link to the server information. In the same time, we can control the monitoring switch of the router as well as the directions of the camera's stepping motor. The second scheme is let the server access the server number of the OTT class application, according to this way we can choose We-chat, an application program developed by Ten-cent. This activation of the alarm is passive, the system occupies the signaling of the communication operator, consume less expense of the discharge and the quantity of the electricity. But the control signal is also need to be compiled in the server, which means trigger actively.

4   The Implementation of NFC-U Disk Encryption Technology

In order to avoid the wrong judgments to influence the user experience, this system should be located an adequate threshold value of the alarm trigger. In some kinds of dangerous radial monitoring area, the threshold value should be set in a relatively small value, which can guarantee the precision of the monitoring, while in other kinds of normal places, the threshold value could be a little bigger.

After the alarm information been triggered, the monitoring device could prompt the consumers by sending the alarm signal to mobile equipments or users' computers. In the terminal end of the monitoring device which usually combined with the server, the ways of warning generally are light blinking and bell calling and etc. But when it comes to the server which there is nobody guarded. Then the system should use the Internet to build the far-end transmission of information. So the system could send the alarm message timely to the person who is in charge of his/her mobile equipment. Therefore this system can help to improve the security of personal information, avoiding the loss of property in a furthest way. The aspect of the mobile equipment uses the APP developed assorted or OTT service to launch every propelling movement [5].

5   Conclusions

This video monitoring alarm system fully integrated the market demand, the router in this system possess the character of open source and stable. As an intermediary of signal transmission, this wireless router cut down the expense on time and space in route planning, which certainly make the cost of target consumers to a minimum. Meanwhile, the usage of thin web client in the video monitoring system only need the support of the browser, there is no need to spend more time and capital on developing a new client software. The design of this system itself has considered the transmission's accuracy of video data current, and it can satisfy the functions which other video monitoring systems do not have, which can use intelligent identify algorithm to handle the exceptional information and alarm the movable termination timely[6]. It is a great improvement to the traditional monitoring solutions

Acknowledgment. This work is supported by the Northeast Forestry University Young teachers in Independent Innovation Fund Project: Research on key technology of face detection and recognition and the Northeast Forestry University Students innovation training project: Intelligent monitoring alarm device based on Embedded Linux.

References

1.   Wang, X., Li, L.: The design and realization of the multi thread video monitoring based on embedded Linux system. Message Communication, 191–192 (2011)
2.   Lu, C.: The research of remote monitoring technology based on VPN. Science and Technology of West China 23, 39–40 (2010)
3.   Yuan, G.: Research on moving object detection and tracking algorithm in intelligent video surveillance. Yunnan University (2012)
4.   Li, F.: The design and realization of the mobile video monitoring system based on Android. University of Electronic Science and Technology of China (2013)
5.   Wang, A.: Research and application of the mobile phone remote video monitoring and early warning system. University of Electronic Science and Technology of China (2013)
6.   Xia, D.: Research on detection, tracking and identification method of the target in intelligent video surveillance. National University of Defense Technology (2012)

The Design and Implementation of an Encrypted U-Disk Based on NFC

Yunhang Zhang, Chang Hou, Yan Chen[*], Qiqing Huang, Honge Ren,
and Zhaowen Qiu

Institute of Information and Computer Engineering, Northeast Forestry University, China
cheny0451@126.com

Abstract. With the popularity of USB3.0, consumers favor a mobile storage which is convenient and efficient. But in the same time, people often pay less attention to the Safety of traditional mobile storage, especially the U-disk, the security of which is almost ineffective. Challenged by severe test of current information security situation, designing a safe and reliable way to protect the user's privacy encryption has gradually attracted both the equipment manufacturers and the user's attention. As we can see among them, as a quick and secure identification technology, NFC gains a great application in mobile payment and authentication field. It also means a great momentum to popularize. Its unique convenience and safety guarantee in the applications has a wonderful prospect in replacing traditional security solutions. This paper discusses a actualization of a new encrypt U disk which is applied with the NFC technology, to understand the principles of NFC technology and its practical application of encryption, this paper can play a guiding significance.

Keywords: NFC, U-disk encryption, authentication technology.

1 Introduction

As a cheap and convenient solution of the mobile storage data, the U disk receive the user's favorite since its birth date. Currently, large-capacity and high-speed U disk can already reached the level of 500Mbps speed, and stability of the U disk data has also been greatly improved.

With respect to the mobile hard drives, the difference of storage media also makes U-disk withstand greater accidental drops and other damage without losing the internal data. However, the security of U-disk data has always been criticized in a long time. With a growing emphasis on person's privacy in the current society, the major manufacturers have also launched their own brand of encryption software for security one after another [1]. It should be noted that the limitations of this approach is giant, generally you only can use this way for some specific brands. Moreover, the encryption method is too simple to be cracked and get the data inside.

---

[*] Corresponding author.

H. Wang et al. (Eds.): ICYCSEE 2015, CCIS 503, pp. 433–438, 2015.

Present mobile storage security solutions in our country is still lagging behind, most solutions essentially relying on encryption software curing in the U disk. And foreign countries already have mature hardware-based encryption schemes, such as biometric locks like fingerprints and Hardware combination lock [2], etc.

Hardware-based combination lock and biosensors identification, these encryption schemes always have a notable disadvantage of their high cost; also they generally applied in military and senior commercial areas. Because of that, the technology is not open.Under these circumstances, designing a solution combined both the hardware and software for civil security has comparatively better market prospects. The combination of radio frequency identification chips and U disk chip can meet this requirement.

Near Field Communication Technology (NFC) based on short distance as well as non-contact radio frequency identification solutions (RFID), is a wireless communication technology. This short-range communications technology is developed jointly by Philips, Sony and Nokia, working at 13.56 MHz. NFC communication theory has both active and passive operating mode, the frequency it used is basically the same as the contactless smart card popular currently. It also works similarly, uses the same communication theory as the non-contact smart card [3].

This paper through the discussion of NFC-U disk encryption technology in the practical applications, studied the feasibility of this safe and convenient way of such hardware level based encryption mode.

2   The Status of NFC-U Disk Encryption

NFC technology is not a brand new technology; it is proposed and advanced by Nokia, Philips and several other giants of communications industry, whose pre- development of the market is basically overseas in other foreign countries.

2011, Google added the support environment for NFC in Android system, and also released Google Wallet application which using the near field communication. Google also cooperate with MasterCard on the program of credit card payments which is based on NFC technology. Foreign countries have earlier research in security field which using the NFC for identification earlier, they have a more mature industry. Among these countries, South Korea is the most representative.82.5% of mobile phone users in South Korea are using NFC phone as a daily payment instrument in their consumptions.

In recent years, the domestic has an increasingly investment on the fields of such smart home and the internet of things, many hardware manufacturers have joined the device's support for NFC. Among them, MI phone and Meizu mobile phone, as a representative the of smart phone manufacturers, studied f foreign models. They enriched peripheral products of NFC, such as the applications of some tag cards, which certainly increased the market's enthusiasm. Currently, the People's Bank of China has reached a basic agreement with the telecommunications operators. They would jointly promote NFC -based payment standard.

In mobile storage security field, Chinese manufacturers are also actively explore the feasibility and constantly improve the technology. These manufacturers launched NFC-U disk storage products which suit the current market. This product mostly use the NFC which is full hardware matches encryption and security kernel , while using stable high-speed Flash storage granules to meet the variety needs of users. Because of the different encryption mechanisms, data transmission requires the use of 16bit ECC error correction.

3   The Implementation of NFC-U Disk Encryption

The basic question to achieve identity authentication and encryption on the underlying hardware of the U disk storage is to address the non-contact RF communications. Here are some main current solutions

(1)   Separate small pieces of flash open area --OTP area to cure a unique number, to have a handshake recognition with the NFC equipments (decoder);

(2)   There is also another way using the unlock APP verification in the Android mobile terminal, this program needs to write the Android program to achieve the aim.

This design uses a unique way curing the unique number of the OTP area. When the users' U disk NFC circuit is activated by external environment, firstly verify whether the external NFC handshake signal can match the curing number or not, if the answer is can, then open the read and write permissions of U disk. After such authorization, users can have the access data operation. It should be pointed out that, OTP area needs algorithm or hardware to realize the feature of logic self-destruct, which prevent the data being cracked in violence and stolen.

In the following section, we briefly outline our design for an NFC-enabled TPM (Trusted Platform Module) architecture. In contrast to existing publications, we propose to integrate the functionality of a low-cost passive RFID interface into the TPM. This allows users to remotely audit the privacy status of the terminal (target) using a conventional NFC device (initiator) by ensuring a direct channel between the user and the TPM.

Figure 1 shows a schematic view of our proposed architecture. It shows the TPM device with its components such as a CPU,I/O interface, voltage regulator, memory, and cryptographic components like SHA-1, RSA/ECC, key generation, and a random number generation (RNG). Additionally, an RFID front-end is connected to the internal bus. It is composed of a digital part which handles the NFC protocol (ISO 14443 or ISO 18092) and an analog part that is mainly responsible to modulate and demodulate the signals of the air interface. Next to the I/O and power interface, the TPM has two additional connections for an external antenna that can be connected or printed on the main board of the terminal.

In general, passive RFID/NFC devices are designed to meet low-resource requirements. They draw their power of the air from the reader and therefore consume only a few microwatts of power to provide a certain reading range. Furthermore, they meet low area requirements so that they can be produced in a large scale with low costs. In fact, most RFID tags can nowadays be produced with costs below 10 cents.

TPMs, in contrast, have an active power supply and provide many resources such as cryptographic engines, CPU, and (in comparison) large memory units. These resources can actually be reused be the RFID/NFC front-end which lowers the requirements of integrating NFC into TPM significantly. The controlling of the protocol, for example, can be handled by the CPU of the TPM and memory can be shared with the RFID/NFC component over the internal bus architecture as well.

By integrating an NFC interface to the TPM, we followed the idea of Parno [7] who recommended to integrate a special-purpose hardware interface to the TPM to establish a direct link between the user and the TPM so that human inter-actors can themselves establish the proximity of the attesting machine. The integration of the NFC interface to the TPM has thereby several advantages.

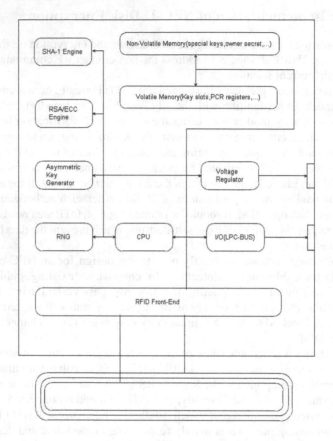

Fig. 1. Schematic of the NFC-Enabled TPM architecture

First, a guarantee is given that the targeted terminal is still in the proximity of the user, as the user performs the touching operation in person.

Second, it gives a guarantee that the TPM is located in the proximity because the NFC module is physically connected and integrated into the TPM and the operational range of NFC is theoretical limited by the coupling property of the magnetic near-field of the reader. Note that RFID standards define a practical reading range of about 10 centimeters. This makes attacks such as machine-in-the-middle attacks much harder to perform.

Third, our proposal of integrating an NFC interface into the TPM allows users to verify the integrity of public terminals with their mobile phone.

By simply touching the antenna of the terminal with their phone, an application can be automatically launched that shows the status of the trust decision on a user-friendly and familiar display output. In addition, the trust decision can be also signalized by a beep, ringing tone, or vibration which makes the application more applicable and comfortable in certain environments. The part of the hardware has the following characteristics:

(1)   NFC module is connected among to the USB HS, encryption master and NAND FLASH:

(2)   The Improved and non-public NFC authentication protocol can meet the needs of high-speed information interaction, which is also compatible with standard protocols, to communicate with the unmodified unlock end;

(3)   By a unique number which is encrypted by algorithm stored in KEY Chip, the authorization mechanism possesses high security;

(4)   The secret key encryption need to be regenerated after each U disk swap;

(5)   By using the superior packaging technology and processing technology, which does not change the volume of U disk fundamentally, in the same time can increase the NFC function and authentication logic;

(6)   It supports AES/SCB2 encryption algorithms, which are encryption algorithms self-defined by High-end customers;

(7)   It have a good support by the system and compatibility [5].

The technical obstacles of the NFC-U disk encryption already not exist currently; the key is the issue of current standardization and user acceptance.

Moreover, there are only a few companies currently having this kind of business; the audiences of customers are still comparably small.

At the same time we need to see that, due to the rise of the current cloud computing, major Internet companies have turned to support of cloud storage , so it means we can Bypassing the problem caused by hardware. The traditional hardware storage devices also expanded the support for cloud storage services on the periphery, its security has been tested by the market. This situation have Greater impact on NFC-U disk encryption scheme.

4   The Generalization Program of NFC-U Disk Encryption

At present, the development of domestic NFC-U disk encryption products below market expectations, a large market capacity. Currently, the development of domestic NFC-U disk encryption products is below the market expectations, it means a large capacity of market.

To promote the encryption security NFC-U disk of this hardware level, we can analyze from the following three aspects, and then begin to solve this problem:

Firstly, technical factors, the traditional storage solutions providers cooperate with NFC security hardware vendors, promote the development of industrial by choosing the suitable technology mode.

Secondly, market factors, the Novelty and convenience of the technology have not been greatly promoted, the public awareness is really low, the improvement of market enthusiasm is not significant, which Causes market differences not been recognized.

Thirdly, the lack of policy support issues. Security privacy and the living, working of people are closely linked; it needs political support for industrial applications and the standard unified planning. Also it can rely on the protocol implementations to grow the market.

5   Conclusions

This paper analyzes the technology model and promotion measures of the NFC-U disk encryption the aim are to have a deeper understanding of NFC-U disk encryption technology.

At present time, several well-known manufacturers of flash memory have launched their own software encryption U disk, but the effect of them all is not ideal [6]

Rely on the base of their users; if we can combine NFC to encrypt and protect the data in the later stage, we can meet the different needs of users by its unique security and convenience. It also possess a great market prospection in the field of data security.

Acknowledgment. This work is supported by the Northeast Forestry University Young teachers in Independent Innovation Fund Project: Research on key technology of face detection and recognition and the Northeast Forestry University Students innovation training project: U disk encryption NFC near field communication technology based on.

References

1. Wang, Z.: The design of Security U disk management tool's software based on particular security chip. Dalian University of Technology (2013)
2. Zeng, K.: The design and realisation of the security U disk based on EZ-USB. University of Electronic Science and Technology of China (2011)
3. Zhao, Y.: The applied research of smart posters based on NFC technology. Zhengzhou University (2012)
4. Chen, T.B.: The New Driving Force of NFC Technology. China Center for Information Industry Development, The Asia-Pacific RFID Technology Association (2009), Chinese International RFID and Smart card Technology and Application Summit Forum Proceedings. Chinese Electronic Information Industry Development Research Institute, vol. 31. The Asia-Pacific RFID Technology Association (2009)
5. Tang, T.: The design and realisation of the Ironkey USB Flash Drive. Journal of Chengdu Electromechanical College 02, 14–16 (2006)
6. Wang, A.: The key integration technology research of the cryptographic chip system. Shandong University (2011)
7. Parno, B.: Bootstrapping trust in a"trusted" platform. In: Proceedings of the 3rd Conference on Hot Topics in Security, pp. 1–6. USENIX Association, San Jose (2008)

An On-chip Interconnection QoS Verification Platform of Processor of Large Data for Architectural Modeling Analysis

Li Qinghua[1], Qin Jilong[3,*], Ding Xu[2], Wang Endong[3], and Gong Weifeng[3]

[1] DTmobile Inc., Beijing 100083, China
[2] AMD Inc., Beijing 100081
[3] National High-performance Computing and Storage Key Laboratory
of Inspur, Beijing 100085, China
qinjl@inspur.com

Abstract. This paper presents introduction for a QoS verification of on-chip interconnection based on the new progress of the industry, which combined with an AMD processor chip design for big data. Some verification experience in architectural modeling and simulation of on-chip interconnection is also introduced in this paper.

Keywords: Interconnect, QoS, Verification, Modeling, Multiprocessor, Computer Architecture, Big Data.

1   Introduction

With the rapid development of internet technology, recently the processor for big data has rise up. From CPU verification point of view, the QoS verification of on-chip interconnection for the scalability and performance of multi-socket is the general trend, due to there are many bottlenecks in big data application. While performance of single core in the CPU chip improves quickly, such as multi-core and multi-thread and the application of turbo boost after the continuous improvement and upgrading is the highlights for big data application. Given the scope and flexibility of the SOC infrastructure, to have verification platform that can deliver pre-RTL scaling of the architecture is important. Without such a tool, we would need to make educated guesses at the correct configuration. This would likely result in overdesign or worse an incorrectly scaled underperforming ASIC. So QoS verification become the key step during the process of processor chip design and later server system design which is called the bottleneck-killer. At present, new generation processor for big data become popular, and has been designed by many company like Intel, IBM and ARM, etc. Figure 1 shows IBM Power8 processor for big data. In big data era processor design and development have become complex SOC design which have multi-core, multi-thread of application and complex data paths, then bottleneck problem appears

---

*Corresponding author.

H. Wang et al. (Eds.): ICYCSEE 2015, CCIS 503, pp. 439–447, 2015.

when a specific application run. An ideal solution would allow quick architectural investigations as well as more detailed architectural modeling with a mix of IP models of many types like C models, TLM models, RTL models and FPGA/Palladium modeling in the verification platform.

Fig. 1. IBM Power8 processor for big data

This paper introduced the QoS verification of on-chip interconnection for a big data processor, which focuses on the analysis of its architectural modeling and interfaces simulation.

2 QoS Verification Platform

2.1 Introduction of Background

In big data era processor design and development have become complex SOC design, and then immediate concern is QoS for the new fabric of SOC chip. Testing QoS requires a subset of the modeling environment. A useful SOC model for performance modeling would need accurate traffic generators in the form of C/SystemC models or trace playback when bottleneck appears.

2.2 QoS Verification Modeling

2.2.1 Verification Platform Overview

Figure 3 and Figure 2 shows a simplistic view of a QoS model and Figure 3 contrasts an architectural performance model. Note that there will be investigations on several QoS possible schemes, Figure 3 and Figure 2 only attempts to show one possibility. Comparing Figure 3 to Figure 2, there add some QoS logic behind the AXI hub QoS management. Here still have generic requestors generating unspecific traffic. The

intent is to test QoS without needing the complexity of accurate IP models. For the
Figure 3, 2 model the following items is necessary:

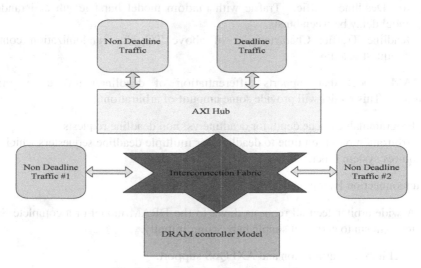

Fig. 2. Basic QoS Verification Modeling

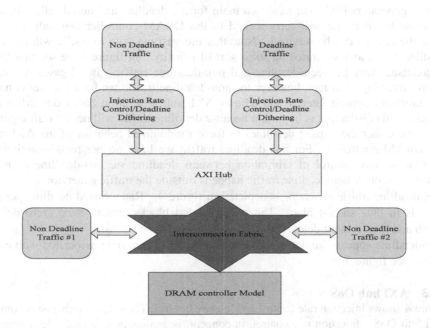

Fig. 3. QoS Modeling with Throttling

— Traffic generators. Traffic distribution needs to be flexible enough to test QoS corner cases.

- Non Deadline Traffic: Traffic with random model burst length and random model delay between bursts.
- Deadline Traffic: Characteristics of above TG plus randomization control of request deadline.

— DRAM model that supports differentiation of deadline/non-deadline traffic classes. This model will provide some amount of arbitration:,

- Programmable queue depth for deadline vs. non deadline requests.
- Arbitration based on time to deadline for multiple deadline sequesters which requires system timer.

— Interconnection Fabric model.

- A wide arbiter feed all requests down to the DRAM model or a complete SDF mechanism to carry out simple hub design initially.

- AXI hub with arbitration and AXI QoS support.

2.2.2 Traffic Generators

Traffic generators (TG) will have two main forms, deadline and non-deadline based. Deadline based traffic will carry a field to the DRAM controller that indicates the latest the request can be serviced. Note that the vast majority of traffic will be non-deadline based and arbitrated on some sort of priority. In figure 3 we see that QoS mechanisms vary between deadline and non-deadline traffic. In all cases we have control over injection rate, however for non-deadline traffic we have another control that monitors request latency and adjusts AXI hub arbitration based on earlier requests. This distinction is important because deadline traffic will need a traffic generator that dither the request deadlines to force randomized behavior of the AXI hub and DRAM controller. For non-deadline traffic we have no per request notion of deadline so any control of arbitration between deadline vs. non-deadline or non-deadline vs. other non-deadline traffic happens outside the traffic generator.

A deadline traffic generator is depicted in figure 4. Randomized deadline per request is in blue shaded block. The green shaded blocks control burst (zero delay) length and inter request delay.

Non-deadline appears similar to figure 4 except the system time associated with each access, see figure 5.

2.2.3 AXI hub QoS

Figure 4 shows injection rate control and latency control. These two mechanisms control AXI hub QoS. Injection rate control, in concept, is analogous to a credit/debit control. For example, if a client has a limit of 4 requests, logic counts outgoing requests and decrements the counter as requests get acknowledged. Acknowledgement occurs for writes when the write goes "clean", for reads the acknowledgement is receipt of the return data.

2.3 Top Level Structure of QoS Verification Platform

Figure 4 shows a high level view of the initial QoS verification platform. TGs are a mix of deadline and non-deadline generators with appropriate QoS mechanisms instantiated. Interconnect could be any number of designs. It also shows an architectural model that adds specific IP models. QoS support is assumed both at the AXI hub level and fabric level. TG means modeling of traffic generators.

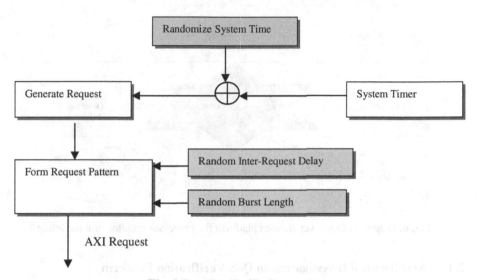

Fig. 4. Deadline Traffic Generator

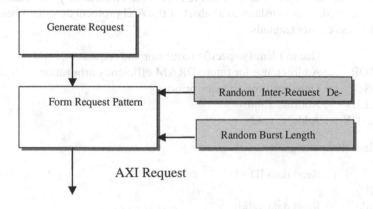

Fig. 5. Non-deadline Traffic Generator

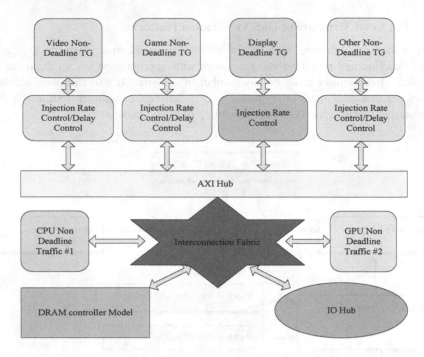

Fig. 6. Diagram of QoS verification platform for processor architectural modeling

2.4   Architectural Investigation on QoS Verification Platform

2.4.1   Subset of the AXI Protocol

The architectural investigation only model read process. Additionally the actual transaction data is ignored. This translates to a subset of the AXI protocol described below:

Read address channel signals:

- ARID   : Use to identify specific requestor and request.
- ARADDR   : Address, use for future DRAM efficiency arbitration
- ARQoS   : Deadline time value or priority
- ARVALID   : Address valid.
- ARREADY   : Address ready

Read data channel signals:

- RID   : Read data ID
- RRESP
- RVALID   : Read data valid.
- RREADY   : Master can accept data.

2.4.2 Monitors on QoS Verification Platform

For monitoring and control of the system , need some method of extracting and inputting information respectively. There will have to be some relationship between TG settings, latency contract and injection rate. Detail is as follows,

Inputs:

— TGs

- Deadline

(a) System time
(b) Burst setting to be randomized in TG
(c) Inter delay setting to be randomized in TG

- Non-deadline

(d) Burst setting to be randomized in TG
(e) Inter delay setting to be randomized in TG

— AXI hub

- Deadline TG
(a) Outstanding requests for injection rate control

- Non-deadline TG

(b) Outstanding requests for injection rate control
(c) Latency contract
(d) Priority range

— Interconnect

Control function depends on the interconnection. For a simple hub/arbiter need not control, but for something like a packet switch may want address range for each stop.

— DRAM/Memory Control

(a) Buffer Depth for deadline and non-deadline traffic
(b) Memory bandwidth

Monitors:

(a) TG (both types) bandwidth/time
(b) Deadline traffic deadline violation
(c) Non-deadline priority/time
(d) Non-deadline latency/time
(e) DRAM/memory bandwith/time

This monitor is tied system timer so that different information can be correlated.

3   QoS Verification Platform Modeling

The QoS verification platform which is architectural modeling with a mix of C models, TLM models, and RTL. Mix of C models, as well as the FPGA/Palladium models and other hybrid modeling, is flexible and able to range of reconfigurable parameters for on-chip interconnection. In the process of the establishment of the platform, EDA tools like SimNow from AMD , many emulators, FPGA platform and Palladium platform is bind together to build this hybrid verification platform, and randomized signal is generated, which bring up successful module optimization of a sample processor chip design as follows.

4   Implementation of Sample Chip

Below is a sample SOC processor, which is implemented in 28nm process Global Foundry, a computing unit consists of four 0.5MB L2 Cache, including the realization of automatic power control & power optimization which obtain 25W low power, suitable for big data computing.

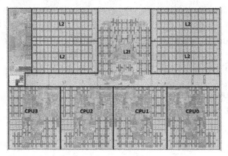

Fig. 7. Micrographs of processor sample silicon die

5   Conclusion

In the era of big data, performance of processor on-chip interconnection will have significant effect on the computing efficiency; this paper introduces a QoS verification platform and its modeling method, and describes the importance of the QoS verification. The performance of the processor obtains the good result based on the interconnection QoS architectural modeling. Based on QoS simulation with margin, the bottleneck does not appear, and modules are easily optimized for the whole server system.

References

1. Dally, W.J., Towles, B.: Principles and Practices of Interconnection Networks. Morgan Kaufmann publications (2007)
2. Xiao, Y., Liu, L., Wei, S.: Design and implementation of Reconfigurable Stream Processor in multimedia applications. In: Communications, Circuits and Systems, ICCCAS 2008, May 25-27, pp. 1382–1386 (2008)

3. Kumar, S., et al.: A Network on Chip Architecture and Design Methodology. In: Proc. ISVLSI 2002, pp. 105–112 (April 2002)
4. International Technology Roadmap for Semiconductors (ITRS), 2003 edn, 2005 edn. Semiconductor Industry Association, San Jose, CA
5. Intel Thurley has Early CSI Interconnect. The Inquirer, http://www.theinquirer.net/default.aspx?article=37392 (February 2, 2007)
6. Borkar, S.: Networks for multi-core chip - A controversial view. In: 2006 Workshop on On- and Off-chip interconnection Networks for Multicore Systems, OCIN (December 2, 2006)
7. Srivastava, M., Chandrakasan, A., Brodersen, R.: Predictive system shutdown and other architectural techniques for energy efficient programmable computation. IEEE Trans. on VLSI System 4(1), 42–55 (1996)
8. Wentzlaff, D., et al.: On-chip Interconnection Architecture of Tile Processor. IEEE Micro., 15–31 (September 2007)
9. Park, C., Badeau, R., et al.: A 1.2TB/s on-chip ring interconnection for 45nm 8-core enterprise Xeon processor. In: 2010 IEEE International Solid-State Circuit Conference Digest of Technical Paper (ISSCC), pp. 180–181 (February 2010)

Paradise Pointer : A Sightseeing Scenes Images Search Engine Based on Big Data Processing

Jie Pan, Hongzhi Wang, Hong Gao, Wenxuan Zhao, Hongxing Huo, and Huirong Dong

Harbin Institute of Technology, China

Abstract. Nowadays, with the rapid development of network, more and more people are willing to share their attractive photos on the internet, especially the sightseeing spot photos. However, there are countless noteless but splendid scenic spots that remain unknown to most people, and it is a pity that one finds a wonderful place but cannot reach it. Therefore, it is meaningful and useful to build an image search website used specially for sightseeing spot images. Meanwhile, we have stepped into the new era of Big Data, the data we need to process is in an explosive growth, including the images. Thus we develop ParadisePointer, a scenery image search engine, which used the processing method on the background of Big Data. In this paper, we are going to introduce the main stages of our system and some key features of ParadisePointer.

1 Introduction

Every day we glance over thousands of images from the internet, and the scenery images are a large share of them ------ We have stepped into a new era of Big Data . Consider a scenario that someone fixes the eyes on a beautiful scenery image and he will get an idea of taking a trip there. A problem faced by him is that he should know the name of the scenery. Unfortunately, most of the world famous image-sharing websites like Flickr and Deviantart [1] do not support "Content-based image retrieval system"i, so users may not have access to know the description of this image if websites do not provide. Of course, many image websites based on CBIR [2] are proposed, such as Google Similar Images, Picitup and Tineye. However, all of them are geared to the needs of general images searching but not specialized for scenery images. Therefore, we get the idea of develop Paradise Pointer, a search engine special for scenery.

Comparing with representative CBIR systems, such as The QBIC (Query by image and video content) system by IBM, IMAGEFINDER by Attrasoft [7], our system has following benefits.

-- **Specification in functions**. Our system focuses on scenery images search and the information about images in other types will not include in our databases. Thus our system is more efficient due to ignoring the needless data.

-- **Meticulous classification**. Images classification is essential in any kind of image searching systems. According to [3], scenery images scenes are divided into two major categories, landscapes and city-shots. Furthermore, each of these two categories

H. Wang et al. (Eds.): ICYCSEE 2015, CCIS 503, pp. 448–452, 2015.

can be divided into multiple parts. We use the hierarchical classification method based on the theory of Chen Chia-Huang and Yasufumi Takamai to build our landscapes classification. As for the city-shot, we will get them classified by compute the proportion of some standard patterns such as straight lines , rectangle, triangles and circle, further discussion can be find in the "Methods" section.

-- **High precision Matching**. Our system integrates multiple features for matching such as texture feature, color histogram feature, edge of images, and reasonable feature invariants. We use four matching methods to ensure the accuracy of matching. More than that, thanks to the Big Data, we can ensure the precision by using millions of images from our massive databases

--**Highly dynamic databases**. Our image-information-databases update periodically through artificial neural networks. We not only acquire data from the internet, but also prompt user to provide more images and labels. Thus the databases can keep fresh and scale up.

-- **Friendly user interface**. We design a simple and elegant web interface for users. A user can simply upload the scenery image, and the results are shown on the same page, including the pictures and text information. If the user is not lucky enough to get the actual information, the system will recommend the user some similar scenery spots. We also allow user to upload their photos to enrich our database.

In this paper, we will introduce the architecture (in Section 2), major methods of our engine (in Section 3), and the demonstration of the system (in Section 4). At last the conclusions are drawn in Section 5.

2   System Architecture

In this section, we discuss the architecture of our system. Our system has two parts. One is search engine. The other is updating module

The architecture of search engine is shown in Figure 1.

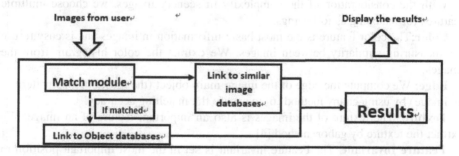

Fig. 1. Architecture

The match module is the main part of our system, which is composed of feature extracting part, database-linking part, and image-feature-matching programs. The feature extracting part is design for extracting the features of the images, which will be stored in our databases. The database-linking module links the image-feature-matching algorithms and our databases. The core part of this module is the image-feature-matching algorithms, these image-processing algorithms will match the data from users matched or find the most similar one. The algorithms will be discussed in Section 3.

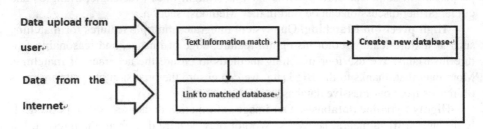

Fig. 2. Update module

The databases updating sub-system is designed to keep our databases up-to-time. The components are shown in Figure 2. We obtain our data from the internet and meanwhile we allow the users to upload their photos to enrich our database. When a user uploads their photos, the system will prompt them to provide exact information of their photos, and the error correction function is also available. When the system receives the data from users, the basic detection module will filter non-scenery images. Then matching and classify module starts. After classification, the similarity sorted module will get the data sorted by similarity.

3   Methods

In this section, we discuss the main methods of the matching module and classification module and similarity sort module.

(1) Match module

With the consideration of the complexity in scenery images, we choose multiple features for matching as followings.

Color: The color feature is the most basic information in images and is easy to use for measuring similarity between images. We extract the color histogram from the images.

Edge: We compute the edge of the image main object (the most obvious pattern in an image) by using canny method [6] to ensure the matching accuracy.

Texture: The texture of the images is also an important feature of an image. We extract the texture by gabor method [4].

Feature Invariant: The Feature invariant is set in the most important position in image matching systems. After careful consideration, we choose SURF (Speeded Up Robust Features) [5], which is a robust local feature detector and effectives in computer vision tasks like object recognition or 3D reconstruction.

(2) Classification module

Obviously, the major shape and color of different types of images is also diverse. So when given an image as an input we wish to classify as one of the following by its corresponding object, such as coast, forest, mountain, open country street, tall buildings.

The color feature is the most basic information in images and is easy to use for measuring similarity between images, so we use the hierarchical classification method for classification[2]. In the first stage, for discriminating night-time view images from others, the value component of HSV color space in the top third region of each image is extracted as a feature. The second stage utilizes image segmentation for obtaining the Region of Interest (ROI). From the ROI, Cb and Cr of YCbCr color space are extracted for discriminating different situations from others. Such that the same image in different weather can also be divided into different types.

We also define basic shape of a scenery images for classification, such as straight lines, triangles and rectangle. For example triangles and rectangles can be easily detected in modern architectures, while the sea and prairie are almost made up by straight lines. So we think it is an appropriate method to classify the images by the proportion one certain or multiple basic shape in images.

4   Demonstration

In this section, we will show the demonstration way of our system.

When a user enters our website, the system will show some example images in different types from our databases in our home page. The system also provides an input interface for users to upload their images, as shown in Fig. 3.

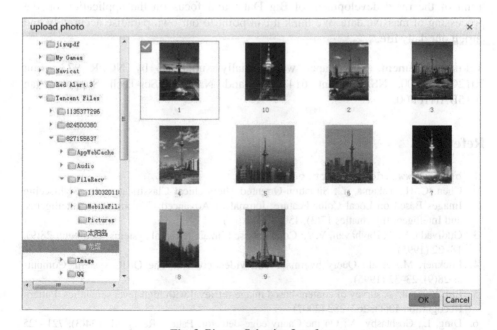

Fig. 3. Picture Selection Interface

Fig. 4. Search Results   **Fig. 5.** Result Details

When a user uploads his own origin image, if system matches some objects, the system will return the information of top-k similar objects as the results (in Fig. 4). If the user clicks one result, the details are shown in the interface as in Fig. 5.

If no matched results are found, return the five most similar objects as the recommend results, and add this image into "unidentified objects database".

5   Conclusions

Our scenery images searching engine is a novel searching engine that specially designed for scenery searching and scenery spots recommending, with a high-efficiency processing methods and friendly user interface. What's more, our engine follow the trend of the rapid development of Big Data, and focus on the application on the processing of massive data. We think it can point to our own paradise accurately and enrich our daily life.

Acknowledgement. This paper was partially supported by NGFR 973 grant 2012CB316200, NSFC grant 61472099 and National Sci-Tech Support Plan 2015BAH10F00.

References

1. http://www.deviantart.com/
2. Chen, C.-H., Takama, Y.: Situation-Oriented Hierarchical Classification for Sightseeing Images Based on Local Color Feature. Journal of Advanced Computational Intelligence and Intelligent Informatics 17(3), 459–468 (2013)
3. Gudivada, V.N., Raghavan, V.V.: Content based image retrieval systems. Computer 28(9), 18–22 (1995)
4. Flickner, M., et al.: Query by image and video content: The QBIC system. Computer 28(9), 23–32 (1995)
5. Liu, Y., et al.: A survey of content-based image retrieval with high-level semantics. Pattern Recognition 40(1), 262–282 (2007)
6. Ding, L., Goshtasby, A.: On the Canny edge detector. Pattern Recognition 34(3), 721–725 (2001)
7. http://www.attrasoft.com/oldsite/brochure.proof.FINAL3.pdf

Chinese MOOC Search Engine

Bo An, Tianwei Qu, Haoliang Qi, and Tianwei Qu

School of Computer Science and Technology, Heilongjiang Institute of Technology,
Harbin, China
ammber83@126.com, haoliangqi163@163.com,
qtw-tony@126.com

Abstract. MOOC (Massive Open Online Courses) has become more and more popular all over the world in recent years. However, search engines, such as Google, Baidu, Yahoo and Bing, do not support specialized MOOC courses searching. The purpose of this demo is to present a vertical search engine designed to retrieve MOOC courses for learner. The demo search engine obtains MOOC web pages by a focused Crawler. And the pages are parsed into structure or unstructure data with a modeling-based Parser. Then the Indexer build index for the data by Lucene. Finally, the extraction MOOC list is made by Course_ranking and Retrieval. The demo search engine is accessible at http://www.MOOCsoso.com.

Keywords: Massive Open Online Courses, MOOC Search, Information Retrieval.

1   Introduction

MOOC (Massive Open Online Courses) is a new form of online education, which is based on the development of courses, educational theory, Internet and mobile smart technology[1] [2]. The New York Times called 2012 "the first year of MOOC"[3] because MOOC gets warming rapidly in global in 2012. The three major MOOC platforms (Coursera, Udacity and edX) provide services successively. Online MOOC courses and users grew by leaps and bounds in the past few years. In 2013, Tsinghua University, Peking University, Shanghai Jiaotong University, Fudan University and other famous universities joined the Edx and Coursera[4].

It is amazing that there is no MOOC search service provided by Google, Baidu, Yahoo and Bing or others in the MOOC age. According to our knowledge, there is no MOOC vertical search service or MOOC search research. Some MOOC platforms do not provide site-search (eg Edx). And the site-searches provided by some MOOC platforms get poor results or unexpected results (such as Coursera). In Coursera we can't extract the courses of " Tsinghua University ". In the future the MOOC will show explosive growth. Therefore it is urgent priority to provide MOOC search service which can help users quickly retrieve the right courses.

This demo shows a Chinese MOOC vertical search engine named MOOCsoso. The rest of this paper is organized as follow: Session 2 presents the architecture of the MOOCsoso and describes the method of each component. Session 3 presents

H. Wang et al. (Eds.): ICYCSEE 2015, CCIS 503, pp. 453–458, 2015.

the interface of the MOOCsoso and show a search application. Session 4 discusses the future development of the MOOC search engine.

2   Architecture of MOOCsoso

2.1   Overview

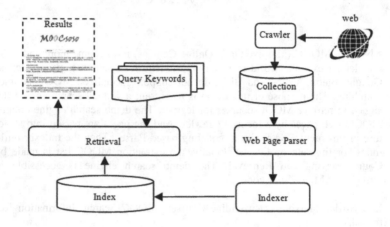

Fig. 1. MOOCsoso Architecture

The architecture of the MOOCsoso is shown in Figure 1. First, the web-based *Crawler* crawl web pages from various MOOC websites. And the web pages are parsed into structured or semi-structured courses texts by the *Parser*. Then the *Indexer* builds the index for the fetched courses texts. When someone submits *Query Keywords*, the information *Retrieval* retrieves optimal search *Results* from the indexed text and the interface shows the results[5].

2.2   Crawler

The *Crawler* of the MOOCsoso crawls more than 13000 MOOC courses. The courses are from the most popular MOOC websites. The URL resources and the courses amount are shown in Table 1. The *Crawler* scans these URLs as seeds. Then the *Crawler* searches their entire Web in first-broad-search strategy[6]. The courses list pages and the course home pages are first obtained. Since the purpose of users is to learn a course or a skill, we crawl the main body instead of the web. The frequency of incremental crawling is once a week because MOOC course construction due to the longer period and the MOOCs update weekly.

Table 1. MOOC Resources

Resource		Courses
Chaoxing	http://mooc.chaoxing.com/	7257
TED	http://www.ted.com/	1731
Sina Open Course	http://open.sina.com.cn/	1037
Coursera	http://www.coursera.org/	753
Netease Open Course	http://open.163.com/	382
Edx	http://www.edx.org/	284
Chinese Universities Video Open Course http://video.jingpinke.com		266
Chinese Online Education Open Source Platform http://www.oer.edu.cn/		245
XueTangX	http://www.xuetangx.com/courses	192
Chinese Open Course CNTV	http://opencla.cntv.cn/	153
Imooc	http://www.imooc.com/	143
Icourses	http://www.icourses.cn/imooc/	110
Khan Academy	https://www.khanacademy.org/	76
Udacity	http://www.udacity.com/	39
Peking University Open Course	http://opencourse.pku.edu.cn/	38
Ewant	http://www.ewant.org/	37

2.3   Parser

The *Parser* parses the web pages obtained by the *Crawler*. The web and their links are restructured to purification text[7]. The MOOC sources are all from some specific MOOC websites. The MOOCs from one website are constructed on the same platform. We create a template for each platform. The template defines the relationship of the web page and the course information. The Parser extracts the course information based on the template. The course information includes course name, university, teacher's name, number of students, descriptions, home link, units, unit content, etc.

2.4   Indexer

The *Indexer* builds index for the parsed text. Lucene[8] is adopted to establish the index of inverted file. The texts are tokenized twice based on the basic dictionary and extended dictionary. The *Indexer* builds index specifically for the key words such as the course name, university, teacher's name, etc.

2.5   Retrieval

The *Retrieval* retrieves MOOCs and outputs the results sorted by the relevance. In the *Retrieval,* we try to analyse the user's intent. If the query is a name of a university, the user intended to search the courses provided by the university. The ranking algorithm in MOOCsoso is named Course_Rank. We set different weight for different fields of the course information. We compute the similarity score between query vector and the course vector. Course_Rank model will be more described in the other paper. Ranking results are the abstracts of the web page including course name, MOOC website, university, teacher's name, number of students and description.

3   MOOCsoso: A Search Engine for MOOCs

Fig. 2. The MOOCsoso Home page

We have built a Chinese MOOC search engine named MOOCsoso. The MOOCsoso is accessible at http://www.MOOCsoso.com. The home page of the MOOCsoso is shown in Figure 2. For example, after we entered the query words "清华大学" (Tsinghua University), the search engine would retrieve the MOOC courses provided by Tsinghua University. The searching results are shown in Table 3.

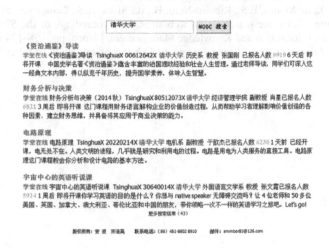

Fig. 3. A MOOCsoso application

4   Conclusion and Future Work

The Chinese MOOC search engine (MOOCsoso) could fix the problems that existing search engine still has no vertical MOOC search. Using the MOOCsoso, Chinese users can search the right MOOC courses efficiently from the web. The MOOC courses are list ranking by the system's recommendation. Not only for the learner, but also for the MOOC creator MOOCsoso is meaningful.

The current version of MOOCsoso can be improved in the following. In the future, we will provide the service of some units or specific skill in a course, video search, and personalized search by analyzing users' behaviors.

Acknowledgement. This paper was supported by the National Science Foundation of China (Grant No. 61370170) and Heilongjiang Education Planning Projects (Grant No. 14G116).

References

1. Sive, H., Sarma, S.: Education: Online on-ramps. Nature 499 (July 2013)
2. Kellogg, S.: Online learning: How to make a MOOC. Nature 499 (July 2013)
3. Pappano, L.: The Year of the MOOC. The New York Times 2(12) (2012)
4. Yuan, H., Xuan, L.: MOOC practice status report and total issues in Chinese universities. Modern Distance Education Research 2014(4) (2014)

458   B. An et al.

5. Shettar, R., Bhuptani, R.: A Vertical Search Engine – Based On Domain Classifier. International Journal of Computer Science and Security 2(4), 18 (2008)
6. Menczer, F., Pant, G., Srinivasan, P.: Topical web crawlers: Evaluating adaptive algorithms. ACM Transactions on Internet Technology 4(4), 378–419
7. Fernandes, D., de Moura, E.S., Ribeiro-Neto, B., da Silva, A.S., Gonçalves, M.A.: Computing block importance for searching on web sites. In: Proceedings of the Sixteenth ACM Conference on Information and Knowledge Management, Lisboa, pp. 165–174 (2007)
8. The Apache Jakarta Project: Lucene, http://jakarta.apache.org/lucene/

HawkEyes Plagiarism Detection System

Leilei Kong[1,2], Jie Li[3], Feng Zhao[1], Haoliang Qi[1], Zhongyuan Han[1,3], Yong Han[1],
and ZhiMao Lu[2]

[1] School of Computer Science and Technology, Heilongjiang Institute of Technology,
Harbin, China
[2] College of Information and Communication Engineering, Harbin Engineering University,
Harbin, China
[3] School of Computer Science and Technology, Harbin Institute of Technology,
Harbin, China
kongleilei1979@hotmail.com

Abstract. The high-obfuscation plagiarism detection in big data environment,
such as the paraphrasing and cross-language plagiarism, is often difficult for anti-plagiarism system because the plagiarism skills are becoming more and more
complex. This paper proposes HawkEyes, a plagiarism detection system implemented based on the source retrieval and text alignment algorithms which
developed for the international competition on plagiarism detection organized
by CLEF. The text alignment algorism in HawkEyes gained the first place in
PAN@CLEF2012. In the demonstration, we will present our system implemented on PAN@CLEF2014 training data corpus.

Keywords: plagiarism detection system, source retrieval, text alignment.

1 Introduction

With the great development of web, especially the abundant online literature sources,
search engine, and the application of machine translation, makes it easier for people to
search, copy, save, and reuse online sources. During the last decade, automated plagiarism detection in natural languages have attracted considerable attention from research and industry[1]. Of these countermeasures, systematic detection using anti-
plagiarism software provides a pertinent and effective method [2].

With the application of anti-plagiarism detection system, more and more high-
obfuscation plagiarism methods, such as paraphrasing and cross-language plagiarism,
begin to be carried out in recent years. The high obfuscation plagiarism means that the
plagiarism is implemented by using sentence reduction, combination, restructuring, paraphrasing, concept generalization, and concept specification and translation. However,
the existing plagiarism detection softwares just perform well in the extract plagiarism or
low obfuscation plagiarism, and the detection of high obfuscation plagiarism problems
keep unresolved. In fact, most of the existing systems fail to detect plagiarism by paraphrasing the text, by summarizing the text but retaining the same idea, or by stealing ideas
and contributions of others[3]. Particularly for Chinese, the system on cross-language
plagiarism detection has not been reported.

H. Wang et al. (Eds.): ICYCSEE 2015, CCIS 503, pp. 459–463, 2015.

To detect the high obfuscation plagiarism, especially the paraphrasing and translation plagiarism, we develop HawkEyes, a plagiarism detection system especially for the high obfuscation plagiarism.

2   Frameworks and Core Algorithms of HawkEyes

2.1   Frameworks

According to Potthast [1], Figure 1 shows the basic plagiarism detection framework of HawkEyes.

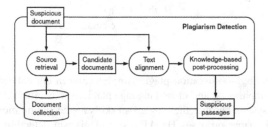

Fig. 1. Plagiarism Detection Framework

The plagiarism detection is divided into source retrieval, text alignment and post-processing by the framework showed in Figure 1. For a suspicious document, the target of source retrieval is to retrieve all plagiarized sources in the source documents corpus which the suspicious document plagiarizes. Given a pair of documents, the task of text alignment is to identify all contiguous maximal-length passages of reused text between the suspicious document and the plagiarism candidate documents. If the suspicious document and the source document are written by different language, this type of detection is called cross-language plagiarism detection. Third, knowledge-based post-processing, where the extracted passage pairs are cleaned, filtered, and possibly visualized for later presentation [4].

2.2   Source Retrieval

The phrases of the HawkEyes' source retrieval algorithms include: (1) chunking, (2) keyphrase extraction, (3) query formulation, (4) search control, and (5) download filtering.

Firstly, the suspicious document is divided into some segments. Commonly, we choose five sentences as one segment. Then, keyphrases are extracted from the suspicious document as the query keywords to retrieve the plagiarism sources. A ranking model based on learning to rank is used to select the query keywords. Thirdly, given the keywords sets extracted from chunks, queries are formulated which are tailored to the API of the search engine used. Fourthly, the search controller schedules queries to the search engine and directs the download of search results. Lastly, HawkEyes filters the results to choose the documents which will be compared in the text alignment phrase.

2.3   Text Alignment

After getting the candidate documents, the task of text alignment is to identify all contiguous maximal-length passages of reused text between the candidate documents and the suspicious document. The detailed comparison is implemented between the suspicious document and each of the candidate document. HawkEyes divided the text alignment into 3 steps: (1) seeding, (2) match merging and (3) extraction filtering.

Given a suspicious document and a source document, seeding is used to look for the matching seeds. These matches are called "seed". Match merging is used to merge the seeds obtained in step (1) to form the longer fragments. The extraction filtering combined fragments to get the final plagiarism fragments according to certain standards. The above text alignment algorithm was described in [5] in detail which got the first place on overall measure in the international competition for PAN@CLEF2012.

In HawkEyes, two core algorithms named Plagiarism Seeding Algorithm based-on Information Retrieval and Bilateral Alternating Merging Algorithm are designed for high-obfuscation plagiarism especially [5]. The evaluation measure Plagdet, Recall and Precision are shown in Ref. 1.

3   Demonstration

The core functions of HawkEyes include: incremental indexing, source retrieval, text alignment, upload and download source documents, user documents management, source documents management, visualization of detection results, etc.

Table 1 shows the time performance of HawkEyes. The average length of source doucuments is 38881 words and the average length of suspicious documents is 13688 words.

Table 1. Time Performance of HawkEyes

Sub-Phase	Time(s)/document	Memory	Processor	OS
Source Retrieval	21.4663	8G	Intel(R) Core(TM) i5-3470 CPU @3.20GHz	win7
Text Alignment	0.43	32G	Intel(R) Xeon(R) CPU E5-2620 @ 2.00GHz	Linux

The detailed framework of HawkEyes System for Plagiarism Detection is described in Figure 2.

The demonstrations for source retrieval and text alignment are shown in Figure 3 and Figure 4. The corpus which the demonstrations used is the train corpus of PAN@CLEF2014.

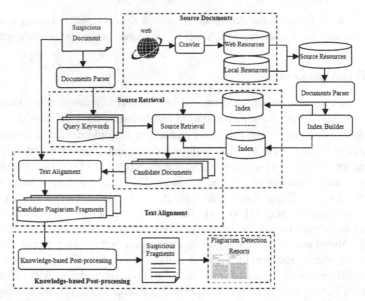

Fig. 2. The Detailed Framework of HawkEyes Plagiarism Detection System

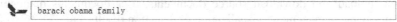

HawkEyes Plagiarisn Deteetion System

Suspicious Document:suspicious-document001-batch1

barack obama family

<u>1332938960</u>

wife.\nWhen President Barack Obama was sworn in President Barack Obama was sworn in yesterday, the image of a first family forfuture ones.\nAs the image of a first family for future ones.\nAs wife...

<u>1229226862</u>

the night of Barack Obama's historic the night of Barack — an African-American family with deep roots in New — an African-American family with deep roots in New the night of Barack Obama&rsquo...

Fig. 3. Source Retrieval Demonstration

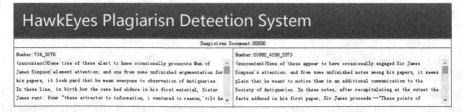

Fig. 4. Text Alignment Demonstration

Acknowledgement. This work was supported by Research Project of Heilongjiang Provincial Department of Education (12541649).

References

1. Potthast, M., Hagen, M., Gollub, T., Gollub, T. (eds.): Overview of the 5th International Competition on Plagiarism Detection, CLEF 2013 Evaluation Labs and Workshop – Working Notes Papers, Spain (2013)
2. Jocoy, C.L., Di Biase, D.: Plagiarism by adult learners online: A case study in detection and remediation. The International Review of Research in Open and Distance Learning 7(1) (2006)
3. Alzahrani, S.M., Salim, N., Abraham, A.: Understanding plagiarism linguistic patterns, textual features, and detection methods. IEEE Transactions on Systems, Man, and Cybernetics, Part C: Applications and Reviews 42(2), 133–149 (2012)
4. Potthast, M., Gollub, T., Hagen, M., Kiesel, J., Michel, M. (eds.): Overview of the 4th International Competition on Plagiarism Detection. CLEF 2012 Online Working Notes/ Labs/Workshop, Italy (2012)
5. Leilei, K., Haoliang, Q., Cuixia, D. (eds.): Approaches for Source Retrieval and Text Alignment of Plagiarism Detection. CLEF 2013 Evaluation Labs and Workshop – Working Notes Papers, Spain (2013)

LRC Sousou: A Lyrics Retrieval System

Yong Han[1], Li Min[1], Yu Zou[1], Zhongyuan Han[3,1], Song Li[1], Leilei Kong[1,2],
Haoliang Qi[1], Wenhao Qiao[4], Shuo Cui[5], and Hong Deng[1]

[1] School of Computer Science and Technology, Heilongjiang Institute of Technology,
Harbin, China
[2] College of Information and Communication Engineering, Harbin Engineering University,
Harbin, China
[3] School of Computer Science and Technology, Harbin Institute of Technology,
Harbin, China
[4] Beijing Institute of Surveying and Mapping, Beijing, China
[5] China Resources, Shenzhen, China
song1230717@126.com

Abstract. Lyrics retrieval is one of the frequently-used retrieval functions of search engines. However, diversified information requirements are neglected in the existing lyrics retrieval systems. A lyrics retrieval system named LRC Sousou, in which erroneous characters are corrected automatically, the mixed queries of Chinese words and Pinyin are supported, and English phonemes queries are also achieved effectively, is introduced in this paper. The technologies of natural language processing, information retrieval and machine learning algorithm are applied to our lyrics retrieval system which enhance the practicability and efficiency of lyrics search, and improve user experience.

Keywords: Lyrics retrieval, information retrieval, learning to rank.

1 Introduction

According to the CNNIC report, by the end of June 2014, the using frequency of network music is up to 77.2%, ranked as the fourth in all network software[1]. Therefore, lyrics retrieval system is an often used function. Lyrics search is asked to retrieve and return the information about the related song (e.g. title, lyric, artist, album, etc.) and then show them to the user after the queries such as the title of the song, a part of the lyric or the artist etc. are submitted. Nevertheless, in practical application, users submit not only standard queries but also nonstandard queries, such as inaccurately remembered lyrics, wrong Chinese spelling, Pinyin used to instead of some forgotten Chinese words, as well as English words which are known only the pronunciation but none of their spelling. In these special cases, traditional information retrieval technologies based on keywords matching are challenged. As a result, users have to change keywords repeatedly to get the satisfactory results. Thus it is useful in improving the user experience by providing diverse, humanized query support and meeting the requirements above.

H. Wang et al. (Eds.): ICYCSEE 2015, CCIS 503, pp. 464–467, 2015.

For this purpose, we implement a lyrics retrieval system named LRC Sousou by using the techniques of natural language processing, information retrieval and machine learning algorithm. The system not only can correct the erroneous characters automatically, support mixed query of Chinese words and Pinyin, but also support the English phonemes retrieval. Specifically, the system implements the function of correcting erroneous characters via the natural language processing technique, realizes the Chinese word and Pinyin mixture retrieval by taking advantage of mixture index. For the sake of homonyms retrieval, English phonemes index and Chinese fuzzy pronunciation index are used. In order to complete the final results ranking, we merge multiple indexes by applying learning to rank algorithm.

2   System Design

There are four main components in the lyrics retrieval system, namely, Crawler and Preprocessing, Build index, Query modification and Retrieval. As is shown in Fig 1.

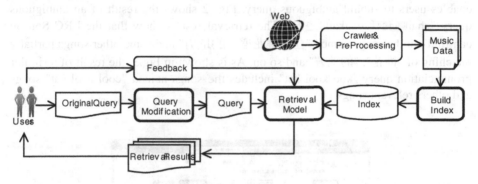

Fig. 1. Framework of the lyrics retrieval system

Crawler and Preprocessing Component crawls the LRC files and the web pages that contain the lyric information from the Internet. We extract information like lyric, artist, album and so on by applying Stanford Named Entity Recognizer (NER)[1].

Build Index Component. In order to meet various retrieval needs, several index strategies, such as word segmentation by Stanford Word Segmenter[2], n-gram, Pinyin and English phonemes based on double metaphone[2], are used to build index respectively on the on the lyric, the title of the song, the artist as well as the album.

Query Modification Component supplies and revises the queries. To conduct the Chinese query modification, we turn the Chinese words into Pinyin and approximate pronunciation firstly, and then use a translation model convert Pinyin to Chinese words. For instance, the original query "潜力值外" is modified to the query

---

[1] http://nlp.stanford.edu/software/CRF-NER.shtml
[2] http://nlp.stanford.edu/software/segmenter.shtml

"千里之外". Moreover, the English words which have similar pronun-ciation are added into the query. The modified query contains the original query, revised Chinese word, Pinyin, approximate pronunciation Pinyin and English.

Retrieval Model Component. The 68 similarity scores between the parts of modified query and fields are calculated respectively by a special vector space model[3]. To combine these features, RankLR[4], a learning to rank algorithm based on logistic regression, is employed to train a retrieval model with labeled data. Besides, the user feedbacks, such as user click and LRC download, are used to adjust the retrieval model.

3   Demonstration

The LRC Sousou has download and index 160288 LRC files from internet by lucene toolkit[3]. Users can use the accurate retrieval option to input precise query to research the artists, song name, Lyrics, album or all of them. And then, smart retrieval option enables users to submit ambiguous query. Fig. 2 shows the result of an ambiguous query "zhai哪套话sengka的地方". The retrieval results show that the LRC Sousou can find out the songs named "在那桃花盛开的地方" and some other songs partially matching of "桃花", "的地方" and so on. As is shown in Fig.3, the result of a similar pronunciation query "koo kool kat" includes the songs named "cool cool cat" songs and others relevant songs.

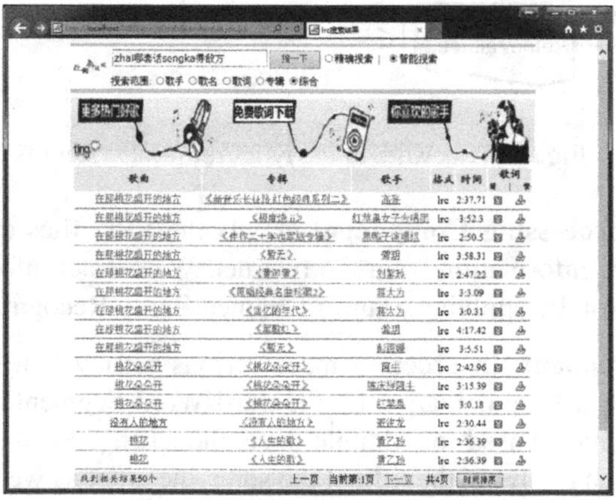

Fig. 2. The example of Chinese ambiguous query

---

[3] http://lucene.apache.org

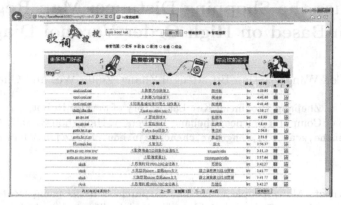

Fig. 3. The example of English similar pronunciation query

Acknowledgement. This paper is supported by Scientific Research Fund of Heilongjiang Provincial Education Department (NO: 12511444).

References

1. ChinaInternet Network Information Center. 34th Statistical report on Internet development Chinese [EB/OL] (July 2014) (in Chinese), http://www.cnnic.net.cn/
2. Phillips, L.: The Double Metaphone Search Algorithm. C/C++ Users Journal (June 2000)
3. Hatcher, E., Gospodnetic, O.: Lucene in Action. Manning Publications (2004)
4. Han, Z., Li, X., Yang, M., Qi, H., Li, S., Zhao, T.: Hit at TREC 2012 microblog track. In: Proceedings of Text REtrieval Conference (2012)

Reduce the Shopping Distance: Map Region Search Based on High Order Voronoi Diagram

Zhi Yu, Can Wang*, Jiajun Bu, Mengni Zhang, Zejun Wu, and Chun Chen

Zhejiang Provincial Key Laboratory of Service Robot,
College of Computer Science, Zhejiang University, Hangzhou 310027, China
{yuzhirenzhe,wcan,bjj,mengnier,chenc}@zju.edu.cn, Zejun.WU@gmail.com

Abstract. Many people would like to purchase items using location-based services to find the suitable stores in daily life. Although there are many online map search engines giving isolated Point-of-Interest as query results according to the correlation between isolated stores and the query, this interaction is difficult in meeting the shopping needs of people with disabilities, who would usually prefer shopping in one single location to avoid inconvenience in transportation. In this article, we propose a framework of map search service using Region-of-Interest (ROI) as the query result, which can greatly reduce users shopping distance among multiple stores. High order Voronoi diagram is used to reduce the time complexity of Region-of-Interests generation. Experimental results show that our method is both efficient and effective.

Keywords: Map Search, Region Search, Voronoi Diagram.

1 Introduction

With the rapid development of the Internet and handhold terminals, more and more spatial data adding to the Internet everyday, and Geographic Information System (GIS) has become a important Big Data Server. Many people rely on GIS to obtain a variety of Location Based Services (LBS) [1]. Especially when people want to shop for something but don't know where to go, the majority of people would turn to online map search engines for help. This demand also promotes the development of map search system.

Similar as other Big Data Servers, GIS and LBS can support only few simple online applications, such as a single store query which cost very little response time. The query results from the current map search engines are based on Point-of-Interest, which are isolated locations ranked by their correlation with the query keyword. People often would desire more shops to choose from when shopping. When recommending isolated stores to users, it is difficult to ensure that they have satisfactory purchase and it is quite often that they have to traverse multiple locations in different regions to get what they want. This will undoubtedly increase the shopping distance for users when recommending

---

* Corresponding author.

H. Wang et al. (Eds.): ICYCSEE 2015, CCIS 503, pp. 468–473, 2015.

shops using map services. The problem is especially severe for people with disabilities, who would prefer shopping in one single region to avoid inconvenience in transportation. While these kind of application will increase time complexity obviously with bad algorithm, which are not suitable for the real online server with such large amount of data.

In this paper, we propose a novel region-based map search method to reduce the expected shopping distance for people with disabilities. Rather than returning isolated POIs, we find dense areas containing multiple POIs as the query result. To improve the algorithm efficiency, we use high order Voronoi diagram to find dense regions containing the query keywords. Our algorithm has a time complexity of $O(n^2)$, which can well satisfy the time requirement of online search.

2   Related Work

Various Geographic Information Systems (GIS) have been developed to provide diversified Location Based Services (LBS) nowadays, include accessibility information service [2]. However, existing works mainly focus on finding Points-of-Interest (POIs) to for a user query by considering both the keyword similarity and geographic distance. With huge amount of data stored in GIS systems, many methods are used to speed up the searching process, such as R-tree and so on [3].

Although it is generally recognized that Regions-of-Interests (ROIs) can better satisfy users search need in many cases that POIs, few systems for ROI-based map service exist because calculating the relevance of an ROI is often too expensive to be applied in practical online service. There are existing researches for retrieving target ROIs from a given candidate ROI set for a user query such as [4]. The results of these methods are sub-optimal and the service quality is highly dependent on the predefined candidate ROI set.

3   Map Region Search

While existing map search mainly focuses on recommending isolated stores for a query, our purpose is to find relevant ROIs to reduce the expected shopping distance for users. We will first define the relevant terms and discuss the measurement for the expected shopping distance.

We assume that most people will finish a shopping task by visiting no more than k stores and we define a set with k adjoining POIs as a candidate *region* for user queries. It is natural to choose the variance of the distances between k stores in this region as a measurement for the *shopping distance*. A region with low variance is the one densely populated k stores, whose expected shopping distance will generally be much shorter than the one with higher variance. Thus the map region search problem can be defined as follows: given a set of n POIs each labeled with multiple query keywords, we want to find the region (which is a set of k POIs) with the minimum variance.

This problem is non-convex in the discrete solution space, meaning no global optimal solution can be found for it generally. The exhaustive search in the solution space would be prohibitively expensive for large data sets (the complexity is $O(n^k)$), which is not suitable for online search services. A naive k-NN search aiming to reduce the search space is to examine the regions generated by a single POI with its $(k-1)$ nearest neighbors. This method can reduce the searching complexity into $O(n)$, but the result is sub-optimal, as shown in the example in Fig. 1:

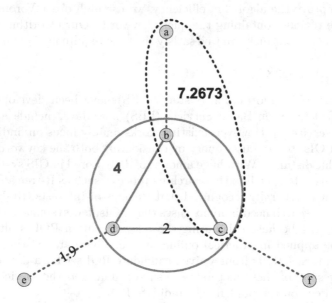

Fig. 1. An example which k-NN method lose the optimal solution

We want to find the optimal region containing 3 stores from 6 stores in Fig.1. The distance denoted by the blue line between stores is 1.9, while red line is 2. k-NN method can only retrieve the purple regions with a variance of 7.2673, which is similar as the set $\{a, b, c\}$, but the optimal region is the green one with a variance of 4, which is the set $\{b, c, d\}$.

In this paper, we use high order Voronoi diagram to improve map region search. Voronoi diagram has been thoroughly studied [5] and widely applied in 2-D space data analysis [6]. Given n points in the 2-D space, a k-order Voronoi diagram partitions the 2-D space into mutually disjoint Voronoi regions. Each Voronoi region is controlled by k points, meaning that for every point in this region, the distances between this point and the k control points cannot be longer than the corresponding distances between this point and other $(n - k)$ points. A case of a 3-order Voronoi is given in Fig.2, in which the blue point is in the Voronoi region controlled by $\{q_3, q_4, q_5\}$; so the distances between this point and each point in $\{q_3, q_4, q_5\}$ is shorter than the corresponding distances between it and $\{q_1, q_2, q_6, q_7, q_8\}$.

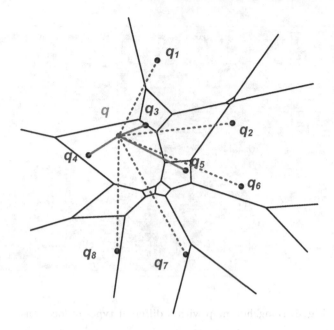

Fig. 2. An example of 3-order Voronoi diagram

It can be easily proven that the optimal k POIs with the minimum variance should control a k-order Voronoi region. Thus our proposed k-Voronoi map region search method contains 2 parts: (1) find out all k-order Voronoi regions as candidate solutions, and then (2) select the best one from them as the query answer. It can be proven that our method always find the optimal solution. In addition, existing works have shown that there are no more than $O(k(n-k))$ Voronoi regions, and the time complexity to find out all of them is $O(n^2)$[6], which ensures that our proposed method needs only $O(n^2)$ time to answer a query, which can be applied in online search services.

4 Experiments and Results

Data used in our experiment is downloaded from dianping.com, which contains location information of 13539 stores in Hangzhou, labeled with 100 distinct key-words. Fig.3 shows an example of the map with 4 different types of stores.

We use this data set to test the performance of the k-NN and the k-Voronoi map search method. In this paper, k is set to 10.

Firstly, we run both of the 2 methods on the No.1 keyword to recommend the top-100 regions separately, and the result is shown in Fig.4. We can see that the variances of regions found by k-Voronoi method are significantly lower than those found by k-NN method.

We also run the two methods to recommend the top-10 regions on all key-words. For each keyword, we calculate the variance in each region and then

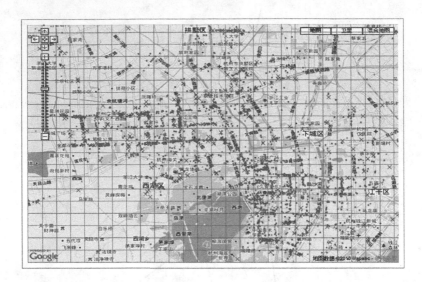

Fig. 3. Hangzhou map with 3 different types of locations

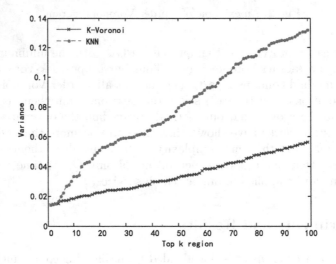

Fig. 4. Result of top-100 regions recommendation on No.1 keyword

calculate the average value of the 10 variances corresponding to the top-10 regions found by the two methods. Finally, we calculated the ratio of the two average value represented by the bars (one for each keyword) shown in Fig.5. All ratios are less than 1 and the average value of them is 0.8035 (represented by the red line in Fig.5), which indicates that our method has 20% gain over the k-NN method.

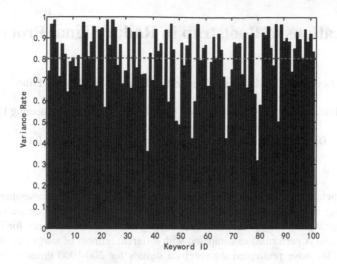

Fig. 5. Result of average top-10 regions recommendation on all 100 keywords

5 Conclusion

In this paper, we present a novel map region search method based on high order Voronoi diagram. When applied in online map service, our method can recommend dense shopping regions with minimum shopping distance. The method is expected to provide better shopping recommendations for people with disabilities.

Acknowledgements. This work is supported by National Key Technology R& D Program (Grant No. 2012BAI34B01).

References

1. Virrantaus, K., Markkula, J., Garmash, A., Terziyan, V., Veijalainen, J., Katanosov, A., Tirri, H.: Developing GIS-supported location-based services. In: Proceedings of the Second International Conference on Web Information Systems Engineering, vol. 2, pp. 66–75 (2001)
2. Miura, T., Yabu, K., Sakajiri, M., Ueda, M., Suzuki, J., Hiyama, A., Hirose, M., Ifukube, T.: Social Platform for Sharing Accessibility Information Among People with Disabilities: Evaluation of a Field Assessment. In: Proceedings of the 15th International ACM SIGACCESS Conference on Computers and Accessibility, pp. 65:1-65:2 (2013)
3. Cong, G., Jensen, C., Wu, D.: Efficient retrieval of the top-k most relevant spatial web objects. Proceedings of the VLDB Endowment 2(1), 337–348 (2009)
4. Fan, J., Li, G., Zhou, L., Chen, S., Hu, J.: Seal: Spatio-textual similarity search. Proceedings of the VLDB Endowment 5(9), 824–835 (2012)
5. Voronoi, G.: Nouvelles applications des paramétres continus lá théorie des formes quadratiques. Journal für die reine und angewandte 134, 198–287 (1908)
6. Aurenhammer, F.: Voronoi diagrams-a survey of a fundamental geometric data structure. ACM Computing Surveys (CSUR) 23(3), 345–405 (1991)

Applications of Bootstrap in Radar Signal Processing

Lei Liu[1], Dandan Fu[2], Yupu Zhu[2], Dan Su[2], and Ming Diao[1]

College of Information and Communication Engineering, Harbin Engineering University,
Harbin 150001, Heilongjiang, PRC
Daqing Normal University, Daqing 163712, Heilongjiang, PRC

Abstract. The bootstrap technique is a powerful method for assessing the accuracy of parameters estimator, that have been widely applied on statistical and signal processing problems. A novel program based on bootstrap for DOA estimation is performed to compared with different number of snapshots in this paper. We have resampled the received signals for 200-1000 times to create new data, therefore the arrival angle is estimated by the music algorithm in the conditions of confidence interval. The demo results show that higher estimation probability and smaller mean square error can be achieved in the situation of fewer snapshots received by passive radar system than that of traditional algorithm.

Keywords: bootstrap, music, DOA, radar signal.

1 Introduction

The problem of DOA estimation of radiating sources is important part in array signal processing, it has been a hot spot in recent years [1]. In smart antenna applications and for channel sounding, estimating the directions-of-arrival (DOAs) is of great interest to obtain high-per-formance mobile communication systems[2].The first DOA estimation algorithms were based on nonparametric estimation. Afterwards, new parametric estimation like the maximum likelihood (ML) or eigenspace-based algorithms, such as the multiple signal classification (MUSIC)scheme was proposed in 1979 by Schmidt[3].For reducing the associated computational load, Ray proposed the Estimation of Signal Parameters via Rotational Invariance Techniques (ESPRIT) [4] within the receiving-array geometry to allow simplified implementations in 1989. The most recent proposal was the covariance-based (CB) DOA algorithm [5, 6]. In order to improve the performance estimation, many "super-resolution" algorithm were proposed. All above results were achieved in the conditions of that we have infinite number of snapshots or SNR. In practical applications like the complex electromagnetic environment, it is difficult to capture the enough radiating source pulses, so the number of snapshots and SNR are often unsatisfactory.

The bootstrap technique was published in 1979 by Efron[7]. Now, bootstrap techniques are available as standard tools in several statistical software packages and are used to solve problems in a wide range of applications[8]. From the independent and identically distributed data (i.i.d) to the unindependent data, many methods like the Model-based, the block (or moving block) bootstrap, the naive bootstrap, sieve

H. Wang et al. (Eds.): ICYCSEE 2015, CCIS 503, pp. 474–479, 2015.

bootstrap and so on were proposed in recent years. Applications of Bootstrap have been reported in radar signal processing [9], Image processing [10], and time-frequency analysis [11]. Although the applications of array signal processing were reported in some articles, but there are very few applications in DOA estimate.

The paper is organized as follows: In Section 2, the signal model is defined, then the music algorithm and the bootstrap method are described. On the basis of the data model, the proposed method is given in Section 3. In Section 4, the demo are performed to estimate the direction of arrival with a large number of simulation experiments. The conclusions are given in Section 5.

2   Signal Model

2.1   Data Model

Suppose that there are N narrow band sources with same wavelength impinging on an ULA (Uniform Linear Array) with M sensors and the spacingbetween the adjacent elements is d. The output vector received by the array at the snapshot t is given by:

$$x(t) = \sum_{i=1}^{N} \alpha(\theta_i) s_i(t) + e(t) \tag{1}$$

Where $\alpha(\theta_i)$ is the steering vector, which is the array response of associated to the i-th signal source; $e(t)$ is M-dimension vector of noise.

The received data matrix in an observed window has decomposition:

$$X(t) = AS(t) + N(t) \tag{2}$$

We assume that noise is spatially white noise, and covariance of received data is given by:

$$\beta_n = E[e(t)e(t)^H] = \sigma^2 I \tag{3}$$

2.2   Music Method

In 1979, R.O.Schmidt proposed the MUSIC [12] method which promotes the development of high-resolution algorithm for array signal processing. It is the most classic subspace algorithm. MUSIC algorithm calculates the covariance matrix array output data firstly, then the characteristics of the matrix is broken down. Get two mutually orthogonal signal subspace and noise subspace, then according to the orthogonality between them can be on the letter Number of each parameter is used to estimate effective.

The maximum likelihood estimate of the covariance matrix is:

$$\beta = \frac{1}{N}\sum_{i=1}^{N} XX^{H} = \beta_s + \beta_n \tag{4}$$

Where $\beta_s = U_s \Sigma_s U_s^{H}$ is the covariance matrix array of the signal, where U_s and Σ_s are composed of singular vectors corresponding to the N largest singular values. $\beta_n = U_n \Sigma_n U_n^{H}$ is the covariance matrix array of the nosie, and $U_n \Sigma_n$ are composed of singular vectors corresponding to the rest M-N singular values.

The spectrum estimation formula is defined as:

$$P_{MUSIC} = \frac{1}{a^{H}(\theta)U_n U_n^{H} a(\theta)} \tag{5}$$

2.3   Bootstrap Principle

The bootstrap principle is very simple,the main idea is resample the data with replacement. It is using resampling sample value instead of real value to estimate parameters.

Suppose that we have data received in $x = \{x_1, x_2,\dots, x_n\}$, which are realizations of the random sample $X = \{X_1, X_2,\dots, X_n\}$, drawn from some unknown distribution F_X. Let $\hat{\delta} = \hat{\delta}(X)$ be an estimator of some parameter δ of F_X, which could be, for example, the mean $\delta = \beta_n$ estimated by the sample mean $\hat{\delta} = \hat{\delta}(X)$, The aim is to find characteristics of $\hat{\delta}$ such as the covariance of $\hat{\delta}$. Normally, the data $(X_1, X_2,\dots, X_n)$ is independent identically distributed (i.i.d.) data . If the sample is dependent data, then we can use the nonparametrically estimation like the residual method or block bootstrap.

3   Proposed Method

According to data model in formula (2) and (4), compute the data covariance matrix β from X, then decomposition of the data β to get the β_s and β_n.

So we can resample the noise subspace β_n to get new residual β_n^* and create new bootstrap data by adding the re-sampled residuals to

$$\beta^* = \beta_s + \beta_n^* \tag{6}$$

Since the noise is supposed the white noise, so we can use the empirical distribution to correspond to the maximum probability density in probability distribution curve.The step of residual method was shown in table 1.

Table 1. The step of residual method

Step 1: Define the received data matrix with L snapshots and βx be the covariance of received data ;

Step 2: Perform SVD of β ,and construct the signal space U_S and noise space U_N;

Step 3: Calculate the residuals βn , create a set of bootstrap residuals βn* by resample with replacement from βn ;

Step 4: Estimate the sample eigenvalues, calculate the direction of arrival with music.

Step 5: Repeat the step 3 to 4 B times to obtain the set of angle.

Step 6: Recalculate angle according to the confidence interval use the empirical distribution.

4   Demonstration

To demonstrate the features of our system, we load some varied data of radar to resample 200-1000 times to compared the traditional algorithm.

The program was based on vc++ and matlab to simulate the demo.

We attempt to demonstrate our system in following steps:

1. Data Load: The radar signal be load into our system.
2. Data set up:We can set up the various parameters of the data includes M、 N、 i、 t and so on.
3. Resample set up:The times of resample or the sample were set up in this step.
4. Method choice:According to the characteristics of the data to choice the bootstrap method.
5. Run: Run the demo.
6. Results comparing: The comparison results was shown after running the program.

Program interface and some results are as follows:

Fig. 1.

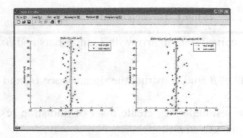

Fig. 2.

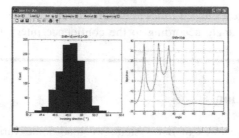

Fig. 3.

5   Conclusion

For the convenience of the DOA research in fewer snapshots case, our system provides the friendly interface to simulate the DOA estimate under the different conditions. Through the image of the graphical interfaces as shown in Figure 2. In the interface, a user could select different data sources, set up the parameters of bootstrap and pick the method. When the run button is clicked, the results are shown in the window. And we can compare to the different results visually. An example for result page is shown in Figure 3.

In short, this demo system can expressed the DOA estimation under different conditions directly. It can help researchers analyze data better and achieves the design expectations. In order to close to the practical applications, we plan to increase data sources and improve algorithms in further demo.

Acknowledgment. This research has been supported by Natural Science Foundation of Daqing Normal University (12ZR24).

Reference

1. Ben-Ari, E., Remez, J.: Performance Verification of a Multimodal Interferometric DOA-Estimation Antenna. IEEE Antennas and Wireless Propagation Letters 10, 1076–1080 (2011)
2. Goossens, R., Rogier, H., Werbrouck, S.: UCA Root-MUSIC with sparse uniform circular arrays. IEEE Transactions on Signal Processing 56(8), 4095–4099 (2008)

3. Schmidt, R.O.: Multiple emitter location and signal parameter estimation. In: Proceedings of the RADC Spectral Estimation Workshop, pp. 243–258 (1979)
4. Roy, R., Kailath, T.: ESPRIT–Estimation of parameters via rotationalinvariance techniques. IEEE Transactions on Acoustics, Speech, and Signal Processing 37(7), 984–995 (1989)
5. Ferreira, T., Netto, S.L., Diniz, P.S.R.: Low complexity covariance-based DOA estimationalgorithm. In: Proceedings of the 15th European Signal ProcessingConference, Poznan, Poland, pp. 100–104 (September 2007)
6. Zoltowski, M.: Novel techniques for estimation of array signalparameters based on matrix pencils, subspace rotationsand total least-squares. In: Proceedings of the IEEE ICASSP, Seattle, WA, pp. 2861–2864 (May 1998)
7. Efron, B.: Bootstrap methods: another look at the jackknife. The Annals of Statistics, 1-26 (1979)
8. Zoubir, A.M., Robert Iskander, D.: Bootstrap methods and applications. IEEE Signal Processing Magazine 24(4), 10–19 (2007)
9. Foucher, S., Farage, G., Bénié, G.B.: Application of bootstrap techniques for the estimation of Target Decomposition parameters in RADAR polarimetry. In: IEEE International Geoscience and Remote Sensing Symposium, IGARSS 2007, IEEE (2007)
10. Chuang, S.C., Hung, W.L.: Image classification using bootstrap likelihood ratio method. In: 2010 International Conference on Machine Learning and Cybernetics (ICMLC), vol. 2. IEEE (2010)
11. Politis, D.N.: The impact of bootstrap methods on time series analysis. Statistical Science 18(2), 219–230 (2003)
12. Schmidt, R.O.: Multiple emitter location and signal parameter estimation. IEEE Transactions on Antennas and Propagation 34(3), 276–280 (1986)

SAR Image Quality Assessment System Based on Human Visual Perception for Aircraft Electromagnetic Countermeasures

Jiajing Wang[1], Dandan Fu[2], Tao Wang[2], and Xiangming An[2]

[1] College of Information and Communication Engineering, Harbin Engineering University, Harbin, China
wangjiajing@ hrbeu.edu.cn
[2] Daqing Normal University, Daqing, China
{601770950,1656975717,113973589}@qq.com

Abstract. In electronic confrontation, Synthetic Aperture Radar (SAR) is vulnerable to different types of electronic jamming. The research on SAR jamming image quality assessment can provide the prerequisite for SAR jamming and anti-jamming technology, which is an urgent problem that researchers need to solve. Traditional SAR image quality assessment metrics analyze statistical error between the reference image and the jamming image only in the pixel domain; therefore, they cannot reflect the visual perceptual property of SAR jamming images effectively. In this demo, we develop a SAR image quality assessment system based on human visual perception for the application of aircraft electromagnetic countermeasures simulation platform. The internet of things and cloud computing techniques of big data are applied to our system. In the demonstration, we will present the assessment result interface of the SAR image quality assessment system.

Keywords: Synthetic Aperture Radar (SAR), image quality assessment system, human visual perception, internet of things, cloud computing.

1   Introduction

SAR has been widely applied to military reconnaissance because the imaging of SAR is not influenced by the weather, geographical location, or time. However, the most serious challenge that SAR reconnaissance mission faces in modern warfare is electronic interference. For this reason, research into SAR jamming and anti-jamming technology has become an important subject in the field of military electronic countermeasures [1]. The assessment of SAR images quality is the basis and the prerequisite for SAR jamming and anti-jamming technology [2], which will promote the precise debugging of electronic countermeasure equipment and have a strong guiding role in actual combat.

The existing SAR image quality assessment methods mainly include subjective assessment methods [3] and objective assessment methods. Subjective evaluation is a very tedious and time-consuming process, which is not recommended in automated

H. Wang et al. (Eds.): ICYCSEE 2015, CCIS 503, pp. 480–483, 2015.

practical applications. The objective assessment methods contain the property indexes of evaluation [4], the structural similarity (SSIM) [5] and methods based on texture [6], and so on. However, the visual perceptual property is not involved in these methods, which might result in low reliability of assessment conclusions.

Therefore, we develop a SAR image quality assessment system based on human visual perception. The novel algorithm extracts the texture feature of the SAR image, and analyzes the texture feature matrix using a discrete wavelet transform which is advantageous to the quality assessment based on multi-channel characteristics of human perception. The final quality is combining the octave feature and the directional selectivity of the contrast sensitivity function (CSF) with the masking effect of human visual perception. The system is applied to aircraft electromagnetic countermeasures simulation platform. The administrators have the ability to interact with the system through a graphical interface that allow them to load the original reference SAR images, add the jamming SAR images, input SQL statements and view scores of all the records from a cloud computing platform. Users are able to obtain the quality assessment results by inputting the information of reference image and changing the parameters of the jamming image.

2   The Principle of SAR Image Quality Assessment System

The SAR image quality assessment system mainly has the following sections: the SAR image data input portion, the image sequence management module, the jamming assessment module and the output portion. Hardware-in-the-loop simulation system is responsible for sending the SAR image data to the SAR image quality assessment system through Ethernet interface. The image data are classified and matched in the image sequence management module, and stored in a database. The jamming assessment module analyzes the image data and evaluates the quality of the jamming image, the assessment results are saved in the database of SAR image quality assessment system with the management technique of internet of things. According to the requirements of user, the image with the corresponding information and the assessment result can be extracted from the cloud computing platform. At the same time, the image data processing server imports the assessment parameters to reporting services process by SQL Server interface and then generates the parameters report. At last, output the final data through Ethernet interface into the simulation demonstration and the later stages.

3   The Algorithm in Jamming Assessment Module

SAR image is a kind of gray image that reflects the backscattering feature of the target [7], it has obvious textural features [8], and hence the novel algorithm uses the texture property which is described as follows in outline:

1. Extract the texture features to obtain the parameter matrix of texture features for both original reference SAR image and jamming image. The contrast parameters extracted from GLCM can be applied to express the texture features for image.

Because the contrast parameters not only can describe different texture features but also can reduce the computational complexity.

2. Decompose the parameter matrix into 5 levels using the 'Daubechies' wavelet. The 16 wavelet coefficients of different frequency bands in different directions are obtained, subsequently. This process simulates the human visual perception model that the visual information of different frequency ranges and the different directions enters into different visual channels.

3. According to the contrast masking effect and CSF of visual perception, the octave weighting coefficient of each sub-band is calculated.

4. Using the angular, spatial and structural similarity of each corresponding sub-band between the original reference image and the jamming image to assess the jamming effect. The final weighted assessment result is given in equation (1).

$$Q = \frac{Q_{LL} + \sum_{i=1}^{N} \sum_{j=1}^{3} \omega_{i,j} Q_{i,j}}{\omega_{LL} + \sum_{i=1}^{N} \sum_{j=1}^{3} \omega_{i,j}} \tag{1}$$

Where Q_{LL} is the quality measurement of the lowest frequency component. $\omega_{i,j}$ is the weighting coefficient for each sub-band, ω_{LL} is the weighting coefficient of the lowest frequency with value 1. $Q_{i,j}$ is the quality of each sub-band, $Q_{i,j} = \beta_1 A_{n_{i,j}} + \beta_2 R_{n_{i,j}} + \beta_3 S_{n_{i,j}}$.

4   Demonstration

Our system uses real SAR image database, the data from this database has been scored by human subjective assessment and the mean opinion score (MOS) has been obtained. When users enter the system, they can select different reference images by querying information of image ID, image name or image location. After confirming the reference image, choose different jamming types and jamming signal ratio (JSR), then press the button of "Assessment Result", the system presents the interface as Fig. 1, it shows the images of reference and jamming, the information about the images, and the quality score of the SAR jamming image which is approximate to the MOS. In order to verify the superiority of the visual perception based algorithm, the system interface also shows the traditional methods of equivalent number of looks (ENL), correlation coefficient (CC), structure similarity (SSIM) and singular value decomposition (SVD).Users can refer and compare the different results with MOS.

The SAR image quality assessment system which is supported by the image parameter database has completed the overall process of experimental data processing and analysis from image acquisition, classification, and matching processing to jamming image quality assessment.

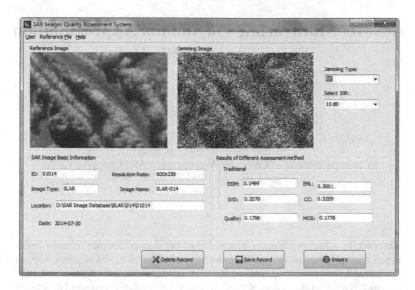

Fig. 1. Assessment result interface of SAR image quality assessment system

Acknowledgments. This work was supported by Research Project of Heilongjiang Provincial Department of Education (12533003).

References

1. Xiaozun, M., Donglin, S., Zhonggan, B.: Simulation on ECM effect based on GMTI using single-antenna SAR. In: 8th International Symposium on Antennas, Propagation and EM Theory Proceedings, pp. 1542–1545. IEEE Press, New York (2008)
2. Yang, L., Daihai, D., Xuesong, L.: An evaluation index consistent with HVS using correlation coefficient for SAR jamming effect assessment. Journal of Electronics and Information Technology 33, 1505–1509 (2011)
3. Shuhong, J., Weisheng, D.: SAR Image Quality Assessment Based on SSIM Using Textural Feature. In: 7th International Conference on Image and Graphics (ICIG), pp. 281–286. IEEE Press, New York (2013)
4. Han, Z., Yu, L., Yi, S.: SAR image quality assessment using coherent correlation function. In: 5th International Congress on Image and Signal Processing (CISP), pp. 1129–1133. IEEE Press, New York (2012)
5. Zhou, W., Bovik, A.C., Sheikh, H.R.: Image quality assessment: from error visibility to structural similarity. IEEE Trans. Image Process. 13(4), 600–612 (2004)
6. Anzhou, H., Rong, Z., Dong, Y.: Perceptual Quality Assessmet of SAR Image Compression Based on Image Content Partition and Neural Network. Chinese Journal of Electronics 3(22), 543–548 (2013)
7. David, S., Adrian, S., Betlem, R., Erich, M.: Geometric and radiometric correction of ESA SAR products. European Space Agency (Special Publication) ESA SP (2007)
8. Gholamreza, A.: A new statistical-based kurtosis wavelet energy feature for texture recognition of SAR images. IEEE Trans. Geosci. Remote Sensi. 50(11), 4358–4368 (2012)

A Novel Multivariate Polynomial Approximation Factorization of Big Data

Guotao Luo[1,*] and Guang Pei [2]

[1]Dept. of Computer, Sichuan TOP Vocational Inst., Chengdu,China
luoguotao@scetop.com
[2]Chengdu Institute of Computer Application, CAS , Chengdu, China
peiguang616@163.com

Abstract. In actual engineering, processing of big data sometimes requires building of mass physical models, while processing of physical model requires relevant math model, thus producing mass multivariate polynomials, the effective reduction of which is a difficult problem at present. A novel algorithm is proposed to achieve the approximation factorization of complex coefficient multivariate polynomial in light of characteristics of multivariate polynomials. At first, the multivariate polynomial is reduced to be the binary polynomial, then the approximation factorization of binary polynomial can produce irreducible duality factor, at last, the irreducible duality factor is restored to the irreducible multiple factor. As a unit root is cyclic, selecting the unit root as the reduced factor can ensure the coefficient does not expand in a reduction process. Chinese remainder theorem is adopted in the corresponding reduction process, which brought down the calculation complexity. The algorithm is based on approximation factorization of binary polynomial and calculation of approximation Greatest Common Divisor, GCD. The algorithm can solve the reduction of multivariate polynomials in massive math models, which can obtain effectively null point of multivariate polynomials, providing a new approach for further analysis and explanation of physical models. The experiment result shows that the irreducible factors from this method get close to the real factors with high efficiency.

Keywords: Mass physical model, Multivariate polynomial approximation factorization.Reduction, Unit root.Binary approximation factorization, Approximation GCD algorithm.

1   Introduction

In actual engineering, especially in big data processing,massive physical models require to build the math model to find solution through analysis. In process of building the math model, massive multivariate polynomials are obtained. As a difficulty in both academic and engineering field, factorization of multivariate polynomials is the first and important step in solving multivariate polynomials. The null point of multivariate polynomials can

---

*Corresponding author.

H. Wang et al. (Eds.): ICYCSEE 2015, CCIS 503, pp. 484–496, 2015.

be obtained through factorization,   providing the basis for a deep analysis of the physical model. From unitary polynomial to multivariate polynomial, from integer field to complex field, the calculation complexity of factorization algorithm of multivariate polynomial is increasing with gradual reduction of accuracy. Since the realization of factorization of unitary polynomial in polynomial time by the use of LLL mesh reduction algorithm[1], Niederreiter(1993)[2] and Ruppert(1999) [3] proposed to establish ordinary differential equation (ODE) to   factorize the unitary polynomial. On this basis, Gao(2002) [4] proposed to establish partial differential equation (PDE) to factorize the binary polynomial, which reduced time complexity compared with the factorization algorithm of the traditional binary polynomial[5]. For the approximation factorization difficulties of the complex coefficient binary polynomial, Gao et.al.(2004) [6] proposed an effective way to use PDE to solve the approximation GCD, which laid a solid basis for studying approximation factorization of polynomial[7,8]. The researchers also had done a great number of work in expanding the factorization of the binary polynomial to the multivariate polynomial. GAO Xuhong et.al.(2013)[9]proposed a new algorithm to draw the sparse factor of multivariate polynomial of integral coefficient. By the use of the knowledge of the finite field, the algorithm reduces the multivariate polynomial of integer field to the binary polynomial, ensuring the stability of the number of nonzero term, factorize the binary polynomial, and finally applies the discrete logarithm method to restore the irreducible binary polynomial to the multivariate polynomial, achieving the factorization of multivariate polynomial of integer field. In terms of efficiency, the algorithm is as good as the traditional factorization of multivariate polynomial, with a high efficiency for drawing the sparse factor of the polynomial.

In the practical engineering, the math model needed to be established is not an integral coefficient polynomial but generally a real coefficient polynomial or a complex coefficient polynomial. Therefore, how to simplify and factorize complex coefficient polynomial is an important problem to be solved in the practical physical model. Thorough study has been made of the complex coefficient unitary polynomial, while the study of the complex coefficient binary polynomial relies too much on PDE[4] and approximation GCD algorithm[6,10,11]. For the multivariate polynomial, the PDE and approximation GCD algorithm are of no use. Therefore, the factorization of multivariate polynomial has always been a difficult issue. Traditionally, according to the irreducible theorem[12]of Hilbert and the irreducible theorem[13]of Bertini, the former theorem reduces multivariate polynomial with unitary polynomial, constructing the irreducible factor of multiple polynomial based on Hensel lifting scheme [14]; the latter reduces the multiple variable to the binary variable, with every variable indicating one platform. These ideas are new, however, in the reduction process, the nonzero term is added to replace the sparse polynomial with the non-sparse polynomial, increasing the difficulty of factorization. But these ideas provide a new way of thinking for researchers: that is the multiple polynomial can be reduced to the unitary or the binary polynomial through the reduction, at last, the irreducible binary factor is restored to the irreducible multiple factor by the use of the factorization algorithm of the binary polynomial. We extend the multiple polynomial factorization of integer coefficient proposed by (Gao, 2013) [9] to the complex field.

This paper presents a novel multiple polynomial factorization of complex coefficient for the multivariate polynomial of complex coefficient in math model. The algorithm reduces the multiple polynomial of complex coefficient to the binary polynomial at first;

secondly, it adopts the approximation factorization of binary polynomial to get the binary irreducible factor; at last, it uses the restoring method corresponding to the replacing to restore the irreducible binary factor to the irreducible multiple polynomial. The algorithm relies on the approximation factorization of the binary polynomial, smartly selecting the unit root as the reducible factor in the complex field. According to the employed reduction skills and restoring method, it can be ensured in a great probability that the irreducible polynomial can get closest to the irreducible factor of the original multiple polynomial. This algorithm can effectively solve the factorization of the multivariate polynomial deduced from the massive physical models , providing basis for a further analysis of the physical model. This paper presents an effective approximation factorization of multivariate polynomials of the complex coefficient in detail on the basis of massive physical models, which indicates the algorithm is an effective solution to the factorization of the multivariate polynomial in massive physical models. We will show you the effectiveness of the algorithm through the factorization of multivariate polynomials of the complex coefficient in the following.

2   Reduction and Restoration Process of Multivariate Polynomials

2.1   Basic Concepts of Polynomials

Considering the square-free multiple polynomial $F \in C[x_1, x_2, \ldots, x_n]$,which is $F = G_1 G_2 \cdots G_r$,in this formula, $G_i \in C[x_1, x_2, \ldots, x_n]$, $G_i \neq G_j$, $i \neq j$, G_i is approximately irreducible in the complex field. If F as weight factor in the complex field, then based on (Gao, 2002)[4] method the weight factor is taken out to make it a square free polynomial.Assuming $N = \{0,1,2,\cdots\}$, N^n indicates the n-dimensional array collection, supporting $\alpha = (\alpha_1, \alpha_2, \ldots, \alpha_n)$, with monomial $x^\alpha = x_1^{\alpha_1} x_2^{\alpha_2} \cdots x_n^{\alpha_n}$. The monomial ordering is defined: 1) there is one and the only one monomial tenable for any $\alpha, \beta \in N^n$, $\alpha = \beta, \alpha > \beta, \alpha < \beta$;   2) If $\alpha \neq 0$,then $\alpha > 0$; 3) For any $\alpha + \gamma > \beta + \gamma$, we have $\alpha > \beta, \gamma \in N^n$. Any multivariate polynomial can be written as : $F = \hat{a}x^\alpha + \sum_{\alpha > \beta} \hat{a}_\beta x^\beta$, here $\hat{a}, \hat{a}_\beta$ is non-zero plural. $\hat{a}x^\alpha$ is the first term of the polynomial F, recorded as : $lt(F)$, x^α is the first monomial, recorded as : $lm(F)$, x^α is the leading coefficient, recorded as: $lc(F)$,   the nonzero term of polynomial F is recorded as: $nops(F)$.

Given $F = \sum c_\alpha x_1^{\alpha_1} x_2^{\alpha_2} \cdots x_n^{\alpha_n}$,the support set of polynomial F is recorded as:

$$Supp(F) := \left\{ \alpha = (\alpha_1, \alpha_2, \ldots, \alpha_n) \in N^n : c_\alpha \neq 0 \right\} \tag{1}$$

2.2 Reduction Skills

For any multivariate polynomial $F \in C[x_1, x_2, \ldots, x_n]$, S is defined as the subset of the finite integer n, recording that $s = e^{\frac{I\pi}{n}}$ is the unit root determined by n, we set $d = 1 + 2 \max_{1 \le i \le n} \deg_{x_i}(F)$ as a positive integer. The reduction process is as follows: the element a_i, b_i of set S is taken randomly to define the reduction method:

$$R_1^{a,b} : x_i \to s^{d^{i-1}} X^{a_i} Y^{b_i}, \ 1 \le i \le n \tag{2}$$

$$R_2^{a,b} : x_i \to s^{2d^{i-1}} X^{a_i} Y^{b_i}, 1 \le i \le n \tag{3}$$

For any monomial cx^α, in which $\alpha = (\alpha_1, \alpha_2, \ldots, \alpha_n), c \in C, c \neq 0$,

$0 \le \alpha_j < d, 1 \le j \le n$, therefore, some monomial is reduced based on formula (2),

we can get: $R_1^{a,b}(cx^\alpha) = R_1^{a,b}(cx_1^{\alpha_1} x_2^{\alpha_2} \cdots x_n^{\alpha_n})$

$$= c\left(sX^{a_1}Y^{b_1}\right)^{\alpha_1} \left(s^d X^{a_2}Y^{b_2}\right)^{\alpha_2} \cdots \left(s^{d^{n-1}} X^{a_n}Y^{b_n}\right)^{\alpha_n}$$

$$= cs^{\alpha_1 + \alpha_2 d + \cdots \alpha_n d^{n-1}} X^{a_1\alpha_1 + a_2\alpha_2 + \cdots + a_n\alpha_n} Y^{b_1\alpha_1 + b_2\alpha_2 + \cdots b_n\alpha_n} \tag{4}$$

Recording: $R_1^{a,b}(cx^\alpha) = cs^{l(\alpha)} X^{a\alpha} Y^{b\alpha}, a = (a_1, \cdots, a_n)$,

$l(\alpha) = \alpha_1 + \alpha_2 d + \cdots \alpha_n d^{n-1}$ $b = (b_1, \cdots, b_n)$

simplified as:

$a\alpha = a_1\alpha_1 + a_2\alpha_2 + \cdots + a_n\alpha_n$ and $b\alpha = b_1\alpha_1 + b_2\alpha_2 + \cdots b_n\alpha_n$.

In the same way, based on formula (3) we can get $R_2^{a,b}(cx^\alpha) = cs^{2l(\alpha)} X^{a\alpha} Y^{b\alpha}$.

Based on the formula (2)-(4), all the monomials of the multiple polynomial are reduced to be the binary monomial, with the original multivariate polynomial being changed into the complex coefficient binary polynomial. The two most critical points for reduction of complex coefficient multivariate polynomial are:

1) After the reduction based on (2)-(4), in a great probability, the number of nonzero term of the polynomial will not be reduced(hat is after the reduction, the following equation will not occur at the same time $a\alpha = \tilde{a}\tilde{\alpha}$, $b\beta = \tilde{b}\tilde{\beta}$).

2) The multivariate polynomial is reduced to be the binary polynomial based on formula (2)-(4), the binary polynomial can not be reduced either.

For condition 1), there is the following nature:We set $S = \{0,1,\ldots,c\} \subseteq N$, the nonzero number of polynomial $G \in C[x_1, x_2, \ldots, x_n]$ is t=nops(G), randomly selecting $a_i, b_i \in S$, then the probability of the binary polynomial $R_i^{a,b}(G), i = 1, 2$ and multivariate polynomial G having the same number of nonzero is:

$$1 - \frac{t(t-1)}{|S|} \qquad (5)$$

Therefore, the value of the set S and the sparseness of the polynomial determine whether the algorithm succeeds or not. For condition 2), from the reduction skills and selection of the reduction factor, we can get that the condition (2) holds up in a great probability (Gao, 2013)[9].

2.3   Reduction Process

Assuming $\hat{R}_1(X,Y)$ and $\hat{R}_2(X,Y)$ are the binary irreducible factors with the same supporting set based on the binary approximation factorization. Now the following operation is performed for the two binary irreducible factors: selecting the monomial coefficient with the same supporting set $c_1 = cs^{l(\alpha)} \in C, c \neq 0$,

$c_2 = cs^{2l(\alpha)} \in C, c \neq 0$, the division operation is performed: $c_2 / c_1 = s^{l(\alpha)}$, the logarithmic with s as the base is taken for the two sides of the equation: $\ln_s(c_2 / c_1) = l(\alpha)$, here $l(\alpha) \in Z$,which is:

$$l(\alpha) = \alpha_1 + \alpha_2 d + \cdots + \alpha_n d^{n-1} \qquad (6)$$

the vector $(\alpha_1, \alpha_2, \ldots, \alpha_n)$ is obtained after the modular operation of the formula (6). The original polynomial coefficient is obtained with calculation of $c = c_1 / s^{l(\alpha)}$.

As the value of set S determines the probability of success of the algorithm, if selecting multiple positive integer n_i in which any two integers are relatively prime, a greater S set can be constructed by the use of the relatively prime nature of integer, which is good for improving the success probability. Correspondingly, the corresponding unit root can be solved according to n_i , more binary polynomials are obtained with the reduction, the Chinese remainder theorem $l(\alpha)$ is applied in the reduction process.

3   Approximation Reduction of the Binary Polynomial

The first section of the paper reduces the complex coefficient multivariate polynomial to the complex coefficient binary polynomial, according to the theorem[9] given by the predecessors, we can get that the reduction method presented in this paper can hold up in a great probability. On this basis, this section illustrates the approximation factorization algorithm of the complex coefficient binary polynomial[4] . This algorithm is a relatively classic approximation algorithm proposed by the predecessors. We will introduce the approximation GCD factorization of the complex coefficient binary polynomial[6,10,11] at first, followed by the introduction of the approximation factorization algorithm of the binary polynomial.

3.1   Algorithm1 Approximation GCD Algorithm

Input : $g, h \in C[x, y]$

Output: nonzero approximation GCD of g and h

Step 1   According to the following two methods, the degree (k) of approximate GCD of g and h is determined.

(1)According to the relationship of $ug + vh = 0$,we construct the resultant $W = W_1(g, h)$,in which $g, h \in C[x, y]$, and satisfying the condition of $t \deg(u) < t \deg(h), t \deg(v) < t \deg(g)$, we calculate the rank of W by the use of the numerical approach.

(2)Value is assigned to some variable of g and h randomly for many times, reducing the binary polynomial to the unitary polynomial and finding the resultant of unitary polynomial for many times, and at last, the degree of the approximation GCD of g and h is determined.

Step2 We set: $t \deg(u) = t \deg(h) - k, t \deg(v) = t \deg(g) - k$,

and restrictive conditions of u and v given by step 1 are met, restructuring $W = W_k(g, h)$, and satisfying the dimensional number of the nuclear space of W is 1.

Step3 Calculating the singular vector corresponding to the minimum singular value of W nuclear space, which is the basic system of solution $[u, v]$.

Step4 According to the least-square relationship $\|h - du\|_2^2 + \|g + dv\|_2^2$, d is found to be the approximation GCD of g and h.

Algorithm 1 is the solution to approximation GCD of binary polynomial of the complex coefficient, through finding solution to GCD of binary polynomials and derivative polynomials.

3.2   Algorithm 2   The Approximation Factorization of the Binary Polynomial

Input: the binary polynomial $f \in C[x, y]$, with the relatively-prime between f and f_x, f is approximation square free with no factor in $C[y]$: $\deg_x(f) = n > 1$, $\deg_y(f) = m > 1$, $\hat{S}$ is a finite subset of C and $|\hat{S}| > mn$

Output: some irreducible factors of binary polynomial f in the complex field

Step 1 Constructing Ruppert matrix

(1)Constructing Ruppert matrix according to the partial differential relationship $\dfrac{\partial}{\partial y}\left(\dfrac{g}{f}\right) = \dfrac{\partial}{\partial x}\left(\dfrac{h}{f}\right)$ and the restrictive conditions of g and h;

(2)Using the singular value factorization of matrix to solve the singular value and rank of Ruppert matrix;

(3) Obtaining $g_1, g_2, \cdots, g_r$ according to the last r singular vector $v_1, v_2, \cdots, v_r$ obtained.

Step 2 Constructing and solving the matrix E_g;

(1)Randomly selecting $s_i \in \hat{S}$, constructing: $g = \sum\limits_{i=1}^{r} s_i g_i$;

(2) Fixed variable $y = \eta \in C$, ensuring this will not change the degree concerning x of the binary polynomial f;

(3) For $y = \eta$, according to the following relationship solving

$$a_{i,j} : \min \left\| rem\left(gg_i - \sum_{j=1}^{r} a_{i,j} g_j f_x, f \right) \right\|_2 ;$$

(4) Solving the characteristic value $\lambda_i, 1 \le i \le r$ of matrix $A = \begin{bmatrix} a_{i,j} \end{bmatrix}$, calculating

$$dis := \min_{1 \le i < j \le r}\left\{\left|\lambda_i - \lambda_j\right|\right\};$$

(5) If dis is close to 0, then turning to step 2.

Step3 Calculating $f_i = \gcd(f, g - \lambda_i f_x), f_i \in C[x, y], 1 \le i \le r$, by the use of algorithm 1.

Step 4 Judging whether the factorization result is right or not.

(1) Calculating $\min_{c \in C} \left\| f - c \prod_{i=1}^{r} f_i \right\|_2 / \|f\|_2$;

(2) If meeting the given error limit ε, then return to factors $f_1, f_2, \cdots, f_r$ and c of the original binary polynomial, otherwise turning to Step1.

Algorithm 2 is the approximation factorization of the binary polynomial with the approximation GCD conditions.

4   A Novel Approximation Factorization Algorithm of Multivariate Polynomial

In this section, a novel factorization algorithm of the complex coefficient multivariate polynomial is proposed   with illustration of detailed examples.

4.1   Algorithm3 A Novel Approximation Factorization Algorithm of Multivariate Polynomial

Input: Any multivariate polynomial of the complex field
$$f = \sum c_\alpha x^\alpha \in C[x_1, x_2, \ldots, x_n] \quad, \quad f \text{ is of square free form, taking:}$$
$d := 1 + 2 \max_{1 \le i \le n} \deg_{x_i}(f)$,taking randomly $n_1, n_2 \in Z^+$ and satisfying $\gcd(n_1, n_2) = 1, n_1 n_2 > l(\alpha)$, in which the form of $l(\alpha)$ is given by (6), we set: $s_1 := e^{I\pi/n_1}, s_2 := e^{I\pi/n_2}$, satisfying $s_1^{2n_1} = 1, s_2^{2n_2} = 1$

Output: The irreducible factor $(f_1, f_2, \cdots, f_r)$ of multivariate polynomial f

Step 1 Reduction

(1) Wet set: $FM := [\ ]$;

(2) Randomly choosing $a_i, b_i \in \{0, 1, \cdots, t\}, 1 \le i \le n$;

(3) Reducing: $c_\alpha x^\alpha, R_1(f_j) = \sum R_{1,n_j}^{a,b}(c_\alpha x^\alpha)$,

$R_{1,n_j}^{a,b}(c_\alpha x^\alpha) = c_\alpha s_j^{l(\alpha)} X^{a\alpha} Y^{b\alpha}$, $R_2(f_j) = \sum R_{2,n_j}^{a,b}(c_\alpha x^\alpha)$,

$R_{2,n_j}^{a,b}(c_\alpha x^\alpha) = c_\alpha s_j^{2l(\alpha)} X^{a\alpha} Y^{b\alpha}$,

$a\alpha = a_1 \alpha_1 + \cdots + a_n \alpha_n, \ b\alpha = b_1 \alpha_1 + \cdots + b_n \alpha_n$,

$$l(\alpha) = \alpha_1 + \alpha_2 d + \cdots + \alpha_n d^{n-1}, \ j = 1, 2.$$

Step 2 Factorizing $R_i(f_j)$ by use of algorithm 2.

Step 3 The irreducible binary factor is reduced to be the irreducible multivariate factor.

(1) Comparing the irreducible factors of $R_1(f_j)$ and $R_2(f_j)$, we take the monomial with the same supporting set ($a\alpha$ and $b\alpha$) where the support and the coefficient are extracted to get:

$$\left[c_\alpha, s_1^{l(\alpha)}\right], \left[c_\alpha, s_2^{l(\alpha)}\right], \left[c_\alpha, s_1^{2l(\alpha)}\right], \left[c_\alpha, s_2^{2l(\alpha)}\right].$$

(2) According to $s_1^{l(\alpha)}, s_2^{l(\alpha)}, s_1^{2l(\alpha)}, s_2^{2l(\alpha)}, n_1, n_2$, the Chinese remainder theorem is applied in calculation to obtain $\hat{l}(\alpha)$;

(3) According to $\hat{l}(\alpha)$ and d, the modeling calculation is used for obtaining $(\alpha_1, \alpha_2, \cdots, \alpha_n), c_\alpha \hat{x}^\alpha$ and the irreducible factor f_{kk};

(4) Judging whether f_{kk} can approximately make exact division of f, if the answer is affirmative, then $FM := [FM, f_{kk}]$, then turning to Step 3.

Step 4 Returning to the irreducible factor set $(f_1, f_2, \cdots, f_r)$.

Instruction of algorithm: selecting two relatively-prime integer n_1 and n_2, calculating the corresponding unit root s_1 and s_2, with which, obtaining $R_i(f_j)$ according to the reduction of (2)-(4). Factorization is performed of the four reduced binary polynomial, selecting the corresponding coefficient and support to be calculated to get the coefficient and support of multivariate irreducible factors by the use of the Chinese remainder theorem. As $n_1 n_2 > l(\alpha)$, selecting multiple n_i can reduce the calculating difficulty of using one n and improve the success rate of the algorithm. In the reduction of Step 3, comparison is made of the binary irreducible factors one by one to obtain the monomial with the same support, where the coefficient and support is taken. If the calculated multivariate irreducible factor is not the irreducible factor of the original multivariate polynomial, it indicates that in the binary irreducible factors, there is more than one factor with the same support, when the comparison should be made again. In the following comparison of support, the monomial already compared should not be considered again.

Algorithm 3 is the main algorithm of the paper, reducing the multivariate polynomial into the relevant binary polynomial for the multivariate polynomial of the complex coefficient, factorizing of the binary polynomial by means of algorithm 1 and algorithm 2 with restoration of the irreducible factor according to the restoration method, which produces the relevant irreducible polynomial factor.

4.2   One Detailed Example

With the complexity of the multivariate polynomial in the physical model and the illumination of this paper, a detailed practical example will be given to demonstrate the effectiveness of algorithm 3, which can effectively solve the arbitrary multivariate polynomial of the actual physical model.

Polynomial f is given: f: = 0.15xyz + 0.94247779607693797155xy + 0.5iz + 3.1415926535897932385i

According to algorithm 3, we set $d = 3$,selecting relatively-prime integ-

er $n_1 = 7$, $n_2 = 9$, correspondingly $s_1 = e^{-\frac{I\pi}{7}}$, $s_2 = e^{-\frac{I\pi}{9}}$. The binary polynomial

can be obtained by the use of the reduction method (By turn $R_1(f_{s_1})$,

$R_1(f_{s_2})$, $R_2(f_{s_1})$, $R_2(f_{s_2})$):

(0.135767176087548417934+0.063775182455346106599i)X12Y9+(0.9338869868
2970452448+0.126962562041509002041i)X7Y5-(0.14968156148667897702-
0.47706962820002442576i)X5Y4 + 3.141592654i
(0.136899767280231021 98+0.061306229036033424026i)X12Y9+(0.9345799276
7150206564+0.12175695102760540970i)X7Y5-(0.14368887699241407782-
0.47890866209399355182i)X5Y4 +3.141592654i
(0.095769681370497522014+0.11544768568573984142i)X12Y9+(0.9082711714
3889345823+0.25161056309906574111i)X7Y5-(0.28563410754739613961-
0.4103817206036381632i)X5Y4+3.141592654i
(0.099887283751752164124+0.11190411317148690093i)X12Y9+(0.9110186887
6359989349+0.24147328020602992514i)X7Y5-(0.27526639135290175225-
0.41741402651463558514i)X5Y4+3.141592654i

The following equation can be obtained by the use of algorithm 2 to factorize equation 4 above by turn:

[2.56913093444233-1.63307178320345i+(0.467945793804887-
0.125585160032648i)X5Y4,   -0.335853001100060-1.412564710609 15i+(-
0.433479672658569+0.0427508995770148i)X7Y5]
[-1.40881525278382-1.61184756465743i-
(0.141039692562378+0.310148051862897i)X5Y4,
1.19419645890214+0.09669701657 23197i+(0.0750488158580470-
0.351509130322330i)X7Y5]
[1.07955705567981-1.21461454550393i+(0.251453929639446-
0.0605098458770287i)X5Y4,
-0.293674078746672-1.03457062542084i+(-
0.322626861902372+0.00204571508754324i)X7Y5]
[-1.34176257168027-0.788149742869556i-
(0.109221050513896+0.222279657161651i)X5Y4,
0.958173781404232-0.454976971835472i-
(0.0582886391133313+0.312828273527952i)X7Y5]

Drawing the coefficient and support with the same supporting monomial, we can

obtain $\hat{l}(\alpha)$ and polynomial coefficient by the use of the Chinese remainder theo-

rem in calculation, which is reduced to obtain the irreducible multivariate factor:

[(0.30000000000000000000-0.12124031600327488101e-20i)xy+i,
(0.50000000000000000001+0.23375639270664146112e-
20i)z+3.1415926535897932385]

The first item is factor f_1 , the second item is factor f_2 , us-

ing $\dfrac{\|f - f_1 f_2\|_2}{\|f\|_2}$ in calculation to get the error: 1.256e-25.

5   Analysis and Experiments of Algorithm

The algorithm selects the unit root as the reduction factor, the unit root is cyclic, which can ensure the stability or small change of the coefficient of polynomial, meanwhile, reduction and restoration can be performed effectively. In the reduction process, the Chinese remainder theorem and modular arithmetic can be applied in calculating the support, with a small amount of calculation, the calculation complexity of the algorithm is mainly determined by the factorization of binary polynomial in the complex field.

In this paper, the factorization of the polynomial is performed on the platform Intel(R) Celeron(R) CPU E3300 @2.5GHz RAM 2.00GB,Digits =20Windows Maple16 for the experiment, which lists the first degree of the polynomial (the highest degree), and coeff error is the absolute error of the polynomial coefficient, backward error is

the integral relative error[6] with the calculation of $\dfrac{\left\|f - \Pi_i \hat{f}_i\right\|_2}{\|f\|_2}$, time refers to the

total time for factorizing the multivariate polynomial.

Table 1. The experimental examples and effect

example	max deg (f)	coeff error	backward error	time(s)
1	3,3,3	0	$1.2e^{-14}$	10.3
2	6,6,10	10^{-5}	$1.35e^{-4}$	687.3
3	12,7,5	10^{-3}	$6.01e^{-3}$	1121.1
4	4,4,4,4,4	10^{-6}	$2e^{-4}$	2124.3

Table 1 is the experimental results by using the approximation factorization algorithm of the complex coefficient multivariate polynomial proposed in this paper. In table 1, example 1 illustrates the multivariate polynomial of one integer in (Corless, 2002)[15], example 2-4 is used for factorizing the sparse multivariate polynomial of complex coefficient with high degree which is randomly produced. We take $d = 1 + \max_{1 \le i \le n} f(x_i)$ in the four experiments, two neighboring relatively-prime integers are taken for n_i according to different requirements, the corresponding s_i is calculated. From table 1, we can see that the algorithm is effective for the factorization of both the multivariate polynomial of integer and the sparse multivariate polynomial of the complex coefficient.

6   Conclusion

In big data processing,for the multivariate polynomial of the massive physical models, this paper proposes a complex coefficient multivariate polynomial approximation factorization algorithm, which can provide effective solution to the factorization of the multivariate polynomial of the complex coefficient in the massive physical models, providing further analysis of the physical model.The algorithm reduces multivariate polynomial to the binary polynomial by the use of the approximation factorization of the binary polynomial at first, secondly, the binary irreducible factor is obtained by the use of the approximation factorization algorithm of the binary polynomial; at last, the binary irreducible factor is reduced to the corresponding multivariate irreducible factor. From (Gao, 2013)[9], we know that the simplification and reduction ideas are effective in a great probability. The first section and the third section of the algorithm are relatively new and simple, with the calculation difficulty mainly focusing on the approximation factorization of the binary polynomial, so we employ (Gao, 2004)[6] factorization algorithm, which can effectively factorize the binary polynomial. The math multivariate polynomials in massive physical models are various, so advanced processing of the physical model should be made, which is good for building and solution of the math model. Meanwhile, the factorization of multivariate polynomial should be optimized, which can improve the efficiency of factorization, providing more effective processing of massive physical models.

References

1. Lenstra, A.K., Lenstra, H.W., Lovasz, J.L.: Factoring polynomials with rational coefficients. Mathematische Annalen 261(4), 515–534 (1982)
2. Chistov, A.L.: Efficient factoring polynomials over local fields and its applications. In: Proceedings of the International Congress of Mathematicians. Math. Soc., Japan,Tokyo, vol. 1, pp. 1509–1519 (1991)
3. Ruppert, W.M.: Reducibility of polynomials f(x,y) modulop. Journal of Number Theory 77(1), 62–70 (1999)
4. Gao, S.: Factoring multivariate polynomials via partial differential equations. Mathematics of Computation 72(242), 801–822 (2002)
5. Wang, P.S., Rothschild, L.P.: Factoring multivariate polynomials over the integers. Mathematics of Computation 29(131), 939–950 (1975)
6. Gao, S., Kaltofen, E., May, J.P., Yang, Z., Zhi, L.: Approximate factorization of multivariate polynomials via differential equations. In: Proceedings of the 2004 International Symposium on Symbolic and Algebraic Computation, Santander, Spain, pp. 167–174 (2004)

7. Kaltofen, E., May, J.P., Yang, Z., Zhi, L.: Approximate factorization of multivariate polynomials using singular value decomposition. Journal of Symbolic Computation 43(5), 59–376 (2008)

8. Hart, W., Hoeij, M., Novocin, A.: Practical polynomial factoring in polynomial time. In: Proceedings of the 36th International Symposium on Symbolic and Algebraic Computation, San Jose, California, USA, pp. 163–170 (2011)

9. Allem, L.E., Gao, S., Trevisan, V.: Extracting sparse factors from multivariate integral polynomials. Journal of Symbolic Computation 52, 3–16 (2013)

10. Lichtblau, D., Polynomial, G.C.D.: Factorization via Approximate Gröbner Bases. Symbolic and Numeric Algorithms for Scientific Computing (SYNASC), 29–36 (2010)

11. Zeng, Z.: The approximate GCD of inexact polynomials part II: a multivariate algorithm. In: ISSAC 2004 Proc., Internat Symp., pp. 320–327. Symbolic Algebraic Comput, New York (2004)

12. Weimann, M.: A lifting and recombination algorithm for rational factorization of sparse polynomials. Journal of Complexity 26(6), 608–628 (2010)

13. Weimann, M.: Factoring bivariate polynomials using adjoints. Journal of Symbolic Computation 58, 77–98 (2013)

14. Lecerf, G.: New recombination algorithms for bivariate polynomial factorization based on Hensel lifting. Applicable Algebra Engeg in Engineering, Communication and Computing 21(2), 151–176 (2010)

15. Corless, R.M., Galligo, A., Kotsireas, I.S., Watt, S.M.: A geometric-numeric algorithm for absolute factorization of multivariate polynomials. In: Proceedings of the 2002 International Symposium on Symbolic and Algebraic (2002)

Author Index